WORLDCHANGING

A USER'S GUIDE FOR THE 21ST CENTURY

EDITED BY ALEX STEFFEN | FOREWORD BY AL GORE

CHANGING THE 21ST CENTURY

DESIGN BY SAGMEISTER INC. | ABRAMS, NEW YORK

STUFF

FOREWORD
Al Gore 11

INTRODUCING WORLDCHANGING
Bruce Sterling 13

EDITOR'S INTRODUCTION 15

HOW TO USE THIS BOOK 25

INTRODUCTION 29
Questioning Consumption 32
Consuming Responsibly 35
Understanding Trade 39
Creating Healthy Homes 47
Doing the Right Thing Can Be Delicious 51
Buying Better Food 53
Better Food Everywhere 57
Eating Better Meat and Fish 63
The Future of Food 66
Preserving Barnyard Biodiversity 71
Cars and Fuel 74
Bright Green Consumerism 81
Designing a Sustainable World 83
Picking Green Materials 87
Craft It Yourself 90
Engineer It Yourself 93
Art Meets Technology 96
Biomimicry 99
Biomorphism 102
Nanotechnology 106
Neobiological Industry 109
Knowing What's Green 114
Producer Responsibility 118
Collaborative Design 123
Open Source 127
Design for Development 131
Bright Green Computers 134

SHELTER

INTRODUCTION 139

Green Remodeling 142

Building a Green Home 147

Living Well in a Compact Space 152

Developing Green Housing 155

Furniture and Home Decor 157

Lighting 160

Energy 162

Using Energy Efficiently 164

Green Power 170

Going Off the Grid 179

Smart Grids 183

Water 186

Conserving Water 187

Thinking Differently About Water 189

Landscaping 198

EcoHouse Brazil 201

Refugees 203

Reinventing the Refugee Camp 207

Transforming Disaster Relief 210

Open-Source Humanitarian Design 216

Land Mines 218

Rethinking Refugee Reconstruction 220

CITIES

INTRODUCTION 225
The Bright Green City 228
Vancouver 231
Portland 234
Retrofitting the Suburbs 238
Big Green Buildings and Skyscrapers 245
Healing Polluted Land 250
Greening Infrastructure 254
Place-Making 259
Urban Transportation 262
Product-Service Systems 268
Chinese Cities of the Future 271
Lagos 279
Megacity Innovations 282
The Hidden Vitality of Slums 286
Leapfrogging 292
ICT4D 296
Brazil's *Telecentros* 300
Leapfrogging Infrastructure 302

COMMUNITY

INTRODUCTION 307
Holistic Problem Solving 310
Education and Literacy 313
Educating Girls and Empowering Women 316
Public Health 323
South-South Science 332
Copyfight 336
Urban Community Development 339
Community Capital 342
Microfinance 346
Social Entrepreneurship 352
Giving Well 355
The Barefoot College 359
Travel and Tourism 363
Global Culture 368

BUSINESS

INTRODUCTION 379
Your Money 382
Creating Business Value from Sustainability 386
Green Marketing 389
Brands 393
Thriving in a Bright Green Economy 397
Seeing the Big Picture 402
Start-Up 101 403

INTRODUCTION 409
Movement Building 412
Networking Politics 418
Amplifying Your Voice 421
Connecting with Others 426
Tools for Talking 431
Demanding Transparency 434
Demanding Human Rights 441
Watching the Watchers 447
Protest 450
Direct Action 456
Nonviolent Revolution 460
Ending Violence 464

POLITICS

INTRODUCTION 473
Placing Yourself 476
Citizen Science 481
Restoration Ecology 484
Ecosystem Services 486
Biodiversity: How Much Nature Is Enough? 491
Sustainable Forestry 494
Creating Rural Sustainability in the Global South 498
Future of the Small Town 502
Local Greenhouse Forecast 506
Climate Foresight 511
A Personal Action Plan 514
Mapping 518
Charting the Deep Oceans 521
Polar Regions 525
The Solar System: Greens in Space 529
Imagining the Future 535

SELECTED BIBLIOGRAPHY 538
CONTRIBUTOR BIOGRAPHIES 547
ACKNOWLEDGMENTS 555
PHOTOGRAPHY CREDITS 557
INDEX 562

PLANET

Foreword

Al Gore

This book is about rising to meet the great challenges of our day.

We find ourselves in a climate crisis of great magnitude. Mankind is literally changing the balance between ourselves and the atmosphere, leading to an unprecedented warming of the earth. Despite the wonderful technological advances of the twentieth century, we still generate power and fuel our vehicles with coal, oil, and gas—and the combustion of those fuels is what is heating up the planet. Meanwhile, more than a billion people find themselves still trapped in dire poverty, hoping for just a percentage of the wealth that we have here in the United States. Millions of children still die from preventable diseases and malnutrition, and throughout much of the world, violence, corruption, terrorism, and oppression are still all too frequently the realities of daily life.

Each one of these problems is serious. Together, they add up to signs that this is a turning point in human civilization, one that requires great moral leadership and generational responsibility.

We have a great challenge. However, because we understand the root causes of these problems, we can also join together to solve them.

We need a new vision of the future. The Bible says, "Where there is no vision, the people perish." Today, facing so many problems, many of us find it very difficult to envision a better future, much less the kind of solutions that might make such a future possible.

That's why this book is so vitally important. *Worldchanging: A User's Guide for the 21st Century* is a compendium of solutions, some little known but well proven, some innovative and new, some bold but as yet untried. This book not only shows what is already possible, but also helps all of us imagine what might be—in our own homes, in our communities, and for the planet as a whole. Taken together, these solutions present a picture of a future that is not dark or catastrophic, but one that is full of hope and within our grasp.

To build that future, we need a generation of everyday heroes, people who—whatever their walks of life—have the courage to think in fresh ways and to act to meet this planetary crisis head-on. This book belongs in the library of every person who aspires to be part of that generation.

Introducing Worldchanging

Bruce Sterling

Suppose I write this: "Introducing this wondrous book is a major challenge."

If you follow my syntax, you may conclude that I have no completed book introduction to offer you. Yet, I'm working on it. I'm deep in a creative process of transition.

Abandoning the theatrical majesty of authorship, I have pulled you backstage into the hairy reality of the book-introduction process. No longer restricted to the classic roles of author and reader, we have become both observers and participants, just like the sagacious individuals behind Worldchanging. We are backstage at the theater, well before the ticket-clutching rubes can attend the show. We can see the curtain pulls, the sandbags, the greasepaint, the unlit limelights, the tremulous, chain-smoking actors. It looks chaotic, tentative, even shabby, but we can already hear the crowd filtering in. A tremendous show is awaiting us. A massively popular hit. Ten billion people are coming, maybe eleven billion. The flimsy walls will shake with their roars of laughter, their tragic sobs.

The future, whether we like it or not, is imminent. It may be unprecedented in shape and structure, difficult to describe with the language of the past, but it will surely come to be, and we will see it. It won't come flying in from outer space, phonily perfect and chrome plated; it will be a quotidian, functional, yet currently unimaginable thing.

"Worldchanging," the gerund-as-verb, is a grouping of online visionaries. They are finding, exchanging, assembling, and discussing information—with one another and with their huge and growing global audience. By their nature, they are a continuous process, rootless yet blooming, a kind of rolling, seed-spewing electronic tumbleweed. They are forecasters of tomorrow's culture. They confidently predict that the tools, models, and ideas for building a bright green future are already here. The pieces are scattered about us, awaiting a slow-gathering combustion of insight, the flash of a cultural gestalt.

The Worldchanging crew, the core people who supply the organizational impetus for this ongoing labor, are a widely scattered global cluster of journalists, designers, futurists, technicians—and very often these renaissance people fulfill all those roles at once. They are true natives of the twenty-first century, with a novel cultural sensibility. They are a particularly insightful and articulate lot. They display a unique ability to turn jagged splinters of information into a coherent portrait of a more advanced way of life. They are eagerly becoming the change they want to see.

"Worldchanging," the gerund-as-noun, is WorldChanging.com, a Web site, and the starting point for this worldchanging book. This book is the result of the thinking behind that Web site, which is a densely annotated creative engagement with the ongoing transformation of our planet. Worldchanging is a hugely exciting and significant endeavor. It functions particularly well. Most highly participatory Web sites have grim, disadvantaged areas—they feel like slums, or feuding arenas, or cranky, viciously elitist VIP rooms. But WorldChanging.com has the distinct look and feel of a vigorous, well-planned, booming community, with swift public transportation, clean air, and an honest city council. Sophisticated, artsy, and multicultural to a fault, Worldchanging is nevertheless full of regional character. It is a vibrant, influential entity that a mere ten years ago would have been unimaginable.

By the earlier logic of the twentieth century, Worldchanging would presumably be edging toward some settled institutional status. A corporate consultancy. A high-powered foundational think tank. Maybe a really with-it global NGO. No longer a gerund, "Worldchanging" would become "The Changed World Group." Guardians of a transformational fait accompli, this brass-plaque institution ("CWG" to its friends and patrons) would have high-rise offices in New York, Seattle, Brussels, Geneva, and Mumbai. It would boast a large staff. Thick plush carpets. Black metal filing cabinets stuffed with forms in quintuplicate.

None of that old-fashioned frippery to be found here, though. Here we are getting a roiling blast of keen inside expertise without the institutional overhead. Worldchanging people are not commissars. They never talk ideology; instead, they talk uncommon sense. They don't spin; they are pragmatic empiricists. They don't do mysticism; they forecast. They don't lobby; they anticipate. They often visibly struggle with ideas and issues that are well nigh unimaginable. That is because our world's future will be, in stark point of fact, well nigh unimaginable—by today's standards, that is. Today's standards are not of great importance, however, because they cannot be sustained. A day doesn't pass when you don't see one of today's standards densely wreathed in black, choking flames. As smoke mounts in the global theater, the Exit signs grow ever brighter. You may not have seen the signs yet, but the hour for panic is already well behind you, as you hold in your hands a book that is practically an encyclopedia of Exit signs.

This book is the canonical version of the Worldchanging vision of tomorrow. It needs no plug and no batteries. It consists of ink and paper, but this dizzyingly comprehensive chunk of treeware is no common, everyday book. It's a search engine. It has hit on a host of critically important, little-known matters that only seem "visionary" at this rapidly passing moment.

Let me introduce you to another world.

Editor's Introduction

Alex Steffen

This book offers ideas about how to change the world. These ideas invite changes in our daily lives and encourage planetary thinking. They show us at once how powerful we are as individuals, and how much we need one another; how imperiled the planet is at present, and how great the future could be.

Planetary thinking is hard. To any one of us, the planet seems almost incomprehensibly vast. Spin a globe, and our hometowns—everything we see and do in our daily lives, all our favorite places and all the people we hang out with—are the tiniest of dots whirling by. Fly over the Pacific at hundreds of miles per hour and we still see nothing below, hour after hour, but empty ocean.

Things are seldom what they seem, however. For humanity as a whole, the planet is small and shrinking quickly.

Because the planet seems so large to each of us as individuals, it's easy to forget how many of us there are (over six billion and counting) and how much stress we collectively put on the earth. Though it's not always easy to see it as we go about our days, our current way of life is unsustainable, and that which is not sustainable does not continue. We are using up the planet, one person, one day, one decision at a time; we're not considering the consequences.

Thinking about sustainability should be like planning for retirement. We save money—build up an investment portfolio, say—and then as we get older, we try to live on the revenue our investments create. In this case, our natural "capital" is a gift we've inherited simply by having the good luck to evolve on such a bountiful planet. And in using that capital, we should leave enough of the earth undestroyed for future generations to meet their needs. In fact, we should really leave the capital alone and just live off the interest.

But we're not doing that. We're living beyond our ecological means, and paying our overdrafts with resources our children will need and pollution they'll struggle to clean up. Those of us in the rich countries of the world, the developed countries, in particular, are using more than our share. What we should be using, in the interest of fairness, is that portion of the planet that we can use without undermining future generations' prospects—divided by the number of people who need to use it.

How much nature is that per person? Well, luckily, some ecogeeks have worked that out for us. They've found a way to measure the impact our lives have on the planet, what they call an "ecological footprint." In an equitable and sustainable world, each person's ecological footprint would work out—in Mathis Wackernagel's equation, which experts consider pretty accurate, if a bit optimistic—to about 4.7 acres (that's 1.9 hectares). In other words, if we divided up the usable part of the world by the number of people who need to use it, we'd each have to find a way to meet our needs sustainably from the bounty of a little less than five acres' worth of resources.

Unfortunately, we're already using an average of 5.4 acres (2.2 hectares) per person, planetwide. To make matters worse, our sustainable share of the planet is shrinking. In part this is a natural result of population growth: divide the planet by more people and you get a smaller piece of land.

The global population has mushroomed from 2.5 billion to over 6 billion in just the last fifty years. And more folks keep showing up at the party. We live on a planet of children, preteens, and teens, where almost 2 billion people are under the age of eighteen. Huge strides are being made toward slowing the population explosion—in educating young women, providing them with economic opportunities, and making sure they have access to family-planning techniques: the three best-proven means of slowing population growth. But the odds are good that most young women will want, as we nearly all do, to have families. This means that even in the best-case scenario there will probably be at least 8 billion of us forty-five years from now—and it could be 14 billion. So even if all goes well, we'll be welcoming another 2 billion to the global dinner table. That's like adding a new city, larger than Seattle, every week, every year, for the next fifty years.

But the planet is shrinking for another reason: we're using it up. We are, as ecologists say, in overshoot, meaning that we're using the planet with such intensity that it is unable to restore itself.

Here too, geeks with supercomputers have gone to work, and what they've found is pretty shocking: we're already using nearly half of the world's "net primary productivity." What that means, for those of us whom math makes sleepy, is that humans are using about 50 percent of all the life on earth—that about half of all the microbes, insects, plants, and mammals on the planet are being sucked into the systems that go to feed our needs. Think of every single living thing on the earth as a river. We're diverting half of that river to suit our needs, already.

This is a pretty clear numerical description of the essential problem, because while we're busy sucking up all that "net primary productivity," there are a whole mess of other critters—from little bacteria and beetles to salmon and tigers—that can't get what they need. Increasing clear-cuts, overgrazed grasslands, eroding farmlands, fishing boats strip-mining the sea, and huge toxic plumes radiating out from our cities: our current overuse of nature is driving species to extinction all around us.

If things go on like this, half the species on earth may be gone forever by 2050. Biologists call this the Sixth Extinction, a die-off much like the last days of the dinosaurs, but spreading much more rapidly.

We need to learn how to live on one planet, because we don't have any others we can live on.

We're also beginning to run out of resources. Some resources—like the solar energy beaming toward us from the sun—are renewable (that is, they don't run out when we use them right). Others, like oil, are not. Unfortunately, our society depends on nonrenewable resources. This cannot continue for two reasons:

First, nonrenewable resources have increasingly severe environmental costs—simply getting them or using them leaves us with less natural capital. Burning oil, for instance, is changing our weather, bringing on a global climate crisis on a scale never before imagined.

Second, we're simply running out of these resources. We may, for example, have already passed the peak of global oil production. Oil company experts debate whether we will effectively run out of oil in twenty years or fifty, but the essential point remains: if you're under thirty, you can expect to see a post-oil civilization in your lifetime.

Meanwhile, our ability to burn oil, to synthesize nitrogen fertilizers, and to brew chemical pesticides, cheap plastics, and alloys is all out of proportion to the limited nature of the renewable natural resources at our disposal. It allows us to support a lifestyle that is both more luxurious and more destructive than we could possibly afford (with our current technologies and designs) if we were reliant on renewable resources alone.

To return to our retirement analogy, if we're living on interest, we're fine. But what if our bills are higher than the checks our interest pays out? What happens if, to cover those bills, we spend not only the interest but also some of the capital? Well, next year, since there's less capital, we'll earn less interest. If our bills next year are higher still, and we have to dip even deeper into our capital, we'll earn even less interest the following year, and so on. Pretty soon, we're locked into a vicious spiral and headed toward bankruptcy. That's a pretty fair description of what's happening to our ecological capital, but the ultimate bankruptcy won't land us in a state-run old people's home—it'll land us in a world of deserts, hunger, and freaky weather.

How long before we're locked into that spiral and off to the ecological poorhouse? No one knows for sure, but the scientific consensus seems to be converging on a figure somewhere around twenty-five years from now. If we haven't stopped hemorrhaging natural capital by 2030, we may not have enough left to choose a different path. Most everyone studying the issue seems to agree that if we haven't made deep and profound changes to our impact on the planet by 2050, we'll have committed our civilization to a series of catastrophes. As *Limits To Growth* author Dana Meadows says, in an era where we seem to be running hard up against the limits of so many natural systems, the ultimate limit turns out to be time.

We need to create one-planet livelihoods, a new model of sustainable prosperity, and we need to do it within our lifetimes. For the future to be bright, it must also be green.

To compound the issue, humanity's appetite for better living is also growing. The Global North and the Global South—the "developed" and "developing" worlds—now live around the corner from each other, mutually dependent. Everywhere in the world, the poor see how the rich live, if not out their window, then on TV. People who live in shanties can compare the material quality of their own lives with that of people who fly over them in jets.

It's very difficult to know that someone out there has a car and a computer, a comfortable office and a beach house, and not, on some level, want those things too, or a version of them that maps to our desires. It would be difficult to find people who willingly and happily choose poverty when they know that others live easily and prosperously. Most of us, nearly all of us, want more than we've got.

For while it's true that our average global ecological footprint is too large, some of us have bigger feet than others. If everyone on the earth lived like the average North American, we'd need five planets to support our lifestyles; if we all lived like the average European, we'd need three planets; while if we lived like the average Pakistani, we'd be using less than one planet.

But there aren't a lot of teenagers around the world clamoring to live like Pakistanis. No, what the kids want, from Cape Town to Caracas to Novosibirsk and everywhere in between, is to live like Americans, or at least Italians: They want stereos. They want refrigerators. They want cars. They want computers. They want better lives. One of the realities of our day is that we live on a young planet, and many of those young people have seen *Baywatch* (at least metaphorically)—they know how the richest among us live, and they want, if not that, at least something better than what they've got. We can be sure that every one of the billions of kids now growing up has their own dreams.

It's worse than wrong to think that we're going to talk them out of pursuing those dreams. In fact, it's hypocritical to think that we should discourage them—especially that those of us in America (the land where the pursuit of happiness is written into our founding documents) should say to the two-thirds of the world living in what we consider dire poverty, "Sorry, some white guys, mostly dead now, set up a system which means that we get PDAs and day spas, but y'all should be happy with a goat and a half-dry well." Where's the fairness in that?

Today's kids want prosperity. This, of course, doesn't mean that they want to be Americans. Korean, Brazilian, Algerian, and Albanian kids mainly want to be Korean, Brazilian, Algerian, and Albanian ... with cars, computers, cool clothes, and nice homes. In short, a huge chunk of our fellow human beings are young, and ambitious for better lives—and they're going to try to get them no matter what we do.

There's no closing the gate behind us. We need to expect that most of the planet will want to live lives closer to those we've established based on our five-planet lifestyles.

We need to create a system that can deliver prosperity to everyone. The future needs to be bright, green, and freely available to all.

When asked how we can create a planet full of prosperous people without destroying the earth, some will articulate a version of the theory that green follows gold. That is, the way to achieve sustainability is to first grow rich. Growing rich gives you the money to invest in more efficient and less environmentally damaging technologies, which in turn gives you a cleaner environment. There's even a technical description of this process, known as the environmental Kuznets curve.

This argument does have some apparent validity. Put simply, we in the Global North got rich essentially by digging and pumping fossil fuels out of the ground and burning them. The industrial age was an age of smokestacks and steam engines, and our head start over the rest of the world essentially came down to getting the engines first and using them to make the rest of world do what we wanted. Along the way, we decided that all that smog and soot and poisonous slurry was cramping our style, and we passed laws banning the stuff (or at least driving it offshore). So now we are less smoggy, very wealthy, and extremely powerful.

Not surprisingly, many people in the rest of the world don't like this arrangement, and they've been trying their best to build their own steam engines and smokestacks ever since. Those who believe in the theory that green follows gold have been heartily cheering them on.

But unfortunately, the model we used to get rich is no longer replicable. As the famous 2002 report, the "Jo'burg Memo," put it, "There is no escape from the conclusion that the world's growing population cannot attain a Western standard of living by following conventional paths to development. The resources required are too vast, too expensive, and too damaging to local and global ecosystems." The "Western model of development" is a one-off.

We need a new model.

We need a new model that allows unprecedented prosperity on a sustainable basis. We need a new model that will let everyone on the planet get rich and stay rich, while healing the planet's ecosystems. We need to create one-planet livelihoods, which are so prosperous, so dynamic, so enticing that the alternative of chasing the old model of green follows gold is revealed as the fool's path it is.

Designing a system that would lead to that kind of sustainable prosperity presents an epic challenge. For that system to work in the real world, it must be rugged and shockproof—because the world's a rough place these days. More than 2 billion people have no access to electricity. About the same number have no safe means of disposing of their sewage. More than 1 billion drink fetid water. Over 1.2 billion don't always have enough to eat, and at least 840 million are suffering from chronic hunger and are only one bad harvest away from mass starvation. Hungry people don't have the energy to work as hard—economists estimate that somewhere between $64 billion and $128 billion is lost annually from developing world economies because of malnutrition. Hungry people are sick people, and sick people in turn slide further into poverty. Common, preventable diseases like childhood diarrhea kill millions each year, and other diseases are growing epidemic in a world where hundreds of millions have no medical care at all. AIDS alone is expected to kill 68 million people by the year 2020, while leaving at least 20 million children orphaned. Some countries, like Botswana and Zimbabwe, will have lost half their adult population to the disease by the end of the decade. Amid these sorts of societal holocausts, all other services—especially education—decline, and uneducated people (875 million people worldwide are illiterate, 60 percent of them women) are in turn less likely to understand good hygiene, to be able to master new farming techniques, or to participate in democracy in any meaningful way (where it exists at all). For the poorest 1 billion people, life has become a series of vicious deteriorations and inescapable traps. And for 2 billion of their neighbors, who are doing slightly better, this poverty creates instability and a nasty back draft, making it harder to make any progress at all. This is part of the context in which the environment is unraveling.

And with increasing regularity, ecological instability flares into outright chaos. James Gasana, Rwanda's minister of agriculture and environment in the early 1990s, told a lesser-known side of that country's tragic past in an article in *World Watch* magazine. We tell ourselves that Rwanda's genocide, in which at least 800,000 people were murdered, was the fruit of ancient and unanswerable ethnic hatreds. But the reality is that those hatreds were fanned into flame by a sharp wind: hunger. With a mostly mountainous terrain, and a population that had grown from 1.9 million to nearly 8 million in just four decades, the tiny country simply couldn't feed itself. The genocide may have been driven by hatred, but it was set into motion by hunger, and the killing was worst in communities with the least to eat. Similar dynamics can be seen in Sudan, where decades of drought has exacerbated long-standing tribal and religious conflicts in regions like Darfur.

But dramatic ecological instability certainly cannot account for all outbreaks of violence. The last decade has shown that even in countries that seem reasonably stable, chaos can spread quickly when pushed by power-mad men. Sarajevo was one of the most enlightened, multiethnic cities in the world . . . and a few years later it lay in ruins, under siege, surrounded by a land of mass graves and rape camps. Liberia was Africa's great success story . . . and a few years later an army of drugged teenagers with automatic weapons were wandering the streets of its capital in wedding dresses and fright wigs, shooting anything that moved.

In fact, thugs, gangsters, dictators, and tribal warlords run much of the world, often at a substantial profit. As Bruce Sterling writes in *Tomorrow Now*, "Outside (and sometimes within) the prosperous bounds of the New World Order is a large and miserable New World Disorder. It includes not only the smoking ground of the Balkans, but the Caucasus, South Central Asia, and vast, astonishing swaths of Africa . . . For the typical New World Disorder soldier, ethnicity and religion are not something you die for—they are stalking horses; useful pretexts for breaking down states and subverting police and governments. The resulting chaos can be structured, made to pay. Revolutionary idealists sometimes begin this process, but once the disorder fully flowers, their doctrines just get in the way. They will generally be rubbed out by greedier, more practical subordinates" (2002).

A far larger chunk of the world is more peaceful, but still rotting with corruption. The debt that chokes many developing countries is largely a legacy of that corruption, willingly abetted by developed-world banks; academics studying development believe that one in every three dollars made in development loans from 1970 to 1990 ended up in secret private bank accounts in places such as Switzerland and the Caymans. Some argue that the figure is much higher, that when you count in kickbacks to bankers, sweetheart deals with multinational corporations, and a long train of local officials and businesspeople siphoning off their cuts, it may be that only one dollar in four was actually used to build anything (and what was built was often shoddy and disastrously inappropriate). The problem seems to be worse now. Global anticorruption groups such as Transparency International have documented an alarming trend over the last decade: while some countries emerged from the Cold War into democracy and transparency, in a far larger number of countries, a facade of democracy masks a level of corruption and influence peddling that approaches open kleptocracy.

Both chaos and corruption make our work more difficult—so much so that any new model of sustainable prosperity needs to not only take them into account, but to actually work to mitigate them. If the answer to our ecological crisis does not also lead to greater security for everyone and help spread democracy, open government, and open business practices, it is in fact no answer at all.

We need a future that is bright, green, free, and tough.

So here we are. We need, in the next twenty-five years or so, to do something never before done. We need to consciously redesign the entire material basis of our civilization. The model we replace it with must be dramatically more ecologically sustainable, offer large increases in prosperity for everyone on the planet, and not only function in areas of chaos and corruption, but also help transform them.

That alone is a task of heroic magnitude, but there's an additional complication: we only get one shot.

Change takes time, and time is what we don't have. And there's always a lag between choice and adoption. In practice, the inertia of that which exists can be massive; the retooling period, even when things are changing quickly, tends to be significant, in many cases lasting decades. Even if we were, universally, as a species, to decide today to completely change our ways, we'd still have to rely on current technologies and practices while setting up the new system. If we all agreed to live more simply, to embrace our best current technologies, and to share fairly and sustainably the planet's resources—to aim for one-planet practices—by the time we all got there, we'd still be in overshoot, because in those intervening decades, we'd have destroyed another huge chunk of the planet's natural systems, and would have, in effect, less planet. So in fact, we'd have to then aim to live with even less impact, which is not only more difficult to get people to accept, but takes even longer to achieve, and by the time we get there, our available natural capital will be even more depleted, and we'll have to live even more humbly, and on and on until we reach stability at a point where the planet is deeply torn up and we're all living on the edge of poverty. Fail to act boldly enough and we may fail completely.

All of this means we have to come up with an answer that takes as a given not the natural capital we currently have, but the smaller pool of natural capital we'll likely have when any proposed change actually takes effect. What's more, any transition will require us to continue doing things as we are for a certain period of time while we retool and redesign, and then to spend a lot of our resources actually rebuilding. Since we are already running an ecological deficit, and since that deficit is getting bigger, there is absolutely no reason to believe that we can try one thing for a couple of decades, and then, if that doesn't work, try something else. The living fabric of the planet, once unraveled, will never come back, at least not for millennia. There are no do-overs on a finite planet.

We don't say it in public, but we've placed a giant wager here on the future of the human race. To win that bet, we must move to a new model, one based on a standard of sustainability even higher than our current one-planet standard, while providing prosperity to billions around the world. And we need to do it in twenty-five years. And we need to get it right the first time. To lose that bet is to endanger everything we love.

That's the bad news.

Here's the good news: we don't have to destroy the planet or impoverish other people to live well. We're often told that there are trade-offs between doing well and doing right, but when you pull back and look at the big picture, those trade-offs usually prove to be illusions. In fact, we're learning more and more that doing the right thing in an intelligent way often pays off handsomely.

That's true in our own lives. We can save energy, and save money. We can live in close-knit communities, cut down on our commutes, and spend more time with our friends and families. If we earn a little extra money, we can invest in companies with smart solutions and often pocket a nice dividend check. We can create lives that go easy on the planet, exploit no one, and yet leave us happier and better off.

That's true in our whole society. We're learning with every passing day that changing the world is a growth opportunity. Vast fortunes will be made by those companies that invest in clean energy and new technologies. Huge savings (and better quality of living) will be realized by those cities that grow smart and that green their infrastructures. Green may not follow gold, but there is gold in green; there are fortunes in fairness.

We can begin now to see the blueprint of a planet that works well for everyone on it, where new thinking, trade, and collaboration yield economic growth and innovations that swell bank accounts while shrinking our impact on the planet.

If we face an unprecedented planetary crisis, we also find ourselves in a moment of innovation unlike any that has come before. We find ourselves in a moment when all over the world, millions of people are working to invent, use, and share worldchanging tools, models, and ideas. We live in an era when the number of people working to make the world better is exploding. Humanity's fate rests on the outcome of the race between problem solvers and the problems themselves. The world is getting better—we just have to make sure it gets better faster than it gets worse.

We don't need a miracle to win this race; we need a movement. We need millions more people who are committed to doing their part to embrace good ideas, find new solutions in their own work, and live and share what they learn.

Each of us has a part to play. Each of us can be a part of that movement to change the world, and the best part is that what's needed is not unflinching loyalty to some supreme leader, or mystical adherence to some cult's belief system, but millions of us doing our best to think for ourselves and share what we know.

That being true, one of the most important questions we can answer for ourselves is, how do we design a personal future that is both bright and green? Small steps are important, but how do we make sure those small steps are leading us down a worthwhile path?

The point of this book is not to tell you that we've got the answers—the point is to share with you some of the solutions that people have tried, so that you can be inspired to find your own. If we were to stop innovating, only following the formulas in this book, we would fail. We need much better solutions, and the only way that will happen is if millions of people take on the pieces of this puzzle and try creating their own version. Everyone on the planet has a piece to contribute—especially those of us who live in wealthy countries, who have experience and skills, wealth and privilege. Suggestions on how to contribute can be found in every chapter of this book, but here are a few guiding principles:

Small steps are great, but look for small steps that influence the big systems to which we're all connected. Cutting our energy use and buying green power, for instance, not only saves us some money (and lets us enjoy the energy we do use in good conscience), but also helps transform the wider energy system. Every time one of us switches to compact fluorescents and wind power, a bunch of good things happen: less coal is burned (meaning healthier air and less damage to our climate); more wind farms are built; and energy companies get a signal that there's a market for energy-efficient products and more clean energy. By designing our lives to be greener, we help nudge the whole economy toward a bright green future.

Individual actions are great, but look for individual actions that will influence others. There's an old saying that living well is the best revenge. That may be true, but living well is certainly the best argument: when we design our personal lives in such a way that we're doing the right thing and having a hell of a good time, we act as one-person beacons to the idea that green can be bright, that worldchanging can be life-changing. One of the biggest mental barriers to changing behaviors is the idea that change will be unpleasant. But we know better, and every time we design a bit of our lives to reduce our impact, support good efforts, and make our lives more comfortable, beautiful, and exciting, we're sending a powerful message to everyone around us.

Good intentions are great, but remember that only passion changes the world. There are more avenues for action than even the most motivated overachiever among us could ever pursue. So we shouldn't try to do everything: we should try to do the right things. When we seize the chance to make changes that are both important and speak to us as people, we transcend good intentions and more meaningfully express who we are. Why be boring? Why follow other people's instructions for designing a better life? The world needs more passionate people, deeply engaged with the business of designing their own lives in ways that speak to them. From passion comes creativity, and from creativity come better answers.

The equation we should all follow is this: do the easy things, then do a few more challenging things that we really believe in and enjoy. If we're home-repair geeks, we should green our houses. If we're policy geeks, we should find the best practices around, adopt them, and improve them. If we're fashion geeks, we should show the world exactly how fabulous dressing green can be. If we're business geeks, we should make our fortunes selling a sustainable product the world really needs. If we're gardening geeks, we should make the yards in our care thrum with life. The world doesn't need our suffering, it needs our shining examples, and every one of us has an example to set.

It turns out that the world is full of both great ideas and passionate people. Since we started the WorldChanging.com site, we've posted thousands of articles about approaches we think are worth watching, and millions of people have visited the site. We've won journalism awards and had people say nice things about us. Much more important, we've heard from thousands of readers who have used the ideas we point to as a jumping-off point for their own worldchanging actions.

But here's the thing to remember: what we've done doesn't say anything special about us. We'd like to think we've done a good job, and we have certainly sought out the smartest collaborators and correspondents we could find, but if we've created something of value, it's because WorldChanging recognizes and works from four ideas as a central premise:

1. We need better tools, models, and ideas for changing the world. Luckily, more are being created every day.

2. The more people know about these tools, models, and ideas, the better their own ideas will get, and the more ideas will become available.

3. Anyone can join the conversation, and the more people do join, the better it gets.

4. The better the conversation gets, and the more people use the tools, the more exciting the adventure becomes, and the more likely its success.

Consider this an invitation to join the adventure. What kind of future will you create?

How to Use This Book

Changing the world is a team sport, and there's a spot on that team for every person on the planet, though finding our spot can be damn hard. Learning what we can do is not easy in itself, but discovering what each of us feels called to do, in a way that only we can do it, is one of the hardest tasks life has to offer. In these times, the question "What will I do?" is one of the toughest we may ever ask ourselves.

Because the question is not easy, this book doesn't offer easy answers. This book isn't about lists of Ten Simple Things You Can Do. It's about providing you with ideas for rethinking your own life, and providing approaches to change. This book is written not by people who know it all (such people are invariably wrong), but by a bunch of your teammates who are working themselves to figure out how we can make a difference together. We do try to tell it like we see it, but our ultimate goal is to provide ideas and stories that suggest ways you can create new solutions yourself. It's not the final answer on anything, but a starting point for deciding how to make your life count.

The solutions you choose to embrace may be comparatively simple (say, deciding to buy clean energy from your power company or to eat organic foods) or they may be more far-reaching (deciding to take your career in a new direction or to mobilize your neighbors to change your community). You're the best judge of how to apply these ideas in your own life.

As you think about what these ideas might mean in your life, we encourage you to treat the information we provide as a set of avenues for exploration. There are a lot of ideas here, and not all of them will work for you: indeed, most of them will work only if you apply your own smarts and adapt them to the circumstances you face. In the long run, some of the ideas will probably turn out to be bad ones. If there were a one-size-fits-all equation for making a better world, life would be easier, but much more boring. Half the fun here is dreaming up new ways to combine, modify, and apply the tools others have created in ways no one else has ever thought of. Changing the world is a team sport, but no two of us play the game the same way.

And no book, not even a six-hundred-page slab of paper like this one, could possibly tell everyone everything they might like to know about every solution, or even cover every worthwhile solution in the first place. So, while we've done our best to bring you insight into an array of good ideas, you'll get the most out of this book if you treat the whole thing as a guide to further exploration.

There's a whole universe of worldchanging people out there, looking to share information and ideas, and their numbers are swelling by the day. Find your allies and heroes, and share what you're learning. One way of connecting to them is to use the WorldChanging.com site. Throughout this book, you'll see Web addresses on the top of the pages. That's our way of reminding you that there's more information about these subjects online. On those WorldChanging.com pages, you'll find access to groups working on these problems, pointers to Web sites exploring the practicalities of these tools in greater (sometimes obsessive) depth, and updates on the stories we tell. You'll even find a glossary of the terms used in this book.

You'll also find ways of telling us about what you've been doing to craft your own answers. We're happy to hear from you too, if you've found any errors or problems in the book, so we can fix them in future editions. Indeed, given the number of ideas covered in this book, and the speed with which we've raced to bring them to you, we almost certainly have made mistakes and missed ideas people ought to know about, so please log on to WorldChanging.com, share what you know, and help us make this book a better and better resource.

STUFF

Our things define us.

What we buy, what we use, what we keep and throw away, what we waste and what we save: the stuff that surrounds us and flows through our lives is a key indicator of the kinds of lives we're living. To be an affluent twenty-first-century person is to float on a sea of material objects—each with its own history and future.

They may be hidden from our eyes, but in practical global terms, those histories and futures tend to be the most important aspects of the stuff we own. The moment we tear the wrapper from a new toy, that toy is already at the end of a sweeping story involving the mining of metals, the pumping of oil, the operation of huge factories, the shipping of cargo containers, the printing of packaging materials, the purchase of advertising, the careful arranging of store shelves, and the final drive home. Simply buying, say, a new laptop connects us with a web of activity that spans the planet.

Another story begins when we throw away our old laptop. It may find itself on a quick trip to the local dump, where it will lie, buried beneath a mountain of garbage, corroding and slowly leaking toxic chemicals for hundreds of years. It might, on the other hand, be shipped off to China, where its circuit boards will be stripped out and where poorly paid workers will extract by hand the valuable metals they contain. Parts of the laptop's titanium body may be sold as scrap and melted down for other purposes. Much of the rest will wind up in an open dump, where children will pick through the dissected electronic remains.

What's true for our new laptop is true for every product we buy: what we actually purchase from the store, as sustainable-design expert William McDonough [see Knowing What's Green, p. 115] points out in his iconic book *Cradle to Cradle*, is just the tip of a vast material iceberg, a gigantic pyramid of extracted resources and burnt fuel, toxic waste, and sweatshop labor. Similarly, our use of a product only marks the start of a new cycle—the product will spend most of its time decaying in a dump somewhere. A Styrofoam carton may only spend fifteen minutes holding the Chinese food we have for lunch, but it could easily spend a hundred years decomposing in some trash heap.

The first problem with the secret life of our stuff is it hides from us the consequences of our actions.

The second problem is, most of the time, those consequences are not pretty.

The mountains of waste that telescope out from us before we buy something and after we throw it away choke the planet with deadly poisons, endanger our health, wreck natural systems [sailors thousands of miles out to sea report gliding through vast floating rafts of trash], and force our fellow human beings to work in conditions many of us would never accept for ourselves. The system today obliges us to take part in creating a mountain of troubles for the world every time we plop down our credit cards.

However, changes are happening; the things we live with every day are evolving, becoming lighter on the planet, fairer to people everywhere, safer for our health.

There are already products on the shelves and in the showrooms that can help us do better, help us turn the mountain of waste we create today into a small hill tomorrow. In this section, we explore how to find those green products and how to make good choices regarding everything from clothes to cars, coffee to computers.

But we don't need to stop there. We know more than ever about how to see the hidden lives of things. We now have access to all sorts of information about products' life cycles—information that gives us the ability to make better choices. More important, as the once arcane information about everything from materials' toxicity to engines' efficiency comes out into the open, a new wave of empowered designers is emerging, and they are bent on shrinking our mountainous impacts into ecological molehills.

The changes rippling through the fields of art, design, engineering, material science, and biology are nothing short of revolutionary. Design tools, especially software, are getting cheaper, more powerful, and easier to use, and collaborative models and new thinking are bringing people together in new ways, enabling them to solve problems using approaches undreamed of a decade ago. Science is unlocking nature's secrets, allowing us to mimic nature's grace, strength, and ecological integrity in everything from product design to industrial systems. Art and technology now inform each other in ways previously unseen, revealing new vistas of possibility. In every field that touches on the conceptualization, manufacturing, and use of stuff, powerful combinations of transparent information, ecological understanding, and advancing technology are unleashing forces for change.

Because of these forces, a more radical evolution in design is under way, and a new generation of stuff is emerging. Imagine things that use minimal energy, that are made with no toxic chemicals, that are completely recyclable, that hurt no one—not even nature—but that perform better and last longer than what we have today.

These products still live mostly on the computer screens of cutting-edge designers, but that's changing quickly. And we have the power to make it change more quickly still. We may not all be designers, but we are all design consumers. By voting with our dollars and showing that we demand smarter stuff—that we insist on quality without the guilt—and that we're willing to walk the talk at the checkout counter, we can help spur this transformation.

We have the power to choose the world we will live in, and through our purchasing decisions, we reveal that world every day. AS

Questioning Consumption

It's a small world, and it's getting smaller. Some aspects of this shrinking are heartening—it's easier to unite social networks, conduct business, and learn about places and people outside our immediate surroundings.

But the planet is shrinking for another, less auspicious reason: we're using it up. Every year we cut more forests, graze more cows, drive greater distances, dump more trash. And since we're already taking more than the planet can give, every year nature has less to offer. To make matters worse, this spiral seems to be accelerating, and the gap between sustainability and everyday practice is widening.

To use less of our planet and become responsible consumers, we have to ask ourselves the fundamental question, how much stuff do we need? The relationship between material wealth and well-being would seem to be a proportional one: when one increases, so does the other. But as it turns out, measures of wealth and health rise together only to a point, and then the pattern shifts. In fact, research tells us that, assuming a basic level of comfort, happiness across most cultures is largely unrelated to material luxury.

Reducing our own levels of consumption saves us money and eliminates unnecessary clutter from our lives. In an era of big-box superstores and viral marketing, most of us go for quantity over quality, lured by the ease of getting more stuff cheaper. When we become conscious about what we buy, we end up with more space for appreciating the objects that surround us, and what we have actually can make us happier.

Ecological Footprints

One way to wrap our minds around the implications of billions of people sharing a small planet is to do some math: divide up the usable part of the globe by the number of people who want to use it. This gives us a sense of what an equitable share would be for each of us. To be fair, though, we should probably not use up everything right away. Our kids and grandkids may want to eat, drink, and breathe, too, so we should probably only take as much as we can while allowing nature to renew itself.

How much nature is that per person? To begin locating the balance point between using the resources we need and minimizing the repercussions of our consumption, we must assess our individual contribution to planetary problems. A leading thinker in analyzing these factors is Mathis Wackernagel, who proposed that we view our personal environmental impact as a footprint. Our "ecological footprint"

represents the ramifications of all the various components of our lives, from food and home energy use to transportation and where we live.

Using a simple online quiz, known as the Ecological Footprint Quiz, we can measure our footprint by inputting details about our lifestyles and habits into a basic calculator. At the end, we get a score that indicates how many acres of land we each require to support our current lifestyle.

According to the Global Footprint Network, the average ecological footprint is 5.4 acres (2.2 hectares) per person. But it's important not to look at the average alone, because some of us have much bigger feet than others. The average American, for example, uses approximately 24 acres (10 hectares), while the average Chinese person uses only about 4 acres (1.6 hectares), and the average Pakistani just 1.5 acres (about 6/10 of a hectare).

The number we score on the quiz is symbolic, of course. The earth will never be sliced up into individual serving sizes, nor could it, since our portions continually diminish as resources dwindle and populations explode. Our own score gives us a very good sense of how far we are from a reasonable level of personal resource use—but we must also consider the fact that the planet can only withstand so many footprints. The giant feet of industrialized nations leave little room for the ever-increasing needs of developing nations.

There are now a number of different variations on the Ecological Footprint Quiz. As a conceptual tool, the quiz can have profound effects on how we make choices and steer our lifestyles toward a more sustainable rate of consumption. AS & SR

Preceding pages: *1000 Trash People*, Lake Stelli, Switzerland, 2003.

Opposite, left: Ukita family with their household possessions, Tokyo, Japan, 2001.

Opposite, right: Brazilian shopper ponders her options, São Paulo, Brazil.

Freecycle

What do we get when we combine the gift economy, sustainable thinking, and craigslist.com? Freecycle.org, a fantastic "reverse eBay" for the stuff we no longer want.

The premise is simple: you join a "freecycling" e-mail listserv for your local community, on which people proffer the things they might otherwise take to the dump, give to charity, or simply leave on the curb. At last count, there were hundreds of such lists set up, in communities large and small.

The Internet is terrific at making markets more efficient, and there is indeed a market for many things we would otherwise throw away. Sure, you *could* post that box of random computer cables on eBay, but your time is frankly better spent doing other things. All you really care about is keeping those cables from ending up in a landfill. If someone is willing to show up and take them away, even if they use just one, the result is still a reduction in waste. Furthermore, if they use just that one cable, then sell the rest on eBay for a profit, all the better for them. Anything to keep the cables from the landfill. This is a stellar example of win-win thinking. AZ

Choice Fatigue

According to happiness researchers (yes, there are experts who measure happiness), we are less happy than we used to be. There are a number of well-studied and well-documented reasons for this, and a few obvious ones (famine, poverty, disease, and war, anyone?), but surprisingly, a leading cause of unhappiness, at least in developed nations, appears to be our overabundance of choice. In the last few decades, rates of depression have dramatically increased worldwide, a curve that corresponds with the upsurge in choice, indicating perhaps that having too many options fosters stress, anxiety, and uncertainty.

"Choice fatigue" seems counterintuitive, because choice is good, right? Well, not so fast. Our consumer culture is relentless, and the more choices we have—the more information we're bombarded with—the more effort we invest in evaluating our options, and the more likely we are to be dissatisfied with the outcome.

The more options we're given, the poorer our decision-making abilities become. Most of us hate making trade-offs and will avoid making choices until we absolutely have to; the decision-making process is fraught with bad feelings from the start. At the same time, most of us are bad at dealing with uncertainty and at estimating odds, especially since we often don't possess enough information to properly calculate probabilities. After spending so much time weighing trade-offs and trying to sift through a deluge of information, our expectations rise so high that we often end up disappointed when the outcome is not as perfect as we had hoped. Consumer satisfaction is nothing more than the miracle of reality matching our expectations.

What's worse, we often adapt to our overabundance of choices by picking things haphazardly and acquiring more than we need. The more we own, the more we get used to all of the stuff surrounding us, and the less special it feels. That's not to say that the only remedy for choice fatigue is getting rid of choice altogether. Rather, we need to find ways to maintain a level head when making choices, and to keep a healthy distance between the destabilizing allure of advertising and ourselves. NAB

Brands in a Choice-fatigued Market

For companies, choice fatigue seems to spell trouble. After all, it would appear very risky for a company to reduce the choices it offers customers when their competitors are piling them on by the dozen. Many new gadgets offer twenty functions in one handheld unit, the idea being that the more things it can do, the better (or at least the more attractive) the product is. But consumers have started wondering, "Why can't I get a cell phone that is *just* a cell phone?"

So why do manufacturers, despite these complaints, continue to develop products that do more and more? Obviously, they aren't taking the consumer's pleas for simplicity as an indication of true preference. Companies want to offer products that emerge from uncharted technological territory, so they hedge their bets and put out the products they believe will entice us with novelty and abundant capabilities. And the strategy works: we continue to buy the latest thing and eagerly anticipate the next. Companies then view our purchases as proof that their convoluted analysis of consumer psychology is valid.

We may be sending mixed messages with our verbal complaints and our spending habits, but sometimes our preference rings loud and clear. The iPod is probably the best example of a gadget with extremely limited functions—and earth-shattering success. The little device is simple, straightforward, and utterly ubiquitous. Nobody wants to be without one, and nobody is asking for an iPod that is also a datebook, a phone, and a computer. In its first iteration, it did one thing, came in one color, and made its way into the hands of nearly everyone who could drop two hundred bucks. SR

RESOURCES

Voluntary Simplicity by Duane Elgin (Harper Paperbacks, 1998)
The notion of voluntary simplicity hinges on the idea that knowing when you've had enough is the key to financial independence and to enjoying your stuff more. It's not about renunciation of wealth or social engagement: it's about focus. If you want to be able to work less, or retire early, the trick is to weigh out what you buy, what you need, and what you'll want in the future: "To live more voluntarily is to live more deliberately, intentionally and purposefully—in short, it is to live more consciously. We cannot be deliberate when we are distracted from life. We cannot be intentional when we are not paying attention. We cannot be purposeful when we are not being present. Therefore, to act in a voluntary manner is to be aware of ourselves as we move through life."

Your Money or Your Life by Vicki Robin and Joe Dominguez (Penguin Books, 1999)
One self-help book worth its salt, *Your Money or Your Life* focuses on simplifying your life by changing your relationship to money, and building financial independence by needing less and investing more.

"Most people in North America engage in the dominant myth of 'more is better' without question, and even good, caring people rationalize excess as necessity. Because of this, so many people, with a helpless shrug, say they need ever more money to meet the demands of 'modern life,' citing a vague boogey-person called 'cost of living.' 'More is better' now means 'more money is better.' This acquiescence to excess then requires putting up thicker and thicker walls between our consciences and the billions of people who live in poverty. If we were selling our time—and perhaps our souls—to a system that truly fed us, that would be one thing. But the economy is not designed for people; rather, people are trained to serve the economy."

Gone Tomorrow: The Hidden Life of Garbage
by Heather Rogers (The New Press, 2005)
"Trash is the visible interface between everyday life and the deep, often abstract horrors of ecological crisis," writes Rogers. Arguing that the consumption so intrinsic to American culture is propelling us toward environmental disaster, she dismisses recycling as a promise meant to normalize growing consumption, examines the multimillion-dollar industry surrounding garbage, and traces the history of waste and waste removal in our country. Rogers argues for a radical change in our everyday treatment and conception of waste, in hopes of reversing the precarious ecological state we face today. As part of a twofold plan, she insists that "trash needs to be addressed in terms of production instead of consumption," and that "industry's inability to regulate itself must be acknowledged and replaced with enforceable environmental measures."

Consuming Responsibly

We all know that buying less stuff makes good green sense. But some of us think this might cramp our style, or at least our fun. We all have needs, and most of us enjoy buying new things, so when we can't buy less, we can at least buy smart. Smart consumption means figuring out who makes the best available products and where they can be found. It also means doing enough research to be able to sniff out deceptive marketing. Whether we're shopping for clothes or food, cars or housewares, or even houses, we can wise up about the goods we buy.

We have the power to bring our values to the checkout counter. We can look for clothes made with renewable fibers; cleaning supplies without toxins and pollutants; and products that don't exploit the workers who made them. As consumers, we can imagine something better than what is already there. With the knowledge we possess about everything from labor practices to design strategies, we can envision the best possible goods. But until our imaginations and the demand we create with our dollars bring those ideal products to fruition, we have to choose the best of what's already around. Here's how.

How to Buy Better Clothes

For most of our immediate needs, we already have the power to make responsible choices. We can buy organic food, use energy-saving lightbulbs, drive hybrid cars. But even if we do all of those things, we are still probably sporting duds that support some pretty abominable practices. Unfortunately, buying an organic cotton T-shirt has less of an impact than buying an organic apple, as there are many more steps between the cotton field and the T-shirt we put on than there are between the orchard and the apple we eat. The process used to mass-produce organic cotton clothing is little different from the process used to produce synthetic clothing—and that process is what's so problematic for the environment. But that doesn't mean our choices don't matter: We can effectively vote with our dollars against unfair labor practices. By trying out alternative fabrics, we can push progress in the industry toward sustainable materials, and that's worth the effort.

At present, two types of "good" clothing are generally available: the gunnysack garments that scream "granola" and a handful of high-fashion (and high-priced) "ultra-green" lines. In a bright green world, every shop would stock a selection of equally appealing, yet far more responsibly created apparel at reasonable prices, much like standard grocery stores that now have an aisle or two of natural foods. But for now, we have to read labels and keep our eyes and ears open for new developments.

Look for the following practices and traits common to socially and environmentally responsible apparel:

Sweatshop-free Labor: It's not enough to simply buy clothing labeled Made in the USA. There are a number of U.S.-controlled regions in other parts of the world where sweatshop laborers turn out garments that can be legally called American-made. There are also sweatshops right here in the United States. The label gives no indication as to the conditions under which the clothes were made or the distance they traveled to hang in your closet. Unless the label explicitly says the garment was made under fair labor conditions, you can't be sure.

Perhaps more than any company before, American Apparel has brought a significant wave of mainstream attention to this issue. The popular maker accomplished this not by selling outrage toward companies who perpetrate poor conditions in their factories, but by creating a real alternative with immense commercial appeal and well-strategized marketing.

Organic or Renewable Fibers: A number of fabrics now on the market do minimal damage to the earth, biodegrade fairly efficiently, and come from renewable resources. And we're sorry to say, the Cotton Board's "Fabric of Our Lives" is not one of them. The conventional cotton that makes up most so-called natural fiber actually comes from the most pesticide-intensive agricultural process in the world. According to the Organic Trade Association, producing a single cotton T-shirt takes approximately one-third of a pound of pesticides and fertilizers—chemicals that permeate the soil, run into the water, and pollute entire ecosystems with heavy toxins. So what can you choose instead?

Organic Cotton: In response to consumers' rising awareness of the ills of conventional cotton, and paralleling the boom in the rest of organic agriculture, organic cotton has been gaining momentum as a reasonable option. When you take the pesticides out of the equation, cotton isn't that bad—no chemicals, better health for field workers, and a quality product you can feel good swaddling your baby in. Of course, farming

Left: Merino sheep, New Zealand.

Opposite: Garment workers, Zhejiang Province, China, 2005.

cotton still takes a dramatic toll on the earth, and organic cotton still costs significantly more than conventional cotton. But it is a preferable alternative, and quite widely available in everything from apparel to bedding, towels to tampons.

Modal and Lyocell: Made from beech wood pulp, these fabrics are incredibly soft, breathable, and pliant—think of the thin, worn softness of your favorite old T-shirt. The textile industry in India welcomed these new materials several years ago as a good replacement for polyester and rayon, because they serve the same function in manufacturing, without taking the same toll on the environment. Because they are made with cellulose (the natural biopolymer found in all plants), they wick moisture away from skin, which makes them excellent for athletic apparel. You'll find them frequently in bedding and clothing.

Bamboo Fiber: Bamboo can be made into yarns of various qualities and thicknesses. One of the most renewable resources for fiber and building materials, bamboo has enormous versatility and durability. Bamboo is commonly woven into fiber blends for everyday garments like socks, sweaters, and stretch pants. So far it is difficult to find fabric that is 100 percent bamboo, but that is likely to change as it gains popularity and processing costs drop.

Merino Wool: Requiring much less processing than synthetic or agricultural fibers, merino wool comes primarily from Australia and New Zealand, where sheep farms abound. People love this renewable resource—which regrows on sheared sheep very quickly—for reasons unrelated to its sustainable qualities. An excellent fiber for outdoor athletic wear, it provides a soft, thin, effective insulating layer that keeps moisture off the skin. Recent advertising promoting organic merino puts heavy emphasis on the importance of raising sheep in healthy conditions and on the benefits of using organic wool for baby blankets and clothing.

Hemp: The most ubiquitous natural fiber, hemp has come a long way since its early days, when it yielded coarse, thick weaves. Hemp can now be made into soft, high-quality fabric as fine as silk. The hemp industry has grown by fits and starts over the years, against varying opposition, but the fiber remains viable for fabric production.

Recyclable/compostable End-of-life Options: Many companies now offer end-of-life return policies that allow you to send back consumer goods once you are finished with them. Patagonia, for instance, runs a return system for long underwear, allowing customers to send back their worn-out garments directly to the company. The company sends the material to Teijin, a Japanese firm that has developed a disassembly and reprocessing technique called ECO CIRCLE. Garments go to Teijin's facilities for recycling and return ready for reuse. Of course, the transportation involved takes its own toll on the environment, but ECO CIRCLE systems are likely to expand into new regions as the technology gains attention. SR

Method

Clean, green genius, the Method Home brand, started by a couple of inspired young guys in San Francisco, is everything that the big cleaning-products manufacturers aren't: their products are biodegradable and nontoxic, and they come in recycled packaging. That's not all. Although they look nothing like the other household products on the shelf, they still cost the same. As a result, Method Home is popping up not in specialty and health food stores, but in the big boxes and supermarkets, where good deals are a big deal.

These products have turned the home-cleaning industry on its ear, and companies like Unilever and Procter & Gamble are taking notice, increasing their budgets for environmental research and development. Putting a hip spin on under-sink staples (and running ads of people doing chores in the buff) has turned Method Home into a mighty and rapidly growing little venture. Consumers who've spent their whole lives grabbing the worst chemical cleansers around now have an easy way to adopt a new Method. SR

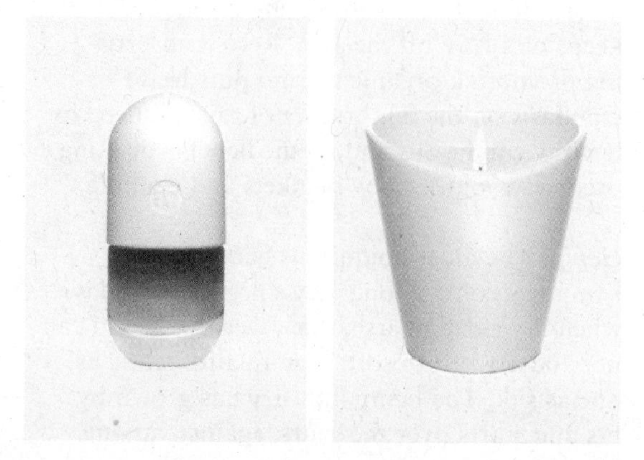

Method Home products.

Greenwashing

As you might surmise, greenwashing is the act of painting a green facade on something to cover up what's underneath. The term emerged during the 1990s, when corporations began to see the value of appearing socially and environmentally responsible despite not yet having brought their practices into line with their new ethical public images. Consumers who saw through green-marketing tactics labeled suspect corporations "greenwashers" in an effort to expose their duplicity.

Now that true sustainable business has taken root, we can demand transparency of supposedly green companies. Those who seem to be covering something up generally don't pass the greenwashing test.

How can you tell if a company's green messages are authentic? Here are some hints. You should be wary of an advertisement or corporate statement for which any of the following ring true:

- The claims are vague—using turns of phrase like "natural," "environmentally friendly," or "good for the planet"—and aren't backed up with specific facts.

- The product is billed as having no environmental impact (pretty much everything humans produce has some environmental impact).
- The company states that by purchasing its products you can "save the earth."
- The claims aren't verified by an independent and credible third party.
- The green products or services represent a tiny fraction of the company's overall business.
- The company doesn't respond to specific questions, whether via phone, e-mail, or a Web site. JM

RESOURCE

The Consumer's Guide to Effective Environmental Choices by Michael Brower and Warren Leon (Three Rivers Press, 1999)
The key word in this book, as is clearly indicated on the cover, is *effective*. In a sea of books about making everyday choices that benefit the environment, almost none gives a scientifically grounded, realistic depiction of how much any one of those choices actually matters in the grand scheme of environmental protection and restoration. What's worse—paper or plastic? Does it even matter? If you eat meat every day but take the bus to work, are you doing better or worse than a vegetarian who drives solo?

According to the authors, "Not all consumption has an equal impact on the environment. For that reason, a 10 percent across-the-board reduction in Americans' consumption would not be the most effective way to reduce environmental damage." Using priority lists, graphs, and tables, the guide offers comprehensive, consumer-level advice for assessing the impacts of our consumption patterns. There's a lot of myth busting and eye opening, and although the book is somewhat dated, many of the points remain relevant and are different from what we normally hear.

Understanding Trade

Your Chinese clock radio sounds, waking you up with news from the BBC, and you slip out of your Egyptian cotton sheets and into the shower. You dry off and put on underwear from El Salvador, jeans from Lesotho, and your favorite blue shirt from Sri Lanka. A cup of Tanzanian coffee, some Brazilian orange juice, and you're off to work in your Japanese car—assembled in Kentucky, powered by gasoline from Saudi Arabia, Nigeria, and Russia. Good morning!

In our increasingly globalized age, you're likely to handle objects imported from several dozen different nations before you even eat breakfast. Advances in transportation and communications technology have made it possible for businesses to source raw materials and products from nearly anywhere in the world and ship them out to consumers around the globe.

The results of globalization go beyond the success that U.S. grocery stores have had selling water from Fiji. You can get a Coke in Fiji, or in any of two hundred other nations. In 113 countries, you can use that Coke to wash down a McDonald's hamburger—or, in India, a vegetarian McAloo Tikki burger. You can watch a Bollywood film on your Kenya Airways flight from Nairobi to London, where the video store has got all the hot new films from Lagos. And you'd probably need to go to Bhutan to avoid seeing ads for the newest Hollywood film.

This increasing integration of global economies isn't an accident: it's the result of policies and institutions put in place at the end of World War II. This system, which included the World Bank and the International Monetary Fund, wasn't just a reaction to the herculean task of rebuilding European cities and German

MSC RITA
PANAMA

and Japanese economies. It was an attempt to ward off the economic slump the planet faced between World Wars I and II, when trade between nations slacked, markets collapsed, and the global economy lapsed into the Great Depression.

After the war, the Allied nations, led by the United States, advocated a system of reduced international trade barriers, convertible currencies, and trade among one another. As their economies recovered, Japan and Germany also bought into this system, and when the Soviet Union fell, newly independent Eastern European and Asian states joined in. With rare exceptions, like Cuba and North Korea, the world's nations have chosen—or have perhaps been coerced—to play by a single set of trade rules.

There's a good argument that this global integration has made the world a better place, even in developing nations. Since the end of World War II, life expectancy in the developing world has doubled. Since 1960, the percentage of the world living with inadequate food supplies has dropped from 56 percent to less than 10 percent. Since 1986, the percentage of people living on less than a dollar a day has been cut in half. In many ways, globalization is working.

Or is it? Critics of increased global trade point to the ways in which developed nations have benefited more than developing nations, and argue that global trade is rigged to further enrich wealthy countries and further impoverish poor ones. Environmentalists point out that shipping Chilean grapes to London entails a huge cost in carbon dioxide that globalization's proponents ignore. Human rights activists worry that the jobs being created in developing nations are jobs for children in sweatshops. And artists and activists around the world are concerned that exports from Hollywood and Tokyo are crowding out local, unique film, music, and fashion.

Global trade is huge, and it's not going away. Few of us would want it to, but most of us would like it to be better—less damaging to the environment, fairer to workers in poor nations, more committed to preserving cultural differences.

Can we turn free trade into fair trade? Can the system of global trade work for the very poor as well as for the very rich? Or does achieving justice and sustainability involve turning away from the systems we've used for the past sixty years? If so, what do we turn to? EZ

Container ship, Halifax, Canada, 2001.

Why China Wins

China is rapidly becoming the world's factory. By one measure, China is responsible for 13 percent of global economic output, making its economy twice the size of Japan's, and third only in size to the United States and the European Union. The United States imported $197 billion worth of goods from China in 2004—11 percent of all U.S. imports. Wal-Mart, the largest retailer in the States, imports up to 70 percent of its inventory from China. Wal-Mart does more business with China than do all of Russia and the United Kingdom.

Wal-Mart buys so much from China for the same reasons the United States as a whole does: Chinese goods are extremely inexpensive, and Chinese manufacturers make a diverse range of products of increasingly good quality.

Goods from China are inexpensive for two reasons: currency valuation and labor costs. For many years, China "pegged" its currency, the yuan, to the U.S. dollar at a fixed rate. During that time, the Chinese economy strengthened dramatically compared to the U.S. economy. If China's currency had floated freely against the dollar, the yuan would have risen—a dollar

深加工
120

would buy 6 or 7 yuan instead of the 8.28 it purchased for more than a decade. So Chinese goods remained cheap for American consumers—too cheap, in the eyes of American manufacturers. Under pressure from the U.S. government, China revalued the yuan in July 2005, but maintained a peg, tying the yuan to the dollar, yen, and euro. Naturally, American manufacturers would have preferred to see the yuan float freely, so that China would be on the same footing with other trading nations, and so that its exports would be significantly more expensive. This would doubtlessly come as a shock to U.S. consumers, who've grown accustomed to low costs at retailers like Wal-Mart.

In addition, China's manufacturing labor costs are incredibly low in U.S. terms—manufacturing jobs generally pay a hundred dollars per month. Workers are able to survive on these modest incomes because employers commonly provide housing in dormitories near the factories, and because the cost of living is extremely low in China. The alternative to factory work is a life of grinding rural poverty, so there are plenty of people willing to fill positions, and an urban job—versus farming and an arranged marriage—is an especially attractive option for young women from rural areas.

China's system of residency permits *(hukou)* makes it illegal for many rural migrants to live and work in Chinese cities, so migrant workers are unlikely to complain about unsafe working conditions, seven-day workweeks, or withheld pay, for fear that bosses will report their work status to authorities. (Undocumented workers in the United States and Europe face similar dilemmas.) In addition, environmental laws in China often go unenforced, making products cheaper, but leading to a massive ecological crisis there [see Chinese Cities of the Future, p. 271]. China does so well in the global economy in large part because its workers (and the environment) pay a steep price so that Chinese factories can keep their price tags small. EZ

The Non-sweatshop Future

Whereas China competes primarily on grounds of low price, other nations are discovering economic opportunity by branding themselves a source of non-sweatshop labor. Lesotho has emerged as one of the world's leading manufacturers of blue jeans, thanks to favorable trade terms with the United States and close collaboration with U.S. retailers Levi Strauss and Gap. Through a combination of factory inspections, community initiatives, and government cooperation, Lesotho has become a model for sweatshop-free garment sourcing. Garment manufacturing represents 90 percent of Lesotho's manufacturing economy and 40 percent of its national economy.

But the garment boom in Lesotho is slowing. On January 1, 2005, the Multi-Fiber Arrangement—a huge, complicated treaty that governs the global garment trade—expired. The MFA limited the amount of garments exported from China and other large textile producers, creating opportunities for smaller nations. Without the MFA, the Chinese share of the international garment market is expected to expand from 20 percent to more than 40 percent. And, unlike China's yuan, Lesotho's currency isn't tied to the dollar, but to South Africa's rand; when

Preceding pages: Chinese workers processing chickens, Jilin Province, China, 2005.

Right: Shipping containers, Montreal, Canada, 2001.

the dollar gets weak, Lesotho's exports to the United States get more expensive, while China's stay relatively cheap. Will Lesotho's garment industry survive? In today's global economy, that depends on Chinese manufacturers and U.S. consumers as much as it does on Lesotho's workers and government. EZ

Understanding Global Shipping

Search Amazon.com for books about the Internet, the global telecommunications network that connects us all, and you'll find more than 23,000 titles. Search for books on shipping containers and you'll find fewer than 200. After all, who wants to read about container ships and the cargos they carry?

Yet these homely metal boxes are the key component in an astounding network of ships and ports that allows goods to flow freely around the world. Before North Carolina trucker Malcolm McLean modified a tanker ship to carry truck trailers in 1955, cargo was moved by armies of dockworkers, who carried boxes and bundles across cargo nets in a model called break-bulk shipping: cargo was shipped in bulk and broken into smaller units when it arrived. Now 90 percent of the world's cargo travels in standard-size metal boxes, 8 feet wide, 8.5 feet tall, and either 20 or 40 feet long (2.4 x 2.6 x 6.1 or 12.2 meters).

The world puts a lot of cargo in these boxes. According to the World Shipping Council, the equivalent of more than 100 million twenty-foot containers (or TEUs—twenty-foot-equivalent units—in the shipping trade) moves from port to port each year. These containers can carry a lot of cargo—up to 53,000 pounds (24,000 kilograms) apiece—meaning that container ships carried roughly 5.3 trillion pounds (2.4 trillion kilograms) of cargo in one year. That's about the equivalent of moving every human being on the planet ... six times.

The vessels that make the global-shipping network function are unbelievably huge. New vessels are classified as Panamax or post-Panamax. Panamax ships are as big as the Panama Canal can accommodate; post-Panamax are larger, typically 1,150 feet (350 meters) long and 130 feet (40 meters) wide. One of these ships can carry as many as 6,600 twenty-foot containers. And ship designers are planning ships at least three times as large.

Container ships don't just move huge cargoes—they move them very cheaply. Shipping a container across the Pacific can cost as little as $500—about $0.02 per 2.2 pounds (1 kilogram). The direction in which a container travels matters a great deal; because the United States imports so much from Asia, and exports primarily waste paper, scrap metal, and empty containers in return, it generally costs three times as much to ship from Asia to the United States as it does to ship the other way around.

Shipping stuff around the world may not cost much, but it weighs heavily on the planet. Sending things around the world can significantly swell their ecological footprint [see Questioning Consumption, p. 32], and the shipping industry itself has not been known as a green standard-bearer, as oil spills, toxic ship-breaking (the process by which old ships are scrapped), and the dumping of polluted ballast water are all commonplace. In addition, transportation is a major contributor to climate change [see Local Greenhouse Forecast, p. 506].

We could green the shipping industry, though. The huge maritime corporation Wallenius Wilhelmsen has brought forward conceptual plans for the Orcelle—what it calls the "green flagship" of the future. The Orcelle would be powered by a combination of sails, solar energy, and fuel cells, using optimal designs to carry the most cargo possible—while requiring no ballast water, using fewer toxic chemicals, and generating no greenhouse emissions. The company predicts that such vessels may be plowing the waves as soon as 2025. EZ

Why Does the United States Still Export Cotton?

That's the sort of question economists like Pietra Rivoli, author of *The Travels of a T-Shirt in the Global Economy,* like to ask. Intrigued when an antiglobalization protester asked her, "Who made your T-shirt?" Rivoli decided to find out. After purchasing a shirt on a Fort Lauderdale street, she traced the origin of the item from a cotton field in West Texas to a factory in Shanghai to a T-shirt printer in Miami—and imagined its likely eventual fate at a textile-recycling facility in Brooklyn and a used garment market in Tanzania. While it wasn't a shock that the shirt was sewn in Shanghai by a woman earning a hundred dollars a month, it was a surprise to Rivoli that the cotton in the T-shirt had been grown in the USA.

It's usually a bad idea to base your economy around the export of commodities. Developed economies can afford to build factories, import raw materials, and export finished goods—which sell for higher prices and aren't as susceptible to price fluctuations. The Ivory Coast exports cocoa beans to Switzerland, which exports high-value chocolate to the world. Highly developed nations tend to export high-value goods, because labor costs to assemble goods are very high.

Frequently, wealthy nations will design and market goods, but will outsource their manufacture to less developed nations. It's hard for American factories to make a profit sewing jeans, so Gap designs next year's line, then pays factory workers in Lesotho to sew the jeans.

Cotton is a commodity that's subject to notorious price fluctuations. It's also hugely labor-intensive to harvest and risky to grow, because bad weather can destroy a crop very easily. Countries that export a lot of cotton usually do so either because they don't have a large industrial sector (Mali, Uzbekistan) or because they've got huge populations of poor rural farmers (Pakistan, India, China).

But neither of those reasons accounts for the United States being the world's largest cotton exporter. Two other reasons do: technology and subsidies. The United States has been a leading cotton-growing nation for two hundred years, first using slaves, then sharecroppers, as a steady source of low-cost labor. In the absence of that cheap labor, cotton farmers in the States had to develop sophisticated technology, with ample help from government-supported agricultural universities. The technology allows them to harvest huge crops with a small labor force, and to turn cottonseeds and waste into profitable commodities.

But the main advantage American cotton farmers have is a cooperative government. The United States pays roughly $4 billion a year to 25,000 U.S. cotton farmers, through a set of subsidies, price guarantees, and insurance policies. One subsidy guarantees U.S. farmers a minimum of 72.24 cents a pound for their cotton. Because the global market price for cotton in 2004 was 38 cents a pound, U.S. taxpayers paid cotton farmers a lot of money. But taxpayers paid cotton farmers even more money thanks to laws requiring American garment factories to purchase a certain percentage of their cotton from American growers. And, just to ensure that cotton farmers don't lose too much sleep, other programs compensate farmers for weather losses, extend them credit, and help them develop further technologies.

High-tech farming and government support make U.S. cotton more globally competitive than it otherwise would be. Because the United States exports so much cheap cotton, the global price of cotton remains low. In her book, Rivoli points to a range of studies suggesting that U.S. subsidies depress the global price of cotton from 3 to 15 percent. Consequently, farmers in the developing world have to sell their crops for lower prices, and make less money. When Africa's leaders say they want fair trade, not more aid, they're talking about cotton: the annual U.S. cotton subsidy is roughly three times greater than the amount the U.S. Agency for International Development gave Africa in aid in 2004. EZ

RESOURCE

Global Exchange
http://www.globalexchange.org/
Global Exchange is one of the most innovative and passionate fair-trade groups around. If you're looking for information on the dark costs of globalization and how we might pursue fairer, more sustainable economies, look no further.

Creating Healthy Homes

No substance is more pure than mother's milk, right? Maybe that was true a few centuries ago, but today breast milk usually contains a whole host of toxins, including contaminants found in things like paint thinner, fungicides, and gasoline. The chemicals of industrial production that lurk everywhere make their way into our babies' bodies with every gulp.

Fear tactics aren't necessarily the best motivation for change, but the ingredients in common household cleaners are downright frightening. The number of toxic substances we generally stock under our sinks and in our utility closets is practically unthinkable, which is just the problem: we choose not to think about them. We also choose not to think about those toxic substances once they swirl down our drains and disappear. But nothing we "wash away" is actually gone—chemical residues remain, floating into the bathwater, and the rest flows into sewers and groundwater, and ultimately cycles back into our lives one way or another.

Fortunately, nontoxic products are increasingly available and affordable. A healthy home and a clean conscience don't have to result in an empty bank account. SR & JF

Ingredients: The Good and The Bad

Kitchen Cleaners: Safety in the kitchen is important to most people. It's a paradox, though, that scouring countertops and sinks with poisonous chemicals gives us peace of mind. Bacteria can be harmful, but toxic cleaning agents do not make our homes safe. Many glass cleaners contain ammonia, which can cause headaches and lung irritation. Phenol and cresol in disinfectants can cause dizziness, and kidney and liver failure. Oven cleaners often contain lye and sodium hydroxide, which can burn the skin, the eyes, and the respiratory tract, and butoxyethanol, which can poison the blood, liver, and kidneys. Scouring powders and cleaning solutions often contain chlorine bleach, the number-one cause of poisoning in American homes.

Homemade or ecofriendly alternatives to these general cleaners include diluted tea-tree oil, an excellent disinfectant; borax, a naturally occurring disinfectant; white vinegar, which can be used to clean floors and other surfaces; and a baking soda mixture, which is great for scouring. We don't have to concoct natural cleansers ourselves—these ingredients can be found in store-bought products as well.

Around-the-house Products: Laundry detergent is one of the worst culprits when it comes to personal-health problems. Detergent additives can absorb into the skin on contact, causing irritation in the short term and cell damage in the long term. Fragrances found in detergents and fabric softeners can contain chemicals that have been linked to cancer and to reproductive harm. To avoid these health hazards, look for soaps and detergents labeled Fragrance Free.

As an alternative, castile soap can be used as a general cleaner anywhere in the home, including in the laundry. Natural oils such as olive, walnut, food-grade linseed, and jojoba make excellent furniture polish and floor wax. (Mineral oil is the more conventional ingredient for wood treatment, but it is derived from petroleum and it off-gases toxins.) Other alternative household cleaners include grain alcohol as a solvent instead of toxic butyl Cellosolve, and plant-oil disinfectants such as eucalyptus, rosemary, or sage rather than triclosan, which is found in everything from detergents and soaps to lotions and mouthwashes.

Bathroom Cleaners: Hydrochloric acid and sodium acid sulfate, both of which are found in toilet bowl cleaners, can burn the skin and eyes or cause vomiting, diarrhea, and stomach burns if swallowed. Other common toilet cleaners contain phosphoric acid, which can cause blindness if it gets into the eyes. Mildew removers often have pentachlorophenol, which can be fatal if swallowed.

Milder alternatives to common bathroom cleaners include grapefruit-seed extract, which can be a disinfectant, and a paste mixture of baking soda and water, which is the all-natural scouring agent of choice.

Shampoos, body washes, and other soaps and detergents are often made from and packaged in petroleum derivatives that present a spate of environmental problems. It's hard to believe, but the average shampoo bottle is made virtually 100 percent from petroleum products, from the soap inside, to the bottle containing it, to the ink printed on the bottle. But we can buy bottles made from recycled plastic, printed with nontoxic ink, and containing natural cleansing agents without spending more money.

Paper Products: Besides the wastefulness and disposability of toilet paper, napkins, and paper towels, there are broader environmental implications associated with these products. Most are bleached with chlorine, which is an irritant to the skin and respiratory system. Discharged by-products from paper mills contain dioxins, which don't easily biodegrade and eventually build up in our air, water, and soil.

Paper products made with post-consumer content are available in ordinary supermarkets, but we need to be conscious of the ratio of recycled content to new content, as many brands contain less than 5 percent recycled material. In general, we should try to buy in bulk to reduce the amount of packaging per product, and to become aware of how many disposable products we buy in the interest of convenience. A staggering variety of single-use cleaning items are now available, and while these products are tempting as time- and mess-saving tools, they are extremely wasteful. SR & JF

The Recycling Debate

Does recycling really matter?

There's no shortage of debate over the net advantage of recycling. Plenty of staunch environmentalists argue that the energy required to recycle bottles and newspapers negates the benefit of creating reusable material. But abundant evidence demonstrates otherwise.

The amount of energy required for recycling depends on numerous factors: what type of material is being recycled, how the material gets collected and delivered to the recycling site, and how processed the resulting material will be. According to the National Recycling Coalition, "it takes 95% less energy to recycle aluminum than it does to make it from raw materials. Making recycled steel saves 60%, recycled newspaper 40%, recycled plastics 70%, and recycled glass 40%. These savings far outweigh the energy created as by-products of incineration and landfilling."

No doubt recycling is an intermediary step: if we truly want to save energy and stem the stream of waste we send to the landfill, we'll have to implement new industrial production systems that generate less waste and fewer disposable components to begin with. (We'll also have to reduce consumption of disposable/packaged goods and find more ways to reuse things.) In the meantime, recycling is certainly a worthy practice, and by simply tossing things in their appropriate bins, we can gain a clearer perspective on just how much stuff we dispose of every week. SR

Dryers

For cleaning our clothes, we have few options that are both ecofriendly and convenient. The washing part can be done with relative ease: we can purchase a more efficient washing machine by seeking out an Energy Star label or its equivalent [see Using Energy Efficiently, p. 164],

Crushed plastic bottles awaiting recycling.

and we can use nontoxic detergents. Washing clothes is still a water-intensive process, but it's getting better. The drying part, however, is much more problematic. After the refrigerator, the dryer sucks more energy from our homes than any other appliance.

Dryer technology has improved so little in the past few years that Energy Star doesn't even offer a label to distinguish the bad from the better. No matter what kind of dryer we own, we're consuming a whole lot of energy and spending a whole lot of money with each load of soggy clothes. What are our options? Well, in dry, sunny climates, an old-fashioned clothesline comes out on top. But if we need to keep our dryers, we can do the following to reduce the amount of energy we use:

- Keep the dryer in a warm part of the house. The machine won't have to work as hard to generate and maintain sufficient heat.
- Dry similar fabrics together. A load of thin, synthetic articles will be done much faster than a mixed load of heavy towels and lightweight sheets.
- Run separate loads consecutively to take advantage of residual heat, and use the Permanent Press setting to complete the tumble cycle with leftover heat.
- Clean the lint filter after every load.

At some point, dryers may make the leap to bright green, but for now, the best we can do is to follow these tips, and—whenever possible—take advantage of a sunny day. SR & JF

Dry Cleaning

When it comes to processes with harmful environmental impacts, dry cleaning takes the cake. Most dry cleaners use a solvent called perchloroethylene, or perc; the process produces waste that the Environmental Protection Agency considers hazardous. Perc is toxic to plants, fish, and marine animals, and toxic to us in large doses. It can potentially contaminate drinking water if it's allowed to seep into the the ground. The substance can enter our bodies through contaminated air and water, and can be stored for prolonged periods of time in body fat, as well as in the plants and soil around our homes.

There are now some green dry cleaners out there that use silicone-based solvents instead of the conventional kind; there are also wet cleaners that use water and detergent instead of solvents. These are good alternatives to traditional dry cleaners, but they are both more expensive than the standard. So just remember that hand washing often does the trick. SR

RESOURCES

Green Clean: The Environmentally Sound Guide to Cleaning Your Home
by Linda Mason Hunter and Mikki Halpin
(Melcher Media, 2005)
Green Clean is a fun, lively romp through the various alternatives to bringing toxic chemicals into our homes. It manages to dance a delicate line—it's ethically concerned without being preachy, grounded in the earth without being crunchy, playful without being vapid: "When it comes to disposing of organic waste, a compost pile wins over a garbage disposal as the best green option. Garbage disposals use a significant amount of water and load sewage treatment plants and septic tanks with organic matter. Composting returns organic matter to the earth. Adding compost to your lawn or garden nourishes plants and improves soil structure and water retention. As Walt Whitman said, 'Behold this compost! Behold it well!'"

Ideal Bite
http://www.idealbite.com/
Ideal Bite offers up a steady stream of tips for greening your daily life. The site tends to focus more on the small steps, rather than the big choices that produce the most change, but sometimes the simple things need attention, too. Generally lighthearted and fun, Ideal Bite is a welcome relief from the daily servings of gloom and doom found on many environmentally oriented Web sites.

Doing the Right Thing Can Be Delicious

We've come a long way from a diet that is sustainable and healthy for our bodies, our communities, and our planet. Although the industrial food revolution of the last century promised abundance and an end to hunger, in many ways it has delivered the opposite. The United States produces more than twice the required daily caloric intake for every man, woman, and child, yet as many people in the United States go hungry as populate the entire country of Canada. Worldwide, 852 million people go hungry every year, and 16,000 children die every day from needless hunger.

The industrial farming revolution of the last century—particularly the introduction of chemical pesticides, monocultural production, and confined animal feedlots—has made farming one of the world's worst polluters. In the United States alone, we blanket the country with billions of pounds of pesticides.

Industrial farming also has the dubious distinction of being one of the world's biggest contributors to greenhouse-gas emissions. Our petroleum-dependent farming eats up oil faster than you can say "Gulf War," using ten calories of fossil fuel for every one calorie we produce. According to journalist Richard Manning, author of *Against the Grain: How Agriculture Has Hijacked Civilization,* the production of just one two-pound bag of breakfast cereal burns the energy of half a gallon of gasoline.

Even worse, as industrial farming pollutes our environment, it also pollutes our bodies. Research from the Centers for Disease Control and Prevention on exposure to environmental chemicals indicates that most of us walk around with a significant "body burden" of chemical residues, many from farm chemicals.

Devouring the fast food, junk food, and fake food that saturates our supermarkets and restaurants has led to a host of health problems as well, from obesity-related diseases that lead to premature death, to certain cancers and neurological and hormonal problems that are associated with the chemicals used in our fields. Our fake-food culture is also largely to blame for nearly 76 million foodborne illnesses, which lead to more than five thousand deaths every year.

Yes, the twentieth century may have seen the fastest revolution in our dietary and agricultural practices in human history, but around the world, citizens (eaters, farmers, policy makers, researchers, and health advocates) have also fostered a different sort of revolution in food and farming, one that holds real hope. Indeed, this new century may see a revolution in food equally startling to the twentieth century's—only this one will be much better for us. Oh, and it will taste much better, too. AL

Slow Food

What ever happened to going to the market for fresh food, cooking at home, sitting down, and eating slowly enough to taste our meal? What happened to gathering together with family and friends to share conversation, trade recipes, and try something new in the kitchen? What happened is fast food. Hardly anyone has remained immune to the instant-gratification food culture we live in, or the breakneck pace of a regular day.

But one person did decide that this grave problem warranted a backlash, and in 1986, in response to an outrageous plan to build a McDonald's on the Spanish Steps of his native Rome, Carlo Petrini issued a rallying cry to citizens worldwide to drag their feet and slow down the acceleration of modern cuisine. Thus began the Slow Food Movement.

The movement encourages us to rediscover and rededicate ourselves to the pursuit of pleasure through food. But it's not just about delighting in all things edible. Slow Food is fundamentally about supporting local farmers, preserving cultural customs, promoting organic agriculture, and teaching people to rely on themselves—not on franchised eateries—for their sustenance.

The Slow Food Movement's snail mascot has made its way speedily around the globe, leaving a trail of motivated slow foodies in forty countries. You can easily join your local association or start one in your hometown and be a participant in the Slow Food Movement's restoration of food culture. SR

Sustainable Farming 101

When we push our carts through the aisles of the supermarket, the piles of organic and conventional fruits and vegetables look nearly identical, but the organic cost more. So what are you buying into when you buy organic?

Well, first and foremost, you're buying a product that has not been grown using synthetic fertilizers or drenched in chemical pesticides. Organic farmers use compost and "green" manure (crops that are tilled into the soil before they mature and help replace soil nutrients) to fertilize crops. They avoid exhausting the soil's nutrients in the first place through crop rotation (in contrast with the "monocropping" of industrial agriculture, in which the same crops are constantly replanted in the same fields, depleting the land). Crop rotation also fosters a thriving population of beneficial microorganisms in the soil, which in turn fortify the crops, to yield more nourishing vegetables and grains. Organic farmers control pests by encouraging helpful predator insects and by shielding crops with protective, lightweight row covers. These techniques are nothing new; organic farmers are simply working with nature's systems instead of against them.

The *sustainable* in *sustainable farming* doesn't just refer to the pesticide-free produce that comes out of the farms' soil. In the Rodale Institute Farming Systems Trial—a 22-year-long study that compared organic and chemically grown soybeans and corn in the United States—researchers found that organic techniques used 30 percent less fossil fuel on average than chemical-dependent techniques, reduced soil erosion, and conserved water. Because sustainable farms grow a variety of crops, they protect biodiversity, which has been severely threatened by monocultural industrial farms. Sustainable farms are simply healthier overall; if the soil is healthier, then what comes out of it will be, too. In addition, plots of organic farms fare better during both floods and droughts, because the soil has a superior ability to absorb and retain water, attributes that will become even more valuable as global climate change escalates. SR

RESOURCES

Fresh Food Fast: Delicious, Seasonal Vegetarian Meals in Under an Hour by Peter Berley (Regan Books, 2006)
Written by the former chef at New York City's vegan restaurant Angelica Kitchen, *Fresh Food Fast* makes good on its title with delicious menus, handy tip lists, and time-saving tricks. AL

Local Flavors: Cooking and Eating from America's Farmers' Markets by Deborah Madison (Broadway Books, 2002)
Fresh-foods maven Madison takes you on a trip through her favorite farmers' markets with recipes that celebrate the delicious fare she discovers. AL

The Organic Foods Sourcebook by Elaine Lipson (McGraw-Hill, 2001)
Lipson's book is sure to become a classic, if it isn't already, among organic-food advocates. Easy to read, clearly articulated, and compelling, *The Organic Foods Sourcebook* drives home the point that organic food doesn't just taste better or look better: it *is* better. With chapters on genetic modification, organic dairy, and everything in between, Lipson goes beyond simply explaining the benefits of organic food, to provide a plethora of resources for further reading on the organic-food movement. In her words: "If there is one thing the reader should take away, it is that we all have the power to create change that is felt at the highest levels. We do it with our choices every day and by sticking with those choices over the long haul. The details of the organic portrait will change with time, but we all must remain dedicated to the vision of a better, healthier way to feed ourselves and live on this planet."

Opposite, left: Organic asparagus, eggs, and quiche.
Opposite, right: Organic vegetable farmer and son, Washington.
Right: Farmers' market, Brooklyn, New York.

Buying Better Food

We all value the modern convenience of one-stop-shopping. But when did it become normal to buy food at office-supply or home-decor stores? No matter what we go into a store for, we can almost always walk out the door with a snack in hand. And that's a problem because we are losing our connection to the source of our food; when we can buy it anywhere, it seems to come from nowhere.

By understanding a few basic things about how our food is grown and processed, how far it travels to get to us, and what happens to it on the way, we can make better choices about what we eat. We'll get fresher, better-tasting food, and in the process spare the environment and support the farmers who work every day to produce the food that sustains us. SR

Buy Local

In the middle of Denver, in the middle of December, you can walk into most any supermarket and buy a ripe mango. This has been true long enough that almost nobody stops to think of the remarkable distance that mango traveled

or of the tree it fell from, which is probably enjoying a balmy tropical day on the other side of the planet. Proponents of eating local food balk at the ubiquitous midwinter mango. Why? Because they think about the baggage that mango flew in with.

The mango carries with it gallons of fossil fuel from the planes and trucks that delivered it, the hard work of laborers who picked and packed it, and likely many pounds of pesticides and preservative coatings that kept it intact during the long journey from the farm to your table. That's a pretty heavy mango.

Now you don't need to feel guilty about eating a mango, but there are a lot of reasons to appreciate food that was grown closer to home. Locally grown produce is fresher, which makes it taste better and ensures that you're getting maximum nutrition for your buck. Additionally, food that's both grown and sold locally skips many steps of processing, packaging, and transporting, sparing the environment and eliminating the dependency on suppliers far removed from the region. It's worth your while to seek out local food sources, and numerous campaigns and community resources all over the country can help you tap into them. A ripe mango tastes delicious, but a peach picked this morning tastes divine.

You can take the decision to buy local as far as you like. The 100-Mile Diet, for example, includes only food that has been produced within a hundred miles of your home. The diet got its start when a Vancouver couple decided to try it for a year and wrote about the results in an online magazine, the *Tyee*. As they discovered, adhering to the diet in a country where nearly everything on the supermarket shelves is imported from thousands of miles away can be costly and difficult. However, with the hundred-mile rule in the back of your mind, you can commit to buying local as much as possible, and when you don't, you'll be more aware of your food's distant origins.

A few decades ago, farmers' markets were scarce in this country. Today, there are thousands of markets scattered from coast to coast. Depending on what you buy, you may pay a slightly higher price when shopping at the farmers' market, but there are two significant reasons to shell out a few more cents: first, you know exactly where that money goes—straight into a farmer's pocket; and second, that farmer's produce is guaranteed to taste better than the grocery variety.

If you can't find a farmers' market nearby, it's not very hard to start one for your community. It may take a while to find and secure an appropriate site and coordinate local vendors, but once it takes off you'll have given yourself and your community an invaluable resource for better food, not to mention a great central meeting place that feels like a weekly neighborhood festival.

Buy Together

For the sake of convenience, many people now buy their food in big-box stores like Costco, or superstores like Wal-Mart, where bulk quantities ensure less frequent shopping trips. These big chains make it very hard for small farmers to break into the bigger markets, and their use of central distribution hubs means that most of the food they sell has taken a world tour before it reaches your pantry. An alternative to the big boxes comes in the form of food co-ops. Food co-ops are community groceries, often structured as worker-owned or member-owned businesses, meaning that the employees and customers have a say in what's on the shelves and where it comes from. The foodstuffs at the co-op usually don't incur a huge markup, because the workers and members have a personal investment in the quality and price of the inventory, meaning it's worthwhile to seek out items that offer optimal value—both economically and nutritionally. Where possible, the best way to do this is to source products from local farmers and vendors, a strategy that guarantees that our dollars support the local economy while buying us the freshest food possible. SR

Buy Direct

The ultimate local food is what you grow in your own garden. But not everyone has a green thumb, much less the time to cultivate tomatoes and eggplants at home. For the next best thing, you can participate in community-supported agriculture (CSA).

Farms that offer CSA programs sell you food directly, delivering a weekly box to your door or to a central location in your neighborhood. All you have to do is sign up, and you can be a part of their regular weekly rotation. With CSA, everyone wins: your local farmers get much better prices for their produce than they would if they sold to a corporate middleman, and you get fresh, wholesome (often organic) fruits and veggies that you can enjoy in the warm glow of having done the right thing. Plus, those boxes of food are usually a great value.

The only problem is that word *boxes*. Often, getting a share in a CSA means getting boxes and boxes of food—mounds of zucchini, mountains of kale—which is great if you're feeding a large household, but not so great for a single person or a couple. Most small households could never eat their way through such a pile of produce.

Many CSA initiatives are starting to respond to the demand for more manageable portions, by offering more flexible memberships. For example, Full Circle Farm, outside of Seattle, allows members to sign up by the week rather than by the season, meaning you only have to budget by the week. This is a much more attractive option for those of us on a tight budget.

Community-supported agriculture programs like Full Circle allow farmers to keep land under cultivation rather than selling to developers, who are likely to turn the farms into more profitable residential developments. And CSA farms located close to cities allow urban dwellers access to fresh or organic produce, as well as the invaluable opportunity to see and understand where their food actually comes from. AS & SR

Buy Fair

Unfortunately, not everything you need grows on your local farm, or even within your state or country. You don't have to give up imported goods to be a responsible eater, but when you do buy them, be sure they are marked Fair Trade.

The fair-trade movement, initiated in the 1980s through collaboration between farmers in developing countries and farmers' rights advocates in Europe, has set the stage for fair-trade-labeling initiatives in twenty countries. Crops such as coffee, chocolate, and bananas traditionally come to us through supply lines riddled with injustice toward workers, contributing to the oppression and poverty of many people in the developing world.

For farmers in developing countries, fair trade addresses one of the barriers to a better life: the unpredictable whims of world-commodity market prices that their livelihoods are often subject to. Sometimes farmers get so little for their wares—and pay so much for supplies and equipment—they end up empty-handed at the end of the year, despite grueling work. Fair trade changes the dynamic of buying and selling in the world market by guaranteeing a fair price to hundreds of thousands of farmers in more than fifty countries.

Fair-trade-certified labels can now be found in dozens of countries. Since 1999, TransFair USA has been certifying fair-trade products in the United States, and the distinctive logo can be found stamped on everything from bags of tea to bunches of bananas. Most coffee shops that purchase fair-trade beans advertise their responsible practices, so be on the lookout for java and other products with a clean history. AL

Greenwashed Food

As a result of the establishment of organic-certification programs in dozens of countries, food packages are now covered in labels to help consumers choose organic provisions. Along with these label initiatives, however, comes the inevitable misappropriation of language by large corporate food companies, which have adopted the word *organic* for marketing leverage and customer manipulation. This is the face of greenwashing [see Consuming Responsibly, p. 38] in the food industry.

In 2002, the United States Department of Agriculture (USDA) implemented the first uniform national standard to regulate the labeling of foods as organic. Under the standard, in order to use the USDA's organic seal, a company must comply with all the following criteria: no use of irradiation, genetically modified (GM) substances, sewer-sludge fertilizers, or synthetic pesticides or fertilizers, and no antibiotics or hormones in meat.

While the regulations represented progress in terms of unifying the organic requirements and bringing organics closer to the forefront of consumers' minds, they also had negative implications for small farmers and food producers. The standards essentially boiled organics down to the lowest common denominator, making it difficult for farmers who already went above and beyond that level to demonstrate the superiority of their product. Even worse, the prohibitive costs associated with obtaining USDA certification made it impossible for many small organic farmers to label their goods "organic."

Large corporate food manufacturers were in some ways granted a free pass, with the ability to dole out funds to create adjunct organic brand names without having to dramatically change their practices. Companies created organic lines that had the illusory appearance of products from small family-operated farms; these include Seeds of Change (M&M/Mars), Boca Foods (Philip Morris/Kraft), and Sunrise Organic (Kellogg). The label designs, the names, even the Web sites give no indication of an affiliation with the umbrella company that manufactures the products. It is a deceptive and worrisome situation for conscientious consumers who want to support organics and not corporations.

How can you be sure your organic food is really organic? And how do you know whose pocket your money ultimately ends up in? The Organic Consumer Association and the Organic Trade Association both have abundant resources that help you learn some simple facts for better navigation of your grocery aisle. Here are a few tips to start you off: Be sure to read the ingredients list and not just the brightly colored ad box on the front of the package to learn which and how many of the ingredients are really organic. At the farmers' market, talk to the farmers and ask questions. Some farmers advertise their products as "transitional" instead of "organic," which means that they are in the process of eliminating pesticides and fertilizers. They may even be producing 100 percent organic goods, but without the stamp of government approval. It's worth it to be curious and to inquire about your food. After all, you are what you eat. SR

RESOURCES

Fields of Plenty: A Farmer's Journey in Search of Real Food and the People Who Grow It
by Michael Ableman (Chronicle Books, 2005)
This exquisite book, part travelogue, part cookbook, and part witness to the crossroads that agriculture and food face today, begins with a series of questions that writer-farmer-photographer Ableman poses while beginning a 12,000-mile (19,312-kilometer) tour of America's farms: "How do we make sure that pure food is available to all, not just those that can afford it? How can we grow food without depending on vast amounts of energy and foreign oil? How can we protect and enhance biodiversity and the natural environment within and around our farms?" Ableman beautifully recounts his quest, which takes him from rooftop greenhouses in Manhattan to the chili-producing deserts of New Mexico.

Organic produce in a supermarket, Poulsbo, Washington.

Farms of Tomorrow Revisited: Community Supported Farms—Farm Supported Communities
by Trauger Groh and Steven McFadden
(Bio-dynamic Farming and Gardening Association, 1998)
This "textbook" for community-supported agriculture, first published in 1990 and then revised in 1997, provides a theoretical overview, real-world examples, and countless resources for anyone interested in reconnecting with the farmer and the land, and knowing exactly where their food is coming from. In the authors' words, "The problems of agriculture and the environment belong not just to a small minority of active farmers; they are the problems of all humanity, and thousands of people are searching for new ways and new solutions." Community-supported farms are undoubtedly one of the most compelling solutions out there for a sustainable future, and this book details the plans, actions, and intentions of the "farms of tomorrow that are needed today."

Better Food Everywhere

To choose healthy food, we have to have choices to begin with. Much of the time, we don't get to pick what lands on our plates. Whether we dine in a school lunch hall, a conference room, a hospital cafeteria, or a restaurant, if we're hungry and the options aren't ideal, chances are we won't choose hunger over an unhealthy meal—and we certainly shouldn't. But if institutional and commercial eateries would offer more healthful meals, we'd all be better off. SR

Where It Matters Most: Schools and Hospitals

Farm-to-school: In most school cafeterias, if the fruits and vegetables don't come from cans, they come from the freezer. Whole grains are as hard to find as a second helping of tater tots. Juice comes from concentrate, and sugary sodas are widely available. What does this mean for our kids? Poor nutrition, yes, but also poor concentration, low grades, and plummeting energy levels. It's no secret that a healthy diet fuels an attentive mind, but when officials cite declining test scores and low attendance rates, how often do they bring up the lunchroom?

Organic produce in a school-lunch line, Olympia, Washington.

Fortunately, a few hot spots around the country have been raising awareness of the severity of this problem and demonstrating innovative solutions. One of the best known is Edible Schoolyard, the Berkeley program founded by one of the icons of American cuisine, chef Alice Waters. The program teaches kids about nutritious food, beginning with the organic vegetable patch at their school. Children take carrots and tomatoes from seedling to salad, learning to cultivate, harvest, and cook all sorts of things. This seminal program has inspired numerous others.

Children everywhere benefit tremendously from having gardens on school grounds. The plots serve as experimental learning labs, and fill open spaces with living, growing greenery and plants instead of asphalt and concrete. Drawing kids into a more intimate connection with the source of their food can give them quite a wake-up call. When a child can pick her own after-school snack from the tree she waters, that's a special thing. She's less likely to munch on foods whose origin she can't imagine.

Farm-to-college: For older students, many more options exist at mealtime. Many high schools and colleges have fast-food joints on the premises—some have entire food courts. While elementary and junior high schools can (and do) ban junk food in the hallways, these kinds of restrictions don't fly for young adults. Luckily, many students know enough not only to reject readily available junk, but to demand better choices. Many colleges have begun to introduce free-range meats, organic produce, and hormone-free dairy products. Some support local farmers by purchasing their products for dining halls and dorms. Among students who are offered a choice, colleges and universities have seen a rapid shift toward a healthier diet. Students who emerge from college already familiar with the advantages of fresh, local, organic food are bound to have an impact on the future of the food industry. In April 2006, the University of California at Berkeley opened the first certified organic kitchen in the nation. By applying for full certification with the California Certified Organic Farmers, the dining hall committed itself to serving 95 percent organic food at all times, and set a new standard for collegiate meal plans.

Farm-to-hospital: You're supposed to go to hospitals to get well, so why do most hospitals serve exactly the fare that makes so many of us ill? In the United States, 38 percent of hospitals have fast-food restaurants inside their cafeterias! In many cases, doctors and residents, some on twenty-hour shifts, can't get a healthy snack on the go. And all those jokes about the repulsiveness of hospital food? They don't exaggerate. But we don't have to stomach it, and many hospitals are getting on board with healthier and tastier fare. The Kaiser Permanente hospitals in California have even set up farmers' markets in their parking lots.

An organization called Hospitals for a Healthy Environment has also begun to turn attention toward the role of food in achieving a sustainable health-care facility. Founded jointly by environmental and health-care agencies, including the EPA and the American Hospital Association, the organization originally focused on reducing pollution and waste from hospitals. But after recognizing the impact of cafeteria opera-

tions on hospital patients and visitors, they also began looking for ways to green the dining areas. Because of the endorsement of this program by major national agencies, the idea is likely to spread. Remember when hospitals had cigarette vending machines in their waiting rooms and ashtrays in their recovery rooms? With any luck, we'll soon look back on Big Macs in hospital cafeterias with the same incredulity. AL

Better Restaurants

For some people, part of the joy of going out to dinner is not having to do any work until a plate of food arrives under your nose. What that also means is that generally, you don't have to put a single thought toward where that food came from. A variety of shocking and horrifying fast-food practices have inspired growing awareness in most people's minds. But all told, there's still a lot we don't know about the food we're being served in restaurants.

Fortunately, some savvy restaurateurs have realized that informing their customers about the food they're eating adds value to every plate. At a handful of fast-food chains and high-end eateries alike, organic vegetables, free-range meats, and local dairy foods are appearing on menus that aren't just the result of chef-inspired top-down food evangelism: it's a shift in what diners want. We're still after the stress-free experience of eating a dish prepared to order—and not having to wash the plate it came on—but we're also craving a little piece of its history. Even if you don't want to maintain a garden or join a CSA [see Buying Better Food, p. 54], you can still eat local and organic by becoming a regular at such an eatery. You'll eat food that's more nutritious—and far more delicious—and you'll also help to spur change.

One of the great pioneers in linking the urban restaurant world to rural agriculture is Blue Hill Restaurant, an award-winning restaurant in New York City's Greenwich Village. When you scan the menu at Blue Hill, you'll find that the dishes center on ingredients that are available from farms in the Hudson Valley and the Berkshire Mountains of western Massachusetts—including owner Dan Barber's own Blue Hill Farm in Great Barrington.

In order to offer the "eat-local" experience to diners outside of the city, Blue Hill at Stone Barns opened in suburban Westchester County, New York. At this Blue Hill, you get more than just a phenomenal meal—you also get to see the local food economy in action. The restaurant includes a working farm that supplies produce and meat to the restaurants, and a classroom where cooking and farming demonstrations give visitors a better sense of their food's origins.

In these tough times for family farms in the Northeast, Blue Hill has risen up as a pillar of support, inspiring diners to care about the food they eat, as well as the local systems that make this kind of eating possible.

At the other end of the spectrum you'll find Chipotle Mexican Grill, a McDonald's-owned franchise that has begun offering natural pork, chicken, and beef—including meat from the well-known Niman Ranch—on its menu of Mexican basics like tacos and burritos. With more than five hundred outlets across the country, and plans to more than double its numbers

Opposite, left: At its salad bar, Evergreen State College provides information on the local farmers who grew the vegetables, Olympia, Washington.

Opposite, right: Volunteers from Evergreen State College harvest leftover crops for use by a local food bank, Olympia, Washington.

Right: At the Blue Hill Restaurant in Stone Barns, New York, diners can look over the fields in which their food was grown.

in the next few years, Chipotle gives those raising ecologically sustainable meats a hearty boost. While naturally produced meats still account for only a fraction of all meat raised in the United States, the rise in demand is steady, and outlets like Chipotle will play a big part in familiarizing people who don't see themselves as "health nuts" with the products.

Taking it further, the new O'Naturals chain bases its entire theme on the goal of providing natural and organic fast food—often sustainably ranched and grown—to people on the go. Founded by Stonyfield Farm president and CEO Gary Hirschberg, O'Naturals has opened a handful of stores in the northeastern United States, with the goal of cranking out meals at the speed of Burger King. It's no small feat to go up against a long-established and relatively homogenous population of cheap restaurant chains—but this one may hit the zeitgeist at just the right moment to really take off. EG

Community Gardens

Few kids who've spent the day cooped up in a classroom will pass up an opportunity to run around and get dirty, so it is no surprise that gardens have proved to be a highly effective tool for introducing nutrition awareness and health education into underserved urban communities. Kids exude contagious enthusiasm, and when they take it home, it infects their family and friends. When a third grader tells his mom about eating a tomato he grew all by himself, the concept of reconnecting with our food literally hits home, and it's a realization that stays with us.

All across North America, you can find community gardens—some as small as a single residential lot, others spanning acres between skyscrapers or factories. Many of the better-established community gardens offer workshops, after-school programs, and employment for young people, and even host their own farmers' markets or contribute produce to a local one. By hosting a hands-on learning environment where community members can watch their own work flourish, urban gardens have found a way to keep people coming back and to get a message of health across in an accessible—and delectable—way. SR

EarthWorks Urban Farm

Once the booming Motor City, Detroit now faces some of the worst dilapidation of any major city in the United States. With high rates of poverty and unemployment, Detroit is beset with a spate of urban problems. But community-based efforts have been making strides toward improving the health and welfare of city dwellers throughout the entire area. One such initiative is EarthWorks Garden, which originated in the back garden of an inner-city soup kitchen and has grown significantly since its inception in 1999. The garden works primarily to educate kids from the public school system about nutrition, biodiversity, and organic agriculture.

Produce from the garden goes toward sustaining the Capuchin Soup Kitchen and local WIC program (Special Supplemental Nutrition Program for Women, Infants and Children). EarthWorks holds open markets several days a week during the growing season to facilitate easy

City Farm in urban Chicago sits on what was once vacant land, and produces organic food and vegetables.

access to fresh, organic products; it keeps prices down and offers coupons to those on financial assistance programs. The volunteer-run markets take place at several WIC clinics around the city, where families can either receive meals prepared with EarthWorks ingredients on-site or use coupons to take fresh produce home.

EarthWorks has its hands in a number of other community projects, mostly focusing on youth education. In addition to teaching the importance of organic and fresh foods, the programs teach the values of mutual respect and community interconnection—a complementary set of lessons that helps solidify basic knowledge and cohere the community. SR

Urban Farming

Historically, many of the world's cities practiced forms of urban agricultural. Paris, just one century ago, grew more than 100,000 tons of crops; it ended up with so much that the surplus was shipped to London. Today, cities from Montreal to Mumbai are experimenting with forms of urban farming. Green roofs [see Greening Infrastructure, p. 256] are increasingly visible amidst the skyscrapers and smog of many cities.

But although programs like EarthWorks Urban Farm are, without a doubt, considered innovative in the Global North, the approach they take is part of mainstream life for much of the Global South. Some 800 million city dwellers around the world grow food, while 200 million earn their living farming in the city. Cities such as Dakar, Sofia, and Singapore produce a significant percentage of their food through urban farms.

Cuba provides us with the gold standard when it comes to demonstrating the value of producing your own food in a volatile world. Following the collapse of the Soviet Union in 1989, Communist Cuba not only lost its biggest buyer (the Soviet Union had paid above-market rates for Cuban industrial-agriculture products such as sugar), but it also found itself faced with a U.S. embargo, and at a distinct disadvantage in the global economy. As a result, Cuba had no money with which to buy oil, fertilizers, or pesticides—the main ingredients of factory farming.

One result was hunger. In 1989 Cubans were consuming an average of 3,000 calories per day; by 1993 that number had dropped to 1,900—the equivalent of skipping one meal. The Cuban response to this crisis, born out of necessity, was to create a system of sustainable agriculture that was not reliant on fossil fuels or global shipping systems.

Urban farming became a big part of this system. This made sense not only because Cuba is an urban nation (70 percent of the population lives in urban areas), but because the lack of fossil fuels made it impossible to continually truck food into the cities from the countryside. The government instituted a program that turned Havana's many vacant lots into farms or community gardens, virtually handing the land off to anyone who agreed to turn it into a viable food source. This scheme was so successful—many neighborhoods were able to produce at least 30 percent of their own food—that it quickly spread to other cities. Today, Havana's crumbling buildings are stitched together with farms and gardens. Forty-one percent of Havana's urban area is used for agriculture, and the city generates 51 percent of Cuba's vegetables.

Today Cubans have regained that lost meal. Best of all, most of what Cuba produces is de facto organic, because the lack of available pesticides and fertilizers meant that scientists and farmers had to devise ways of protecting and controlling crops using only what nature provided. ZH

RESOURCES

On Good Land: The Autobiography of an Urban Farm by Michael Ableman and Alice Waters (Chronicle Books, 1998)
When we think of Southern California, what we tend to envision is not the avocado, cherimoya, and fig trees that grow so contentedly in the Mediterranean climes, but rather the sprawling pastel-colored suburbs and long stretches of six-lane freeways. In the latter half of the twentieth century, as those hard bits of suburbanization crept over most of the area's undeveloped land, the farm that Michael Ableman calls home somehow escaped defilement.

In *On Good Land*, Ableman tells the story of growing and preserving his family's farm in the midst of encroaching tract houses and highways. His own learning process, and the education his farm has offered his kids, his farm-hands, and his neighbors, lends a poetic note to a compelling message; as a farmer situated in the middle of an increasingly urban and disconnected world, he wants us to know that the land matters more than we can imagine. "If you eat," Ableman writes, "soil is your business."

Fast Food Nation by Eric Schlosser
(Harper Perennial, 2002)
Investigative journalist Eric Schlosser tells you everything you ever wanted to know (and probably more than that) about the "dark side of the all-American meal." You will never look at Mickey-D's the same way again—and that's a very good thing.

Super Size Me directed by Morgan Spurlock
(2004)
In this documentary, Morgan Spurlock literally sacrifices his body to illustrate the destructive power of fast food. A news story about two teens who tried to sue McDonald's for allegedly making them obese inspired Spurlock to find out exactly what a steady diet of fast food would do to the human body. The parameters of his experiment were simple: for one month he would only eat things that were offered on the McDonald's menu; he had to try everything on the menu at least once; and any time a cashier asked him if he wanted to "Super Size" his meal, he had to accept. To further replicate the experience of the average American, he did not allow himself to exercise—he even limited the amount of steps he could take in a day.

Spurlock's quick physical decline is irrefutable—midway through the film, a previously straight-faced doctor is so alarmed by the results of Spurlock's blood work that he practically begs him to end the experiment prematurely.

Spurlock's video diary—sometimes humorous (he appears for a weigh-in in a red-white-and-blue Speedo), sometimes nauseating (one Super Size meal makes him vomit)—is stitched together with interviews with a series of experts exploring the rise of obesity in the United States. However, overall he seems to understand that you'll catch more flies with humor than with didacticism.

Eating Better Meat and Fish

Meat has been a fundamental part of the human diet since we came down from the trees. The problems with it are not inherent; rather, they lie in the many ripples of impact that meat production—and fish production—send out into the world. Both large-scale livestock farming and the massive fish farms that dot our coasts destroy the environment: they cause air and water pollution, deforestation, and depletion of fossil-fuel resources.

Some people choose vegetarianism as a way to take a personal stand against destructive factory ranching and farming practices, but for most people, a meat-free lifestyle has no appeal; meat tastes good, plays an integral role in traditional cuisine, and makes up a significant portion of some basic diets. In reality, we could reduce all of the harmful cycles of livestock farming and still have enough meat for a summer-long BBQ. We just need to be smart about our own choices, and encourage smart practices from ranchers and fishers.

The best way to do this is to scout out meat and fish raised or sourced in ecologically friendly, humane, and healthful ways, and take charge of what we put on our plates. We can't yet find these in every grocery store, but they're becoming more common and more affordable. Sustainable meat and seafood appear on menus in all sorts of restaurants, and because their quality is a selling point, eateries generally do their utmost to let us know we're getting something good. In the supermarket, various labels may indicate how the animals and fish were raised and which name brands are associated with responsible ranching and fishing.

The benefits of buying and eating better meat and fish start with improving our own health and stretch to encompass the health of entire ecosystems. We can choose meat that comes from diverse populations of sustainably farmed livestock breeds, and fish that come from protected and respected populations. It's a perfect example of the power of the consumer to steer an industry. When we make smart choices at the meat counter, we let producers know that responsible practices yield better profits.

Sustainable Ranching

Our factory farms dump a shocking amount of animal waste into our waterways and soil every year. Farm machinery churns a significant portion of greenhouse-gas emissions into the atmosphere—and on top of that, cow flatulence adds abnormal quantities of methane, compounding the greenhouse problem before the equipment even enters the picture. But if we don't want to give up burgers—and we don't have to!—how can we remove ourselves from this cycle of environmental degradation? By buying sustainable meats.

What happens on a sustainably run ranch? Responsible ranchers regard their animals as part of a whole system, in which cows are more than steaks on the hoof: they serve as catalysts in restoring and maintaining healthy land. Allowed to graze wherever they like, cows can have a concentrated impact that quickly degrades pastureland. But with managed grazing, a technique that alternately grazes and rests pastureland, the impact is distributed and can actually facilitate increased plant growth, biodiversity, and fertility. Plus, wandering free makes for happier cows. EG

Buffalo Commons: Preserving the Great Plains

The vast farms that span the central United States are dying by their own hand. Decades of shortsighted farming practices have resulted in eroded soils and depleted aquifers. With environmental damage compounded by job loss and steady out-migration, America's heartland faces a bleak future agriculturally.

But it's not too late to turn things around, and ideas abound—like the Buffalo Commons, a suite of ideas for ecological and social restoration of the Great Plains that's been circulating since the late 1980s and continues to carry revolutionary promise. The idea? Restore native grasslands of the Great Plains and bring back herds of buffalo. The Great Plains Restoration Council envisions a decades-long effort, combining community building with the restoration of a continuous wildland corridor, extensive and spacious enough that populations of buffalo and other prairie wildlife will be able to roam freely once again. Reestablished buffalo herds could also be managed sustainably to supplement the beef industry with another kind of red meat. Bison burger, anyone? EG

Cow Power

Dairy cows produce copious milk. They also produce copious poop. It's not something most people think of as environmental pollution, but cow poop is a real problem when allowed to run into waterways and seep into the ground. But we don't have to let cow waste go to waste. If processed correctly, it can become a power source. That's just what Central Vermont Public Service (CVPS) has been doing with its Cow Power program, which promises to provide "renewable energy one cow at a time."

In order to reap power from poop, farms install an anaerobic digester which, over a period of twenty days or so, breaks down some of the collected poop's solids into acids, which feed bacteria, which in turn digest the manure and produce biogas. The gas is then pushed through a pipe into a modified natural gas engine, and electricity generated by burning the gas is fed into the CVPS system. The digester also produces a low-odor slurry that makes a fertilizer that is safer than raw manure.

Cow Power is a gracefully circular way for cattle to give a bit of energy back to the systems that support them. Participating dairy farmers get an additional source of income, almost literally turning waste into gold. And Vermonters who sign up for bovine-generated electricity support renewable energy *and* the state's traditional dairy-farming industry. EG

Fish for the Future

For centuries, it seemed that the ocean's vast bounty knew no limits; coastal cultures that subsisted on fish were some of the world's richest and healthiest societies. But population growth, increasing wealth, and the establishment of highly industrialized fishing techniques have led to an unsustainable demand on the undersea food supply. Today, fishing fleets must legally abide by strictly regulated quotas, and advances in fishing technology help control the size of each catch and reduce the capture of undesired species. Nevertheless, many conduct illegal operations outside the relatively weak confines of international law.

The result? A report published by the United Nations Food and Agricultural Organization in 2004 stated that "28 percent of fish stocks worldwide are either overfished or nearing extinction"; another 47 percent are "near

the limits of sustainability." Farming fish and seafood through aquaculture has taken off, but fish feedlots generally take their cue from cattle feedlots—they are over-crowded, soaked in chemicals, and polluting.

Plans are being implemented to recover fisheries in many parts of the world. Scientists and fish advocates are calling for the establishment of protected marine reserves, international conservation agreements, and quotas. Simultaneously, businesses are springing up based on principles of better and more sustainable seafood farming and harvesting practices.

Change will likely come slowly to the fishing and seafood industry. Meanwhile, what can we do? For now, eating only sustainably raised or harvested seafood is the best way to avoid contributing to the crisis. Ultimately, a growing market for such seafood will drive fishing and aquaculture practices—meaning that our dollar does carry real power for change. EG & GF

Sustainable Fisheries Certification

Unlike the Department of Agriculture's Organic Standard label or the Environmental Protection Agency's Energy Star label [see Using Energy Efficiently, p. 166], a standard, overarching sustainable fisheries' certification has not yet been established for restaurants, retailers, and fishers. Right now, a proliferation of systems, protocols, and evaluation schemes, as well as conscientious distributors like EcoFish and CleanFish, are dedicated to identifying and supplying sustainable seafood.

In 1995 the United Nations Food and Agriculture Organization developed a code of conduct for fisheries, which has since been widely recognized, fostering the formation of the Marine Stewardship Council (MSC). This international nonprofit has worked to establish a universal certification and labeling process. The MSC guidelines call for accountability through reliable and independent auditing and the establishment of transparent standards based on good science. If the fish on our dinner table has been certified by the MSC, we can be sure that we're supporting good work toward the restoration of the ocean's wealth. EG

Seafood Watch

Seafood Watch, a program launched by the Monterey Bay Aquarium, has created a concise, informative list to help fish lovers keep track of which species we can grill free of cares, and which are ecological no-no's. The California aquarium offers this information to visitors on a wallet-sized reference card. It can also be downloaded or requested from the aquarium's Web site, which features comprehensive information on fish classifications, plus ways to get involved in protecting ocean life. Updated regularly to reflect improvements and declines in fish populations, the searchable online Seafood Guide has detailed information on the ecological status and nutritional value of different species of wild and farmed seafood. This is a great resource full of constructive solutions that anyone can use. EG

RESOURCES

Keep Chickens! Tending Small Flocks in Cities, Suburbs, and Other Small Spaces
by Barbara Kilarski (Storey Publishing, 2003)
Imagine your own flock of chickens living peacefully in your own urban oasis. In *Keep Chickens!* Barbara Kilarski guides the chicken enthusiast through raising and cultivating chickens in small spaces. Providing advice on everything from choosing breeds to gathering eggs to complying with city ordinances, the book is interspersed with smart tidbits about the joys of raising chickens in the city and illustrated with vintage

Opposite, left: Organic dairy farmer, Farmington Township, Minnesota, 2005.
Opposite, right: Fisherman hauling in his catch, Newport, Rhode Island, 2001.

chicken advertisements. As Kilarski notes, "Keeping a small flock of chickens, even two chickens, is an enjoyable and rewarding pastime that, incidentally, makes us a little more self-sufficient and brings us closer to the slow, measured beat of nature."

Ocean Friendly Cuisine: Sustainable Seafood Recipes from the World's Finest Chefs by James O. Fraioli (Willow Creek Press, 2005)
This book features recipes that emphasize tilapia, farmed shellfish, and other sustainable seafood; information on how to figure out where the fish at your local market or restaurant comes from; fisheries issues; and more.

Cod: A Biography of the Fish That Changed the World by Mark Kurlansky (Penguin Books, 1998)
Mark Kurlansky recounts humanity's shared history with the now-collapsed cod fisheries off North America's North Atlantic coast.

Song for the Blue Ocean: Encounters Along the World's Coasts and Beneath the Seas by Carl Safina (Owl Books, 1999)
Safina intertwines politics, science, and a travelogue to expose the precarious ecological state of the world's oceans.

The Marine Stewardship Council
http://www.msc.org/
The Marine Stewardship Council's Web site teems with good information—for shop and restaurant owners and consumers alike—on how to learn more about sustainable seafood, where to buy it, and how to prepare it. It's a good resource, especially for those trying to better understand fishing as an industry, and our power to change it.

The Future of Food

In the past half century, farming has, for the most part, been transformed from the sort of family enterprise depicted in children's books into full-blown industrial resource extraction. Preserving the world's soils and the biological diversity of crops is key to maintaining healthy global ecosystems, curbing poverty, and ensuring a peaceful future for billions of people. Possible solutions abound. Instead of turning the soil every year, we can grow perennial plants that need no tilling. Instead of growing square miles of one kind of corn, we can assemble a whole community of crops based on native plants that can support each other. In other words, we can build a farm the way nature would. AS

Understanding Soil

Soil is the skin of the earth. It's the first point of contact between the planet and the atmosphere. The highly fertile top layer of soil—the uppermost twenty centimeters or so—is known as topsoil. Like the air we breathe, this layer of earth is so ordinary and ever-present that it is easy to take for granted. But it is absolutely essential to our lives, health, and prosperity.

By maintaining larger and larger farms, using giant machinery that tills the topsoil and kills all but the planned-for crops, planting giant monocultures (identical crops), and soaking the land in pesticides, weed killers, and chemical fertilizers, industrial agriculture strip-mines the soil.

As cultivated topsoil is scoured away by rain, or blown away by wind, what remains is less fertile. What's more, the steady chemical beating topsoil across the United States has taken over the last fifty years has killed off many of the microorganisms that keep soil alive. Dead soil is no longer soil: it's just wet dust. In what were once the thriving farm towns of America's heartland you can see what happens when topsoil is destroyed. You can see it in the fields themselves, in the dust clouds that blow through the region, and in the coffee-black run-off that swells the rivers with every serious storm. AS & EG

Prairielike Farms and Smart Breeding

To put a stop to the degradation of our topsoil, we must change the way we farm. Today, we grow most crops on mass monocultural farms of annual plants. These farms require a huge amount of labor—including yearly plowing, which causes soil erosion—and huge quantities of chemical fertilizers and petroleum-based pesticides. To move beyond this type of farming, which locks us into constant service to compromised and degenerating land, we need to think of farms the way we think of prairies.

Wes Jackson, founder of the Land Institute, knows this. A fourth-generation Kansas farmer armed with a PhD in genetics, Jackson has devoted his life to demonstrating that "we have to farm the way nature farms" (Benyus, 2002). The edible prairie would be a complete revision of the present industrial agriculture we now employ, replacing the perpetual depleting of resources with a self-regenerating process more akin to nature. One of the first steps is the renewal of perennial crops. Most modern crops were bred from perennial wild relatives, and Native Americans in the Ohio and Mississippi valleys "farmed" wild knotweed, maygrass, marsh elder, and little barley prairie plants we now regard as weeds. Now, too many crops are annuals, which disallows the possibility of a farm with continuous lifecycles. A prairielike farm would be able to start anew in spring without massive amounts of human and mechanical labor.

There's one big problem with prairielike farming, though: time. Plant-breeding programs take many generations to perfect, and experiments in building plant communities don't show results for years. Because our farms are in such bad shape, we don't have that kind of time—but we could use new technologies to breed the new hybrids faster.

Genetic modification of crops has become a notoriously controversial and widely protested science. But the same insights that have introduced little-studied and potentially dangerous crops into our food supply can yield safe and beneficial plants that may help rehabilitate agriculture.

Scientists have begun to discover, study, and develop the dormant characteristics that are an inherent part of the natural heritage of plants. The process of awakening these characteristics in plants—known as smart breeding—is akin to taking age-old agricultural techniques such as crossbreeding and hybridization and combining them with the highly refined genetic science of the twenty-first century.

Smart breeding produces plant varieties in much the same way traditional breeding does, but at a greatly accelerated pace. To create a new strain of tomato, farmers in the past would cross various existing tomatoes and see which hybrids and mutations grew best. That hit-or-miss approach is slow, since each new generation of plants must be grown out and its seeds harvested before the next generation can be planted. What's more, because mutation strikes randomly, a farmer's best, juiciest tomato may suddenly lose a gene critical to its juiciness. History shows that traditional breeding, while often successful given long time frames, is full of dead-ends, lost opportunities, and mistakes.

Smart breeding starts with the same plants but applies genetic analysis to figure out exactly which genes are responsible for the desired traits. Sometimes, the process takes the form of traditional breeding—keep this seed, toss that one—based on more precise knowledge. Other times, the process involves turning on or off specific genes in a plant's DNA—a goal achieved only at random in traditional breeding. Still other times, sequences of genes are lifted from one variety of a plant and introduced into another variety—something entirely possible on a traditional farm, but done by smart breeders in a way that doesn't introduce an element of uncertainty (how will that tomato's gene for frost resistance express itself in this tomato?).

But even when it involves cutting and pasting genes from one tomato into another, smart breeding more closely resembles nature than does the gene-splicing many of us fear, because there's no chance of crossbreeding two species that would not interbreed in the wild. Smart breeding and transgenics are as different as apples and anthrax. Even many biotech skeptics think smart breeding poses minuscule risk to human health or the environment.

The pay-offs, however, could be huge. Before factory farming, most crop species appeared in hundreds of varieties that were often admirably adapted to local soil types, weather patterns, and growing conditions. Replacing these many "heirloom" varieties with a few industrially bred ones not only allowed for the creation of the monocultures that corporate farms spray so vehemently to defend, it also drove many heirloom varieties into extinction. As we move away from petrochemical-based agriculture, we need to both rediscover and reinvent heirlooms.

Smart breeding could do even more, though, than help great-grandma's tomatoes make it through a scorching summer: it could deliver the kinds of plants we need to regenerate farmland. One result might be "super-organic" food crops bred with green biotechnology. Each plant of today and yesterday contains the genetic potential to thrive on the soil-conserving, biodiverse, prairielike farms of the future. AS

The Benefits of Bioengineered Crops

Transgenic genetically modified (GM) crops are a risky work in progress. The health effects of eating, for example, a tomato into which genetic material from a salmon has been introduced, are simply not known—the impact on the human body may be minimal, but we can't yet speculate. The environmental risks of farming GM crops are significant—genetic engineering introduces into the environment self-reproducing species that haven't existed long enough to be studied in depth. The race to develop GM crops is largely an attempt by large corporations to monopolize factory farming, and raises huge social and economic questions. Corporate research into GM crops is not geared toward redesigning crops for free distribution to the world's poorer farmers but toward design-patented products that prop up agribusiness and generate wealth for a powerful few.

Nonetheless, genetic modification is not inherently evil, and when applied with wisdom, it can have positive results. An excellent example is New Rice for Africa (NERICA), a strain of rice that may succeed in bettering health in West and Central Africa, restoring agricultural sustainability there, and improving the economics of food importation in the regions.

The great benefit of NERICA is that it mixes African rice *(Oryza glaberrima)*—which is highly resistant to drought and local pests but which has a very low yield (triggering widespread slash-and-burn farming)—and Asian rice *(Oryza sativa)*—which has a very high yield per plant but is much more sensitive to environmental conditions (triggering increased use of pesticides). These two species of rice do not cross naturally or through traditional hybridization techniques—the genetic differences are just too great—but biotechnology has produced more than three thousand NERICA lines, allowing farmers to choose those that best fit their regional needs.

In making use of 1,500 varieties of African rice that were facing extinction, the NERICA initiative has helped to preserve genetic lines at risk as farmers shift to higher-yield Asian varieties. And NERICA reaches beyond Africa: 42 million acres (17 million hectares) of rice in Asia and 9 million acres (4 million hectares) in Latin America grow in conditions similar to West Africa's. AS & JC

The Future of Aquaculture

Dr. Martin P. Schreibman has been growing tilapia for years in tanks in his lab at the Aquatic Research and Environmental Assessment Center of Brooklyn College. "You could set a tank up in your basement and grow enough fish to pay your rent," Dr. Schreibman stated in a *New York Times* article, and it may be true. He envisions a day when fish farming throughout New York City—using systems scalable to tight urban spaces—will replace resource-wasting importing as a ready source of local seafood.

Farmed fish, shellfish, and crustaceans represent almost one-third of the seafood we eat today. With worldwide demand for seafood on the rise and most wild fisheries going under, aquaculture has become a lucrative business. The industry and its methods have their critics, but not all aquaculture is bad.

Shellfish aquaculture can actually have a positive impact on the environment. Creatures such as oysters, clams, and mussels eat by filtering plankton from the water, needing no external food supplements. Since their harvests must come from clean waters, shellfish farmers often make staunch advocates for coastal ecological preservation. Much of the shellfish on the U.S. market today is farmed.

Fin fish farming in coastal waters can be more problematic. Farmed salmon, raised by the thousands in net pens, produce a corresponding load of water-polluting feces. Diseases can spread quickly through the crowded pens. Antibiotics used to treat these diseases can then leak out into the water, where they can help disease-resistant organisms develop. And it all adds up to less-than-healthy salmon steaks on your plate.

Many researchers and environmentalists believe that the solution lies in removing fish farms from areas bordering wild waters. Although Dr. Schreibman's tanks are still largely experimental, his work shows that sustainable aquaculture is possible—even in a Brooklyn basement. EG

RESOURCES

The Essential Agrarian Reader: The Future of Culture, Community, and the Land
edited by Norman Wirzba
(Shoemaker and Hoard, 2004)
There's no doubt that the last century has been witness to a concerted movement away from agrarianism and toward a globalized, industrialized planet—with avocados from California, rice from Thailand, coffee from Colombia. Concerned that fewer and fewer people actually know or care about their food's source or its production, and alarmed by the new global order, Wirzba expounds the modern agrarian movement's push for responsible action in the interest of lessening the divide between food's production

Opposite, left: Researchers testing New Rice for Africa (NERICA) seed samples, Kindia, Guinea, 2002.
Opposite, right: Farmer learning about NERICA rice from local agricultural experts, Koromayah, Guinea, 2002.
Left: Intensive work is being done to restore Olympia oysters to Puget Sound in Washington State, 2003.

and consumption. "However much we might think of ourselves as post-agricultural beings or disembodied minds, the fact of the matter is that we are inextricably tied to the land through our bodies—we have to eat, drink, and breath—and so our culture must always be sympathetic to the responsibilities of agriculture," Wirzba writes. His book is, in many ways, an homage to the work of farmer-poet Wendell Berry, and it teaches important lessons about the American agrarian past and the possibilities and necessities of an agrarian future.

The One Straw Revolution: An Introduction to Natural Farming by Masanobu Fukuoka
(Gardners Books, 1992)
From a small village on the island of Shikoku in southern Japan comes a story that brings together the classic elements of Taoist innovation: observation, insight, and a cheerful disregard for conventional wisdom. While walking past an old, unplowed field one day, Masanobu Fukuoka noticed healthy rice seedlings growing among the weeds. Emulating the conditions that these seedlings grew in, he stopped plowing his farm or flooding his fields to grow rice, as conventional practice would dictate. Instead, Fukuoka pioneered a method of natural farming that has resulted in his fields' being unplowed for twenty-five years, yet producing yields comparable to other Japanese farms. His methods require no machines or fossil fuels and create no pollution. Drawing on a faith in natural cycles, the Fukuoka method demonstrates a way of farming that means interfering with nature as little as possible. Fukuoka's innovative method of farming, dubbed the One Straw Revolution, has won supporters and adherents around the world, where his methods have been adapted to local conditions. ZH

Meeting the Expectations of the Land
by Wes Jackson, Wendell Berry, Bruce Colman
(North Point Press, 1985)
A classic in the field, this book of essays about the nature of good farming still speaks sense twenty-odd years after it was written, by reminding us that all agriculture depends on certain fundamentals: soil, sun, and water—certainly—but also care, intelligence, and a respect for legacy.

The Land Institute
http://www.landinstitute.org/
Wes Jackson's Land Institute does much-needed work on changing farming practices to work with nature. Those with a serious interest in the subject will find the institute's reports and studies invaluable.

Hungry Planet: What the World Eats by Peter Menzel and Faith D'Aluisio
(Ten Speed Press, 2005)
Sometimes the best way to gain a wide-angle perspective on a global situation is to piece together a number of close-ups. Authors Menzel and D'Aluisio do just that in *Hungry Planet,* profiling thirty families from around the world and combining stunning photographic essays with descriptions of each family's weekly food intake. *Hungry Planet* is a worldchanging resource that blends the emotional power of a family photo album, the revealing content of a documentary, and the educational value of world food statistics—presenting them with clarity and simplicity:

"Obesity and its health consequences are just collateral damage. Growth-seeking companies cannot imagine a better place to find new buyers than the emerging economies of developing nations ... Compare the diets of the two Chinese families, one urban and one rural. Both are in a transition from a diet of poverty to one of affluence ... The foods on display in the [rural] family portrait are mostly raw or minimally processed ... This rural family buys cubes of chicken stock as a luxury; others are beer, cigarettes, and, as is common nearly everywhere, Coca-Cola ... In contrast, the urban family can afford—and has access to—a great many more foods of different kinds. It buys many of the same basic foods as the rural family, but adds ... baguettes, sugar-free gum, Häagen-Dazs Vanilla Almond ice cream, and takeout from KFC. Such is the global food market in the early 21st century."

Preserving Barnyard Biodiversity

Over the last three thousand years, humans have bred an amazing variety of tasty things: tomatoes and turnips, chickens and cows. Because these species were domesticated in many different areas, wide variation emerged within plant and animal families. Yet we generally only eat a tiny sampling of the thousands of varieties that have evolved around the world. Wheat and rice, for example, dominate our grain sources to the exclusion of numerous nutritious ancient grains. As the poet Gary Snyder reminds us, there's no such thing as a post-agricultural civilization. Our agricultural prospects depend on improving, not minimizing, crop diversity.

Futuristic farming technologies may sustain food production as agriculture undergoes dramatic shifts, but even if techniques like smart breeding [see The Future of Food, p. 67] live up to their promise, we still need diverse genetic material to work with. As we experiment with cooking up recipes for the foods of the future, we need an assortment of ingredients. We can find the genetic variety in heirloom plants and in barnyard biodiversity. Our ability to grow bright green farms in the future depends on both. AS

Seed Banks

Most home gardeners know all about seed preservation—they meticulously save seeds from their best plants for use in next year's garden. Large-scale seed-saving operations, however, aim to achieve much more than another sweet batch of cherry tomatoes.

Seed banks aim to preserve humanity's agricultural heritage, acting both as a resource and a fail-safe. They help farmers actively promote diversity by reviving traditional crops and preserving the variety and abundance of domesticated plants—guarding against possible disasters. Seed banks are humanity's insurance against disaster, a promise to the future that farming, and thus civilization, will go on. EG & AS

Native Seeds/SEARCH

Corn, beans, and squash are known as the "three sisters" of southwestern Native American agriculture. In the early 1980s some elders from the Tohono O'odham Nation wanted to revive one of these sisters, a variety of squash they remembered from their childhood. Enlisting the help of biologists and ethnobotanists, the elders were able to track down and obtain a handful of seeds. The hunt for these heirloom seeds led to the founding of Native Seeds/SEARCH (NS/S), a repository of crops traditionally grown by Native American nations from that region.

Today, NS/S grows the seeds of more than two thousand crop species are grown there, more than half of them relatives of the "three sisters." The other half includes grains, chilies, dyestuffs, and melons. The program has brought about a hundred crop species back from the edge of extinction.

Not only has NS/S helped to revive traditional Native American farming in the region—which had all but disappeared—and to preserve the biodiversity of the Southwest, it has boosted the health of the Tohono O'odham Nation's people. The nation, like many other indigenous North American groups, has suffered from endemic diabetes since adopting the typical Euro-American high-fat, high-sugar diet. Many of the native plants that Native Seeds/SEARCH has helped reintroduce to the region are specifically well suited to controlling diabetes; prickly pear paddles, for instance, are a great source of nutrients and of soluble fiber, which slows the rate of digestion, keeping the body's glucose levels more stable. Other crops simply serve as nutritious, high-protein staples that cost very little to cultivate.

Native Seeds/SEARCH has proven that traditional crops have a future.

International Center for Agricultural Research in the Dry Areas
Farming dry, arid land poses complex challenges, and the crops we've bred over generations to grow well in places that have little water are both a heritage and a toolbox. The International Center for Agricultural Research in the Dry Areas (ICARDA), located in northern Syria, banks seeds from 131,000 varieties of plants, gathered from arid regions across the Middle East, Central Asia, and North Africa. In protecting and enhancing native staple crops, ICARDA works to alleviate poverty by boosting agricultural productivity. Samples from the seed bank have also been used to reintroduce crops into war-torn Afghanistan. EG & AS

Global Crop Diversity Trust
The Global Crop Diversity Trust (GCDT) is both the net and the umbrella of the world's seed banks. Many seed banks are lost as a result of catastrophe or a lack of funding. Grants from governments, foundations, and private corporations allow the GCDT to secure consistent funding for the preservation of crop diversity around the world. The endowment enables people with a close connection to agriculture and the land who otherwise have few or no resources to ensure a livelihood for future generations. The GCDT makes seed banks a global priority and provides them with the funding they need to survive. EG & AS

Sangams
Small farmers also play a significant role in protecting seed diversity. The Centre for Indian Knowledge Systems (CIKS) in the state of Tamil Nadu, India, works with hundreds of small farmers to form groups, called *sangams*, wherein farmers work together on programs to maintain organic farming practices, or sell biopesticides to supplement their incomes. These *sangams* also manage community seed banks. The center provides the initial funds needed to construct the storage facilities, after which *sangam* members themselves maintain the seed banks, making small monthly contributions to a communal bank account, and electing officials to oversee the borrowing and replenishing of seed stocks.

Navdanya, a related organization based in New Dehli and Uttaranchal, in northern India, has been running a rigorous indigenous seed-bank program that has saved and stored hundreds of varieties of seeds, including two thousand varieties of rice in forty seed banks in thirteen Indian states. Some 70,000 small farmers belong to the organization, and their efforts are proving that local seeds and local knowledge, just as much as science and technology, hold the keys to creating the agriculture of the future. ZH

The Polar "Doomsday" Vault

A reinforced concrete vault in a mountain cave on a frozen Norwegian island could someday become one of the most important resources for sustaining life on earth. According to plans, this "doomsday" vault will hold two million seeds, representing an extensive selection of food crops and other plants from around the globe. Curated with the help of the Global Crop Diversity Trust, the bunker will be able to withstand disasters and threats of all kinds, and the island's permafrost will help keep the seeds

frozen, ensuring their longevity. When completed, the vault will be one of the most comprehensive seed banks to date, yet it comes at a very low cost because natural conditions keep it consistently cool and highly secure. Unlike other seed banks, which share their contents with farmers today, the doomsday vault is meant to protect the world's food supply in the future from catastrophes like nuclear war, asteroid strike, or environmental disaster caused by accelerated global warming. Call it civilization's insurance policy. SR

Nikolai Ivanovich Vavilov

The world owes a debt of gratitude to Nikolai Ivanovich Vavilov. Well ahead of his time, Vavilov recognized as early as the 1920s the critical role that protecting agricultural biodiversity could play in human survival. His work was the origin of today's seed banks. An avid traveler, Vavilov amassed what was then the world's largest collection of seeds while researching ways to develop more productive strains of vital species.

In the 1930s and '40s, Vavilov faced extreme opposition from the Stalinist Soviet government. Trofim Lysenko, the head of agricultural affairs for the Communist Party, opposed Mendelian genetics, as they countered his own theories, and in 1940, Vavilov was thrown in jail, charged with sabotaging agriculture and promoting "bourgeois pseudoscience." As Russia's agricultural legacy was systematically destroyed and Vavilov languished in prison, fellow scientists vigilantly guarded his seed collection. During the Siege of Leningrad by the German army, they chose to go hungry rather than pillage the food sources in the storehouse; several starved to death while watching over the seeds. Vavilov ultimately starved to death as well—in prison.

Today, the N. I. Vavilov Research Institute of Plant Industry remains one of the largest seed banks in the world, though it faces financial troubles. The institute received a grant from the Global Crop Diversity Trust in 2003 to ensure that its unique contents are properly regenerated and preserved.

Building a Market for Heritage Breeds

"When someone buys a $199 turkey from us, they're not buying a turkey," says Patrick Martins, cofounder of Heritage Foods USA. "They're buying a story."

The tale the pricey poultry tells is the comeback story of the Bourbon Red Turkey. Popular in the early twentieth century, the breed has teetered on the edge of extinction for years. Martins first became interested in rare, heritage breeds while raising and distributing 1,500 Bourbon Reds through Slow Food USA's [see Doing the Right Thing Can Be Delicious, p. 51] Heritage Turkey Project. The undertaking proved that it takes only a modest boost in the market to save a species: even this relatively small surge in demand improved the Bourbon Red's chances for survival; the breed's status went from "rare" to "watch" on conservation lists. Heritage Foods now works with small farmers to offer an ever-growing selection of rare types of pork, lamb, bison, and poultry.

Opposite, left: International Center for Agricultural Research in the Dry Areas (ICARDA) seed bankers with drought-resistant lentil varieties.

Opposite, right: ICARDA scientist entering seed data into a handheld computer.

Right: Heritage turkeys on S&B Farm, Petaluma, California.

The extinction of heritage breeds would be a profound loss to agricultural biodiversity—and to the taste buds of meat lovers everywhere. Breeds such as Berkshire Pork, Barred Plymouth Rock Chicken, and American Bronze Turkey definitely taste different—and in many cases, better—than the stuff you've been getting at the supermarket year after year. EG

RESOURCES

Seed Savers Exchange
http://www.seedsavers.org/
Seed Savers Exchange is a network of people committed to collecting, conserving, and sharing the seeds of genetically unique heirloom varieties of vegetables, herbs, and other plants. These plants possess qualities specific to the ecological niches in which they've been grown—qualities that have been passed down through generations. Seed Savers publishes *Seed to Seed*, a well-regarded guide to growing and saving heirloom seeds.

Enduring Seeds: Native American Agriculture and Wild Plant Conservation
by Gary Paul Nabhan
(University of Arizona Press, 2002)
An eloquent argument for the preservation of agricultural biodiversity, *Enduring Seeds* reflects perspectives that Nabhan has gained through his experience in desert ecology, Native American farming, and seed conservation.

"Descending out of the Texas Hill Country, I was supposed to be on my way to Houston. But the story of the *Zizania texana*, Texas wild rice, had pulled me off course and into a landscape too manicured and manipulated for its own good. As I drove along, idly listening to an Austin country-and-western radio station in my rented car, I hummed a sad song that summed up my pilgrimage. It was 'Needle-in-Haystack Time Again,' looking for a plant so rare that I'd need a world's expert to take me to it."

Cars and Fuel

For many North Americans, the car has become both a necessity and a shrine. Sprawling suburbs and bad urban planning have made it nearly impossible for us to get anywhere without driving. Even when we do have other options, many of us are reluctant to give up our cars, because our cars—and the idealized "open road" they call home—have become part of our identities. As discussions of a future oil crisis and the role of oil in the Iraq war heat up, so do the efforts of some manufacturers and politicians to convince us that our cars are the very essence of freedom.

But the truth is our cars are just tools—and we need better tools now. The auto industry was built on a seemingly endless supply of gasoline, but it is now becoming increasingly clear that the end is, in fact, in sight.

The concept of "peak oil" is pretty simple: Oil is a limited resource. Whenever we're dealing with finite resources, there comes a point when dwindling supplies make it more expensive to extract more of that resource, and we start extracting less. The term for that point is *production peak*. Without a doubt, at some point cheap oil will start running out, and less and less will be available.

The critical question is, when? Some geologists argue that the production peak is imminent; others say that we have ten to twenty years, and still others say we have the entire twenty-first century to figure out a new energy future. Opinions may vary, but automakers are of one mind on the issue, and have already begun to plan for the end to cheap oil.

Even if we *did* have decades before we hit peak oil, we can no longer ignore the

damage our cars do to our planet. Finding, pumping, and burning fossil fuels is changing our climate, polluting our seas, and involving us in resource wars. Peak oil or no peak oil, there are no two ways about it: we have to shed our dependence on oil *now*.

How do we kick the oil habit? Many auto companies are working on short-term solutions: high-profile hybrids (quickly becoming Hollywood's trendy new accessory); more efficient gasoline engines; cleaner-burning diesel engines; and biofuels and synthetic fuels to substitute for petroleum. Right now, our choices matter more than ever—every time we go to the showroom and drive a hybrid instead of an SUV off the lot, we send a message to automakers to keep the new solutions coming.

The long-term solution is likely electric, whether the electricity is provided by a super-efficient battery or by a hydrogen fuel cell. Researchers and automakers are looking to electrify the drivetrain, powering vehicles only with electricity and motors, rather than with petroleum fuel and combustion engines. Several years ago, most people would have said that hydrogen-powered cars were the only solution—and exciting new developments from manufacturers like Honda have restored faith not only in hydrogen cars but also in a future hydrogen economy. However, hydrogen development is more challenging than had originally been anticipated, and will clearly be slow to yield workable results. We can't assume that hydrogen-powered cars will be perfected in time to head off the potential shortages and disastrous environmental consequences of peak oil and global warming.

There may not be one right answer to the car question. Every piece of the puzzle could be important—from saving fuel now, to supporting new vehicles that run on clean energy, to backing legislation that will pave the way for the future of cars and fuel. The road ahead demands a different kind of car. MM & AS

Drivers' Ed

In the immediate term, no simple solution to our fuel and emissions problems presents itself. There is no magic rabbit coming out of the technology hat. The best thing we can do right now is to choose wisely and behave responsibly.

There are two basic and related approaches to shedding our dependence on oil: substitute and conserve. On the substitution front, we can use biofuels and synthetic fuels whenever possible. On the conservation front, we can swap our gas-guzzlers for cars that are more fuel efficient: smaller, newer, diesel (powered by biodiesel, of course), or hybrid-electric. And we can choose not to buy SUVs.

When it comes to conserving fuel, it's not just what you drive, it's how you drive it. Abiding by the following tips will help you conserve:

- Avoid idling. If you need to warm up your engine on cold winter days, keep idling time to less than thirty seconds.
- Drive slower. When Congress imposed the 55 mile (88.5 kilometer) per hour speed limit in response to an earlier oil crisis, the result was an estimated 2.5 billion gallons (roughly 9.5 billion liters) of gasoline and diesel conserved in 1983 alone, according to the Natural Resources Defense Council. That figure works out to 163,000 barrels of fuel per day—the output of a midsize refinery.
- Drive steadier—jackrabbiting gobbles fuel. Use cruise control when possible.
- Use an overdrive gear if your car has it. This mode reduces your engine's speed, which saves gas.
- Keep your car clean and well maintained. Dirty air and fuel filters reduce gas mileage, as does low tire pressure. Also, refrain from using your car as a storage unit—the more weight you're carrying around, the lower your gas mileage will be.

No matter what you drive, you can benefit from the above suggestions. But if you own a hybrid, you can squeeze out even more outrageously good mileage by knowing how to best drive your vehicle, as follows:

- Accelerate quickly from a stop. Unlike in a standard car, this is actually more efficient for a hybrid's electric system.
- Coast or use cruise control whenever possible.
- Brake gently whenever possible—this will recover braking energy to the battery.
- Use the real-time miles-per-gallon display to monitor your car's activity—you'll likely be able to determine other patterns. MM

Preceding page: The Trans-Alaska Oil Pipeline snakes 800 miles across Alaska.

Biofuels and Synthetics

Biofuels (fuels made from renewable resources such as soybeans) and synthetic fuels are bound to play an important role in the evolution of sustainable transportation worldwide. Biofuels, like hybrid cars and rooftop solar panels, are a kind of bridge technology, helping us get where we need to go without cutting us off from our existing systems.

Ethanol, an alcohol produced in a variety of ways from cereal grain or biomass waste, is used in gasoline blends with up to 85 percent ethanol (in a mixture known as E85). Vehicles that can burn either gasoline or ethanol blends are called flexible-fuel vehicles, and there are millions of them on the highways in the United States and around the world. From an automaker's point of view, creating a flex-fuel vehicle is relatively simple: you use a sensor and engine-control software that will determine what type of fuel the car is using and then adjust the fuel injection and combustion accordingly. The biggest stumbling block to E85's becoming a fuel of choice is that a limited number of gas stations are outfitted to provide it. Efforts, however, are under way to increase the number of E85 stations. Ford Motors, for example, is working with VeraSun, an ethanol provider, to do just that. Ethanol can also be used in low-ratio blends with diesel.

Biodiesel, a chemical substitute for petroleum diesel, is created by processing vegetable oil—either freshly pressed or recycled. Biodiesel can be used as a direct substitute for petroleum diesel, either in blends, or neat (called B100). Diesel engines require no modifications to run on biodiesel. A diesel engine running biodiesel is efficient and clean. Co-ops, producers, and distributors around the nation sell high-quality biodiesel for anywhere from $2.50 to $3.90 per gallon.

Straight vegetable oil (SVO) can be used as a fuel substitute in a diesel engine. Unlike biodiesel, SVO is not processed and transformed before use. But SVO does require system modifications: a second tank in the vehicle, and a heater that warms the oil before it enters the engine.

Fischer-Tropsch (F-T) synthetic fuels are the product of the thermochemical transformation of a hydrocarbon gas. The process can turn gasified coal, natural gas, methane from landfills, and gas derived from biomass into synthetic diesel, gasoline, and aviation fuels. The resulting F-T fuels are high performing and clean burning, and can be used in blends or neat.

The Department of Defense is keenly interested in the expansion of the United States' use of F-T fuels, which could provide a non-oil (read: politically secure) fuel source to meet the country's 300,000-barrel-per-day burn rate. MM

A Guide to Making Your Own Biodiesel

Although biodiesel stations are becoming more common, knowing how to whip up your own supply can be empowering and cost-effective.

In essence, mixing up biodiesel is not much more complicated than baking a cake, but a badly mixed cake won't ruin your engine. USE EXTREME CAUTION IF YOU UNDERTAKE THIS PROCESS! Just like in any chemistry experiment, these substances carry inherent risks, such as flesh burns, blindness, and fire. Wear safety goggles and rubber gloves. Prepare your first batch under the supervision of someone with experience. Work slowly and be precise in your measurements. Follow these instructions at your own risk.

You can use either store-bought vegetable oil or waste cooking oil from a restaurant. Generally, restaurant owners will be delighted to have you haul away a few gallons of their grease, because otherwise they pay a removal service to get rid of it. Note that if you are gathering waste oil from various sources, each source will have a different pH, so you'll have to make sure to get a mean pH reading, representative of the whole batch.

Ingredients and Supplies:
5 grams lye (KOH)
petri dish
1 liter distilled water
3 beakers (one 20 mL, one 1,500 mL, and one 500 mL)
3.8+ liters filtered waste vegetable oil
1 12-oz. bottle isopropyl alcohol (rubbing alcohol)
1 bottle phenolphthalein solution (pH indicator, available at pool/hot-tub suppliers)
graduated syringe or eyedropper
1 bottle methanol (you can also use ethanol)
1 kitchen blender (which you can never use again for smoothies!)

Step 1 — Titration:
Titration helps you determine how much catalyst you must add, by indicating the acidity of your oil.

Measure out 1 gram of lye into your petri dish. Dissolve into 1 liter distilled water. In the 20-milliliter beaker, dissolve 1 milliliter vegetable oil into 10 milliliters isopropyl alcohol. Swirl or warm very gently to dissolve the oil into the alcohol until the solution is clear. Add 2 drops of phenolphthalein solution to this mixture, and swirl to dissolve. Using your syringe or eyedropper, add 1 milliliter (only!) of lye–distilled water solution to the alcohol and oil solution. Continue adding 1 milliliter of lye–distilled water solution at a time, swirling continuously, until it turns hot pink and holds its color for at least 10 seconds. The number of milliliters of lye solution used, plus 3.5, equals the number of grams of lye you'll need per liter of oil.

Step 2 — Brewing:
Measure out 1 liter of filtered-waste vegetable oil into your 1,500-milliliter beaker. Measure out 200 milliliters of methanol into your 500-milliliter beaker. Measure out into your petri dish the number of grams of lye that you determined during titration.

Pour the methanol into your blender. Add the lye. Blend at low speed until fully dissolved. This reaction creates sodium methoxide. Because of rapid evaporation, the rest of the process must be done immediately. Use extreme caution! Do not inhale or ingest this stuff!

Pour filtered-waste vegetable oil into the sodium methoxide solution in your blender and blend for 15 to 20 minutes continuously. After blending, leave the mixture alone to settle for at least 8 hours, at which point you will have two layers: glycerin on the bottom and biodiesel on top. Both of these substances are nontoxic. The biodiesel goes in your tank, and the glycerin can be turned into soap. You've reincarnated grease into two phenomenally useful substances. SR

Tomorrow's Hybrids

Even if we do succeed in creating hydrogen-powered cars relatively soon, we're not going to jump right from a few gas-electric hybrids to all-hydrogen autos.

The best strategy for bridging the gap between our present overconsumption of oil and the future establishment of a new auto industry is to upgrade our hybrids so that they are even cleaner and more efficient. The next generation of hybrids will be grid-charged plug-ins that will use an advanced and efficient combustion engine combined with a robust electric motor and grid-chargeable battery pack.

Endowed with such a system, this plug-in hybrid would operate in its electric mode for longer ranges of time and at higher speeds than do current models. It would also be able to burn biofuels, synthetics, or blends in the engine to further reduce petroleum usage.

There are a few challenges, of course, such as the current limitations in the kinds of energy storage (batteries, ultracapacitors) that would be necessary for providing enough power and driving range (the distance the car can be driven before it must be recharged) to substitute for combustion engines. And anytime electricity is involved we have to consider the source of the power and the environmental impact of producing that electricity. MM

Diesel-electric Hypercars

A startup company in Carlsbad, California, has designed what could be called the anti-SUV. The Aptera from Accelerated Composites is a compact (two-seater) diesel-electric hybrid made of high-end composite materials.

Fuel efficient? Oh, yeah: the Aptera gets an estimated 330 miles per gallon at 65 miles per hour. That's not a typo. The combination of a super-streamlined shape, ultra-low-weight materials, and high-output supercapacitors (instead of batteries) gives the design incredible efficiency. And because the composite production process developed by Accelerated Composites is faster and more efficient than previous methods,

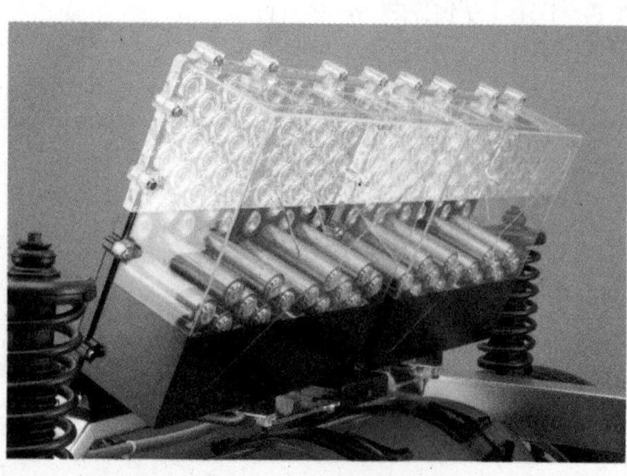

the overall cost of the vehicle could be startlingly low for a hybrid—about $20,000.

Another perk of the Aptera is that it's a three-wheel vehicle, which means (along with its sub-1,500-pound [680-kilogram] weight) that it qualifies as a motorcycle instead of a car. Safety isn't an issue, however, because the passenger compartment is built to formula race-car crash-cage specs.

Before we get *too* excited about this, we need to bear in mind that Accelerated Composites hasn't yet finished a prototype version—the Aptera currently zooms across the landscape only on computer screens. A variety of factors could reduce the overall fuel efficiency of the vehicle: it may need more weight for wind safety; real-world streamlining may not match computer estimates; passenger weight could reduce efficiency.

That doesn't mean that we should dismiss such designs as novelties that will never have a real place in the consumer world. In fact, we should demand that automakers produce some of their prototyped but never released high-mileage diesel-hybrid designs. Forget hypercars, even diesel versions of the typical gasoline-electric hybrids would be very valuable.

A couple of decades ago, diesel meant sooty smoke belching from big-rig trucks, and foul smells from cars. More recently, modern diesel-engine design, coupled with wider availability of much cleaner types of diesel fuel, make diesel a more attractive option. Diesel-hybrid vehicles are a surprisingly ecofriendly choice, combining modern clean-diesel engines, Toyota Prius–style serial hybrid-electric systems, and biodiesel/vegetable oil fuels. They could deliver amazing mileage, cleaner air, and vastly reduced petroleum dependency. Comfortable, powerful sedans could get upwards of 80 miles per gallon and be carbon neutral. That's a cool ride. JC

Opposite, left: The diesel-electric hybrid Aptera could get 330 miles per gallon.

Opposite, right: Honda's FCX fuel-cell stack could power the first commercially viable hydrogen car.

The Hydrogen Economy

Hybrids and biodiesel cars are fine first steps, but they are only first steps—and baby steps at that. In order to avoid the troubles that we'll face when we hit peak oil, and to reduce emissions enough to ensure a bright green future, we need a whole new auto industry, one that is not dependent on fossil fuels. Hydrogen is widely considered to be the fuel of the future, and may well power the next generation of ultraclean cars. The technology roadmaps of today point toward electrification of our vehicles as the answer; hydrogen fuel-cell vehicles would use onboard fuel cells to generate electricity from hydrogen gas, a renewable energy source.

Hydrogen cars, and an entire economy based on hydrogen-generated energy, sound great in concept. In reality, hydrogen vehicles carry a long list of unknowns. Foremost are the cost and performance of the hydrogen fuel cells themselves. Currently, hydrogen fuel cells require complex manual assembly, and therefore production is slow and costly. Furthermore, fuel cells have consistency, quality, and performance issues, including materials' degradation and the inability to work in subfreezing temperatures.

Another major issue is finding a way to store enough hydrogen on board the car to support a driving range that is equivalent to that of current gasoline-fueled vehicles, without compromising the passenger and cargo space. Finally, there must be an infrastructure to support the delivery of hydrogen to refueling stations for retail sale.

All of these issues, and the perpetual promise that hydrogen fuel-cell vehicles are "just a decade away," have, to some extent, diminished the excitement over the idea of hydrogen cars. The concept has started to feel dated, like talking about freezing your head after you die—a vaguely embarrassing symbol of a particular era of futurism.

But Honda may change all of that with a hydrogen fuel-cell vehicle slated to begin production in Japan within the next three to four years. The FCX line has been Honda's fuel-cell vehicle prototype for a few years now, but the first iteration was ill suited to regular use. Tiny, somewhat underpowered, and saddled with a driving range equal to about half that of a typical gasoline-fueled car's, the old FCX simply wasn't an attractive option. The new FCX design, however, changes all of that, and manages to induce the kind of auto lust that previous hydrogen fuel-cell vehicles couldn't muster.

Sleek, roomy, and built upon Honda's latest-generation fuel-cell system—a fuel cell stack (similar to a battery) providing 134 horsepower (that's a respectable 100 kilowatts of power) and a hydrogen storage tank allowing more than 350 miles (560 kilometers) range—Honda's FCX looks like a viable contender.

More important, Honda has simultaneously developed the latest generation of its Home Energy Station (HES). The Home Energy Station uses regular natural gas to create hydrogen to fuel the FCX at home. But the same system can also supply electricity to homes—the heat produced from refueling can even provide hot water. If it works as claimed, the FCX/Home Energy Station combo will not only slash carbon emissions by 40 percent, it will also cut combined electricity, gas, and fuel costs in half.

A car that doubles as a home generator—that's more than an advanced form of transportation. In the bright green world we're trying to build, distributed energy technologies and smart power grids [see Smart Grids, p. 183] may make home power generation a common part of our lives. The point's not to go "off the grid" but to hook into the network, helping make the clean energy we need to power our society into the future—even as we drive—with clean consciences. JC & MM

RESOURCES

The Hydrogen Economy by Jeremy Rifkin (Tarcher, 2003)
Rifkin refers to hydrogen as "a promissory note for humanity's future on Earth." *The Hydrogen Economy* is the best primer out there on the coming oil crisis and the new economy that is poised to emerge from the harnessing of hydrogen as an alternative fuel source. In the book's final chapter Rifkin delves into the thorny questions of who will end up controlling this future fuel, and the potential implications of this new economy on our sociological frameworks: "Hydrogen, because of its universality, offers the prospect that we might be able, at long last, to democratize energy and empower every human being on Earth. But, while the opportunity exists, there is no guarantee that hydrogen will, in fact, be fairly and equitably shared among all people. Much depends on how we come to 'value' hydrogen. Will we see it as a shared resource, like the sun's rays and the air we breathe, or as a commodity, bought and sold in the open marketplace, or something in between?"

Biodiesel Power: The Passion, the People and the Politics of the Next Renewable Fuel by Lyle Estill (New Society Publishers, 2005)
"Biodiesel poses a problem for the EPA because their procedure calls for hooking rats up to pure emissions and calculating how long it takes them to die. In the case of biodiesel, the rats don't

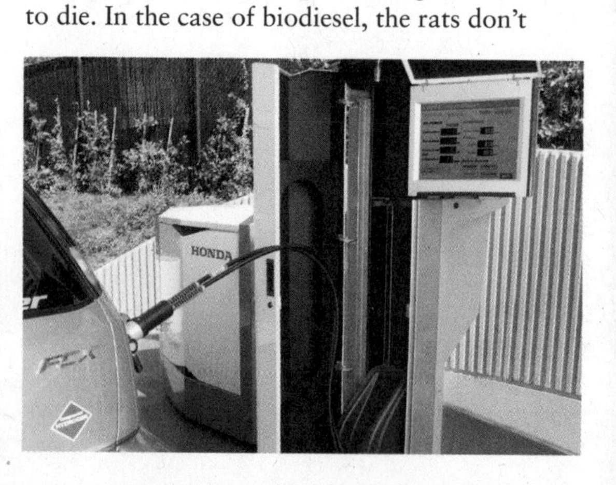

die." For this and many other reasons, Lyle Estill stands as a staunch advocate of B100 (vegetable oil fuel unadulterated by petroleum). Estill's *Biodiesel Power* weaves a basic education in biofuels together with a series of anecdotes from his years entrenched in the biodiesel community, which has been moving swiftly into the public purview in the light of skyrocketing gas prices and fear of peak oil.

Green Car Congress
http://www.greencarcongress.com
Mike Millikin's Green Car Congress is the best single source for news on hybrids, hydrogen cars, alternative fuels, and related topics. It's one-stop shopping for the automotive eco-geek. AS

Opposite, left: Hydrogen refueling station.
Opposite, right: Honda's experimental home fueling station could provide heat and electricity for a home as well as fuel for a hydrogen-powered fuel-cell vehicle.

Bright Green Consumerism

If we want to change the world, it helps to learn how to see systems and look for places in those systems where small, strategic actions pay off in big results.

Take the things we buy: the products that wind up on store shelves are part of a vast system of research, design, manufacturing, and marketing. No matter how brilliant a designer is, if the company she's working for doesn't think her designs will sell, they're history.

Fortunately, this system comes with its own essential leverage point: our buying power. Already, we can buy a number of greener, more responsibly made goods, but those of us who are really serious about building a better future can go a step further. We can work to send signals to the market researchers, who can give designers the green light to turn more visionary designs into revolutionary products.

Let's imagine an example: The CEO of a company that makes golf ball–cleaning machines. The head of his design department has been bugging him for a year, talking about new technologies and design approaches that could yield a machine that costs more but uses much less energy, while delivering even shinier golf balls. This tempts the CEO; he'd like to do the right thing, but the company's investors are clamoring for better quarterly returns on their money, and this year doesn't seem like the right time to try something risky.

Then the CEO's marketing director brings him some research showing that an increasing number of people are willing to pay a premium for high-performance green products (which is already true, as the popularity of hybrid cars shows), and that some country clubs are already responding to their members' requests to use fewer pesticides, less water, and more native plants on their golf courses (which is also already true). Realizing that many golfers may also be the kind of people willing to pay more to go green, the CEO agrees that a limited run of specialty ecofriendly golf ball–cleaning machines could make sense. It's worth a try, and if it succeeds, it'd be good PR.

The designer gets the go-ahead. The next summer, a small number of ecologically friendly golf ball–cleaning machines hits the market.

Everything hinges on what happens next. If enough golf course managers buy the green machines, the company will manufacture more; the more they make, the cheaper the machines will become; as the price drops, the green version will become more competitive with other golf ball–cleaning machines, until, eventually, no one will be able to remember why anyone bought the old ones in the first place.

This is how innovation spreads in a capitalist system. But how do we get enough golf course managers to buy green machines in the first place? That's the critical question.

Often, one person can be the trigger. Let's imagine a small circle of golf course managers who've been friends and golf buddies for years. One of them believes strongly that links should be not just green, but bright green, and over time, though they kid him relentlessly about being a hippie, his friends have more or less come to share his views.

Now imagine this world-changing golf course manager finds himself at a golfing convention, and comes across the booth where the green golf ball–cleaning machine is being demonstrated. At that moment, he has a tremendous opportunity.

Because he writes the checks at his club, he can buy a couple of the new machines himself. That's a pretty good start. As with any new gadget, the best way to sell it is to create a buzz. When his manager friends come play on his course, they get to try out the machine and see how great it is. Eager to stay on the cutting edge, they might buy one for their own clubs; eventually it will end up circulating in online golf forums, where it might get picked up by an editor at a big golfing magazine.

It might not work. The green version might still sink into the murky depths of corporate obscurity (though not without furnishing the manufacturer with some ideas for future product development). But it might become a smash hit; indeed, the actions of that first golf course manager might be the tipping point for selling enough of the green machines to convince the CEO to produce more machines the next year, and to spend more money advertising them. Once that happens, the whole cycle of innovation and investment kicks in, and before we know it, duffers everywhere are polishing their Titleist with clean consciences.

It doesn't stop there. If sales of the machines take off, the manufacturer's CEO will certainly brag to his friends about it. Some of them are sure to run their own companies, and they might start designing ecofriendly bowling pins and tennis nets. In a short time, the entire recreational equipment industry could be a model of sustainability.

Now in a system like this, a number of people play a key role: the designer has to create a better product; the head of marketing has to have her eyes open to emerging trends in responsible business; the CEO has to have enough vision and courage to give the green version a chance. But the single most important person in this system is the golf course manager.

By using both his purchasing power and his influence to speed adoption of better golf ball–cleaning machines, the course manager is leveraging the entire system. He is changing the world at the place where his power will do the most good.

Here's the kicker: we're all golf course managers. No matter what we do or where we live, we all have the power to use our money and influence on others to support companies that are making better stuff. Often, once we know something about green design, we find that the equivalent in our own lives of a better golf ball–cleaning machine already exists, or is on its way, and is waiting for us to embrace it and change the world. AS

Designing a Sustainable World

The critical issue—for people, organizations, and governments alike—is knowing where we want to be. The imaginary, an alternative cultural vision, is vital in shaping expectations and driving transformational change. Shared visions act as forces of innovation, and what designers can do—what we all can do—is imagine some situation or condition that does not yet exist but describe it in sufficient detail that it appears to be a desirable new version of the real world.

John Thackara, *In the Bubble*

For better or for worse, the material stuff that surrounds us shapes our lives. Products have brought what is arguably the zenith of human comfort to those who can afford them. We're knee-deep in useful things such as refrigerators and quality footwear, yet we're also laden with the detritus of the last generation of objects. The periphery of our comfort zone is lined with waste.

The fact that all stuff—every ballpoint pen, every pair of flip-flops—was made with intention is almost as astounding as the sheer number of things around the world. Some designer ensured that it would take scarcely any thought to use our

TranSglass, designed by Emma Woofenden and Tord Boontje, is glassware made from recycled wine and beer bottles.

coffee percolator. While we weren't paying attention, the designer also made sure that it would look embarrassingly out-of-date as soon as possible. This habit of designing for obsolescence, using centuries-old manufacturing technologies, has created a huge set of challenges. We could resolve them by collectively renouncing all but the most basic of material comforts. Alternatively, we could accept the status quo. But while one approach seems retrograde, doomed to failure, the other is simply unthinkable. Perhaps our ticket to a better, more sustainable future is to do what human beings do best, given the chance—design our way out of the conundrum.

The inventor Edwin Land once referred to creative acts such as design, as a "sudden cessation of stupidity." The twenty-first century has already seen a huge wave of such moments, and we have had the opportunity to make designed things more sustainable. We're not lacking in creative acts, ideas, or strategies: we have them in spades. Our greater challenge lies in knitting all of these together.

Product design isn't merely architecture for small things: it's a field in which a whole set of dynamic and unpredictable factors must be considered. We manufacture consumer objects by the thousands, and we release them into the world like flocks of birds. Today's product designer often has little control over where these objects go, how they are actually used, whether they get hacked, axed, or modified, and how they're disposed of when they break or wear out. Still, most of an object's ecological impact is determined at the design stage, so in this seeming chaos is a vast, often untapped opportunity for smarter, more effective design.

One thing that product designers have that architects lack is speed: things can be cranked out in a fraction of the time it takes for a building to be developed. The field of product design has also evolved quickly. In 2000, the well-known science-fiction author and renowned design visionary Bruce Sterling released his "Viridian Design Manifesto," in which he called for a completely new approach to industry, design, and "social engineering." The manifesto appealed for "intensely glamorous environmentally sound products," goods that would be irresistible to consumers for their sheer gorgeousness, that would establish a market in which buying unsustainable products would amount to fashion suicide. Less than a decade later, Sterling's vision is coming to fruition.

Demand for truly ecofriendly products is now growing so fast designers can't keep up. Well into the 1990s, ecologically responsible furniture amounted to little more than globs of recycled plastic melted into the shape of chairs and sofas; today there's a good chance that a sleek, top-of-the-line office chair might be the most ecologically responsible choice.

Most of the green products on the shelf today are mere half-steps, metaphorical references to sustainability. A solar cell-phone charger should be more than

a friendly, guilt-absolving talisman with an ecosensitive sheen. It should be made, used, and retired with biological cleverness and the lightest of impacts. Tiny, hesitant improvements are a terrific way of perpetuating a broken system, but many of the components for fully overhauling this system are already here, waiting to be assembled. Sustainability can be applied to anything that's made.

We must bring about a full-scale convergence of sustainable approaches. Humans are relentless tinkerers; production and industry are central obsessions. Since the mid-1990s, crafty ecodesign has been seized upon, commodified, made bright and clean. We now know how to turn unneeded grass into cabinets, to weave flame-retardant cloth that will compost in a field, to send electronics back to their makers to be disassembled and made anew. We can power our toxin-free laptops by teasing energy out of the sun.

In trying to create this change, designers can spend a lot of time fussing about technical details that are often outside their purview, thereby failing to play to their greatest strengths. The liberation of sustainable design will mean changing the way we compose and conceive of our material world, piece by piece. Designers may be able to come up with great solutions for the complex challenges that humans create, yet they alone can't solve the ecological design problem. Businesspeople decide what gets made. Governments make rules. The global market rolls on. And at every moment we all make decisions to buy, to demand, to repair, or to opt out.

Design guru John Thackara writes: "We've built a technology-focused society that is remarkable on means, but hazy about ends. It's no longer clear to which question all this stuff—tech—is an answer, or what value it adds to our lives." If we step back from the surfeit of stuff, we can see the systems for change orbiting around us, but only if we get involved will they be able to maintain their momentum. We decide whether to share things with our neighbors or hoard them in the attic. We're the ones who can alter our clothing, customize our furniture, and choose to use things for years longer than is expected. Rather than waiting for green products to appear—stamped, sanctioned, and ready—we can demand them, or create them ourselves. DD

RESOURCES

In the Bubble by John Thackara
(MIT Press, 2005)
John Thackara has been tracing the cutting edge of design and sustainability long enough to see how our cultivated obsessions with technology actually operate on a grand scale. Rather than microscopes, he contends, we need "macroscopes" to see the patterns and implications of all our small design decisions. Design exists, after all, to serve human ends, and some of Thackara's most compelling and critical ideas focus on the concept of designing less for technology and more for people.

Opposite, left: These 100 percent recyclable Dalsouple rubber flooring tiles are produced in a system that recycles almost all of the waste generated during manufacture.
Opposite, right: These hay and resin garden chairs designed by Jurgen Bey for Droog Design can be composted at the end of their lifespan.
Left: This David Hertz-designed coffee table is made using Syndecrete, a composite made from natural minerals and recycled materials.

"Although many people perceive design to be all about appearances, design is not just about the way things look. Design is also about the way things are used; how they are communicated to the world; and the way they are produced. The dance of the big and the small entails a new kind of design. It involves a relationship between subject and object and a commitment to think about the consequences of design actions before we take them, in a state of mind—design mindfulness—that values place, time, and cultural difference."

Eternally Yours: Time in Design by Ed van Hinte (010 Publishers, 2005)
Durability and endurance create interesting enough challenges to have preoccupied the Eternally Yours Foundation for eight years. Their work is characterized by radically innovative engineering, as well as the sort of delicious Dutch product design that edges into conceptual art. They celebrate the objects we want to keep in our possession, and they wrestle with the paradoxes of longevity: long-lasting things can mean less consumption—or the preservation of mistakes.

This book is the result: a luxurious, distinctive little publication bound in embossed gold foil, with an exquisite binding that exudes care and preciousness. You'll want to keep it—which is exactly the point.

ecoDesign: The Sourcebook by Alastair Fuad-Luke (Chronicle Books, 2002)
Although as of yet there is no such thing as a truly sustainable product on the market, Alastair Fuad-Luke's *ecoDesign (The Eco-Design Handbook: A Complete Sourcebook for the Home and Office* in its second edition) is a pragmatic look at the current approaches, including everything from multifunctional furniture to innovative materials, solar gizmos, and energy-efficient refrigerators. His is the one and only sourcebook to catalog the various attempts at making sustainable products—as well as the materials, organizations, and designers responsible for the prototypes. Heavy on pictures and light on theory, *ecoDesign* celebrates any and every approach to sustainable product design, even if the results are marginal. Though it clearly illustrates how much farther we have to go, the sourcebook is an important early indicator that there are hundreds of designers and engineers out there working hard to reimagine products, using an entirely new set of criteria.

The Total Beauty of Sustainable Products by Edwin Datschefski (Rotovision, 2001)
What would a day in a totally sustainable society be like? What would you eat, what clothes would you wear? If you went out today, would you know how to buy—much less create—sustainable products?

As Edwin Datschefski writes, "Only one in 10,000 products is designed with the environment in mind. Can a product really represent the pinnacle of mankind's genius if it is made using polluting methods?"

Datschefski is a real enthusiast for sustainable alternatives—as well as a passionate collector of examples. After years spent tracking products at the forefront of green design, he happily casts aside a minefield of jargon in discussing the way things are made. His simple rules—that things must be cyclic, solar, and safe; that an object's total beauty must not be undermined by hidden impacts—are a refreshing reminder that products need not be complicated to be effective.

Casting a wide net in his estimation—over objects as well as buildings and food—Datschefski contends that almost all environmental destruction is related in some way to unsustainable products. This doesn't stop him from uncovering a wealth of well-considered alternatives, each approaching sustainability from a slightly different angle. None are perfect, but they're all real, here, and pointing energetically toward the future.

Opposite and following page, left: Among the renewable materials catalogued on Material Explorer's searchable Web site are the Woven Wood panel by Marotte, Lama Concept's Cell-carpet, and Kinnisand's three-dimensional Wave carpet, which comes in up to eighty-one different colors for endless design combinations.

Picking Green Materials

Materials are ingredients of human creativity and designers are always looking for new ingredients to cook with. While one engineers shining plastics doomed for the landfill, another tries to figure out how to make cars from hemp and kenaf fibers.

What if we remade the world out of unbleached cotton and recycled bottles? Would it really emerge as something more reasonable and uncomplicated, smelling delightful, bathed in virtue? It's important to remember that just because something *looks* sustainable doesn't mean that it *is*. We don't need to stop engineering things with wires and metal or casting products in plastic; rather, we need to realize that we're fully capable of being smarter about cooking things up, and more responsible about managing them once they're released into the world. We're unlikely to see global material culture overcome by such restrictive aesthetic simplicity anytime soon, and besides, working from a tiny, spare palette of materials would be catastrophically impractical. Sure we could make a mobile phone out of soy plastic, but when the ants start climbing into our pockets to eat our gadgets, we'll become all too aware of the drawbacks of narrow approaches to green design.

All good intentions become null and void when things cease to work, so objects such as electronics need to be stable enough to function—to say nothing of surviving an onslaught of hungry insects. Crushed sunflower shells can make great coffee tables, but they make really terrible solar panels. Without plastics, we'd have neither pacemakers nor drip irrigation systems.

The truth is, material itself cannot make a product sustainable, because there's no such thing as a "sustainable material." Certainly, there are some materials that should not be used at all, but sustainability is a matter of what we do with the materials we choose. All materials are extracted, heated, and cooled; we pour still more energy into them every time we shuttle them about. Even recycled glass bottles eat up more energy than we care to admit. By using what we have, and wisely choreographing the flow of materials, we can start to eliminate the damage done along the way. Durable things can conserve water, save lives, and generate green energy.

Databases and Online Libraries: The Software of Materials

Sourcing materials involves so many tangled interdependant networks, it's overwhelming to try to understand the behind-the-curtain elements of the goods we buy. Having more organized, centralized information could be the key to understanding and gauging the

impact of our products. There are a number of huge databases emerging that catalog the ingredients designers have available to them. When we, as consumers, can look up materials and learn about their origins and applications, we will have a lot more power at our fingertips (and satisfy a world of curiosity).

Among the best tools for uncovering unheard-of ingredients are the growing number of online resources such as Material Explorer and the international library at Material ConneXion. There can be such a thing as too much information, but both of these sites are curated and constantly updated to home in on the new and the strange. This saves hours of browsing aimlessly through search engines looking for, say, injection-moldable potato starch, or ultrathin recyclable packaging that mimics the chemistry of eggshells. Material ConneXion is the world's largest resource of information about new materials—from yarn made of seaweed and cellulose to sheets of crushed almond shells—with libraries in major design centers like New York, Bangkok, and Cologne. Its most compelling industrial curiosities aren't always space-age: the Italian company Corpo Nove makes a stinging nettle fiber that updates Napoleonic technology. Using textile technologies reintroduced during the cotton shortages of both world wars, the company can turn the de-stung hollow fibers of stinging nettles into a naturally insulating fabric that varies in temperature control according to how the fibers are twisted or spun. Because nettles are perennials that also grow in nitrogenous or overfertilized soil, they might be able to offer an interesting solution to some of Europe's agricultural problems.

We know that it takes a lot of energy to ship reclaimed lumber from Melbourne to Montreal, so one of the most useful things about any database is the localized, place-specific information we can find there. Material ConneXion and Material Explorer connect us to suppliers of horsetail-hair fabric in Hong Kong, holographic tiles in the Czech Republic, and salago-shrub papers in the Philippines. It is easy to envision, then, how this type of database could connect us to local suppliers.

Software is also the key to analyzing materials' sustainability. This is traditionally the stuff of engineers: what amount of energy and resources really did go into our crushed-almond countertops? The world is still awaiting the invention of a database for comparing materials' suitability in a given situation, or breaking down materials' chemical composition and biodegradability. PRé, a Dutch company that makes SimaPro software, does just this kind of evaluation—though not yet for every material under the sun. By analyzing the life cycle and environmental impact of a product, SimaPro connects us to a wealth of information about real impacts, like the amount of energy needed to make a bathroom tap out of steel. It's not a stretch to imagine a locally based information system that supplies details like this—empowering us to get past the seemingly green and attain the truly green.

RESOURCES

Transmaterial by Blaine Brownell (Princeton Architectural Press, 2005)

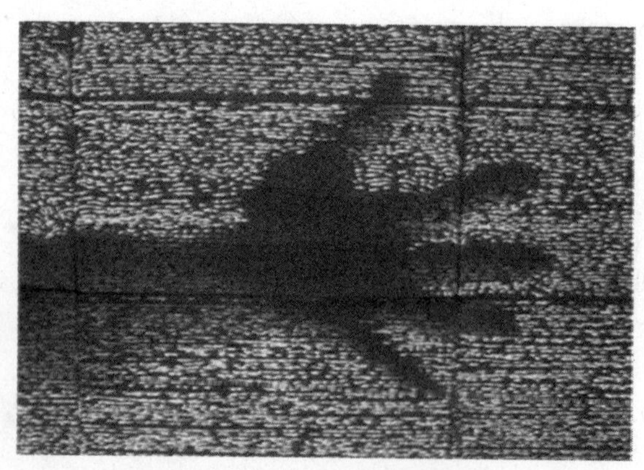

Indoors, soft forms cover the walls to clarify voices in a room. Outdoors, Australian-engineered concrete quietly sucks carbon dioxide out of the air, and biomimetic [see Biomimicry, p. 99] paint—reverse engineered from lotus petals—repels rainwater on the building facades. Plants snake up recycled steel trusses, covering whole walls in greenery while keeping brick and mortar from being devoured by roots. When the sun goes down, photovoltaic [see Green Power, p. 172] windows stop collecting solar energy and start displaying graphics. Interior windows glow with curtains of electroluminescent vines embedded in silk.

Recombining matter in stranger and more varied ways than ever, materials science has been innovating like mad. *Transmaterial*, Seattle architect Blaine Brownell's collection of some of the best new ingredients for architecture, is a material enthusiast's dream. The picture catalog and guide to all things possible tidily classifies and lists materials, processes, and products by supplier. While combing the world for the most interesting substances, Brownell found such oddities as concretes that are translucent, aerated, recycled, and light reactive, neoprene for wrapping horses' joints, and discarded sorghum stalks that have been turned into smoothly finished cabinetry. Like wood that bends in several directions at once, the most exciting materials are unexpected, defying their natural properties.

Some of the most interesting applications featured in *Transmaterial* first saw the light of day in artists' installations. Sachiko Kodama and Minako Takeno created their *Protrude, Flow* sculptures with magnetic fluids that dynamically respond to temperature, sound, and light, changing shape as viewers move around them. Thom Faulders lined a listening room with NASA's memory foam, a material long neglected because it failed to withstand extreme temperatures in space thirty years ago. Down here on earth, our feet leave deep, clear footprints across Faulders's room.

And what about natural daylight, essential to the happiness of building occupants? Besides skylights and light tubes, a roof can be lined with translucent Aerogel. The lightest substance on earth, it works as a champion insulator, while filtering daylight. There are even ways to bring natural light into the unlit rooms of older or much larger buildings, by channeling real sunlight from rooftop collectors into dark interior spaces many stories below. We can sit in a patch of sun in a basement room and watch as the light shifts with the cloud cover.

Though *Transmaterial* focuses on small, improved material and energy efficiencies, it touches briefly on almost every entry's environmental qualities. Alulight, for example, is a strong and rigid foamed aluminum that's five times lighter than pure aluminum. Working on the same principle, Axel Thallemer's inflatable Airtecture structures replace matter with air. Airtecture supports whole rooflines and bears a building's load with air pressure, fabric, and various engineering heroics.

A reminder that there's far more invention out there than we can even begin to keep track of, *Transmaterial* is a guide to the possible. DD

Design and Environment: A Global Guide to Designing Greener Goods by Helen Lewis, John Gertsakis, et al. (Greenleaf, 2001)
Designing for sustainability means building

Opposite, right: Litracon transmaterial is a light-permeable building material made from optical fibers and fine concrete.
Right: Alulight energy absorbers, used to protect car passengers during impact, are made from lightweight metallic foam.

bridges between disciplines and presenting solutions in languages that are foreign to most designers. So there is a critical space for books on sustainable design that fill that realm: annotated, numbered, cross-linked, and brimming with frameworks. *Design and Environment* is, somewhat paradoxically, a deeply practical academic book, targeted at the businesspeople and managers who make so many of the decisions about our material culture.

"Designers are at the significant point of conjunction between technological and cultural worlds. They are therefore in a privileged position to capture and act on signals for change."

Craft It Yourself

The abbreviation DIY, which stands for "do-it-yourself," also stands for empowerment, creativity, and access to information. At this point, DIY can rightfully be called a movement—one made up of people who would rather make it themselves than pay for it, design it themselves than accept it as is, and know for themselves that it is one of a kind.

The DIY movement makes the design process accessible to everyone. It changes our relationship to commercial products, and our perspective on mass production. Certainly, we can't make everything we need, but simply by viewing ourselves as active participants in the fabrication of our belongings, we shake up the dynamic between consumer and producer.

Though the DIY concept has been around for decades, it has gained momentum since the late 1990s, largely as a result of our greatly enhanced access to tools and information. We can now easily track down the basic instructions for doing everything from rigging our own surround-sound speaker system to building a captain's bed.

Unlike in its earliest days, the DIY idea today has infiltrated a very main-

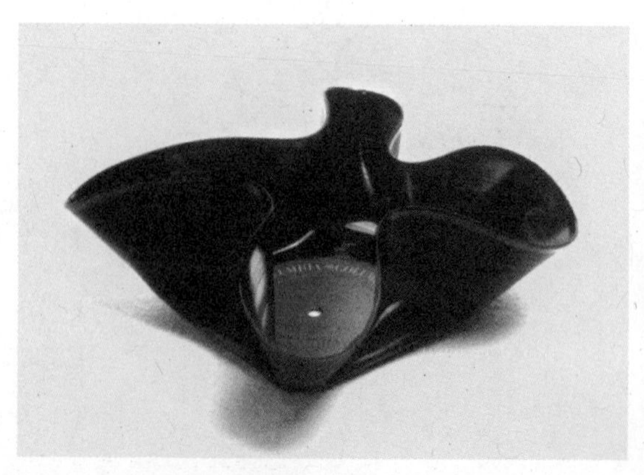

stream audience that could easily afford to pay for home improvements or new clothes, but would rather have a hand in making something unique. Executive knitting circles have popped up all over, because making something yourself feels good, if nothing else, and that's reason enough to do it.

Another phenomenon that makes DIY hot is the desire for uniqueness in the things we own. Having one-of-a-kind commodities matters enough to the mainstream market that mass-produced items are now created to appear flawed and handmade. But millions of identical sweaters with the same stray strand of yarn don't really cut it.

DIY culture has given rise to a number of popular publications, an entire cable TV network, and thousands of blogs and Web sites. With each experiment yielding a different result, and plenty of media outlets for sharing experiences, the wealth of resources has become a self-sustaining treasure chest growing at an exponential rate. SR

DIY Culture

Do-it-yourselfers do not see limits, they see sprawling potential, order drawn from chaos, hope. Outside of mainstream culture, many think this way out of necessity. They live in conditions that force them to invent solutions to everyday problems: the Delta dweller who builds a raft out of plastic water jugs, the desert trekker who sews a sun shade out of scraps.

Cooking-with-wax Bowls: You can buy them by the dozen in thrift stores: old LPs—most too scratched or warped to listen to—make great raw material. Spread a little peanut oil on both sides, pop them in a 350-degree oven for half a minute, then slide on your mitts and mold the vinyl into bowls.

Laundry Lamp: Don't those Cheer-y bottles deserve more than a few weeks in the laundry room? Empty them of their suds, fit them with a corded light socket, drill in some feet, and poof! You've got yourself a "bright-whites" lamp.

Curiously Strong Sound System: Those little tins have to be good for something besides delivering fresh breath. Make a set of iPod speakers out of two Altoids tins, two playing cards, and a set of headphones.

Sod Sofa: Build a living, biodegradable couch in your backyard. How much oxygen did your furniture produce today? In our version of the future, the things we loaf around on inside will be as beneficial as the stuff we grow out back. In the meantime, sculpt lawn furniture from the lawn itself. Unlike your standard-issue sofa, this lush green sofa is totally organic, requires no synthetic finishes, and can be brought to life, Golem-style, from salvaged dirt. Note: Couch may require mowing.

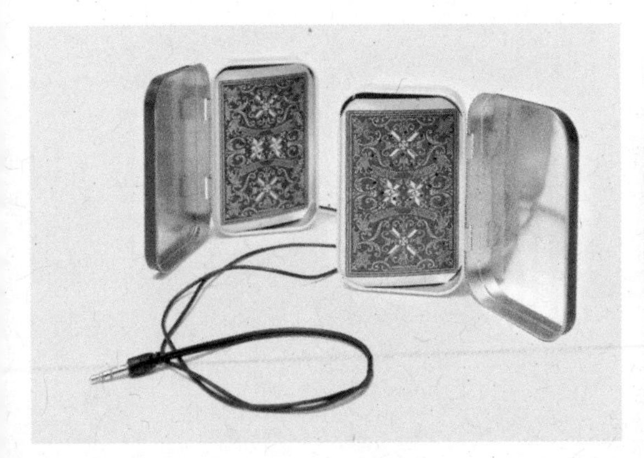

In cultures of leisure, do-it-yourselfers have other reasons for making things. They may see what surrounds them as a set of broken parts in need of fixing, or simply want to build something of use that bears a human stamp. Whatever the motivation, those who make things themselves make a mark on the world, creating a thing of beauty where there existed only the ghost of an idea. Global consumer culture produces enough cast-off stuff every year to build a ladder to the moon, but DIY reusers aim to save the world from drowning in its own junk. When put through the reuse mill, the water bottle becomes a chandelier, the worn-out sweater a brand new scarf.

The do-it-yourself design concept is the brainchild of Marcel Duchamp, the surrealist who coined the term *readymade* to describe artworks made from everyday wares (a urinal, a bicycle seat). Today, the people who make things from cast-off objects do not mean to launch an art movement—pop art or a new Bauhaus. What they do claim, as do-it-yourselfers, is a belief in making things work by some mash-up of imagination and sheer force of will. Since every reject of consumer culture can be reinvented in a thousand ways, the ideas here are intended to be inspirational—a pit stop in the creative process, not a final destination. It's your job to get behind the wheel, floor it, and go places we never knew existed. SB

RESOURCE

ReadyMade: How to Make (Almost) Everything
by Shoshana Berger and Grace Hawthorne
(Clarkson Potter, 2005)
This "do-it-yourself primer" by the cofounders of *ReadyMade* magazine includes projects that range from creating an Eames-style bookshelf out of discarded dresser drawers to fashioning a messenger bag from those wasteful wrappers that newspapers come sealed in when they land on our doorsteps. Informative sections detail the history and environmental legacy of our most commonly used materials—paper, plastic, wood, metal, glass, and fabric.

The book is written in *ReadyMade's* signature casual, humorous style. Die-hard DIYers might find it a bit too jokey at times—features like "How to Avoid Plastic Surgery" and "Iron Man: A Look Back at the Origins of Heavy Metal" are a questionable use of the (recycled) paper they're printed on. But the projects presented here are clearly illustrated and well explained, and will provide plenty of fodder for first-timers and reuse veterans alike. As the authors put it, "Perhaps you will try making a thing or two. Perhaps you only imagine the day when you will make something. Either way, we hope that the projects you see here have a catalytic effect, inspiring you to rethink the purpose of an old telephone book or bicycle wheel." CB

Engineer It Yourself

The do-it-yourself (DIY) movement extends into realms more radical and technical than art and craft. People who enjoy pushing the boundaries of home-cooked science and engineering are hacking high-tech mechanical devices for computing, transportation, even flying, and have published step-by-step instructions for other industrious trailblazers to do it themselves. Along the way, they've helped to beat a path forward for technologies that benefit us all. SR

The Democratization of Making Things

The United States used to be a nation of tinkerers and inventors—what happened? Did they all go away when the megamarts moved in? Turns out, they're still out there, and they're closer than you think. Quietly, over the last few years, a small army of makers has been publishing discoveries, connecting with one another, and using the Internet to "make versus take." Some of us have been making things ourselves for years, but until now it was never possible to go so smoothly from idea to research to meeting other makers to producing a finished collaborative project that could be shared around the planet. The collection of tools, their ease of use, and the way they all work with one another is turning anyone with curiosity into an engineer. The physical objects around us are beginning to resemble computer code—they are describable and replicable. People are sharing, people are making things—and there is no stopping them. We can all vote with our ideas by bringing them into reality. On the Instructables Web site (instructables.com), anyone can share their creations with the world by posting how-tos, recipes, step-by-step guides, and annotated photos.

We're seconds away in Internet time from being able to literally fabricate our ideas almost immediately. Imagine designing a cellphone cover and printing it out on a 3-D printer just as you would print a document on a paper printer. Already—from a company called eMachineShop—you can download free software, design an object, send off the file, and have the part sent to your door. For $20,000 you can put together your own "Fab Lab," a fabrication laboratory of open-source computer-controlled design and manufacturing equipment. Just a few years ago this would have required a multimillion-dollar investment; soon it will be cheaper than a high-end TV. The Center for Bits and Atoms at Massachusetts Institute of Technology (MIT) is dropping Fab Labs around the world, with the goal that they will become self-sustaining.

When anyone and everyone has the means to solve their own puzzles and problems, to share their ideas and works, and to invent new things, the creative human spirit will express itself in so many ways that is it impossible to predict just what we'll build together as citizen engineers. PT

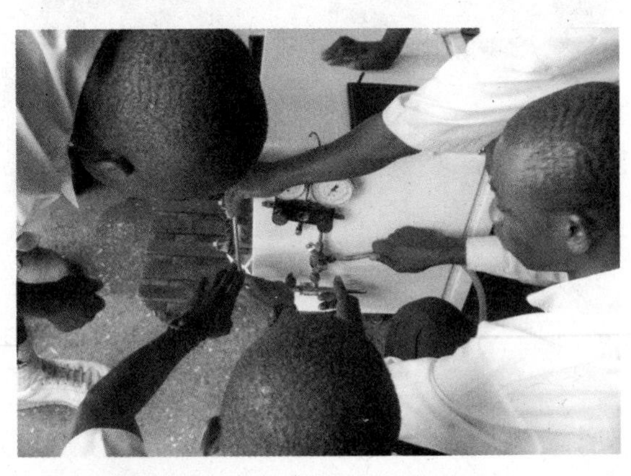

These Fab Lab students in Ghana are working on a vortex refrigeration system, which could, inexpensively, keep everything from food to vaccines cold, 2005.

Instructables

Instructables.com, the brainchild of half a dozen or so MIT PhDs, is an online collaborative repository of how-tos that anyone can access to learn how to make just about anything, and to share their own inventions. Instructions are linear, similar to recipes, and they're available for everything from a 3-D chocolate printer to home-canned applesauce to self-replicating robots. Makers join the site, upload photographs of their projects, tag photos with expanded information, then invite the world to comment or improve upon their designs. Take a look at the following Instructables' samples:

Make Your Own Pedal-powered Air Compressor: What can you do with an old electric motor and a bicycle? Make a pedal-powered air compressor to build other things, of course. Just grab an old motor mount and install an axle with one fixed cog and belt pulley. Welding the bike frame and handlebars and rigging a chair yield a Mad Max–like device that can reach up to fifty pounds per square inch in about fifteen minutes of pedaling.

Etch Your Own Circuit Boards: If you have an itch to design your own electronics, you'll inevitably need a circuit board. Countries around the world stamp these out by the hundred thousands, but you can make your own. Use a laser printer or copier to print out your electronic design on transparency paper, then transfer the toner from the paper onto a copper board by heating it with an iron. Next up, use some etching solution to remove the copper from the portions of the copper board not covered by the toner. After giving the board a quick scrub to remove the toner (your design will remain on the board in copper, which was protected from the etching solution by the toner), you're ready to drill some holes in the board for wiring. PT

RESOURCES

Make Magazine
http://www.makezine.com/
Make is what you'd get if your school's science fair was run by hackers, home energy nuts, and mad inventors in Day-Glo lab coats. Every issue bends your mind. Even for those of us who couldn't solder a circuit if the fate of the world hung in the balance, it's great fun. AS

Radical Simplicity: Creating an Authentic Life
by Dan Price (Running Press, 2005)
This fantastically accessible book, filled with photographs and hand-drawn diagrams of underground houses, handmade saunas, and all-natural domes, is Dan Price's DIY guide to a simpler life. The book bears witness to the author's insatiable drive to design and hand build structures, and along the way makes the convincing case for personal sustainability and connection to the places we inhabit. Examining structures from kivas built into hillsides to underground sleeping quarters, Price reveals the potentials of an alternative lifestyle, one in which shelter is accessible, portable, personal, and constantly changing. As Price states, "nowadays you can stand ... nearly anywhere in America and be assaulted by the over-building, the sprawl,

The folks at Squid Labs deploy a candle-powered hot-air balloon. SpokePOV is an open-source electronics kit developed by Adafruit Industries for making artwork out of your own bike wheels. Both are available on Instructables.com.

the general messiness and ugly weed-like growth of our highly industrialized society," and his book aims to illustrate the alternatives to this troubling trend.

Art Meets Technology

We live in a world where mobile technologies, genetically modified organisms, and virtual realities are commonplace, and increasingly define our lives. People from a vast range of disciplines now use their skills to create objects that not only push the edge of innovative technology, but invite us to question our relationship to it, to comment on it, to use it for something other than its expected purpose, or to hack it ourselves.

The way artists use and misuse emerging technologies in their work can prompt deeper reflection about our society and its relationship to technology than a two hundred–page report written by eminent sociologists can. But what really sets such work apart is its frequent exploration of issues that are immediately and achingly relevant. Scientific advancements, technological change, or notions about the environment fall into sharp relief when artists use familiar electronic detritus as raw material. Traditionally, there has been a divide between the mastery of hard technologies and their use as a medium for commentary, but that's changing fast.

Right: KnitPro is a free Web application that helps users create their own subversive needlework patterns.
Opposite, left: The Hug Shirt delivers virtual hugs long-distance.
Opposite, right: *Disaffected!* challenges players to navigate the perils of Kinko's employment.

Imagine a cell phone that monitored your location using networked surveillance sensors within the phone, and sent text messages to help you keep track of others. This is the premise behind LOCA: Location Oriented Critical Arts, a Finnish project that explores the use of consumer gadgets as nodes for monitoring networks. The idea could be both sinister and constructive; LOCA's projects test both sides of the fence.

If the idea of being tracked stirs uneasiness, there are other great applications for sensing technologies. Myriel Milicevic's Neighbourhood Satellites allow users to playfully monitor the conditions of their environment through a handheld gadget that measures air quality, light, and cell-phone reception.

On another plane altogether, microRevolt's knitPro is a Web application for generating knitting patterns as protests against sweatshops. You can upload digital images of your choosing, and the application will generate a pattern, on a scalable graph, for knit, crochet, needlepoint, or cross-stitch projects.

Cultural artifacts such as these use technology to throw light on themselves. Their authors, however, often don't regard themselves as media artists but as engineers, architects, designers, or hackers. Those who do define themselves as artists usually operate outside contemporary art circles; you won't find their work in contemporary art exhibitions, magazines, or galleries.

Regardless of context, these works accomplish what art should accomplish: they trigger new experiences that transform our perceptions of what is and what could be. It's a whole new realm where art meets technology. DD & RD

Hug Shirt

The Hug Shirt, developed by Cute Circuits, allows you to feel the physical closeness of a distant loved one by generating the sensation of being hugged. How is this possible?

Embedded sensors and electronics in the shirt are able to pick up signals such as a heartbeat and body temperature from a loved one at the other end of a mobile phone. When the sensors process the signal, embedded mechanisms in the shirt re-create the physical pressure and warmth of a real hug. RD

Disaffected!

A company called Persuasive Games has created a line of video games—"anti-advergames"—designed to encourage players to rethink their relationship to consumption and to engage in corporate critique. The game *Disaffected!* satirizes the infamously disordered and endlessly frustrating customer service inside a Kinko's store.

In *Disaffected!*, the player controls the employees behind the counter. As customers file in and form a line, the player must find the customers' orders while coping with a host of

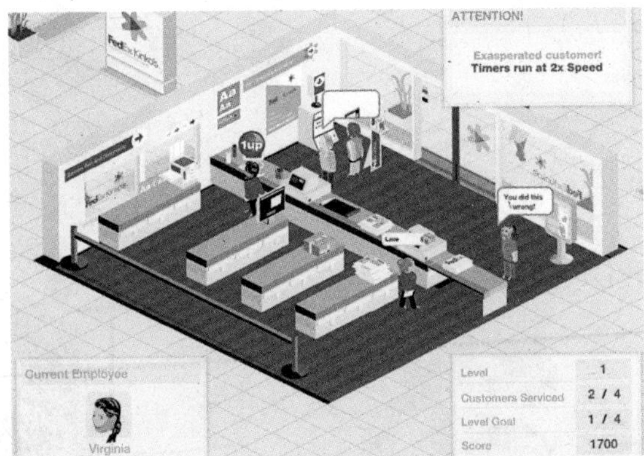

problems and staff ineptitude. Obstacles such as employees who refuse to work or to abide by systems infuse the game with an undeniable element of comedy. They also demand critical thinking from the player, opening up a whole new level of engagement, not only with the game but with the world at large. RD

Wearables for Environmental Awareness

Imagine having the ability to create a sonic map of your experience of walking through a city. Just as an adept deejay can layer a seemingly infinite array of sounds into a singular audio experience, new recording technologies can now enable people to compose music in real time as they walk through noisy surroundings.

The Swedish Viktoria Institute and RE: FORM Studio developed "Sonic City," a system that records information about the environment, and user activity, and maps it to the audio processing of urban sounds. The result is music that you can hear almost simultaneously through headphones. When you wear the sensor-equipped jacket, your physical movement and the activity of the city around you creates a "personal soundscape"—an absolutely unique audio record of your urban adventures.

Also designed for city dwellers, Miki Yui and Felix Hahn's "Acoustic Survival Kit" counters the stress caused by exposure to urban noise, with sound-emitting devices embedded in clothing. The garments emit subtle sounds that fuse with the sounds of the surrounding environment, protecting you from overstimulation and teasing you into more active interaction with your urban environment. RD

"Sonic City" creates unique soundtracks for urban experiences.

Carbon-sniffing Robot

We live in a world of flows and systems that remain largely opaque to us. Making invisible things visible opens a window onto how our world actually functions—and sometimes suggests tactics for improving the world. We can't see the air, but that doesn't mean we're breathing free and clear. From off-gassing paints to factory emissions, potential health hazards hang thick in our air.

In response to this silent threat, artist Sabrina Raaf has created a carbon-sniffing robot. This squat automated vehicle patrols the periphery of a room armed with a green crayon. Every few inches, the robot takes a reading of the carbon dioxide level in that particular spot and makes a vertical mark—corresponding with the CO_2 concentration—on the wall. The accumulation of green lines begins to resemble grass, its growth visible to—and entirely determined by—gallery visitors. The end result is a chart of the air quality in the space throughout the installation, and a reminder that, although we can't always see it, our lives are changing the air itself. RD

RESOURCES

Telepresence and Bio Art: Networking Humans, Rabbits, and Robots by Eduardo Kac
(The University of Michigan Press, 2005)
Recognized worldwide for his interactive Internet installations, bio-art pioneer Eduardo Kac documents in this landmark book the evolution of his field—art that bridges the divide between biology, technology, and innovation. Using examples from his own cadre of world-changing "events," Kac argues, "telepresence works have the power to contribute to a relativistic view of contemporary experience and at the same time create a new domain of action, perception, and interaction."

Creative Capital Foundation
http://www.creative-capital.org/
Creative Capital is a different kind of arts funder: the foundation provides rigorously screened artists not only with sizable grants, but with professional development opportunities and critical connections. It supports both artists' projects and the artists themselves over an extended period of time, making a point of picking artists whose work has the potential for real impact in the world—artists like Sabrina Raaf, Red-Dive, and Critical Art Ensemble. If you want to know where to look for the next generation of worldchanging artists, keep an eye out for who Creative Capital is funding.

Ars Electronica
http://www.aec.net
The Ars Electronica annual festival (and the companion organization the Ars Electronica Center) consistently presents the most cutting-edge examples of work springing from the intersection of art and technology, to the extent that attending their annual show has become something of a pilgrimage among certain crowds. There are many people exploring the fringes of media arts and digital culture, and to some very real degree, they all orbit around Ars Electronica.

Biomimicry

Biomimicry—usually called bionics in Europe—is design inspired by nature. Velcro, for example, was inspired by the way burrs stick to fur—the scratchy side of Velcro acts like burrs, the soft side acts like fur. When well done, biomimicry is not blind imitation, but inspiration for transforming the principles of nature into successful design strategies. Biomimicry can be applied at any scale, from tape that imitates a gecko's skin to achieve super-adhesion, to a high-rise that imitates a termite mound to achieve passive air-conditioning.

We humans have gotten ideas from nature for as long as we've been around. But our application of those concepts has often been haphazard and imprecise. Buckminster Fuller [see Biomorphism, p. 104] was the first to propose that nature's teachings offered the perfect tool for green design. Janine Benyus, with her book *Biomimicry*, was the first to bring that idea to the world at large. Now designers are learning how to make biomimicry an actual methodology, rather than just occasional serendipity.

Nature is inspiring not because it's perfect, but because it's prolific. There are some drawbacks to nature's design: natural products need continual maintenance and/or rebuilding (though this can be an advantage in products meant to biodegrade or become obsolete), and organisms can't borrow designs from one another; they have to evolve from their existing designs. Evolution requires each solution to be better than the last; it's not a testing ground for new strategies that might get worse for a few generations before getting better. However, everything we see in nature has been field-tested for thousands or millions of years, and the earth has come up with

countless clever solutions we might never have dreamed of. JJF

Biomimicry in Action

Biomimicry's core idea is, as Janine Benyus says, treating nature as model, measure, and mentor. Using nature as model, we can get ideas from organisms to solve our problems. Whatever we are trying to do, there are usually several organisms that have evolved successful strategies (like burrs sticking to fur) to do it. Applying nature as measure, we can look to the natural world to see what is possible. Spider silk, for instance, is stronger than steel and tougher than Kevlar (which itself is purported to be five times stronger than steel), but the "factory" that produces it—the spider—is smaller than your little finger, and employs no boiling sulfuric acid or high-pressure extruders. Taking nature as mentor, we are able to recognize that we are part of a larger system, and that we should treat nature as a partner and teacher rather than as a resource to be exploited.

Biomimicry can be achieved on different levels: the *form and function* level, the *process* level, and the *system* level, according to Benyus. Biomimetic *forms and functions,* such as Velcro, are the most common manifestations of biomimicry. Biomimetic *processes* manufacture products as nature would. The self-assembling coatings developed at the Department of Energy's Sandia National Laboratories are an example of biometric process—they grow from a solution the way seashells grow in seawater. These incredibly strong, transparent coatings, which could be quickly grown in laboratories, could revolutionize all kinds of finishes for everything from cars to contact lenses. Biomimetic *systems* feature closed-loop lifecycles [see Neobiological Industry, p. 109] that recycle the outputs and by-products of one process as inputs for another. As William McDonough and Michael Braungart note in *Cradle to Cradle,* "waste equals food" in a biomimetic system. And today we can also add a fourth level: the *design* level. Biomimetic design processes—in forms such as genetic algorithms and iterative design—can produce cheap, green results never seen in nature. Nature's students are learning quickly. JJF

"Evolved" Antennae

Designing spaceship antennae is difficult: the components and their interactions are maddeningly complex, and trying to come up with workable designs that do everything they need to can eat up a lot of NASA scientists' time. Their latest antennae, however, were *evolved,* not designed. They were produced using genetic algorithms—search algorithms that find optimal solutions by combining and recombining bits of code (like genes in an organism) and testing these virtual organisms (each one a possible solution) against set criteria; over many generations of mixing DNA and culling the best performers, the algorithm finds new and better solutions. These antennae look nothing like traditional devices, because the algorithms try things that humans never would. They simply keep tweaking randomly, allowing designs to evolve through a kind of "natural selection" based on criteria set

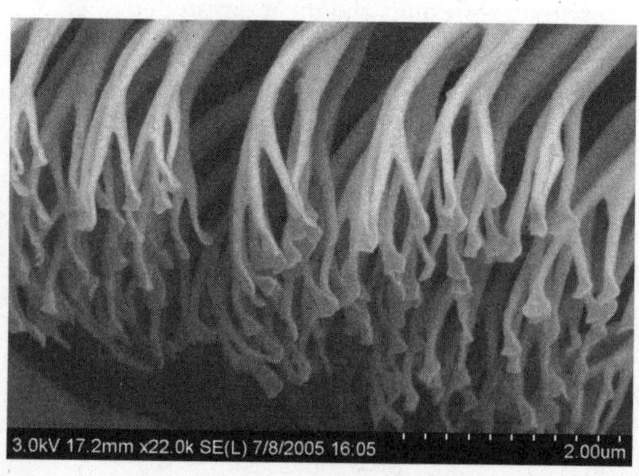

by the engineers. In this process, the criteria for a "healthier" antenna included better amplification and a better signal-to-noise ratio.

NASA's evolved antennae, though, are really just signals of a larger trend. Increasingly, designers of all kinds are programming software that lets generations of possible designs grow and evolve in virtual spaces, prototyping those designs that best fit the criteria they're looking for and seeing which work best in the real world. These simulated bits of biomimicry—computer programs that do everything from finding the best shape for a machine part to telling delivery companies the quickest route to get a package from point A to point B (by following rules based on ants' foraging patterns)—can help designers and engineers come up with solutions we couldn't have dreamed of a decade ago. And since those solutions often are more efficient and work better than the previous ones, evolutionary design can be a powerful tool for making the world more sustainable. JC & AS

Gecko Tape

What makes a gecko's feet so sticky? They are equipped with complex microstructures that have incredible adhesive power and that work on virtually any surface, and even underwater. The microstructures leave no residue, and even clean themselves during use. A tug at the correct angle allows a gecko to peel its feet off a surface without solvents or other glue-removing mechanisms. Scientists at Lewis and Clark College and the University of California, Berkeley, can now synthesize materials that imitate gecko feet to create the wonder adhesive of the future—not just a replacement for tape or glue, but for nails, screws, and a lot more. At the same time, being able to simply peel apart components—with no chemical residue—would enhance our ability to design for disassembly, an important part of Cradle to Cradle thinking [see Knowing What's Green, p. 115]. JC

Lotusan

Scientists at the German company ISPO wanted to find a way for buildings and products to clean themselves. Since the lotus flower grows in swamps yet is always pristinely clean, they studied it to see why it stays spotless. They found that the surface of a lotus leaf, under a microscope, appears craggy, and that dirt particles cannot keep a grip on the tiny peaks. The "crags" also cause rainwater to ball up and roll off rather than sticking. Thus, dirt sits loosely on the surface of the leaf, and whenever it rains, water droplets stick to the dirt and roll off, carrying the dirt away. ISPO has made several products for use on buildings and automobiles—including roof shingles and paints—based on the "Lotus Effect," under the Lotusan brand name. The paints dry with a lotus-like surface structure, and get cleaned every time it rains. Water repellency also makes the paints highly resistant to mold, mildew, and algae, and the company claims the paints will last 40–100 percent longer than normal exterior paints.

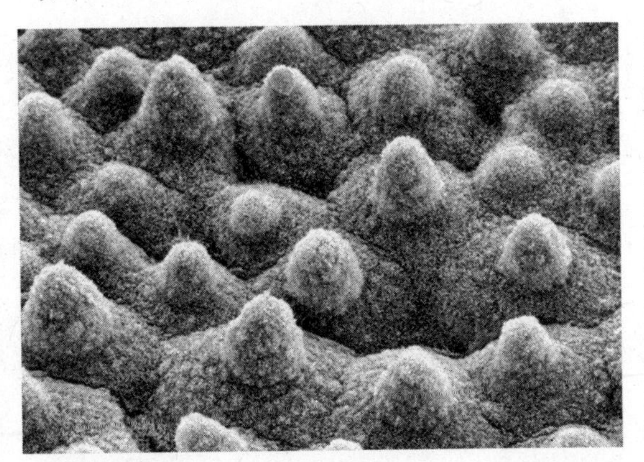

RESOURCES

Biomimicry: Innovation Inspired by Nature by Janine Benyus (Harper Perennial, 2002)
What is ecologically sustainable design but design in accordance with the natural world? As Janine Benyus puts it, "The core idea is that nature, imaginative by necessity, has already solved many of the problems we are grappling with. Animals, plants, and microbes are the consummate engineers. They have found what works, what is appropriate, and most important, what lasts here on Earth. This is the real news of biomimicry: After 3.8 billion years of research and development, failures are fossils, and what surrounds us is the secret to survival."

Cats' Paws and Catapults: Mechanical Worlds of Nature and People by Steven Vogel (W. W. Norton & Co, 2002)
What do spiders' legs and cherry pickers have in common? Ever thought of shells as the technical basis for corrugated cardboard? Nature and man use the same cadre of building materials, yet their designs diverge. In *Cats' Paws and Catapults,* Steven Vogel compares these two mechanical worlds and introduces the reader to the field of biomechanics, a fascinating exploration of humankind's improvements on nature's genius. Writing without jargon or pretense, Vogel hits the nail on the head when he argues that "perhaps the best encapsulation, if a trifle trite, is that nature shows what's possible ... Sure, nature is wonderful. But bear in mind that we do what she doesn't."

Page 100: Insight into the gecko's ability to cling to nearly any surface has inspired a new kind of adhesive tape. Seen here are a laser-illuminated microsphere footprint (left) and an electron micrograph (right) of the spatular tips on a single gecko seta.

Page 101: The unique surface structure of a lotus leaf is revealed under a microscope (left). Lotusan paint uses the structural properties found in lotus leaves to shed water and dirt (right).

Biomorphism

Evolution sculpts. The shapes we find in nature are not random, but quite often the perfect form for the job. Because design nearly always involves decisions about form, learning to follow the forms found in nature is, in itself, a powerful design strategy. That strategy—known as biomorphism—also has a profound and unexpected side effect: the results are usually beautiful. Unlike biomimicry, which endeavors to improve an object's performance or efficiency by modeling its functional design on natural principles, biomorphism is all about the form.

Buckminster Fuller once said, "When I am working on a problem I never think about beauty. I only think about how to solve the problem. But when I have finished, if the solution is not beautiful, I know it is wrong." If the twenty-first century has a governing design aesthetic, it is shapes and colors, patterns and flows that remind us that nature knows a thing or two about blending form and function elegantly.

The designs nature creates are an unlimited source of stylistic inspiration; civilization will come to an end before we run out of brilliant ideas drawn from

Opposite, left: This Biothing Reticulars ceiling fixture was generated algorithmically based on the mathematics of wave interference.

Opposite, right: This Front Design lamp traces the path of a fly circling a light.

observing nature. What's so wonderful about biomorphic design today is that, while the patterns that inspire it may have originated many millennia ago, they are being applied to electronic devices, textiles, housewares, furnishings, and buildings that ride the cutting edge of high-tech invention. More and more, biomorphism not only expresses beauty in concert with sensible design, but is also a manifestation of the tech nouveau.

Biothing / Genware

Sometimes we don't perceive the mathematical foundation of artistic endeavors: we merely love the intricacy of a pattern or the symmetry of a shape. But Columbia University's research and design lab, Biothing, headed by professor Alisa Andrasek, puts algorithms front and center. Biothing uses genetic algorithms [see Biomimicry, p. 99] to generate designs for everything from textiles to buildings, using a computer program—created by the lab—called Genware.

Genware creates integrated "pattern intelligence" that can become a part of the manufacturing process, allowing complex patterns to be replicated at varying scales and used in the creation of interior accessories and other functional objects. Genware mimics natural design in that the code is set, like DNA, but the outcome is variable and unpredictable as a result of "preprogrammed" natural selection. This way each new product has its own unique characteristics, like a living thing.

One of Genware's products, Reticulars, demonstrates the design process in the form of a densely woven curtain that can be hung in front of a window or used as a screen. Two different algorithms generated the materials for the curtain, which are composed of a "responsive" fiber that can react to environmental conditions like light and sound. The design transforms what we normally expect to be a static object into something alive and reactive. The periphery of a room ceases to be forgotten space; the curtain ceases to be merely a utilitarian piece of fabric. Interior spaces become interactive, dynamic environments that change according to the time of day and the activity in the room.

Front Design

Though not particularly green, Front, a Swedish design firm, practices biomorphism in some odd ways, and is bright as a flash-bang grenade. At Front, rats, snakes, and flies are part of the team. Many of the firm's products and experiments are dictated by letting critters do some artistic alteration of their own—rats gnaw on wallpaper until a pattern of holes and tears emerges that allows the old wallpaper to show through in a most unique way; a motion-capture camera records a fly's seemingly random flight around a lightbulb, and then that flight pattern is molded into a beautiful lampshade.

Front also uses the greater forces of nature to create household items. By setting off a few sticks of dynamite in the snow, the design firm created a mold for a lounge chair. By examining the way different qualities of light trans-

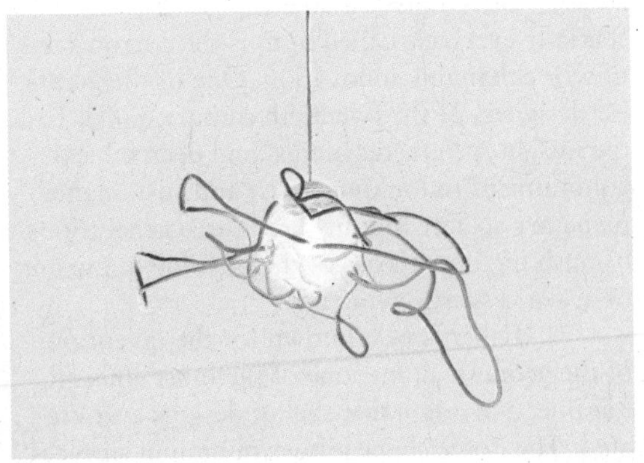

form a room every day, the team came up with a UV-sensitive wallpaper whose pattern waxes and wanes with the sunlight, changing a living room's decor continuously.

Buckminster Fuller

If success or failure of this planet and of human beings depended on how I am and what I do, how would I be? What would I do?
R. Buckminster Fuller

Buckminster Fuller (or "Bucky" as essentially everyone called him) is the patron saint of worldchanging innovation. One of the greatest designers of the twentieth century, Fuller had the insight, prophetic genius, and deep ethical commitment to building a just and sustainable planetary society that inspired three generations to push the boundaries of the possible in pursuit of a world worth living in.

Fuller is best known for the invention of the geodesic dome, one of the most efficient, durable, and adaptable shelter designs ever created. The dome shape allows minimum surface area to enclose maximum interior space, thus reducing material requirements, conserving heat, and facilitating unobstructed airflow throughout the space.

But Fuller's work touched on nearly every aspect of life. In an era when environmentalism was only the concern of a few fringe tree huggers, he advocated designs that allowed us to do more with less: he invented a "World Game" to encourage students to think globally and find a way to "make the world work for 100 percent of humanity in the shortest possible time, through spontaneous cooperation without ecological damage or disadvantage to anyone"; he built a prototype green vehicle, the Dymaxion car; he even coined the term *Spaceship Earth* in an effort to remind us that this one planet is all we have.

Above all else, Bucky Fuller stands as an icon representative of all that one intelligent, committed person can accomplish when he or she decides to make a difference, and of the kind of openhearted, humble, long-term thinking we need if we are to build a better future. AS

Biomorphic Architecture

The Core

The Core is the new education center for the Eden Project—a renowned sustainability center and ecovillage in Cornwall, England. It is one of the greatest large-scale examples of biomorphic construction. The photovoltaic panels that cover the building have been arranged according to the Fibonacci sequence—the natural mathematical pattern we see in the arrangement of a sunflower's seeds or in pinecones. The panels spiral outward, articulating the petals of a flower. The building is meant to mimic the form and function of a tree, with a trunk in the center and overhanging roof canopies that collect sun and provide shade.

In the Core's first four months of operation, the structure's enormous photovoltaic system generated enough electricity to power two three-bedroom homes for a year. At this rate, the Core is projected to save as much carbon dioxide each year as twelve trees could absorb in a century.

The Milwaukee Art Museum

The new wing of the Milwaukee Art Museum (MAM), the Quadracci Pavilion, looks like a bird in flight. Designed by the renowned architect Santiago Calatrava, the expansion features a soaring, winglike rooftop known as the Burke Brise Soleil, which has a "wingspan" wider than a 747. The overhead structure acts as a sunshade and can be moved (or flapped) to control heat and light in the building. Made from steel fins of varying lengths, with curved windows and high ceilings, the Quadracci Pavilion makes the whole museum seem airborne.

From Calatrava's pavilion, you can walk along a suspended pedestrian bridge leading from the museum's waterfront site all the way to downtown Milwaukee, or vice versa. Because this pathway hangs aboveground, it serves as a symbolic transition from the bustle of the city to the serenity of the museum. Like many other museums that have been transformed by world-famous designers, the MAM now draws visitors as much for the building itself as for the art contained within it.

Kunsthaus Graz

If you visit Austria's Kunsthaus Graz at night, you are likely to see one of the most innovative high-tech displays of multimedia art anywhere in the world. The rounded exterior of the building has a glass surface embedded with nearly one thousand fluorescent lights. Each bulb can be individually manipulated for the display of animation and live-action film, or the creation of a moving light array that essentially transforms the facade into a giant computer screen. In this way, the exterior becomes an interactive, communicative skin between the museum's interior and the public space surrounding it. The building itself has a decidedly extraterrestrial appearance, with large-scale curved sides that float above a glass-walled ground floor, and supports what look like protruding antennae along the top.

Completed in 2003 on the occasion of the European Cultural Capital events in Graz, the Kunsthaus is a dramatic biomorphic architectural feat, made even more remarkable by its juxtaposition to the adjoining old clock-tower

Opposite: The Milwaukee Art Museum's new Quadracci Pavilion has birdlike wings that move to shade the building's interior.

Left: Austria's Kunsthaus Graz takes its shape from nature, while being entirely contemporary.

building. UK architects Peter Cook and Colin Fournier designed the Kunsthaus with the intention of highlighting the contrast between traditional European architectural and cultural styles and those that will define Europe's future. SR

Nanotechnology

There are many definitions for nanotechnology or nanotech, but the one we like the best is simply this: engineering functional technologies at the molecular scale. To the extent that we think about nanotech at all, we tend to associate it with science-fictional possibilities: self-organizing swarms of self-replicating nanobots, moving through the world reorganizing matter at will, perhaps getting loose and turning the planet into "gray goo," a term coined by nanotech pioneer Eric Drexler in *Engines of Creation*. These images have about as much to do with the nature of nanotech as artificial skin grafts on burn patients have to do with the bioengineered replicants in the movie *Blade Runner*.

Nanotech as it's actually emerging is prosaic, practical, and profound. With serious adoption of nanotechnology, we will begin moving from a "heat, beat, and treat" industrial era— jokingly summarized in the motto "If brute force doesn't work, you're not using enough of it"—to an age where nano-engineers talk about "nudging molecules into place," and "enticing" carbon atoms to "bond cleanly." The mindsets are as similar as those of a pile driver and a watchmaker.

Nanotechnology involves the creation and use of molecular-scale technologies. Miniature machines, each the size of a speck of dust, might one day be dispersed to perform individual tasks.

Ultrafine manufacturing means that we're getting closer and closer to building machines designed to near-zero tolerances (making practically frictionless engine designs possible), producing raw materials with near-complete purity, even creating new materials that can be put together into stronger, lighter, and more flexible parts. Since it is theoretically possible to eliminate all but the most benign waste, manufacturing could be made almost completely nontoxic, and resources could be extracted from stockpiles that are inexpensive and widely available.

Advances in nanotechnology and other micromanufacturing don't just apply to one industry: they apply to everything that can be rendered in bits, including living molecules. Our ability to work at finer and finer scales applies to DNA as well as to digitized plastic prototypes. Some researchers, for instance, think that the ability of strands of DNA to "compute," to link up in specific combinations under specific circumstances, may one day give us the ability to build living computers, which would use biology instead of electricity to drive computation. AS

Nanotubes: Wonders and Risks

Nanotubes, one form of a particular type of carbon molecule called a fullerene, may well end up being one of the most important materials used in twenty-first-century manufacturing and design. They are nanotechnology's ultimate multitaskers, acting as conductors, semiconductors, or insulators, depending upon how they're shaped. They're also amazingly strong: by weight, they have fifty times the tensile strength of steel, and in theory could be up to a thousand times stronger; at the same time, nanotubes can be very flexible ("deformable") without losing resiliency. When used in a composite material, they can increase its toughness, change its electrical behavior, and allow it to store energy. Nanotubes can be used to make sensors, light-emitting diodes, even computers.

The main drawback of employing nanotubes has been the expense and difficulty of their production. In mid-2005, however, researchers at the University of Texas, Dallas, and the Commonwealth Scientific and Industrial Research Organization in Australia came up with a way to make strong, stable macroscale sheets and ribbons of multiwall nanotubes at a rate of 7 meters (23 feet) per minute. The team described potential applications of the process as including transparent antennae, high-quality electronic sensors, supercapacitors and batteries, light sources and displays, solar cells, artificial muscles, tissue-growth scaffolding, and much more.

As exciting as carbon nanotubes may be, it's important to remember that tiny particles can be highly toxic, even when made of otherwise innocuous materials. Some research has shown that nanotubes can irritate and cause inflammation of the skin and, when inhaled, particles the size of nanotubes appear to make asthma worse. Does this mean that carbon nanotubes are too dangerous to use? Maybe not. As it turns out, under real-world conditions, normal handling of carbon nanotube–based materials doesn't result in dangerous levels of nanoparticles. And researchers are finding ways to reduce the toxicity of the most dangerous nanotubes dramatically: a recent minor modification reduced the cytotoxicity (the dose at which 50 percent of affected cells die within forty-eight hours) by more than 10,000 times, making the modified nanotube essentially nontoxic. JC

The Precautionary Principle

Writers of bad science fiction and easily startled doomsayers love the idea of out-of-control nanomachines—machines that self-replicate and devour everything in their path.

Though that scenario has been largely refuted, it is still very plausible that nanotech and other new sciences could produce unforeseen and detrimental results; therefore, nanotechnologists need a managing principle akin to the Hippocratic oath. A reinterpretation of the Precautionary Principle—the idea that any actions whose outcomes are unknown or potentially negative should be avoided—is probably the best tool to keep those nanomachines in check. Chris Phoenix and Mike Treder of the Center for Responsible Nanotechnology have reinterpreted this principle to both protect the public and allow scientists enough flexibility to fully explore a technology that could have planet-saving implications. Their version of the Precautionary Principle calls for "choosing less risky alternatives when they are available" and for "taking responsibility for potential risks."

Rather than asking—like traditional risk-assessment methods—"how much harm is acceptable?" the Precautionary Principle should ask, "how much harm is avoidable?" Dale Carrico, who specializes in the history and philosophy of technological development, puts particular emphasis on the last part of the principle. He argues that the Precautionary Principle, when it is open to broad participation, has the best chance to both protect us from hazardous results and encourage innovation and experimentation. As he explained in a column on BetterHumans.com, "Even expert knowledge is most useful when it is answerable to multiple and contending stakeholders ... rather than imposed unilaterally by an organized authority ... When the Principle places a burden of justification on those who propose we undertake a risky development, this is not the creation of a new and arbitrary burden, but the fairer distribution of the burden of development onto all its stakeholders."

We have a need for rapid innovation, but we also have the right to demand that innovators behave responsibly. JC

Green Nanotechnology

One of the cool things about nanotechnology is that the more advanced it gets, the greener it gets. Nanoscale production techniques will let us dramatically reduce the amount of waste involved in manufacturing. As we get more adept at recycling and reusing materials, nanoscale production will let us do much more using less energy, less new material, and fewer scarce resources. Nanotech can also help us move toward a cleaner planet in the following ways:

- Nanoscale materials are already being used as environmental sensors that are able to detect minute quantities of toxins and hazards. The same kinds of nanomaterials can also be used to filter contaminants such as metals, bacteria, and even viruses, from water.
- Nanotubes can make the process of producing hydrogen from water twice as efficient, so they may play a big part in making hydrogen-powered cars possible.
- Photovoltaic cells made from nanotubes are more flexible, lighter in weight, and potentially less expensive than traditional silicon solar cells. In theory, we'll be able to do a lot more with them: embed them in fabrics, wrap them around curving structures, and paint them on walls and roofs.
- The development of nanoscale wires able to turn heat into electricity and vice versa will mean refrigerators that can operate without pumps or chemicals, solar panels that are able to extract power from heat or light, and even vehicles that can draw power from the heat of engines. JC

Molecular Manufacturing

In the not-so-distant future, molecular manipulation and rapid prototyping will combine to produce a tabletop nanotech manufacturing system—a nanofactory. This sounds prosaic; it will be anything but. Molecular manufacturing has the power to transform the way we make things.

Consider the process of designing a butterfly-sized self-guided microairplane today. It would take many hours to find lightweight materials, and many more hours to construct the device by hand. The apparatus could carry only a tiny battery and a few bits of electronics. Building and testing each new design would probably require several weeks, and in the end, the microplane would not be able to do very much.

With a nanofactory, the picture would be very different. Building with molecules, nanofactories could develop supercomputers smaller than a grain of rice, motors smaller than a cell, batteries more efficient than the energy-storage mechanisms of real butterflies. Rather than spending days of labor building each new design, a nanofactory could produce it directly from blueprints in minutes. In fact, a tabletop nanofactory could build hundreds or thousands of butterfly microplanes in parallel. All the different material properties required could be obtained by rearranging molecules. All the nanoscale machine systems would have computer-controlled shapes, built directly by computer-controlled nanofactory machines. The nanoscale components could be assembled in a straightforward process into products the size of a butterfly, a car, or even a jumbo jet or spaceship.

Rather than taking several weeks to build and test one new design, a nanofactory could design and test several butterfly-sized self-guided microairplanes in one day. It would be more like software engineering than hardware research and development. Consider how fast the dot-com companies appeared, and how diverse they were, and you can imagine how rapidly new products could appear.

In fact, some designs could be developed even more rapidly by a semiautomated process. The designer would provide an outline for a design with a few parameters that could be tweaked. The nanofactory would build hundreds of variants of the design in parallel. The variants could then be tested in parallel, and those that worked best would be used as templates for the next round of variants—suffice it to say that designs could be optimized quite rapidly.

In a word, clamoring for better, greener, fairer designs, and for the ability to move quickly from idea to prototype to tested product, can help us build a better future. CP & MT

Neobiological Industry

Two hundred years ago, as the Industrial Revolution gained momentum, it didn't much matter how we made things, because manufacturing was little and the earth was big. Today, however, we have a global population of more than six billion, and almost everything in our lives is a product of industrial manufacturing—now it matters a lot. The manufacturing process has fallen ill; escalating demand and hunger for profit have led to rapidly depleting resources and the degeneration of the environment. It can't go on, and it doesn't have to. We already have the tools to create healthier industry—we just need to use them.

Conventional industry is an ecological disaster because it makes things using a "heat, beat, and treat" [see Nanotechnology, p. 106] approach. Most of our materials don't come out of the ground ready for use; they have to be smelted, distilled, or reacted with toxic chemicals at high pressures, a wasteful process. The few materials that need no manipulation (and most of those that do) usually have to be cut into different sizes or shapes before becoming products, which creates even more waste.

Early on in the industrial age, nature's bounty was far more plentiful, and thus cheaper, than human labor. When it became clear just how many uses oil had, industry embraced the energy-dense resource and went into overdrive exploiting its tremendous range of applications—from fuel to plastics to pharmaceuticals. In a short time, our dependence on fossil fuels has become perilous to our existence on earth, not only because of the environmental damage wreaked while obtaining them, but because we simply don't have a lot left.

Today, to make the objects we use in daily life, we employ tens of thousands of chemicals—many known to cause cancer and mutations—and vast rivers of raw materials and energy. Extracting those natural resources has torn up the planet, and industrial waste has polluted our air, water, and soil and is changing our climate. This way of working is drawing to an end.

We can make some improvement by working *with* nature, but we can spark massive change by working *as* nature. This "neobiological industry" blurs the line between the born and the made—maybe to the point of meaninglessness. In addition to the revolutionary use of biomimicry and biomorphism [see Biomimicry, p. 99 and Biomorphism, p. 102] in design, there is a growing role for what Janine Benyus labels "bio-utilization" and "bio-assistance."

Bio-utilization is the use of parts of organisms as raw materials—whether it be the use of wood for a house, or the use of horseshoe-crab blood for a cancer drug. Biological materials are being employed in many new ways. Plant oils are being used to form biodegradable plastics, and waste fibers from crops have been used to strengthen concrete. Even the art world is experimenting with bio-utilization—turning flowers into stereo speakers and grass into photo canvases.

Bio-assistance is the domestication and use of organisms—whether it be herding sheep for wool, or growing a virus for building a battery. This often involves biotechnology, which most of us think of in terms of either medical breakthroughs or genetically altered Frankenfoods. But there is a third kind of biotechnology that may play a much larger role in our future: industrial biotechnology. The term evokes images of ultra-high-tech gene tweaking and the creation of new kinds of organisms, but in what is probably its most useful environmental application, bioengineering will help us find more effective uses for naturally occurring organisms.

As we gain a better understanding of nature's systems, more and more of the work of our industrialized civilization looks like clumsy, even clownish, aping of work that nature does with precision, ecological health, and beauty. We're beginning to realize that our industrial technology is nothing next to the power living things have to digest, filter, grow, sense, and even compute—and that the true masters of manufacturing on this planet are microbes.

We can build industry that is far more sustainable using biotechnology to replicate nature's systems and designs. With the help of biotechnology, we can create pools of hacked bacteria that spit out hydrogen, tanks of tweaked fungus that convert garbage into methane, and vats of tame microbes that allow us to design machines and structures with natural materials that resemble shells and spider silk.

Real concerns remain about the safety of bioengineered organisms, but the best way to realize the sustainability benefits of biotechnology while protecting the planet and ourselves is for all of us to understand the technology and to demand the best practices possible. As Stewart Brand of *Whole Earth Catalog* fame wrote in a recent article in *Technology Review*, "The best way for doubters to control a questionable new technology is to embrace it, lest it remain wholly in the hands of enthusiasts who think there is nothing questionable about it."

Tomorrow's industry will seek to produce objects that work as well as those in nature, through processes that run on sunlight and treat waste as energy. Tomorrow's industry will eat, digest, and secrete the things we need not just in imitation of living beings, but through the actual cells of living beings. Industry will not just be biomimetic, it will be neobiological. JJF & AS

Technical Nutrients

Sustainable architecture and design expert William McDonough [see Knowing What's Green, p. 115] maintains—and many agree—that recycling as we know it will never solve our massive waste problem. Running some glass and aluminum through another incarnation does divert it from the landfill, but ultimately does nothing to change the process by which these problematic items come to be floating around our commercial sphere. As one example of a deeper solution, McDonough's design firm, MBDC, designed an upholstery fabric from wool and cellulose that can be tossed into a compost pile at the end of its life to completely biodegrade. McDonough has dubbed the material a "biological nutrient" —when the "wasted" object is discarded it becomes food for cultivating more wool and cellulose, beginning the entire process again from its birth.

In contrast to the "biological nutrient," which has limited uses in the production of consumer items since it eventually loses its structural integrity on the way back to the earth, "technical nutrients" have a repeated lifecycle within industry. Well-designed synthetic materials can be reprocessed in their entirety, coming back to life as products of equal value or function. It's a process that improves upon standard recycling, a process in which the recycled material has little real value except as a reconstituted composite of used components. In the new closed-loop system of material reuse, everything that reaches the end of its life cycle gets reabsorbed, either harmlessly composting into soil or seamlessly finding another life in the industrial process. AS

Kalundborg and Industrial Ecology

"Closed-loop" thinking is already starting to spread. Industrial systems that employ the concept work like nature, according to MBDC: "Instead of designing cradle-to-grave products, dumped in landfills at the end of their 'life,' the new approach transforms industry by creating products for cradle-to-cradle cycles, whose materials are perpetually circulated in closed loops. Maintaining materials in closed loops maximizes material value without damaging ecosystems."

The best proven example of this kind of industrial ecology is the Kalundborg industrial park in Denmark. At Kalundborg's center is a coal-burning power plant, which not only generates electricity but supplies waste steam to run a nearby pharmaceutical factory and oil refinery. Waste heat from those facilities is in turn used to heat 3,500 homes in the area. The refinery's waste water cycles back to the coal plant to provide more steam. Fly ash from the coal plant is turned into concrete at another factory, and so on. The whole industrial park is piped together into an "industrial ecology." Obviously, coal power plants and oil refineries are not sustainable businesses, but because it has been designed along industrial ecological lines, Kalundborg as a whole is more sustainable than it might otherwise be. Industrial ecology mitigates the bad effects of today's industry: it may well supercharge tomorrow's. AS

Hydrogen-producing Algae

Many eco-futurists promote hydrogen as a cleaner replacement for fossil fuel. But making hydrogen currently takes a lot of energy, or requires *using* fossil fuel. The University of California, Berkeley, and the National Renewable Energy Laboratory (NREL) are working on making hydrogen in a new way: with algae.

The process works in the lab, and it does not require a genetically altered organism—the scientists identified a certain common green alga *(Chlamydomonas reinhardtii)* that excretes hydrogen gas under certain conditions. They describe the alga's hydrogen-making capacity as a "molecular switch" they can turn on or off.

Professor Tasios Melis of UC Berkeley said in a university press release that the discovery is "the equivalent of striking oil," but qualified his statement by adding that the process is still in the lab. "While current production rates are not high enough to make the process immediately viable commercially ... it is conceivable that a single, small commercial pond could produce enough hydrogen gas to meet the weekly fuel needs of a dozen or so automobiles," Melis said.

Before the process is scaled up to industrial levels, the scientists want to improve its efficiency—the algae need to get at least five times better at producing hydrogen to be competitive with fossil fuels.

There may be another interesting use for algae: getting the critters to eat carbon dioxide (the greenhouse gas emitted when we burn fossil fuels like coal and oil) and then turning them into biofuels. Chemical engineer Isaac Berzin has developed a method of capturing the carbon dioxide from smokestack emissions using algae. His process, based on technology he developed for NASA in the late 1990s, captures more than 40 percent of emitted CO_2 (on sunny days, up to 80 percent). The algae produce biodiesel effectively enough (a single acre of algae ponds can produce *15,000 gallons* [56,781 liters] of biodiesel) that full conversion to biofuel for transportation might be achievable. Berzin's company, GreenFuel, has multiple test installations under way, and expects to have a full-scale plant up and running by 2008 or 2009.

Of course, the carbon dioxide doesn't just magically go away—it is released when the cars burn the biofuel—but Berzin's process does transform CO_2 that would otherwise simply be pollution into a (temporary) resource, at least making the carbon dioxide perform double-duty. And being half as bad to the climate would be a whole lot better than the status quo. JJF & JC

Bioplastics

Experimental corn-based plastics have been around since the 1930s, but industry has found it difficult to make plastics that are easily extracted from and returned to the natural ecosystem without harm. Nylon 11, for example, which is produced from the oil of castor beans, still requires a great deal of heat and pressure and several toxic chemicals in processing, and when disposed of, it ends up no more biodegradable than other plastics.

But a new generation of "bioplastics" is pouring out of industrial research facilities. While replacing petroleum-based plastics is, in and of itself, a move toward sustainability, what's truly exciting about these new materials is the window they give on the future: a whole array of toxic synthetic materials may one day be replaced by more natural alternatives.

NatureWorks

Cargill's NatureWorks PLA is a biodegradable corn-based plastic that has come the closest of any plastic to using sustainable resources and having harmless waste. Researched and produced since the 1980s, it has faced serious stumbling blocks: it is more expensive than the petrochemical plastics it replaces, and some environmentalists have been reluctant to promote it because the corn used as feedstock is not limited to organically grown, non-genetically-modified plants. Restricting the type of corn used would raise the plastic's price even further. The benefits of its biodegradability have been limited by the lack of municipal composting programs in the United States. Time will tell whether NatureWorks can overcome these obstacles; if not, corporate investments in bioplastics are unlikely in the near future.

Bio-PDO

DuPont and Tate & Lyle are using corn to produce PDO, a key ingredient in the production of some polymers used for clothing, carpeting, plastics, and other products. According to the companies, producing Bio-PDO consumes 30–40 percent less energy per pound than producing

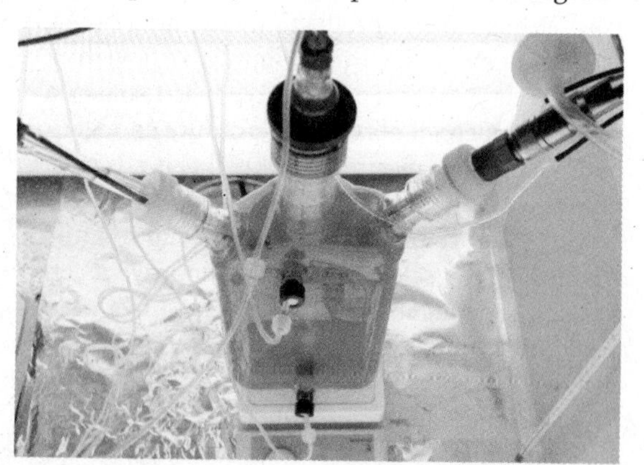

Hydrogen may one day be produced by specially engineered algae, such as these at the National Renewable Energy Laboratory.

petroleum-based PDO; if used as planned, its manufacturers say, it will save the equivalent of 10 million gallons (38 million liters) of gasoline per year. In 2003, the Environmental Protection Agency presented DuPont with its annual Presidential Green Chemistry Award in recognition of its work in developing the PDO process.

Orange-rind Plastic

Cornell chemists have invented a plastic made from orange rinds and carbon dioxide (CO_2). The resulting polymer is similar to polystyrene—not usable for structural applications like polyvinyl chloride (PVC), but good for packaging and cloth fibers, similar to NatureWorks PLA. The researchers do not mention whether the plastic is biodegradable, but even if it isn't, it could reduce oil dependence and sequester CO_2 at the same time. Other scientists are investigating the possibility of making plastic from the cellulose in common agricultural wastes such as stalks, leaves, and husks. This would not only reduce oil use and avoid taking land away from food crops, it would give farmers a much-needed source of additional income. Some chemists are even investigating bacterial digesters as a means to convert the cellulose, instead of heat and pressure-based refinement. This would make plant-based plastics even more of a neobiological solution. JJF & JM

Synthetic Biology

Engineering life is hard. It's made even harder by the law of unintended consequences: living beings are incredibly complicated, and they mutate quickly and in unexpected ways, and we simply don't yet understand them all that well.

Enter biological engineering, or synthetic biology. Biological engineering uses bits of DNA ("BioBricks") to build pseudo-organisms that can grow and act (even replicate) in precisely controlled ways. These "machines" are not quite like anything found in nature, and yet clearly, in most ways that matter, they are alive. Biological engineering is not ordinary biotechnology or genetic engineering. It is the application of engineering principles to the construction of novel genetic structures; in contrast, genetic engineering is often a trial-and-error process, with numerous unanticipated results. Many of the reasonable concerns about genetically modified foods and animals come from this hit-or-miss aspect of biotechnology. Biological engineers take a more systematic approach, using an increasingly deep understanding of how DNA works to make microorganisms perform narrowly specified tasks. (The Massachusetts Institute of Technology now has a major in biological engineering, the school's first new major in decades.)

Standardized, cheaper genetic engineering tools will lead to a better understanding of biological processes. This should allow scientists to avoid the unanticipated system interactions feared by knowledgeable opponents of biotech. Even if the technology is misused, the standardization of the process will make it easier to recognize and respond to accidents or malice.

As long as biological engineers approach their work with an appreciation for its ultimate effects, standardized components and rigorous design methodology could actually make the broader use of biotechnology a safer prospect. AS & JC

Viruses Making Batteries

What if we could grow rechargeable batteries in vats? Dr. Angela Belcher of the Massachusetts Institute of Technology (MIT) thinks she may have found a way to do just that. Along with her team at MIT, the 2004 MacArthur "Genius" established a technique that combines both biomimicry and bio-assistance, using viruses to literally produce battery material based on the way the red abalone generates its shell—but much faster.

The viruses are not engineered, but are put into an environment they normally would never inhabit; their reaction to the surrounding materials turns them into a template that grows the desired material. Belcher and her team have also developed other viruses and yeasts that grow a variety of materials for engineering and medicine.

It is still just research, but it shows promise, and this kind of engineering will increasingly dominate manufacturing, because of its inherent advantages. As Belcher told *Discover,* "An advantage over chemistry is that you have directed evolution on your side. So instead of being limited to a chemical that you pull off the shelf to build a material, you can evolve the biomolecules to be better and better at doing what you want them to do." JJF

DNA Computers

Silicon Valley, the California epicenter of the dot-com boom, was named for the material that engineers use to create microprocessors, the mechanisms that made the boom possible. But just as the dot-com era saw an eventual decline, so silicon may approach its limits in satisfying the ever-increasing demand for speed in computer processing.

Can anything outrace silicon and endure rising expectations for performance? Believe it or not, our own DNA proves to be a more viable competitor than silicon. DNA molecules have the capacity to perform rapid simultaneous calculations—much faster than the fastest computer—and they clearly don't seem to be nearing an inevitable demise.

Scientists are now investigating ways to integrate the astounding abilities of life's genetic material into nanocomputers that will be able to perform much faster, and store vastly more data, than the machines we work with today. These DNA computers, embedded with "biochips," herald a future that used to seem like pure sci-fi, and that now appears to be around the next corner: the era of biological computation.

Knowing What's Green

How do we know if a product is green? It's almost always hard to tell—as the story of the Swiss bag company Freitag makes abundantly clear. Freitag puts out innovative, trendy, and ecoconscious bags by recycling the printed tarps that cover the cargo carried by commercial trucks. Each one is unique, with strong graphics cut from monumental advertisements. You can custom design a bag on their Web site by picking out sections of truck tarp. The bags are assembled under fair labor conditions by Swiss and Moroccan workers, using friendly, nontoxic processes. But the ironic beauty of these one-of-a-kind objects has attracted counterfeiters, who make copycat knockoffs unsustainably, in dubious circumstances, from new material—effectively nullifying the intentions of the original concept. On the surface, there's no knowing that the knockoff is any less green than the Freitag bag [see Consuming Responsibly, p. 38].

Of course, some items on the shelf are dressed up in packaging that references nature but merely wears the idea of sustainability. We can be reasonably certain that a knockoff handbag isn't radioactive, but there can be plenty of other hidden en-

Freitag bags, made from reused shipping tarps, have become an icon of eco-chic fashion.

vironmental costs, all made worse by their invisibility. Setting standards for measuring and communicating those costs is complex, and it requires a transparent flow of information about where things come from and where they go. The early stages of that effort are emerging all over the world.

We like to think, at the bare minimum, that we aren't inadvertently going home with a fistful of mercury when we buy an alarm clock. This is a pretty reasonable assumption. The government standards and regulations that ensure this are some of the least inspiring, yet most important, steps toward making products sustainable. Manufacturers must meet those standards before they put a product on the market. So while rules can never supplant innovation and creativity, they can at least ensure a minimum level of consumer safety and restrict destructive practices.

As it turns out, not all manufacturers make reducing our cancer risk their first priority. Outside the European Union, which recently mandated new restrictions on hazardous substances, most governments are doing very little, and moving very slowly to do it. This makes the flip side of legislation, the certification scheme, so interesting: it is the carrot to legislation's stick.

At the moment, product vigilance is an anarchic sort of realm inhabited by numerous competing agencies awarding certifications and labels for meeting certain standards relating to off-gassing, recycled content, and toxins—and each certification comes with its own regional rules and boards and bureaucracies. But they're all there, growing and multiplying. Selling to a global market increasingly means that there are fewer places to hide: the things manufactured in one country have to satisfy the requirements of another.

The Cradle to Cradle Protocol

Furniture that won't off-gas, toys that are nontoxic, cars that are completely remanufacturable—in most countries, including North America, no standards exist to ensure that we get these types of products from manufacturers. Companies are operating totally within their rights when they make appliances in toxic soup and sell them to us on the cheap. But while regulations set a low minimum threshold that products *have to* meet, certifications are the standards that products *get to* live up to. Instead of focusing only on preventing the very worst manufacturing practices, what if we reframe the problem by focusing on creating the very best things possible?

William McDonough is mainly known for shepherding green building into the American public mainstream, but his partnership with German chemist Michael Braungart has been helping to change the way small things are made. Their Cradle to Cradle (C2C) protocol, outlined in their book *Cradle to Cradle,* is the most thorough set of guidelines available for the voluntary certification of sustainable products, and among the best set of guidelines for making things that are high performing, efficient, and harmless to delicate ecologies in the air and the human body.

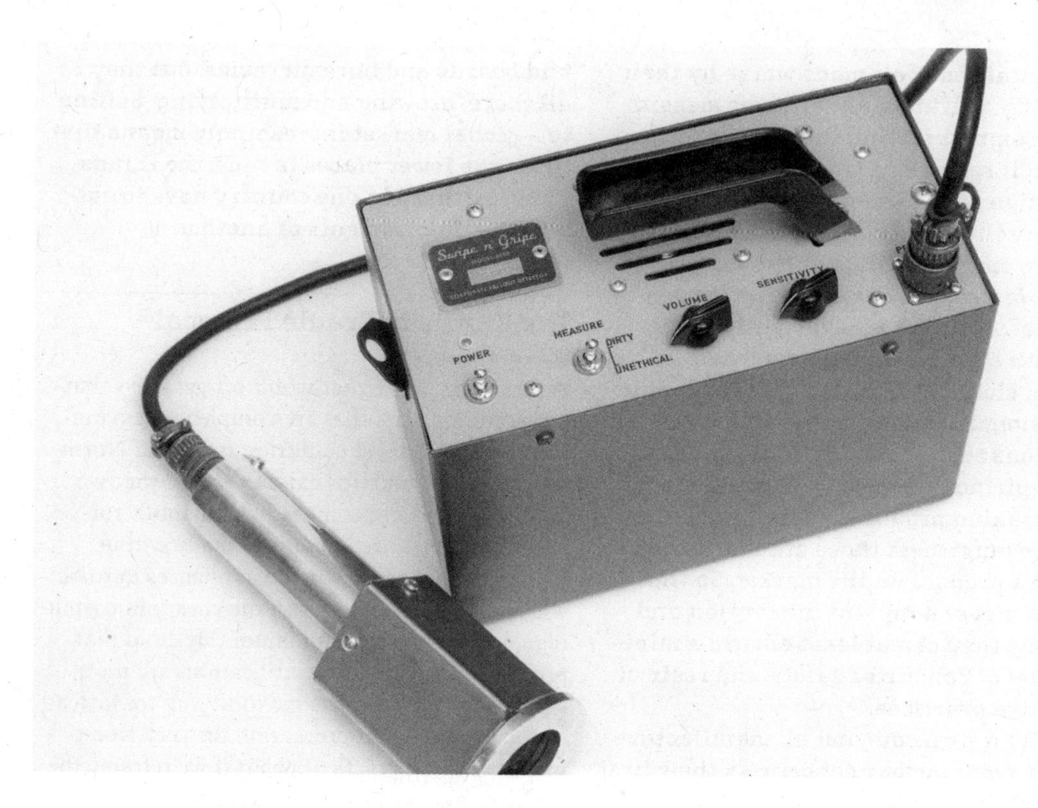

When you buy a toaster, you probably don't know or care precisely how much energy was used to make it. But it would be great to know that an honest, strict third party had stamped it as environmentally sensitive—neither carcinogenic, energy hogging, nor an offense to climate policy. Making a Cradle to Cradle product isn't easy, but it's based on a fairly simple notion: that everything we own should be either recycled, remade, or buried in the ground to compost.

Like the label that guarantees that our carrots are organic and pesticide free, the Cradle to Cradle platinum rating cannot be applied to, say, an office chair unless the workers who made it are being treated and paid well, and the chair isn't full of poisons. Some chemicals are simply blacklisted for reasons of human and environmental health. Herman Miller, a company that helped pioneer ways to use Cradle to Cradle principles in design, applies this thinking to every product it makes. It simply won't do business with material suppliers who won't provide their recipes for proprietary chemicals.

The protocol is an unusual private certification, so C2C maintains control over its high standards. But it would take years or decades for most companies to duplicate McDonough and Braungart's research. What makes Cradle to Cradle protocol so compelling is that, for the first time, someone has set a remarkably high standard for genuinely sustainable design. DD

The Corporate Fallout Detector

The Corporate Fallout Detector (CFD), designed by James Patten, is a blend of performance art, ethics, and technology. When the CFD scans a barcode, it crosschecks it against several online databases that catalog manufacturers' environmental and ethical misdeeds; the databases get their information from various sources, such as *Ethical Consumer* magazine and the European pollution database. The CFD emits a series of clicks—kind of like an old Geiger counter—to indicate a product's reputation. The more clicks you hear, the worse the practices of the company that made it.

While the CFD isn't actually available to the public, it does point the way for other tools that may soon help us know the real stories behind the products we buy. DD

Using Embedded Identification Tags for Good

Say we buy a pair of running shoes, and when we buy them, someone gives us a whole set of instructions: "These shoes come apart," we're told, "and the fabric areas can be taken out and turned into soil. They're also made of very well-considered, nontoxic plastics, easily disassembled, and used to make new shoes just as soon as you bring these back to the shop." "Too much information," we think, and promptly forget. They're good shoes, but within a season they're worn out, and off to the dump they go.

For things to get recycled, they need to make their way back to the factory instead of going to the dump. The missing trick to closing the loop on objects is information tracking. Even when we release recyclable, "disassemble-able" things into the world, one question remains: whose problem are they when someone has long forgotten what they were made of? How do we take advantage of all the clever thinking that went into them, and make sure that their materials get cycled back into a closed system?

One of the answers might be in the way we embed information into products. Radio frequency identification (RFID) is a tiny, chip technology that uses an antenna and silicon chip to transmit information to anyone with the proper equipment to read it. That information can reveal an object's source, ingredients, or serial number. The simplest, cheapest RFIDs are passive ones that can be read with a transponder, and are also the most likely to find their way into clothes—after they're done helping Wal-Mart run a supremely efficient warehouse. What if we could wave a gizmo at our closet and get information about our running shoes—which were conveniently tagged with data about their history, composition, or intended resting place? Once we can discern that much about an object, recyclers can easily sort things and send them back to their maker.

In his design book *Shaping Things,* Bruce Sterling sketches out a scenario in which such information becomes ubiquitous: "This whirring, ultra-buzzy technology can keep track of all its moving parts and, when its time inevitably comes, it would have the grace and power to turn itself in at the gates of the junkyard and suffer itself to be mindfully pulled apart."

Because they raise privacy concerns, RFID systems are somewhat double-edged. They're great for tracking wildlife for research and conservation purposes, yet they could become authoritarian tools (if tags remain active even after a product is purchased, they could be used for surveillance), which is why people are working to develop standards for their ethical use. Many chips are laced with heavy metals, so wide adoption of standard RFID chips could mean salting the planet with low-level toxins. But there's a real difference between using toxic tags to track our movements and embedding recycling instructions into the next generation of organic, printable, nontoxic RFID chips. Used for our collective benefit, these "spychips" could be the unexpected key to making our relationship with things radically more sustainable. DD

The Corporate Fallout Detector sniffs out bad corporate ethics and environmental practices, much as a Geiger counter detects radiation.

RESOURCE

Cradle to Cradle: Remaking the Way We Make Things by William McDonough and Michael Braungart (North Point Press, 2002)
This now-classic book on design and sustainability is a window into a radical framework for changing the way all things are made, and is one of the most accessible books on design and sustainability yet written. McDonough and Braungart sweep aside the slow work being done to marginally improve efficiencies, and argue that nothing less than an ambitious, full-scale—and positive—vision is required to create a

sustainable economy: "Should manufacturers of existing products feel guilty about their complicity in this heretofore destructive agenda? Yes. No. It doesn't matter. Insanity has been defined as doing the same thing over and over again and expecting a different outcome. Negligence is described as doing the same thing over and over even though you know it is dangerous, stupid, or wrong. Now that we know, it's time for a change."

Producer Responsibility

Our everyday life is affected by our relationship to things. But if we limit our notion of this relationship to materials and energy choices alone, sustainable design will not take us as far as we need to go. The greatest promise for gains in sustainability lies not merely in designing *things*, but in designing the services and systems that take into account our interaction with our products and enable us to take responsibility for objects and waste.

We want to carry mobile phones in our pockets and make coffee at home; yet phones and coffeemakers are thoroughly disposable objects. Expensive to repair and cheap to throw away, their design quickly becomes obsolete, and their materials are rarely considered valuable enough to remanufacture. But as the sheer volume of waste that we produce becomes crushing, and new resources harder to capture, companies are trying to figure out how to redirect even the smallest extinguished products back to their manufacturers. DD

Extended Producer Responsibility

Cars have always been disposable, peeled apart for scraps or left as dry husks on the landscape when they die. But scrap doesn't confer any worth—there's a downcycling that goes on, a loss of value. Steel gets weaker, colors duller. Other parts—seats and glove compartments and seas of tires—are politely ignored. In the early 1990s, Germany decided to change the rules, requiring automakers to take ultimate responsibility for the cars they sell. Aside from getting on car companies' nerves, the dictate meant that engineers and designers had to figure out a way to make those dead cars valuable and easy to take apart. As with any new development, the mandate initially cost companies time and gave them headaches—but then it started saving them money. It turns out that cars that are easier to disassemble are also easier to assemble.

The European Union adopted Germany's approach with their End-of-Life Vehicle Directive, a governmental regulation based on a simple concept: manufacturers are held responsible for the things they sell when they get old, broken, or obsolete. Previously, these same companies might have offered pure product–service systems—upgrading, servicing, and taking back appliances—as buyer incentive, yet the choice was business-driven. Taking stuff back as a matter of regulation is a different game. It's not purely for profit: it's an issue that's bigger than a single company.

Most electronic waste is sent to Asia to be disassembled by hand—the only known technique for recovering metals in products designed exclusively for their utility to the consumer. New, expanded regulations in Europe now require that manufacturers design for the electronic waste their products generate, and these guidelines are being closely followed by China. When refrigerators stop working in Europe, the company that produces them can no longer leave Europeans, their communities, or rural Asia buried under the weight of the defunct machines. SC & DD

Designing for Disassembly

We buy things and take them home. They're lovely: they take care of our need to open bottles, protect our feet, light the bathroom, keep our books off the floor. And yet even after breaking, they continue to take up space, as if waiting for another step that never comes.

What's the best future that we can imagine for our disused stuff? We can design things to be disassembled, but the ultimate strategy would use excellent materials, easy to recover—and ward the landfill off in perpetuity. The goal is the closed loop, in which objects are continuously remade, spiraling upward in quality and harmlessness. This is the trash of the finest ingredients, reflecting the fact that things can't last forever, and that fashions, in fact, change frequently. Think of the avocado-colored stove. Most designed objects date themselves by their color or their shape. We keep creating new avocado stoves. And colors, shapes, and technology just shift and keep shifting.

Dead objects are such a shame. It can take close to forever to figure out how to take apart a filing cabinet—and then what do we do with it? But it takes five minutes to take apart Steelcase's Think chair, after it's spent a good fifteen years rolling around an office. Making things easy to take apart also makes them easy to repair. Things that disassemble are designed for a dual human purpose, one that accounts not only for the time a fixture spends adorning our ceiling but also for the time it takes to unscrew that fixture until we're left with a little pile of metal and plastic—waiting and full of potential. DD

Pop-apart Cell Phones

Today most cell phones and other small electronics are shredded instead of being taken apart for recycling, because it costs more to pay someone to disassemble the phone than the phone's recyclable parts are worth. Well, what if the product disassembled itself?

Nokia has prototyped a cell phone that pops itself apart in two seconds, as opposed to the two minutes normally required for manual disassembly. The phone uses special alloys and polymers that change their shapes when heated with a laser—unthreading the screws, pushing apart the case, and popping off the circuit board.

The temperature needed to disassemble the phone is 140–300 degrees Fahrenheit (60–150 degrees Celsius), cool enough to not melt the surrounding plastic, but hot enough to not be triggered accidentally.

In a world of pop-apart electronics, recycling becomes a snap. JJF & AS

Refill or Recycle

Collecting obsolete products for recycling can be tough in North America. Instead of insisting that folks recycle their lifeless toys, companies have to entice people into doing

so. These days, Hewlett-Packard (HP) makes well-considered, easily disassembled hardware that just keeps getting better. In the meantime, the company is working at getting back all the older machines that languish unused in people's homes, each containing a wealth of heavy metals difficult to recover. The fact that take-back laws don't apply throughout the United States hasn't stopped HP from recognizing that people will pay to have their old computer peripherals recycled. Responding to our environmental concerns, the company offers to pick up our old printers at the door, and entices us with the promise "We'll take away your old PCs, printers, and guilt."

Absolution may be enough incentive for some people, but HP has their eye on capturing back all of their raw materials, so the return process has to be easy for the consumer. In a long-running global effort to recycle old printer cartridges, HP started including return mailing packets with its new cartridges. It turned the tide: at last count, the company was flooded with 140 million pounds (63.5 million kilograms) of old cartridges and electronics a year.

Designing for Repair

Ultimately, most objects are disposable. In spite of all our efforts, they still break down, or unexpectedly drain themselves of their early seductive sheen. First our cassette players became outdated, then they broke. But what if we paid not for the machine itself but for the use of a portable music player? What if it was the company's job to fix it, replace it, upgrade it—and eventually recycle it?

The UK-based Sprout Design, one of the few ecodesign studios in the world, comes up with products that are repairable or replaceable. Sprout developed, for Japanese company Toginon, a range of extremely high-quality knives, which come packaged with two blades. When one blade gets dull, which should take some time, the other blade can be used and the dull one sent back to Seki City—one of the last remaining sword-making towns in Japan—where Toginon's employees will sharpen it and send it back.

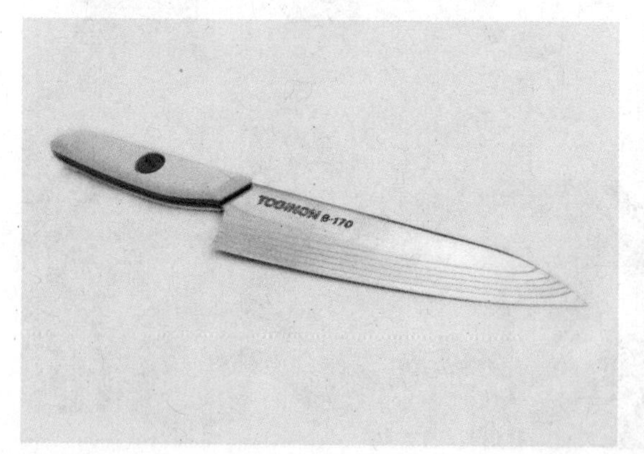

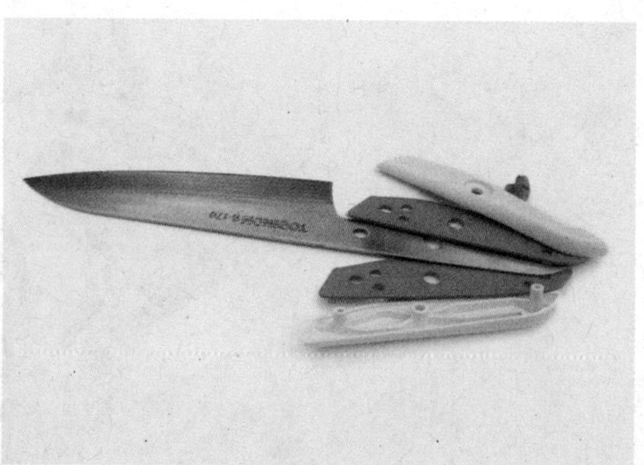

Collaborative Design

Most of the newer objects in our lives started out as pure data. The curves on a Prius automobile share a digital handprint with eyeglasses and ballpoint pens. While still in its virtual, 3-D form, digitized design is just information: we can e-mail a half-formed 3-D object to Singapore, and have someone manipulate it and pass it back. With the right technology, we can test and even prototype the thing on demand. This is truly remarkable: digital product design is still a new field, and it wasn't long ago that a standing army of machinists and draftspeople was needed to crank out detailed, hand-machined prototypes in model shops. Designers now have the freedom to refine and discard quickly and digitally.

Take your toothbrush. A designer used complex, powerful 3-D computer-aided design software to make it, fitting its pieces together and spinning them around in virtual space. There's nothing inherently green about this process, with the small exception of all the prototypes that didn't need to be machined by hand. Software, however, is a light medium: easy to share, easy to distribute. The company Blender 3D makes a free, open-source [see Open Source, p. 127] version of 3-D software and Google's free SketchUp is remarkably easy to use. Software like these versions of 3-D software can transform data into real objects through quick fabrication ("fabbing") technology, which is getting cheaper by the minute. Access to a team of skilled model makers used to be a huge barrier to making complex things, and now that barrier is gone. Product designer Ronen Kadushin makes his work available online as "Open Design," an extension of the open-source-software model.

Design is, and always has been, a collaborative field. Green design—which involves even more people, including recyclers, disassemblers, and chemists—is even more so. But if one of the barriers to sustainable design has been that we, as users, haven't been able to make the rules, the democratization of design tools means that more people can enter the game.

Ecodesign's Killer Application: Alias

It's unlikely that the designer who diligently crafted your toothbrush with pixels and light gave a thought to sustainability. Her main concerns were more likely the way the object looked, how it felt, and how it worked. She probably used a 3-D tool that let her do this more easily than ever before. If ecological impacts were ever part of the equation, it was likely someone else's problem.

But what if today the same designer wants to make a toothbrush that's indigo blue, matte, and rubberized—as well as nontoxic, easy to disassemble, and molded in a wind-powered factory outside Grand Rapids? What if she doesn't have time to wait for an engineer to analyze how much energy is required to make it? What if, as she draws the toothbrush handle, the 3-D program does her ecological calculus in real time, showing the relative impacts of choosing one supplier or one material over another?

Alias, a star among 3-D software developers, makes tools for industrial designers; William McDonough [see Knowing What's Green, p. 115] has rewritten the guidelines for sustainable product design. Their collaboration is yielding

Preceding pages: Workers recycle electric motors, Fengjiang, China, 2005.

Opposite: This Toginon knife is designed to be easily disassembled for repair, resharpening, and recycling.

what could be, for large companies, sustainable design's killer application: a custom 3-D tool that could inform the designer about the ecological impacts of each of her decisions.

One of the toughest challenges in the creation of sustainable products is making sustainability relevant and important right from the beginning—the defining moment in a product's life cycle. Alias's Tom Wujec believes that this challenge can be met by making information about the impact of design decisions visible. Despite having so much influence over what gets designed, managers never ever see life cycle as one of their priorities, but Alias is working on powerful tools for visualizing every element of a product's life. That way, anyone involved in creating or marketing a product can see ecological impact as pervasive and compelling. "If we visually and graphically represent sustainability," says Wujec, "you can change the way people think."

For designers, this is revolutionary. A tool that makes sustainability visual and persistent as they work is the first manifestation of what could be a hugely important ingredient in our technological destiny. Technologies like computer-aided design are never solutions unto themselves: they're simply tools that enable people to make the best decisions. Because Alias is building an interactive, visual connection between the object and its invisible footprint, it hands the power of designing green back to the person who sets the system in motion. More fundamentally, it educates designers as they work, letting them do what they do best: come up with great solutions to the challenges that humans create. DD

ThinkCycle

Several years ago, a group of graduate students at the Massachusetts Institute of Technology (MIT) Media Lab set up an open, online structure so that they could work together on design and engineering projects. Their collaborators were based all over the world and included academics and technology-minded engineers, inventors, and researchers. Taking a hint from the open-source approach to creating software, in which widely distributed groups of people build on one another's work, ThinkCycle opened projects up to ongoing peer review. The diverse challenges ThinkCycle set out to tackle ranged from biomimicry for sustainable design [see Biomimicry, p. 99], to rural neonatal care, to arsenic remediation. ThinkCycle's biggest success was an innovative treatment device for cholera, designed using input from people in the medical field and taking inspiration from other disciplines. A standard IV can cost $2,000; the devices ThinkCycle came up with go for $1.25.

By focusing expressly on designing for sustainability and development, ThinkCycle has carried on the legacy of Victor Papanek, the UNESCO designer who, throughout the 1970s, made high-performing, inexpensive devices—such as a transistor radio fashioned from a used can—for the developing world. Papanek refused to patent any of his works, being far more interested in creating a public domain of form and function. To Papanek, a design was most successful when it cost only pennies. Papanek was uncommonly gifted in his ability to see simple, hybrid high-tech/low-tech solutions to tough problems. In this sense, he had a rare

The Portable Light demonstrates how high technology and collaborative approaches can improve the lives of people in remote communities.

genius—but what ThinkCycle and its followers can do is allow us to spread that genius around. ThinkCycle provides a most compelling model of open-source design, painting a picture of how we might best organize, and capitalize on, the genius of the many.

Lars Eric Holmquist, an expert on ubiquitous computing, once criticized designers for their habit of developing concepts that depended on imagined, unrealized technology. As he stated at a 2002 Doors of Perception conference, "You essentially build a fetish object that has the appearance of a real artifact, but it doesn't actually do what you have promised it will do. What you're doing, in effect, is praying to the gods of technology to come and sort out all your real problems." Even teams of people can fall prey to this kind of thinking, but distributed peer-reviewing can prove the viability of design concepts, just as it can prove scientific theories and computer code. We often lack the technological or contextual knowledge to effectively solve design challenges; by bringing together complementary knowledge bases, ThinkCycle created a brilliant, pragmatic model for conducting reality checks on visionary concepts and designs. DD

Portable Light

For nearly as long as it's been available, electric light has relied upon heavy infrastructure. To light our homes we must link into a network of wiring that brings electricity and provides a physical connection to power plants far away. On or off the grid, access to electricity requires serious investment. On top of that, most lighting systems are fragile, and most solar photovoltaic systems cumbersome. Until very recently, both were best suited to static use in buildings.

In Mexico's Sierra Madre mountains, the seminomadic Huichol (or Wirrarika) travel four hundred miles through rugged terrain during their annual pilgrimage, often on foot. During the wet and dry seasons, they return to the Sierra Madre to farm. Yet without electricity, the Huichol have limited access to education and employment. After sundown they can't see to read, work, or study, and many of them end up leaving for the cities to join the industrialized economy. Paradoxically, it's not electricity, but the lack of it, that poses a threat to Huichol culture. There are no grid connections in the mountains, and the means of delivering electricity are too impractical and expensive to serve the needs of the Huichol.

So it goes with older technologies. But the last few years have seen new developments in both light-emitting diode (LED) [see Lighting, p. 161] technology and flexible, lightweight photovoltaics. Both are light and durable, and when combined, they're essentially self-powering. The use of high-brightness light-emitting diodes (HBLEDs) means that extremely bright, efficient light can be produced by a single miniature diode powered by small solar panels. When architect and interdisciplinary designer Sheila Kennedy and her colleagues traveled into the Sierra Madre, they unexpectedly saw the potential to apply these new, transformative technologies to mobile light sources. Kennedy, who has embedded LEDs into fabric, conceived of a way to allow the Huichol distributed, decentralized access to light, enabling children to read and adults to work while still preserving their nomadic culture.

The resulting project, Portable Light, separated lighting and solar technology from permanent structures to create something entirely new. Kennedy's materials-research firm MATx first consulted and collaborated with Huichol community leaders to discern their needs. University of Michigan students then devised a whole series of self-powering prototypes based on a flexible skeleton of lights, batteries, and photovoltaics; the units are easy to wear or carry, and provide a range of power capacity. After five hours in the sun, the modular candles on the portable light unit, the Community Bag, can provide bright light to several people; the illumination can be extended by exchanging used lights with charged ones. Huichol are well known for their beading and sewing, so the team also created a Portable Workshop to allow artisans to work on craftwork in the evenings. Portable Light Reading Mats are covered with fabric that illuminates a school desktop, and are integrated into a larger Reading Stool, which folds into a backpack. Students can group their stools together to form a learning station. The stool itself is a traditional Huichol design, and therein lies an essential ingredient in the project: the technology is brought into a new context and modified, customized, and embedded into the local aesthetic. Kennedy's objective is to create Portable Light "light kits" so that textile artists can weave flexible solar panels and lighting directly into fabric.

The prototypes that the Portable Light team brought to the Huichol were not fixed; like the technology itself, they weren't static. Much of the project's strength actually lies in its lack of a single, finished design. Besides being stunningly simple, the Portable Light system is reconfigurable: with enough units, a community's systems can be aggregated, knitting a network of lights into a larger power array for community activities, or even minor surgeries. In the Sierra Madre, the Portable Light project is carving out a space where new technology is used to preserve indigenous culture. There is a growing need worldwide for tough, self-powering, portable sources of light. Untethered to the electricity grid, portable lighting could have huge potential beyond the realm of the Huichol. DD

RESOURCES

Design for the Real World: Human Ecology and Social Change by Victor Papanek (Academy Chicago Publishers, 1985) and ***The Green Imperative: Natural Design for the Real World*** by Victor Papanek (Thames and Hudson, 1995)
Victor Papanek was industrial design's original critic, famously calling his profession one of the most harmful in existence. But he was also a brilliant advocate for design in the interest of human need, and brought an unusual reverence and weight of responsibility to the field. In language suffused with the history of design, he describes the Inuit of Greenland and northern Canada, whom he lived with for a time, as the best designers in the world—all of their decisions, he suggests, are perfectly in balance with their environment. We rarely talk about design in these terms.

The Green Imperative follows up the classic *Design for the Real World*, and includes more than enough cautionary tales, but its ultimate focus is on solutions. Papanek isn't nearly as interested in clever technologies as he is in appropriateness—and merit: "The exploitation of cuteness, convenience or fashion manipulates people's reactions and emotions unscrupulously; it represents the engineering of desire. To put it differently: convenience is the enemy of excellence; fashion is the enemy of integrity; cuteness is the enemy of beauty."

Open Source

Most big software corporations treat their computer programs as a trade secret. The "source code," the fundamental workings of the software that drives the programs they sell, is protected both technically and legally from the prying eyes of the public. One result is that no one can build on proprietary software without explicit (and usually costly) permission. One consequence of that, in turn, is that new ideas, security patches, better ways of working, and improvements in the software are often stymied. Another inadvertent consequence is that these companies have a vested interest in preserving the marketing image of their software—people can be, and have been, sued for pointing out flaws in these sort of proprietary programs. Despite large budgets and high profits, much commercial software has tended to get more and more bloated with each passing year, and less and less stable and secure.

But an alternative to proprietary software exists: open-source software. Open-source software is called that because the basic workings of the programs, including the operating system, are hidden from no one. Anyone with an interest and some programming skills can take a look under the hood and see how the programs work—which also means that anyone can propose ways to make them work better. Their proposed fix will then be weighed by others working on the project, and if it is found clever, most likely it will be incorporated into the next version of the program. If not, the person who proposed it is free to take the open-source software and start their own version—as long as anyone who wants to can look at and propose changes to their code.

This collaborative and nonproprietary approach to building software has caused a technological upheaval in the last few years. It turns out that a whole bunch of smart people collaborating in an open and noncommercial manner can do things that even huge corporations and government agencies have a hard time doing. The individual contributions of each participant can be relatively small—if there are enough of them, the combined experience, creativity, and intelligence of the folks involved can make even complex tasks easy. With enough eyes, as the open-source proverb goes, all problems are shallow.

What makes the open-source model so powerful is that it turns competition toward collaborative ends. Open-source programmers still compete to be more clever and skilled than their colleagues, but when one makes a breakthrough, the nonproprietary nature of the code means all benefit. It also makes growth explosive. Some heavy-hitting industry analysts are already predicting that the open-source Linux will pass Microsoft Windows to become the most popular operating system on the planet by the end of the decade.

Open-source software would, by itself, be an important tool, but the real revolution of open source is the model itself. All around the world, people are putting the principles of open collaboration to work on all manner of projects, which far transcend the world of software. What started as a way to make better and freer software is unfolding into a model for making a better and freer future.

Linux and Its Lessons

Linux is a completely free computer operating system, gradually built up through the efforts of many programmers and volunteers, and developed into a workable free operating system for PCs in the 1990s. From there, programmers could bootstrap their way up on existing code and build ever more sophisticated applications. Countless free programs have been derived using Linux, ranging from the fundamental operating system that tells a computer how to do basic tasks to esoteric applications of every imaginable variety.

Since Linux is free to modify and redistribute, many different "distributions" arose including popular ones like Red Hat, SUSE, and Debian. A few years into the twenty-first century, Ubuntu ("Linux for Human Beings") was founded by Mark Shuttleworth of South Africa, with the goal of being robust, easy to use, regularly updated, and free of all licensing charges. Its Web site states: "'Ubuntu' is an ancient African word, meaning 'humanity to others.' Ubuntu also means 'I am what I am because of who we all are.' The Ubuntu Linux distribution brings the spirit of Ubuntu to the software world." Ubuntu has since become the most popular Linux distribution for personal computers.

Today, Linux is just one of a vast number of open-source software projects. You can find high-quality open-source projects for videoconferencing, Web browsing, programming, word processing, making spreadsheets, modeling in three dimensions, mapping, online collaboration, and so on. These are the cultural riches of the twenty-first century—a beautiful and functional heritage of technology open for all to use, learn from, and build upon.

As in other open cultural arenas where anyone can contribute or start new projects, the material produced using open-source software varies in quality. Yet the best open projects are world-class. How? First and foremost, on the best open projects a core team of bright, dedicated people coordinate and filter contributions. The team is motivated by the desire to solve a personal problem, gain a reputation, or make tools. At their fingertips, the team's members have a vast and continually growing library of previously created, high-quality code, free for the using. By writing and distributing the debugging process among a number of different users, the team can usually maintain a high level of quality control.

Because they are open—and often created in a nonprofit spirit—popular open-source projects sometimes receive substantial financial and in-kind contributions. Open-source software also presents opportunities for making money; skilled personnel supply customers with setup, assistance, training, or customization. As one bit of folk wisdom goes, in an open-source world, bricks may be free, but it still sometimes makes sense to hire a mason. The funding and financial potential behind open-source projects inspires quality work.

The open-source concept is no longer limited to software—cultural and intellectual goods of many kinds are increasingly easy to find in free versions. The growing open-journal movement is a great example; instead of paying exorbitant fees to profit-oriented journal publishers, academics are creating their own alternative venues online, and funding them with a small percentage of their research funds. Once research is published in an open journal, it's free for anyone to use, forever—a logical move considering how much research is publicly funded in the first place. The virtual Public Library of Science has received top-quality rankings for its free journal offerings, and is being joined by numerous other innovative publishers.

Society at large benefits from the growing pool of public goods: a body of knowledge about software, culture, and science; a physical pool of libraries, schools, hospitals, universities, and transportation infrastructure; a social pool of communities, nongovernmental organizations, and networks. The more knowledge we have access to, the richer we are. Imagine expanding our measures of wealth to include "free GDP" (gross domestic product)—the transactions we *didn't* have to pay for, but got for free from our shared heritage of public goods. Every time you enjoy public art or music, every time you use free software, every time you read Jules Verne or Shakespeare, you are better off. Shouldn't we count that as wealth? Contributions to the pool of public good can raise free GDP tremendously. When we start keeping track, we'll see that pairing collaborative work with free distribution is an essential element in building a bright green future. HM

Distributed Collaboration

No official planner decides how much fruit to bring into a city, how many sweaters must be knitted, or when a new restaurant should open—yet all these functions, and many more, work pretty well most of the time. Structures like markets, price signals, legal systems, and contracts combine with the decisions of informed consumers to channel people's efforts into mutually beneficial activities.

Similarly, collective intelligence is now being enhanced by online tools and clever social structures. Most Internet users are familiar with Wikipedia, the open-source encyclopedia that allows anyone to edit entries or download content. More than a million entries have been created in the English version of the site, many of high quality. But this didn't happen by chance: the creators put guidelines in place to catch vandals, resolve conflicts, store all versions of the content, and discuss contentious or unclear issues. This combination (discouragement of unproductive behavior and encouragement of productive contribution) lies at the heart of distributed collaboration.

Another way of working together is through sharing resources, whether physical or virtual. ClimatePrediction.net uses volunteers' computers to model the earth's climate, with the goal of better understanding the uncertainties in climate projections. Each volunteer's computer simulates a different possible scenario; the results are sent back to ClimatePrediction.net's central computers, which analyze all the scenarios to get a better estimate of the likely range of future climate change. Though each individual computer makes a modest contribution, when many thousands of them are combined, they can outclass the fastest supercomputers.

A further example of distributed collaboration is Project Gutenberg, which provides for free the text of tens of thousands of out-of-copyright books. Typically, books are scanned in by dedicated contributors, and a rough version is created through character recognition. Next, each page is assigned to at least two volunteers for proofreading, which is done online in a matter of minutes. Most of the contributors have never seen one another, yet they slowly create more and more together, ratcheting up the quality and quantity of free material. Thousands of volunteers thus happily donate their idle moments in the interest of building a collective literary monument. HM

Reputation Systems

If you have a bad experience with an auto repair shop, the shop probably doesn't care too much—you're just one customer, after all. But if you happen to be a talk-show host or a newspaper writer, the repair shop treats you differently, since you could scare off other customers by spreading word of your bad experience far and wide. Reputation systems extend this power to all of us. By creating publicly available information via Web sites where anyone can share their service experiences, positive or negative, these systems help us to filter out the bad and highlight the good.

Tagging systems are the easiest reputation systems to understand. Users create virtual lists of favorites, and share their lists with others over the Internet. CiteULike is a great little project that applies this idea to academic literature, offering scholars a means of keeping up with the increasing pace of discovery and publication without reading every article out there. Standard ways of evaluating literature online, although useful, indicate only how often other articles cite a particular article; CiteULike allows direct access to experts' opinions. By browsing the favorites of people whose previous recommendations you've liked, you can find relevant articles in a variety of fields.

Such opinion-sharing systems have great potential to increase trust among people who would otherwise be strangers. Those who rely on this kind of "social credit check" benefit from judgments made collectively by groups of peers rather than judgments made by a few remote agencies. Reputation systems have been used by travelers sharing travel tips or simply endorsing or warning against places they visited, and by hosts offering those travelers lodging. Why shouldn't they be used for sharing rides, or tools, or civic contributions—in addition to information and knowledge? A successful reputation system must filter out ignoramuses and shills, survive legal challenges from parties who rate poorly, and account for reasonable differences of opinion. But done right—with the interests of users in mind, with the goal of shining a spotlight on our better options—they foster good work and good deeds. HM

Open Learning

Every year, millions of students confront the stark reality of calculus. Think of how much easier it would be for them to learn if they and their teachers had access to the accumulated experience and wisdom of other students and teachers. Envision a way to record the learning experience of students, to analyze what works in overcoming stumbling blocks, to pinpoint the applications that motivate students to care about the topic. Imagine a system that watched how different people learn, observed what teaching techniques work best for different topics and types of students, and then made all this information available—on top of a vast library of open-source teaching material.

This is a challenge for the next wave of open source; meeting that challenge will require a marriage of technical advances and social insight. Every good teacher has some understanding of how students can learn effectively, and most students can identify what tools or techniques help them learn. Capturing even a fraction of this knowledge would be immensely useful. What if a student could easily find answers to questions like these: "What ideas would help me understand climate change better?" "Why should I care about calculus?" "Who could I read to better grasp the current political situation?" What if the answers combined the guidance of a mentor with source material from some of the world's best minds?

That world is coming. Students already have direct access to many great thinkers and doers through initiatives like MIT OpenCourseWare [see Education and Literacy, p. 315], and ResearchChannel. It's not too much of a stretch to picture combining the content provided through such online courses with an online educational resource that's as easy to contribute to as Wikipedia—but one that incorporates many adaptive, personalization, and data-mining algorithms that help customize and direct an individual's learning process. If such an initiative is to be universally accessible, it should be based on free software. This new, global public good would help students, policymakers, businesspeople, and citizens become informed and empowered to positively change their world. HM

RESOURCES

The Success of Open Source by Steven Weber (Harvard University Press, 2004)
"Open source is a real-world, researchable example of community and a knowledge production process that has been fundamentally changed, or created in significant ways, by Internet technology." So argues Steven Weber in *The Success of Open Source,* which takes a close, sociopolitical look at the success of open-source code in subverting assumptions about how businesses are run and products created. He goes on to say: "Open-source software is a real, not marginal phenomenon. It is already a major part of the mainstream information technology economy, and it increasingly dominates aspects of that economy that will probably be the leading edge over the next decade." Open source, he argues, is more than software—it's a way of organizing production so that it can be used for works for public good, and that is definitely worldchanging.

The Cathedral and the Bazaar by Eric Raymond (O'Reilly Media, 2001)
Available as a free download at http://catb.org
Raymond's classic work on the nature of open-source enterprises remains an excellent summary not only of why the open-source movement arose, but what open, collaborative projects can achieve that centralized, heirarchical projects cannot: "Linus Torvalds's style of development—release early and often, delegate everything you can, be open to the point of promiscuity—came as a surprise. No quiet, reverent cathedral-building here—rather, the Linux community seemed to resemble a great babbling bazaar of differing agendas and approaches (aptly symbolized by the Linux archive sites, who'd take submissions from *anyone*) out of which a coherent and stable system could seemingly emerge only by a succession of miracles. The fact that this bazaar style seemed to work, and work well, came as a distinct shock."

Design For Development

While sitting with a group of students, Japanese designer Isao Hosoe pulls out a carefully folded map of Venice. The paper is precisely creased into a field of tiny hills, perfectly regular, which collapses into a pocket-size package. As he opens and closes it, the map folds with a single movement like an accordion. It has no moving parts, is made of common materials, and is embedded with no electronics. But this map, says Hosoe, is an ideal illustration of high technology: an object that is perfect for its context. Those of us who want sustainable solutions for the world's poorest could learn a lot by taking this simple notion to heart. Combined with local knowledge and a willingness to listen, a sense of context is paramount if design is to succeed in the Global South.

Not all design for development incorporates the high technology of mobile phones, computers, or solar panels [see Green Power, p. 172]. In extremely rural areas, high tech may refer to deceptively simple but extremely ingenious designs that don't require specialized materials or equipment to build or repair. For example, the international design and consultancy firm Arup created, for a school on the Tibetan plateau, a latrine that uses the simplest of technologies to achieve a clean, low-odor washroom. The Ventilated Improved Pit (VIP) latrine is a waterless toilet with a solar-operated flue that sucks fresh air through the latrine, down the pit, and out a vent. The mechanism is ideally matched to its environment, requiring no power or water infrastructure to be completely effective. It's a model of appropriate humanitarian design, influenced by the needs of the users, not solely by the ideas of foreign designers.

Most things manufactured for use in the Global North are what designer Enzo Mari would call "designs that seem": they lure us in by telling a story of their profound usefulness, so we eagerly buy them. For the most part, such objects collect dust. But simple, straightforward "designs that are" truly serve the needs of the poorest.

Most industrial designers can only dream of developing a life-changing product. But most professional designers don't work in rural Kenya; they lack the experience and insight to see where solutions are hiding; and they don't know the relevant communities well enough to see the nontechnical barriers to successful design. Their well-intentioned efforts become mere theoretical exercise.

Truly amazing success stories, though, are beginning to emerge. The best designers take a hybrid approach: they still bring in expertise from the outside but they co-create solutions with the people of the Global South. Their innovations can change the game for people, economies, and communities all over the world. DD

Amy Smith: The Best Solutions Are Hiding in Plain Sight

In Africa, the women are the farmers. Women invented domesticated crops. If you're talking to the right people, they should be a group of elderly women with their hair up in bandannas.

Amy Smith

Ever since she was a Peace Corps volunteer in Botswana in the 1980s, Massachusetts Institute of Technology's (MIT's) Amy Smith wanted to work with people in the Global South to develop tools that would improve their lives. She's since brought engineering knowledge to communities that were formerly unable to test or purify drinking water or efficiently grind grain into flour. Whether they involved passively purifying water with the sun or fueling cooking with agricultural waste instead of trees, Smith's collaborative solutions have achieved high-tech brilliance with simple ingredients.

Take her Phase-change Incubator, for instance. In many Ugandan communities, lab workers can't access a reliable source of electricity. But to test for bacteria (and therefore provide the correct medication), they must incubate samples at a precise temperature for twenty-four hours. Smith solved this problem by using fire instead of an electrical current, heating a chemical compound within the incubator to create a "phase-change" reaction that would maintain a steady temperature for long periods of time. For Haiti, a nation that is 98 percent deforested, Smith's D-Lab developed a cooking fuel made from sugarcane remnants, an agricultural waste product. The fuel reduces the need to use the nation's remaining trees for wood charcoal. For Honduras, D-Lab created a water-testing kit that costs twenty U.S. dollars, a fraction of the thousand U.S. dollars conventional apparatuses cost.

Smith and her D-Lab have overcome fundamental challenges that had been long neglected. With the D-Lab, Smith takes engineers into the field and teaches them how to work with people in the developing world to overcome their obstacles. Together they work to uncover small solutions to big problems, solutions that are always elegant in their simplicity.

"Looking at things from a more basic level, you can come up with a more direct solution, and a lot of people go, well, duh, that's really obvious!" Smith told *Wired* in 2004. "But that's what you want: people saying it should have been done that way all along. It may sound small in theory, but in practice it can change entire economies ... A lot of people look at where technology is right now and start from there, instead of looking at the absolute functionality. If you go back to the most basic principles, you can eliminate complexity. The stuff I do is just very

Above and opposite, left: People in Ghana making clean-burning charcoal briquettes to replace the need to burn firewood in deforested areas.

Opposite, right: Whirlwind's cheap, rugged wheelchairs are changing lives in the developing world.

simple solutions to things, which is critical when you are developing applications for the Third World."

Many of the significant daily challenges faced by the developing world's communities are unknown to those outside the particular community. Making just over two pounds (one kilogram) of flour from grain takes about an hour of labor. A typical hammer mill, which cuts this time to just over a minute, employs a tiny replacement screen that commonly breaks down, rendering the machine useless. Smith created a screenless hammer mill that cleverly uses airflow to separate hulls from grain. Like many of the best designs for people living in the Global South, the hammer mill isn't patented, allowing wide access to the technology. ▫▫

Mobility: Ralf Hotchkiss and Whirlwind Wheelchair International

While *confined* is a word often heard in reference to wheelchairs, for persons with mobility disabilities a wheelchair is a liberating tool. Unfortunately, it's a tool unavailable to millions of poor people in need of one around the world. Engineer and longtime mobility advocate Ralf Hotchkiss has pointed out that of the 20 million people worldwide who need wheelchairs, 98 percent don't have one. For people injured by disease or war, a wheelchair can be a life-changing gift.

Hotchkiss has been working to bring mobility to such people in developing nations since 1980. He began building his own custom wheelchairs after a severe injury left him with paraplegia; at the time, American wheelchairs were prohibitively expensive, and it only made sense to him to build them from scratch. In his travels to Nicaragua, Hotchkiss discovered that wheelchairs were even more expensive there. However, the local people he met who used wheelchairs had insights that hadn't occurred to him. So Hotchkiss began working with people in the developing world to design wheelchairs appropriate to their location.

Of the organizations that are working for mobility, Hotchkiss's organization, Whirlwind Wheelchair International, is distinguished by its bottom-up approach; it focuses on designing and training over donation. Since each community comes with different terrain and with unique limitations, Whirlwind works on the ground, designing for a local context—and is committed to producing with local materials and skills; community building and local pride come with collaboration. Whirlwind compares its design development process to the open-source-software model; it promotes open access to knowledge about evolving wheelchair designs.

This kind of thinking is essential to sustaining a community. Having worked in more than forty-five countries, Whirlwind has found that local differences—terrain, even the size of people in one community versus another—vary considerably. Whirlwind's rugged, innovative chairs are cheap, they work on unpaved roads, and they can be repaired with local materials. They're also lightweight, foldable, and able to navigate steep slopes without tipping—a problem with some conventional wheelchairs. Whirlwind has also created wheelchairs for very small children. Experience and support have allowed

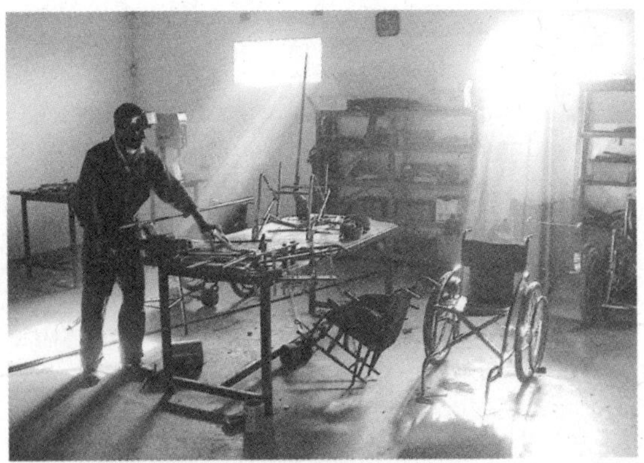

the initiative to help communities scale up their production so that they can produce chairs in larger volumes, to reach more users. Fundamentally, though, Whirlwind's work is about access. As Whirlwind designer Marc Krizack has written on the Whirlwind Wheelchair International Web site, "Providing wheelchairs is not about wheelchairs. It is about providing people with the one thing they need to move out into their own communities—to go where the action is. It is about integrating people with disabilities into their society." ▫▫

Bright Green Computers

Computers move fast. They slip through our houses, obsolesce, and cycle away, wearing out as fast as a pair of shoes. The modern world's greatest tool is among our most disposable and resource-heavy. Judged solely on performance, computer design has progressed staggeringly well and astonishingly fast. But to judge it from a green perspective, we have to take whole systems—and people—into consideration. When our laptops die and we toss them, they either rot in landfills or children in the developing world end up wrestling their components apart by hand, melting toxic bits to recover traces of heavy metals. Did someone forget to design for these kids? With so many products coming out in green versions, why do our computers still only come in the standard, havoc-wreaking model?

Paradoxically, computer designers, engineers, and inventors have more tools than ever to make long-lasting machines that can be continually upgraded or safely discarded and easily recycled. Our computers can transcend their shortcomings. We can eliminate the toxins and emissions generated during production and after disuse. We can buy a computer without the

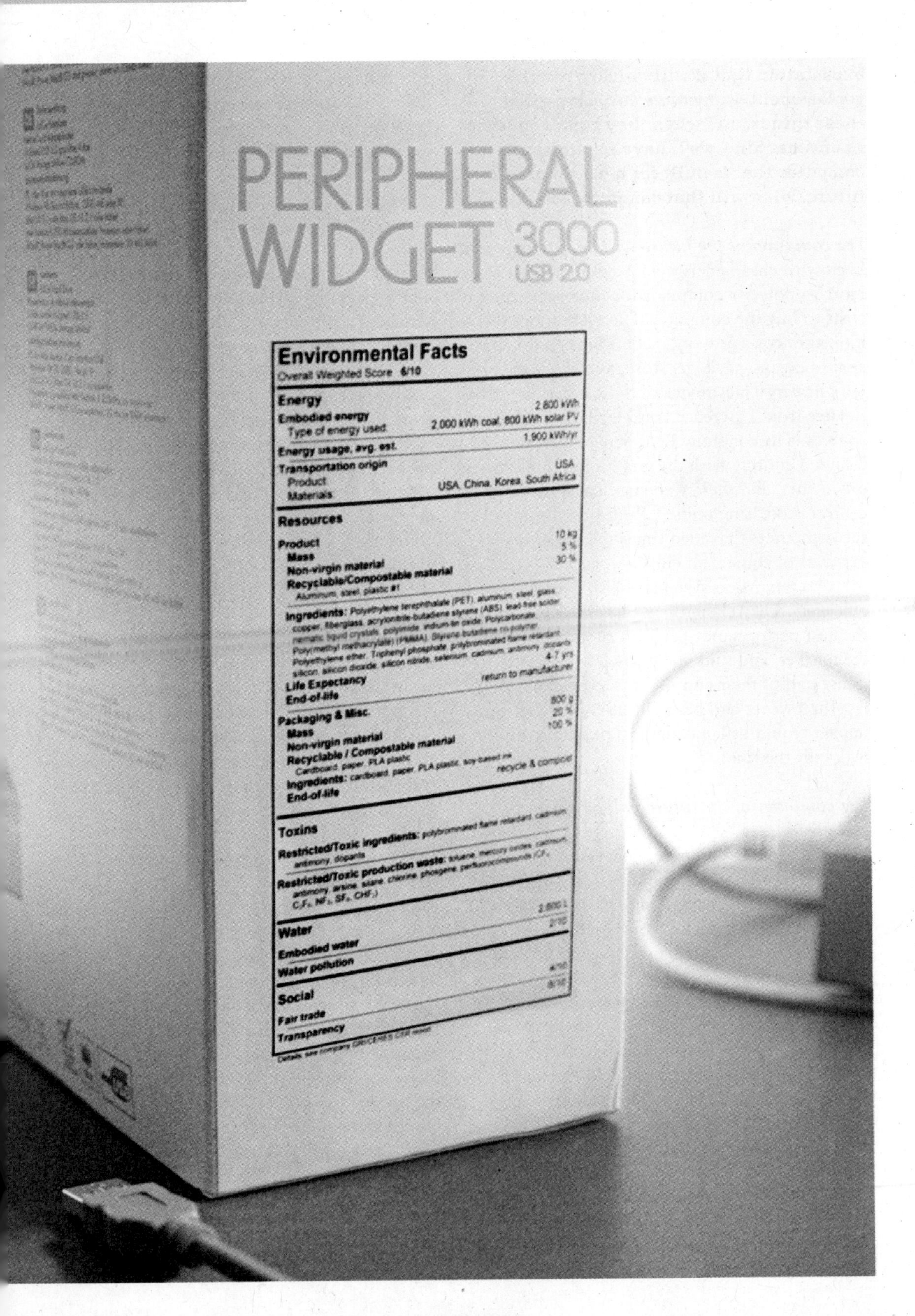
PERIPHERAL
WIDGET 3000
USB 2.0
Environmental Facts
Overall Weighted Score 6/10
Energy
2,800 kWh
Embodied energy
Type of energy used
2,000 kWh coal, 800 kWh solar PV
1,900 kWh/yr
Energy usage, avg. est.
Transportation origin
USA
Product
Materials
USA, China, Korea, South Africa
Resources
10 kg
Product
5 %
Mass
30 %
Non-virgin material
Recyclable/Compostable material
Aluminum, steel, plastic #1
Ingredients: Polyethylene terephthalate (PET), aluminum, steel, glass, copper, fiberglass, acrylonitrile-butadiene styrene (ABS), lead-free solder
4-7 yrs
return to manufacturer
Life Expectancy
End-of-life
800 g
20 %
100 %
Packaging & Misc.
Mass
Non-virgin material
Recyclable / Compostable material
Cardboard, paper, PLA plastic
recycle & compost
Ingredients: cardboard, paper, PLA plastic, soy-based ink
End-of-life
Toxins
Restricted/Toxic ingredients: polybrominated flame retardant, cadmium, antimony, dopants
Restricted/Toxic production waste:
Water
Embodied water
Water pollution
Social
Fair trade
Transparency

expectation that it will quickly require replacement. Indeed, we can have all of these things, and when they come together in one machine, we'll have in our hands a computer that is built for a bright green future. What will that computer look like?

The computer of the future will be superefficient

Even with cheap energy, it's beginning to cost more to power a computer for four years than it costs to buy the computer. There are more than monetary costs at work, too: it takes a pound of coal to create, package, store, and move every ten or twenty megabytes of data, according to estimates from Lawrence Berkeley Labs and other sources. The company P. A. Semi was recently founded entirely with the goal of making efficient processors, and their Web site boasts that their central processing units (CPUs)—the brains of the computer—have ten times the performance-per-watt of equivalent chips.

Sun Microsystems is also betting big on efficiency. They claim their UltraSPARC T1 gives the best performance-per-watt of any CPU on the market, and said in a 2005 press release that if just half of their entry-level servers from the last three years had been UltraSPARC T1s, the impact would be equivalent to pulling a million SUVs off the road.

The computer of the future will be clean and green

Manufacturing a computer generates a tremendous amount of waste, some of it hazardous. Over the last few years, companies like VIA and NEC have removed lead from their fabrication process—but this can be pushed even further. Circuit boards could be created with a fraction of the waste produced now. Computer chip fabrication processes could be radically cleaned up, as well; today they use deadly gases such as arsenide, silane, phosgene, and chlorine, and greenhouse gases such as CO_2 and nitrous oxide—not to mention the few thousand gallons of water and kilowatt-hours of electricity it takes to fabricate a computer's worth of chips. These are solvable problems. In 2004 Texas Instruments worked with Rocky Mountain Institute to design a greener domestic fabrication plant that saved so much money in water, power, and construction costs that it was cheaper than manufacturing overseas. That success simply involved better facility design. Imagine if we redesigned the whole process.

In comparison to reformulating the computer chips, reformulating the shell that surrounds the innards is a snap. Sleek aluminum cases are more robust and far more recyclable than plastic ones; the shining, coated plastics many computers have now are cobbled from a stew of materials that are unrecyclable and expensive. NEC's 2002 PowerMate eco was the first computer to have a fully recyclable plastic case, monitors without harmful gases, and no toxic flame retardants. The jury is still out on whether old electronics cases that companies purport to be biodegradable will actually become the compost pile the companies claim they will, but NEC has recently created a mobile phone with a compostable case, made from a plastic derived from corn and natural fibers, and HP has prototyped a printer with a corn-based plastic case.

Although tossing a computer within a year or two of purchase remains a seriously unsustainable practice, it's still vital for these machines to contain fewer toxins, for the sake of the land into which their insides gradually bleed, and the hands into which their salvageable parts end up falling, somewhere thousands of miles from where they began.

The computer of the future will be recyclable, its parts reusable

After years of packing hazardous chemicals and heavy metals into their gadgets, and having the whole pile of materials speedily obsolesce, more and more companies must now consider the scenario of seeing their old goods dropped back into their laps. Requiring manufacturers to take back their own products when customers discard them is a simple but powerful force for changing the production process. Faced with disassembling parts and cycling them back into the fabrication process means making some very careful decisions about how those parts are assembled in the first place.

In Europe, companies have been legally obliged to take back their products. Though they initially balked at the idea, it was a profound motivating force for change in their production process. Many have found that redesigning products for disassembly has actually helped their bottom lines, increased efficiency in labor and materials, and thus saved them money. A Panasonic 1984-model television had 39 plastic parts made from 13 different plastics, and took 140 seconds to disassemble. The company's 2000-model television contained a mere 8 plastic parts made from 2 types of plastic, and took 78 seconds to take apart (with a corresponding advantage in the time it took to assemble).

Hewlett-Packard, which already offers recycling for old machines (both their own and other companies'), is looking forward to this new style of producing electronics. But until fully closed-loop products begin to trickle back to the company, HP reuses what it can, downcycling and disposing of the rest. Although it's already happening here and there, there's a clear consensus that when all companies have to take back their products, much more design effort will go into creating self-perpetuating cycles of components and materials.

Better still—and cheaper—are the machines that can be reused, reconfigured, or upgraded. Most brand-name manufacturers are laggards here; the vanguard is small computer service shops and resellers—such as Immaculate Computers in Berkeley, California—making what are called white-box systems, cobbled together out of generic components, or souped-up older PCs. When a machine is easy to upgrade, it is also easy to take apart and swap parts in and out of—the same features that make it easy to recycle.

The computer of the future will be about services

Of the $250 billion spent every year on powering computers worldwide, only about 15 percent is spent computing; the rest is wasted idling. Why have that idle horsepower at all? Because we want it to be there when we need it. One way to have even more horsepower, and only pay for the time we're using it, is to switch to a "thin client" or "dumb terminal"—a service that connects a client computer to a big industrial-strength server somewhere else. The machine then has a fast Net connection to the server, a screen, and a keyboard, but its brains are kept elsewhere. Sun Microsystems is trying to push the world toward systems like this, since existing thin-client systems use a tenth of the power and reduce raw-material usage by a factor of 150. Sun's model involves centralized server farms running proprietary hardware and software, but there's no reason thin-client systems couldn't be established running open-source software on millions of generic machines scattered across living rooms around the world.

Bringing it all together

Each of these approaches is an intervention, a tweak in a system. Just incorporating one of them really isn't good enough. While it's still rare to see a product that avails itself of more than one solution, the industry now has the knowledge and capacity to design a convergence of approaches. Devices use less and less power, while renewable energy gets more and more effective. One machine uses a tiny bit of energy, and eventually it may gently generate its own. The greenest computer will not be a miracle in a box: it'll be the product of a sustainable system just now being imagined. Designers are now capable of making many small changes converge into a full rethinking of the industry, into directions as yet unknown. And when they do, we will have on our desk a machine that connects us to the world through information technology; that links us to a cycle of manufacturing that spews very little waste and reabsorbs what it does release; that is put out by companies mindful of the people who make their computers and, more important still, the people who take them apart—and all in a product that is lighter, sleeker, and more elegant than any we've seen yet. DD & JJF

Page 134: Breakthroughs in engineering that have resulted in the NEC PowerMate eco (left) and the PWRficient processor (right) are making computers less toxic and more energy efficient.

Page 135: Tofu and potato chip manufacturers alike have to come clean about their ingredients. What if every object had its ecological costs displayed as clearly and scientifically as calories on a nutrition label?

SHELTER

We think of our homes mostly in terms of memory and emotion—a magnificent meal prepared in the kitchen, a child's first steps across the living room, a basement that always gives us the shivers. The longer we live in a home, the more the physical structure fades into the background. Our lives unfold within it, but we tend to forget about all the ways we continue to interact with and depend on the services our home provides.

At its most basic, a house is a tool for living, but most homes today are blunt, poorly designed tools. They waste energy and water, lack safety and comfort, and are often made from materials whose production has huge environmental and social consequences. According to the Worldwatch Institute, "People can live in a typical house for 10 years before the energy they use in it exceeds what went into its components—steel beams, cement foundation, window glass and frames, tile floors and carpeting, drywall, wood paneling or stairs—and its construction."

But better homes are on the way. A confluence of new technologies and approaches is beginning to allow us to create buildings of a whole new stamp: buildings that are light, airy, comfortable, and stylish; buildings that sip water and make all their own energy; buildings that are better for us and, over time, better for our pocketbooks.

These bright green homes are emerging from the same design explosion that is putting sustainable products on store shelves. Because building a home demands so much more of an investment than, say, making a toaster, the innovations are slower to arrive on your block. At the same time, because architects, designers, and engineers are learning so much so quickly, and increasingly sharing what they find, the year-to-year leaps in performance we're beginning to see in bright green buildings can be astounding. The best structures being built today would not have been possible ten years ago: the best structures we'll be able to build ten years from now may not even be imaginable today.

Better still, these design breakthroughs don't just benefit those of us who are affluent. The same kind of open, innovative approach that's growing bright green buildings in the Global North offers the potential to transform the lives of billions in the Global South. Indeed, many of the emerging solutions for making life better in comfortable neighborhoods are making life itself possible in troubled places.

Solar panels are simply smart if you live in the right areas of the Global North (and, with breakthroughs in design, more and more places are the right areas), but if you live in the Global South, solar panels can bring refrigeration to store needed vaccines, and lighting so kids can read at night. In the Global North, using the rainwater that falls on your roof is a good idea. In parts of the Global South, it can mean avoiding sickness and death.

Nowhere can we see more clearly the potential that radical new design has to change lives than in refugee camps. The world already has millions of refugees. If environmental conditions worsen and large-scale problems like war and famine continue to go unaddressed, some experts believe, the world will have hundreds of millions of refugees by the second half of this century. The refugee problem presents an unparalleled challenge, but it can be met by new models for providing housing, food and water, medicine, and energy to people who have lost everything.

No matter where we live, the same forces are in play: people are working together more effectively than ever to create more powerful solutions than ever.

Homes are tools. We are building better ones. And when we build better homes, we build a better world. AS

Green Remodeling

We might be the most conscientious recyclers on the block, dragging our bins diligently to the curb each week and going back inside with a guilt-free glow, but—surprisingly—if we remodel our kitchen or bathroom just once, according to the Portland (Oregon) Office of Sustainable Development, we are likely to generate the equivalent by weight of four years of weekly curbside recycling.

Realistically, waste falls low on the priority list of most homeowners when they are planning to remodel. The process of dreaming up our ideal homes—and the effort to save money—easily eclipse our potential guilt over sending our old drywall and two-by-fours to a landfill. None of us wants to incur unnecessary costs when we remodel, but it doesn't hurt to find out how spending a little more up front can buy long-term savings for our wallets, our health, and our planet.

Sustainable building materials have historically been more expensive and more difficult to acquire than traditional building materials, but this has been changing rapidly in the last few years, as knowledge and demand have increased. Green building has begun to shed its early "granola" aesthetic, giving way to a much wider variety of styles and designs. This means that the sustainable choice and the preferred choice may be the same choice. We can now remodel with materials that create less waste, look better, and keep a healthier home. SR & JF

Flooring

Reclaimed Wood

Every home tells a story. Some chapters of the story were written before we moved in, but others are ours to compose. When selecting materials and decor for our home, we contribute to the tale that our house inevitably tells to our children, our guests, and the next owners. If we choose to line our floors with tropical hardwood, we may be writing a story we'd rather not tell.

In a hundred years, we have lost more than half of the world's rain forests. Most of this devastion comes from logging tropical hardwoods such as teak, mahogany, and rosewood. The demand for exotic timber flooring drives deforestation—this is especially tragic because it's unnecessary. We can obtain the same gorgeous, solid, natural-wood flooring in a sustainable way—and for less money than rain forest wood—by reclaiming and reusing discarded hardwood.

People who know wood understand that the best-quality timber comes from old-growth trees, whose wood is much harder, and generally has a smoother and more uniform surface. However, in this case "old-growth" translates roughly into "precious and endangered."

Over the last few centuries, builders used ancient hardwoods in numerous buildings that now stand abandoned. Instead of pulling

trees from forests, we can pull wood from these places—demolition sites, warehouses, and old mills. In cities, too, municipal crews cut down a huge number of trees from streets, backyards, and parks that could easily go into our homes as beautiful new floors, instead of going into the landfill. A number of small business and entrepreneurial individuals are now saving and reselling this salvaged wood for interior use.

Like any antique, a reclaimed-wood floor not only looks beautiful and unique, it holds the character of its past. Think of the story hidden within a wood floor made from the support beams of an old granary or a barn roof. Pioneer Millworks, one of several companies that salvages and resells usable wood from fallen trees and industrial sites, can tell us exactly where our wood came from—whether its an old Buster Brown shoe factory in Missouri or an abandoned Royal Canadian Air Force hangar. Visible evidence of the wood's history, such as nail holes and streaking, give the reclaimed timber a vintage aesthetic, a great complement to modern and antique decor alike. When it comes time to remodel our homes, it wouldn't hurt to spend a little time thinking about how we'd like its next chapter to sound.

Forest Stewardship Council Certification for New Wood

If we must use new hardwood, we can turn to the Forest Stewardship Council's (FSC's) reliable, widely honored monitoring-and-certification system for tracking the responsible maintenance of the world's forests. Buying FSC-approved wood guarantees that the timber came from a responsibly managed logging operation, and that using it does not contribute to the destruction of the world's forests [see Sustainable Forestry, p. 494].

Bamboo

If you've ever walked through a bamboo forest, you know the clean sound of stalks clicking together in the breeze, and the uncomplicated lines of thousands of vertical poles growing close together. Bamboo grows with astonishing speed—up to twelve inches per day—making it one of the most renewable resources used in building and product design. Though not technically wood, bamboo, when used for flooring, functions in the same way that hardwood does, with a very light, bright tone and smooth finish. This stable, durable grass (that's right, it's a grass, not a tree) epitomizes "ecofriendliness" and costs less than many of its sustainable couterparts. SR & JF

Non-wood-flooring Options

Wood doesn't work everywhere. When it doesn't, we have the option of using a number of new or widely unknown materials in place of common choices like synthetic tiles and conventionally manufactured carpet. These materials are perfect environmentally friendly flooring alternatives.

Linoleum

Made from linseed oil, wood dust, rosin, and jute, quality linoleum contains only natural ingredients, and takes well to natural pigment dyes. You can buy it in sheets or tiles, and now also in modular units called Marmoleum Click, which fit together without glues or adhesives and allow you to design your own unique floor pattern and color scheme.

Preceding pages: A sustainable, energy-efficient housing design by Integrated Architecture in Grand Rapids, Michigan.
Opposite, left: A green remodeling job by Velocipede Architects, Seattle, Washington.
Opposite, right: The FSC logo identifies products that contain wood from well-managed forests certified in accordance with the rules of the Forest Stewardship Council.
Left: Bamboo forest, Maui, Hawaii.

Carpet
Some people love wall-to-wall carpet because of how soft and forgiving it makes a floor. It's great for kids who like to tumble—except that regular old carpet comes drenched in toxic chemicals that go flying into the air each time our feet jostle the fibers. The chemical gases, as well as the particulates from the fiber, can cause all kinds of respiratory problems for us and our kids.

A classic example of a major corporate overhaul intended to foster sustainability is the story of Ray Anderson and his legendary carpet company Interface, Inc. Interface took responsibility for its petroleum-intensive manufacturing some time ago. It has since become a recognized leader in environmentally responsible business. While still not perfectly green, Interface uses a number of different strategies in the pursuit of that goal. Its modular carpet tiles reduce waste and employ biomimetic [see Biomimicry, p. 99] design, and several of its lines also utilize recycled material. In the past ten years, Interface has reclaimed almost 60 million pounds (27.2 million kilograms) of carpet, and it currently has research in progress to develop completely recycled and recyclable carpets. SR & JF

Recycled-glass Tiles and Countertops

Glass is an amazing material. Not only is it smooth, strong, long-lasting, and light-conductive, but it can be recycled infinite times. Other materials, such as plastic, gradually deteriorate over multiple uses, but glass can be melted down and turned into something else over and over again, without ever compromising its quality. Post-consumer glass containers now make up the second largest source of consumer waste (after paper), so glass is nearly unbeatable as a recyclable material. We can do our part to conserve this great resource by recycling glass containers and by supporting industries that recycle and use recycled-glass products.

A good place to start is with interior design. Since the mid-1990s, architects and material designers have begun to realize that the unique qualities of glass make it an ideal material for building—and not just for using in flat-paned windows and doors. Recycled glass now shows up in everything from kitchenware to bathroom tiles to glass-aggregate floors and countertops.

Probably the most stunning architectural use of recycled glass is Vetrazzo, a ceramic aggregate material. A strong, stable material made from 85–90 percent post-consumer glass, Vetrazzo has an attractive marblelike surface. It is usually used in countertops and tables but can also be used in floors and walls. The material comes in a wide variety of colors, and can be custom ordered in any combination of colors and aggregate sizes. SR & JF

Reuse Centers

So much of the material that gets thrown away during building renovations and demolitions could be reused instead. Most of it is still viable and functional, and there is no reason for it to go to a landfill. Why does it get thrown away? Because it's easy for contractors to fill up dumpsters, and it's hard for homeowners to find

Right: Vetrazzo's recycled-glass countertop.
Opposite: Doorknobs and electrical outlets await reuse at the ReBuilding Center, Portland, Oregon.

exactly what they want by picking through gutted interiors.

Reuse centers have alleviated the burden on both sides by collecting used building materials, organizing and storing them, and reselling them at a reasonable price. The service these centers offer saves an enormous amount of solid waste from the landfill, and saves money for everyone during the remodeling process. On top of that, reuse centers are the best places to find unique, original decorative components that can go into a beautiful, personal home.

When you walk into the ReBuilding Center in Portland, Oregon, you have at your fingertips a vast selection of readily available material for creating the perfect interior space. The ReBuilding Center is the largest not-for-profit reuse center in North America, containing 60,000 square feet (5,574 square meters) of ever-changing inventory that yields new treasures by the hour.

The ReBuilding Center makes its social and environmental aims very clear: to contribute to environmental sustainability through the reuse of otherwise landfill-bound materials, to make materials accessible to all income levels by offering affordable prices, to create local jobs with living wages, and to educate the community about the benefits, both environmental and social, of reuse.

According to the center, on average, it diverts over 2,250 tons (2,000 metric tons) of reusable building materials from the landfill annually. Compared to the approximately 250,000 tons (226,796 metric tons) of construction and demolition debris generated annually in Oregon, this number is relatively small, but the combined savings from all reuse centers eliminates a thick layer of garbage and redistributes it as usable ingredients for somebody's nest. SR & JF

RESOURCES

Green Remodeling: Changing the World One Room at a Time by David Johnston and Kim Master (New Society Publishers, 2004)
If you want to bore down into the underlying principles of green building and really know that you're making the right choices in your remodel job, Johnston and Master's book is an essential resource. There are the normal to-do lists and tips here, but there is also a depth of carefully explained research into why certain decisions are greener than others. You don't need this book if all you're doing is replacing the flooring in your kitchen, but if you're trying to truly make a green home, you'll probably find yourself dog-earing pages and underlining passages ... and learning a lot along the way.

As Johnston and Master explain, "The energy that buildings require starts accumulating long before the buildings and homes are even in existence. The energy required to extract, transport, manufacture, and then re-transport materials to the point of use required a substantial amount of energy at a significant cost to the environment. The sum of all the energy required by all the materials and services (including the costs of upkeep and maintenance) to go into constructing a building is called the *embodied energy* ... For example, stones excavated from a nearby hillside for a new patio have lower embodied energy than stones that must be transported from another state."

Green Home Remodel
http://www.ci.seattle.wa.us/sustainablebuilding/greenhome.htm
These online PDF guides are the best, currently available, free resource on sustainably remodeling your home—period. Covering painting, roofing, landscaping, plus salvage and reuse and working with a contractor, these guides are simply indispensable—if you're thinking about making some changes to your crib, start here. You'll get more excellent advice like the following:

"Initial price gives only a peephole view of the true cost of a product or design over the lifetime of your home. A low purchase price may mean a good deal, or it may signify a lack of quality or durability. Or it may mean that some environmental, health, or social costs are not included in the price. A higher purchase price can mean a better deal in the long run: you can actually reduce the cost of living in your home by choosing resource-efficient fixtures (lowering monthly utility bills) and durable materials (requiring less frequent replacement). Lenders are beginning to recognize the value of ongoing savings to the homeowner. The savings from a more efficient home can cover and even exceed the incremental addition to your mortgage payment, meaning the improvements pay for themselves, and then some."

G/Rated
http://www.green-rated.org
Portland's G/Rated is a service of that Oregon city's Office of Sustainable Development, and offers an excellent set of online guides to building with the planet in mind. Though some of the information applies only to the local area, much is solid and useful no matter where you live. Every city ought to have a service like this.

According to the site, "Green building is growing in popularity with good reason—it makes sense. Building green is also about adventure and exploration—finding those one-of-a-kind pieces of salvaged wood or lighting fixtures that add character to your living room, or learning new ways to save energy and save water. Your decisions help drive sustainability in the marketplace, ultimately creating more cost effective green product choices."

G/Rated Home Remodeling Guide: Designing and Building a More Sustainable Home
(G/Rated, 2005)
ecotrust.org
If you're more of a paper person, you can get much of the best information from the G/Rated program distilled into a handy little book, available for purchase online. The guidebook promises, "Whatever your budget or personal preferences, you can meet your goals while minimizing your impact on the environment."

Building a Green Home

Most of us own a small arsenal of appliances that make our lives more efficient. Microwaves, laundry machines, electric toothbrushes, portable audio players—we value these things for the functions they serve, and we select them based on how well they work for us. Paradoxically, although we fill our homes with life-improving devices, we don't often regard the house itself as a tool for a better quality of life.

Architects and designers are starting to think more about the relationship of a home to its inhabitants: How can a home more effectively self-regulate energy consumption? How can it monitor potential hazards and dangers so that its inhabitants don't have to? How can the house become a flexible accommodation that will permit us to stay in the same place as our families grow? When a home is designed with these thoughts in mind, it starts to act more like an appliance than a container—what Le Corbusier called a "machine for living" (1923)—responding to our needs and evolving with our lifestyles. The homes of tomorrow are flexible, adaptable, and interactive, melding technological advances with our changing values.

Solar technologies are merging with highly evolved technological tools to produce more compact and functional fossil fuel–free power sources. Remote digital programming has converged with home systems and appliances to create "smart home technology" that allows you to turn off lights and appliances when not at home, or to detect hot spots before a fire breaks out. Factory production strategies have been applied to housing design, resulting in the rapidly growing prefab movement. From all angles, we are witnessing a sea change in the role of the home and our relationship to it. SR

A Zero-energy Home

We don't often think of our homes as gas-guzzlers, but buildings account for 50 percent of energy consumption in the United States and for more greenhouse-gas emissions than automobiles.

We can burn less energy without sacrificing comfort. We can heat and cool our homes with nothing but sun, shade, and breeze. In fact, it is possible to "zero out" our energy consumption so that we generate as much power as we use, and sometimes even achieve a surplus that can be fed back into the power grid and used by someone else [see Going Off the Grid, p. 179].

Architect Zoka Zola's Zero Energy House on Adams Street in Chicago is a terrific example of a stylish, energy-balanced dwelling. The building is completely self-sufficient, generating 100 percent of its own power on-site through solar and wind energy.

The Zero Energy House emphasizes passive techniques for generating energy and maintaining indoor comfort. "Passive" solar energy makes use of existing conditions and natural methods like conduction and radiation to heat a building. ("Active" solar energy, by contrast, requires pumps and motors to circulate heat and power collected by solar panels and shingles.) In the case of the Zero Energy House, the building's structure has been designed with passive solar in mind—the house's orientation, overhanging rooflines, strategically placed shade trees, and carefully installed insulation allow it to capture sunlight and heat. Thermal mass, another passive technique, uses exposed surfaces to absorb heat in daytime and radiate heat at nighttime or during low temperatures. In the winter, the Zero Energy House's south-facing windows allow ample sunlight into the house, heating the concrete walls inside. In the summer, shade trees act as passive means of cooling, blocking hot sun rays from the house.

During warmer months, natural ventilation keeps temperatures stable by allowing cross-room circulation between open windows and skylights. This approach can eliminate the costs, noise, and environmental problems associated with air-conditioning and fans. To make cooling even more effective, plants surround the building,

offering insulation, shade, and protection from UV exposure. Ivy shrouds the building's exterior, and other foliage covers the building's roof [see Greening Infrastructure, p. 256]. The green roof absorbs rainwater and provides extra space for planting a diverse range of native and edible greenery. It can be used as leisure space and it protects the building from the damage normally caused by temperature fluctuations and sunlight.

It's easier to implement these types of energy-efficient building strategies during the construction process than it is to add them to existing buildings. Because the Zero Energy House was designed and built from the ground up, the architect was able to study the natural cycles at the site in order to establish a symbiotic relationship between the house and its environment. For example, the architect mapped out the location of the sun (and corresponding shade) at various points during the year to ensure that the house would always be efficient, regardless of weather or season.

Though the Zero Energy House is a large-scale project, building to take advantage of renewable energy doesn't have to cost more, at least in the long run. Building a new home with well-planned orientation, good circulation, and proper insulation makes for an inherently less energy-intensive dwelling. Once these passive design strategies are in place, you can watch the savings pile up—proof positive that ecofriendly construction is also wallet friendly. SR & JF

Smart Home Technology

If your home had a brain, how much easier would your life be? Smart home technology can reduce our environmental impact and increase home efficiency, home safety, and personal independence. With digitally programmed systems, homes can take care of themselves, making it easier for us to monitor and manage our living environment. Whether it comes in the form of windows that open remotely to allow overhead ventilation, electronics that shut off when not in use, or laundry machines that offer custom settings, technological innovation can conserve both the energy our homes use and the effort we spend to keep them safe and comfortable.

Automation works well for digital security systems, fire and flood monitors, and irrigation systems. We can now lock our doors, activate security lighting, and water our yards with the push of a button. Just as computers have grown cheaper with increased demand, these types of in-home technologies will become more accessible as their financial and environmental benefits become apparent.

Smart home technologies also drastically enhance self-sufficiency for older people and people with disabilities. "Assistive technologies" permit people with physical limitations to control the systems in their homes with minimal effort. The priceless sense of independence that results can dramatically increase quality of life and general well-being for people who otherwise have to rely on nurses and aides.

Geothermal Pump

A project at Duke University called SmartHouse could raise the bar for home systems. The most exciting innovation the students involved in SmartHouse have been developing is an automated geothermal heat pump designed to increase energy efficiency. Geothermal pumps measure the difference in temperature between the constant level below the earth's frost line and the varying levels inside a house. The pump operates according to that difference, cooling indoor temperatures when they rise above underground temperatures. Think of it as all-natural air-conditioning.

A Prefab Home

A prefab renaissance is sweeping through the housing market. Prefabrication has been around for decades, but boxy little homes were long considered a lowbrow architectural style. Today's prefab home, however, can be cutting-edge, modern, and minimalist, and in the best cases, ecologically conscious.

Like any other industrially produced product, prefabs can be easily modified to incorporate specific features, such as sustainable materials and green design. In this way, we get to bypass much of the tedium and inconvenience of the building and contracting process. We can skip straight to ordering exactly what we want in a prepackaged container.

Prefab is currently sitting pretty in the spotlight of fashionable architecture. The designers who pioneered these new styles have had a big influence on public acceptance of prefab's revival. Not all of them have designed homes with mass appeal; still, they have shifted the general mindset about prefab housing. The simple fact that compact, efficient homes can beat out oversize, energy-intensive homes as the new symbol of material comfort shows that some homeowners have become more conscious about the space they need.

Prefab also has benefits outside of the high-end home market. Mass customization, mobility, and versatility in building materials make prefab homes ideal for low-income communities and areas that have been struck by disaster, such as the Gulf Coast and Pakistan, which respectively were hit in 2005 with Hurricane Katrina and the Kashmir earthquake.

As consumers get hip to the importance of green design, prefabs are begining to incorporate more ecofriendly features. Many of the existing prefab companies already offer their homes with optional green elements such as solar

Opposite: A computer rendering of Zoka Zola's Zero Energy House, Chicago, Illinois.
Left: An architectural rendering of the Duke University SmartHouse, a live-in research laboratory built by Duke's Pratt School of Engineering.

shingles, tankless water heating, bamboo flooring, and energy-efficient appliances. The manufacturers emphasize that off-site fabrication reduces waste and energy expenditures during construction, and that the footprint of a prefab home has less impact on the building site than does the foundation of a traditional stick-built house. For consumers, it's a great situation—we can get exactly what we want, customized down to the last detail, cheaper and faster than if we went with conventional construction. In other words, a growing proportion of homeowners can now afford a simple, efficient, stylish dwelling. SR

A Bright Green Home

Most architects don't, or simply can't, take their buildings' longevity into consideration. They may hope a structure will last, but once built, it's generally all downhill. Most homes are in their best condition, and are best suited to their inhabitants, when they are brand new. As life inside changes—families grow, kids move away—the house simply ages, and remodeling only offers moderate improvements. When we begin to feel our house no longer serves us, we pull out outdated components or tear down the house entirely to build a more appropriate one. Not only does this incur huge cost and enormous inconvenience, it generates an astonishing amount of waste, most of which goes straight into our exponentially growing landfills.

Australian architect Andrew Maynard recognized this problem and set out to fix it. He designed a home that incorporates a dimension new to housing: time. Constructed like the children's toy blocks that are connected by hinges, Maynard's Holl House stacks, unfolds, and changes shape easily. Shifting needs of the house's residents and rapid advancements in technology do not date the house as a relic of an earlier time; rather, the house adapts. Over time, with this kind of arrangement, a home no longer acts only as a shelter, but as a service system, keeping up with its inhabitants' demands for a comfortable and convenient space. This not only allows compact living to be much more efficient, with units that can serve as kitchen, closet, and sitting room in one, it gives inhabitants a new kind of freedom in their living space. SR

RESOURCES

U.S. Green Building Council
http://www.usgbc.org
"Build green. Everyone profits." That's the motto of the USBGC, which works as the central hub to the entire American green-building industry, offering professional training and accreditation, sharing information about new green-building innovations, and overseeing the LEED rating system. Though they have limited information for the casual worldchanger, if you're even thinking about embracing green building in some way in your business, this should be your first stop.

And if you're new to the field, USGBC's Emerging Green Builders is a program you should know about. Emerging Green Builders links together young architects, designers, builders, and planners who have a passion for sustainability, through an e-mail listserv and through cool events in a number of cities.

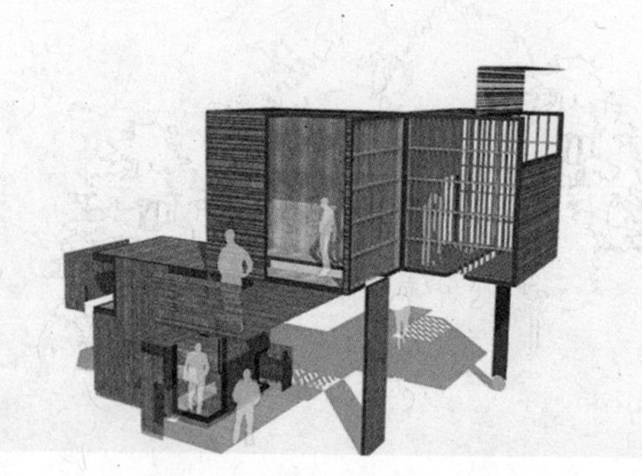

A Place of My Own: The Education of an Amateur Builder by Michael Pollan (Delta Book, 1997)
Many of us cherish the dream of one day building our own dream homes. And the writers among us tend to obsess about a particular subset of dream home: the writing studio. Writer Michael Pollan went a bit further than dreaming, though: he not only imagined a better place to write, but he set about building it, and then shared his experiences in this little gem of a book. And in every step of the process—from sketching out his writing cabin to doing the finish work—he found insight into the meaning of home and building history.

"The development of clear, leaded glass in 1674, followed a century later by sheet glass made with iron rollers, coincided with—and no doubt helped to promote—important changes in people's attitude towards the world beyond the window. Beginning with the Enlightenment, people were less inclined to regard the world outside as perilous or profane; indeed, nature itself now became the site of spiritual sanctity, the place one went to find oneself, as Rousseau would do on his solitary walks. Nature became the remedy for a great many ills, both physical and spiritual, and the walls that divided us from the salubrious effects came to be seen as unwelcome barriers."

A Good House: Building a Life on the Land by Richard Manning (Penguin Books, 1993)
We've been hoping to find places where we can live in some more balanced way with the land around us almost since the Industrial Revolution began. Richard Manning decided to spend a year building a green home on a rural lot in Montana, and chronicled not only the construction process itself but the ways in which the work and the thought he put in helped him rebuild his own life, after he lost his job as a muckraking environmental journalist. Technically, this book is a bit out-of-date, but the journey Manning takes us on couldn't be more timely:

"My environmental reporting, my investigations, the thing that had smashed my career, had taught me that there is enormous consequence to the way each of us lives our lives. Distilled to its simple forms this means a house takes forests from mountains. A house takes coal from the hills. A house takes life from the planet. Yet each of us must have a house. For those of us who value nature, this is the primary contradiction of our lives.

"It occurred to me that I was uniquely situated to investigate this matter from one end to the other, not by reading or interviewing or thinking but by asking the question in a more basic manner, by asking the questions of each board and wire in my house. Could I build a house in such a way as to ensure my happiness and still do minimum damage to the earth?"

Dwell
Rejecting the conventions of most home and architecture magazines, where design and interiors are portrayed devoid of a human element, *Dwell* set out to highlight the energy, diversity, and charming imperfection of well-lived-in spaces: "At *Dwell*, we're staging a minor revolution ... We think that good design is an integral part of real life. And that real life has been conspicuous by its absence in most design and architecture magazines ... we want

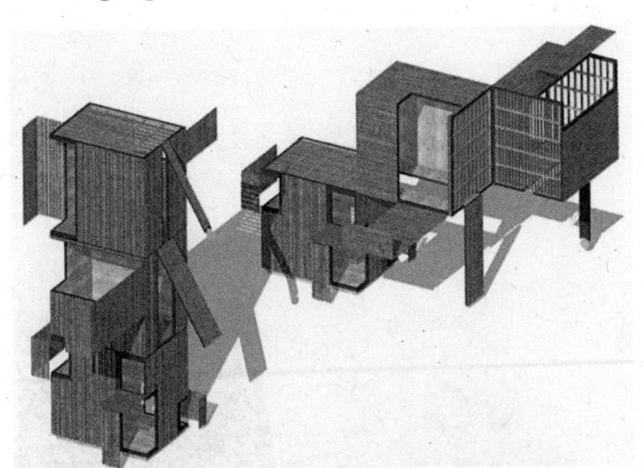

Opposite, left: Assembling the Sunset Breezehouse, a prefab design by Michelle Kaufmann.
Opposite, right, and left: Architectural renderings of Andrew Maynard's shape-shifting Holl House.

to demonstrate that a modern house is a comfortable one."

Dwell (particularly editor in chief Allison Arieff) has also been largely responsible for the renaissance of prefab home design, which has played a significant role in the modern perception of what makes a home green, emphasizing that compact spaces and off-site fabrication are key elements of sustainable living today.

Living Well in a Compact Space

Urban living is better for the planet [see Cities, p. 225]. So why does living in a city so often evoke a sense of resignation? Probably because so many compact living spaces are neither comfortable nor sustainable.

Urban dwellers deserve to feel good about their compact spaces. Done right, small spaces can be wonderful, some of the most pleasant living environments around. Efficient living is green living—a compact living space is intrinsically less wasteful and less expensive than a sprawling space. Simplicity and efficiency in a living space are a statement of a truly empowered existence.

With this in mind, designers are turning our most sustainable dwellings into spaces we love to live in. In North America, this process started nearly a century ago with the invention of the Murphy bed; the only multipurpose, space-saving item that can compare is the futon. But we've come a long way since the Murphy bed; flexible systems that range from easily customized, modular shelving to all-in-one pods that are practically stand-alone apartments are the wave of the future. SR

Right: Abito's elegant and compact apartment.
Opposite, left: Andrea Zittel, *A to Z Management and Maintenance Unit: Model 003*, 1992. Steel, wood, carpet, plastic sink, glass, and mirror, 86 x 94 x 68."
Opposite, right: A Brave Space Design Tetris corner unit.

Abito

Imagine living in a 347-square-foot (32-square-meter) studio that contains all of the accoutrements of a three-room apartment and still manages to feel spacious.

Abito, a UK-based company, has developed a high-rise apartment template that accomplishes all that, while looking magnificently futuristic. The key to this marvel is a stand-alone central pod that serves all the functions that usually go on in multiple rooms. One side of the pod contains a bathroom behind double doors, and a second side has a cupboard with washer/dryer hookups. A third side contains a two-person wardrobe and a utility console, and a fourth contains a kitchen unit with an integrated fridge/freezer, a two-burner range, an oven/microwave/grill combo, and a waste sorter. The bedroom is more like a wall with a fold-up bed, and is positioned opposite the pod's wardrobe. On top of the pod, there is space for an extra bed or additional storage (though there is a remarkable abundance of storage elsewhere).

The Abito apartment is elegant, with white doors and stainless steel finishes. Each apartment has 10.5-foot (320-centimeter) ceilings and floor-to-ceiling windows, which helps make the minuscule space feel deceptively larger, while providing cash-strapped consumers with features generally found only in luxury buildings.

Priced at just over U.S.$155,000, these little apartments cost significantly less than most one-bedroom apartments in Manchester, where Abito is putting up its first building. Abito is currently only planning buildings in the United Kingdom, but it seems like an obvious recipe for success among urbanites just about anywhere. SR

Andrea Zittell's "A-Z Living Units"

In a small space, the more we can fold, tuck, and trundle away our things, the better. We may think our apartments look better after we've shoved the laundry under the bed, but imagine how streamlined our spaces would be if we could tuck away our appliances, stovetops, and bathroom sinks.

Designer and artist Andrea Zittel makes this possible with "A-Z Living Units," freestanding compartments that contain everything from full minikitchens and lounge areas with bookshelves to utility rooms—all packed in a box.

The units, which can be custom-made, are not only brilliantly compact, but they're also great looking: modern, clean, and simple. They're perfect for renters who want to be able to move from place to place while keeping things consistent and efficient. Sharing a house with multiple roommates gets a lot less stressful when every roommate essentially has a mini-apartment in their room. SR

Convertible Kitchenettes

The DoubleSpace Kitchenette from Vestal Design is a life-size jigsaw puzzle. The unit transforms from an armchair to a stovetop in a matter of seconds.

It's an odd combination, but then again, we'd probably never need to cook while lounging,

or vice versa, and the unit proves that two mutually exclusive domestic activities pair excellently in a single piece of furniture. The countertop and dual electric burners stow safely beneath the cushioned seat until we need them. When it's time to cook, the cushions disappear beneath the stovetop to protect them from spills. Best of all, we can roll the units around our spaces as much as we want, constantly rearranging the layout of our homes. SR

Tetris Shelves

Tetris Shelves show that nostalgia is good for something besides retro revival. The shelving units are shaped exactly like the pieces in the classic video game, and fit together in myriad ways, stacking vertically, horizontally, or diagonally, to work with almost any room layout. Of course, the original version (from Brave Space Design) comes with a designer price tag, but given the simple geometry, building one's own version seems like a pretty straightforward DIY project for any industrious, space-seeking individual. SR

RESOURCE

Space: Japanese Design Solutions for Compact Living
by Michael Freeman (Universe, 2004)
Combating the idea that small spaces impose limitations and curb possibilities for interior design, *Space* presents a compelling argument that, in fact, tiny living quarters invite tremendous creativity, largely through the use of flexible and multifunctional parts such as sliding doors and foldaway furniture. Through short profiles and mesmerizing photographs, Michael Freeman leads us through the various elements that comprise a well-conceived compact space. "Material success allows bigger homes, and this expansion of personal, proprietary space is rarely questioned as a good thing," Freeman points out. However, with increasing urban density and rapidly depleting resources, massive, energy-intensive homes are not only becoming an unquestionably *bad* idea, but they are in many places simply not possible. These Japanese urban minihomes are a testament to "less is more" thinking—evidence that compact houses can feel airy and spacious with the right planning.

Developing Green Housing

If building green houses is good, then building green housing developments is even better. When you combine sustainable design with densely planned neighborhoods, you get future-friendly city living. Considering the breakneck speed of urban growth around the world, dense, green housing complexes may be the only way to shelter everyone without compromising the planet.

A handful of forward-thinking cities have been fervently installing solar panels and planting rooftop gardens in the last few years [see Vancouver, p. 231], but overall, developers have been slow to go green, assuming that the extra cost of sustainable building isn't worth the investment. The truth is, green building actually gets cheaper as builders scale up the size of their developments: a thousand-unit complex faces only marginal cost increases above traditional construction, and a five thousand–unit complex can incur no additional cost at all. In addition, green housing is much more marketable than conventional housing, and will only gain appeal as green features take on more importance to homeowners.

This idea is so simple to understand that you would think large-scale developers would embrace green building, recognizing the many benefits and appreciating the roughly equivalent costs. But despite the apparent gains, developers are slow to change their ways. Still, there is no doubt that the trend is approaching a tipping point, and those who still build conventionally may soon find themselves in the minority. In the meantime, developers who have pushed ahead with their green dream buildings despite the extra costs are proving that these complexes can not only meet a dizzying array of low-impact goals, but can also be exciting spaces to live in. SR

BedZED

An ecofriendly housing complex has turned a former sewage plant into South London's hippest address. The Beddington Zero Energy Development (BedZED) is a high-design, low-impact complex with eighty-two occupied units and a very long waiting list. It also happens to be the best model of sustainable urban development out there today.

BedZED hits all of sustainability's high notes. It's a great example of brownfield remediation [see Healing Polluted Land, p. 250] turning a wasteland into an attractive, thriving community full of green space. The developers used renewable and/or recycled materials, sourced from within a thirty-five-mile radius whenever possible.

BedZED is a "zero-carbon" development, meaning it's powered exclusively from on-site energy sources, with no carbon dioxide emissions. All its homes have solar-power panels to meet some energy needs, and the complex's main source of power, a heat-and-electricity unit, is fueled by waste timber, which would normally be sent to the landfill. South-facing window banks that provide natural heating and light, and triple-glazed glass that provides superior insulation help to diminish energy needs, while rainwater harvesting and on-site water recycling dramatically reduce overall water consumption. To boot, BedZED residents aren't automobile dependent, thanks to easy access to public transportation, local schools, and shops—all of which are within walking distance—and to a car-share program that counts at least half the complex's residents as members.

BedZED opened for sales in 2000, several years ahead of the curve of sustainable building's popularity. By 2003, all its units had been sold, not so much because the market for *green* homes had grown, but because the market for *good* homes was still strong. BedZED's being an emissions-free development wasn't enough to sway a new generation of homeowners by itself. Living here isn't just greener, it's better: homes have abundant natural light, private gardens, and rooftop terraces. These features, in addition to the shared community transport programs, all lured prospective buyers and tenants.

BedZED is the brainchild of Bill Dunster architects, BioRegional Development Group, and the Peabody Trust for charitable housing. On the coattails of BedZED's success, BioRegional plans on building a kindred development, Mata de Sesimbra, near Lisbon, Portugal. This project will be much larger than BedZED and will include a nature reserve, a cork-forest-reforestation project, and a tourist village. The goal is to surpass BedZED's environmental achievements by implementing a twenty-year zero-waste plan, 100 percent renewable energy, and a commitment to buy 50 percent of the food sold in on-site shops and restaurants from local growers and producers.

Meanwhile, Bill Dunster and his team at ZEDfactory keep cranking out the innovations: they've published "ZEDupgrade," a guide to retrofitting existing homes with more ecologically responsible features and systems, and they've developed and published "ZEDstandards," a green-building code inspired by BedZED and intended for adoption by industry partners as a benchmark for zero-carbon-emission developments.

One of the most inspiring things about the BedZED team is that they are not seeking to create a monopoly or to compete with potential imitators. Rather, they hope to encourage wide use of their approaches. This is the true spirit of green innovation: a shared desire to encourage better building practices and a willingness to support new endeavors by freely sharing expertise. SR

RESOURCE

BioRegional Solutions for Living on One Planet by Pooran Desai and Sue Riddlestone (Green Books, 2003)
BioRegional may just be the single coolest environmental group on the planet. It's the group's recent work—like the BedZED development and their greening of London's 2012 Olympics—that is particularly exciting, but they got their start by

asking how local sustainable-development efforts can thrive in a global economy. Much of their early work explored ways of meeting resource needs through local economies: creating local paper cycles, promoting hemp clothing, connecting local farmers to local households, finding markets for locally grown lavender. These and many other of the group's projects are featured in *BioRegional Solutions:* "We have been keen to engage with the market on its own terms and to link ourselves back to the local environment and the earth's natural nutrient and energy cycles. We value technology and the marketplace, but recognize that these can only bring long-term benefits when they are linked to natural cycles—i.e., when we work with, rather than against, nature."

Furniture and Home Decor

Once we've decorated our countertops with recycled glass tiles and laid the Marmoleum flooring [see Green Remodeling, p. 142], the real fun begins. Buying furniture may be the most enjoyable part of designing our personal spaces.

But we often fall off the bright green wagon when shopping for furniture. In an attempt to get the style we want, we forget to find out where the material in our dining room table came from, what impact the stuffing in our new couch may have on our health, and how much landfill space our entertainment center will take up when we redecorate the next time.

The home decor industry isn't about to help us out—it thrives on our desire for brands, our ever-changing whims, and the fact that we buy home items under the assumption that we'll eventually upgrade, redecorate, or otherwise toss them out. With this in mind, one of the best things we can do is to think long-term—buy something that has durability not only with respect to its structure, but also with respect to our tastes. Before we buy, we should ask ourselves, "Is it really what we want? Is this actually comfortable? Is it sustainable,

Opposite: BedZED, London's famous bright green housing development.
Right: IKEA showroom. Most furniture from less conscientious companies is designed to quickly obsolesce.

durable, and will it wear well over the next five or ten years?"

We also need to understand that very few home furnishings are sustainable and many are downright hazardous to our health. A lot of people don't realize that most low-cost wood furniture is actually made out of particleboard, usually pressed together with formaldehyde—a known carcinogen. Such particleboards and plywoods emit small traces of carcinogenic gases in our homes, contributing to health problems. Furniture with foam-filled cushions, such as sofas and loungers, poses another threat: foam is usually treated with fire-retardant chemicals called polybrominated diphenyl ethers (PBDEs), which are particularly harmful to fetuses and can even cause brain and reproductive disorders in adults.

These problems are not exclusive to cheap furniture. Designer plastic and metal items often have toxic finishes or other material additives. In addition, luxury "exotic" woods like teak and mahogany, used in expensive handmade furniture, usually come from endangered old-growth forests in Myanmar or Brazil.

Fortunately for the green-design movement, sustainability is climbing up the priority list for consumers and manufacturers alike. The demand for sustainable home furnishings has increased over the past few years, and designers have happily responded with an abundance of stylish ecofriendly furniture that incorporates the playfulness of contemporary design—finally giving sustainability the sexiness it requires to gain mass appeal. From cork and bamboo to recycled waste products pulled from dumpsters, to new materials such as fiber-optic threads and Homasote (a hard material made from recycled-paper composite), sustainable materials [see Picking Green Materials, p. 87] are allowing designers to promote environmental sustainability while furthering the evolution of design.

Currently, the biggest obstacle for most bright green consumers is the cost of such design—demand hasn't quite reached the point that ecofurniture is affordable to all. But larger-scale furniture manufacturers like IKEA and Herman Miller are already turning out mid-range to truly inexpensive pieces that are as good to look at as they are good to the planet. SR & JF

Recycled Furniture Grows Up

Not everyone loves the idea of dumpster-dive decor, but as urban scavengers find new ways to turn reclaimed or salvaged materials into better-than-new home furnishings, it's getting harder and harder to tell when something's been rescued from the curb.

Scrapile, a line of furniture from Brooklyn-based designers Carlos Salgado and Bart Bettencourt, may come from salvaged wood scraps, but it is a far cry from shabby chic. The designers drive a truck around the Northeast gathering discarded material, which they then haul back to their shop, where they stack and refinish the wood to create the multitoned, striated pattern that characterizes their work.

The end result is a refined and unique piece infused with the history of the many layers that comprise it—a coffee table could include both an imperfect piano top from the Steinway factory and planks from a suburban home's demolished deck. SR & JF

Renewable Woods

If you have your heart set on brand new furniture, you can buy pieces made from renewable fiber and timber sources, such as coconut palm, sorghum, and bamboo. Check out the following:

Durapalm: Coconut-palm wood, Durapalm has a unique natural pattern and the dark tone of an exotic wood. Coconut-palm plantations produce a tremendous amount of waste in the form of "retired" trees—the trees produce nuts for eighty to one hundred years, after which planters cut them down and replace them with younger, more productive trees. In an effort to eliminate the wasteful disposal of perfectly good palm wood, timber distributors now source this wood, and inevitably some crafty designers turn it into furniture.

Kirei: You don't often see sorghum on the coffee bar next to the sugar packets—it sometimes does show up in the form of a molasses-like sweetener, but most North Americans don't eat it. Elsewhere in the world, though, sorghum is one of the most important cereal crops. It's also a big producer of worthless by-products; after harvesting, its stalks are either burned or thrown out.

Kirei Board uses sorghum waste in a new way, by heat-pressing stalks with a nontoxic, non-VOC (volatile organic compound) adhesive. Aside from reducing landfill mass, decreasing air and water pollution, and creating jobs for rural farmers, Kirei also offers the benefit of looking completely distinct from wood. With a highly textured and variable surface, it makes unique and beautiful pieces, from chairs and tables to flooring, cabinetry, and wall covering.

Plyboo: Plyboo? That's right: plywood from bamboo. Bamboo's expansive root system sends up profuse, rapidly growing shoots throughout its lifetime, making the grass a highly renewable source.

Plyboo is simply made from laminated bamboo strips that have been boiled in order to remove their starches.

Designers, builders, and carpenters have already widely adopted Plyboo, since it offers such a straightforward alternative to ubiquitous plywood. You'll find Plyboo furniture in a wide range of prices and styles. SR & JF

Office Chairs

Whether self-employed or office-bound, most people sit in a chair to work, at least part of the time. Cheap office chairs are uncomfortable; comfortable office chairs are exorbitantly priced; and almost all office chairs end up as a huge amount of waste—they are tossed out and replaced with even more frequency than employees are.

Herman Miller, the champion of environmentally sound chair design, released the first Aeron chair in the mid-1990s. The chair sparked a sea change in seating design, introducing principles that are now fundamental to industrial design: ergonomic precision and scalability. The chair's curves mirrored those of the human body, and its adjustable settings offered comfort to users of all heights and weights. In addition, the chair consumed minimal natural resources and was easily disassembled and recycled.

Herman Miller's more recent creation, the Mirra chair, greatly improves upon the Aeron. First, the Mirra adjusts more easily. Second, it

Opposite: The furniture shown includes a table made from salvaged wood by Scrapile, a table and chair made from Kirei Board by AAA Design, and an end table made from Plyboo by Plywood Office.

accommodates people even in the highest height and weight percentiles, so its design is even more flexible and adaptable than the Aeron's. Third, it costs nearly one-third less than the Aeron. And fourth, the Mirra passed the McDonough and Braungart Cradle to Cradle [see Knowing What's Green, p. 115] product certification, which guarantees that it was created with an eye toward its entire life cycle—from the material sources that went into creating it to its end-of-use destination. SR & JF

Paint

Painting our homes is a great way to personalize our spaces. But you know that new paint smell? That means we're huffing volatile organic compounds (VOCs). These compounds can cause physical discomfort such as respiratory irritation, headaches, and nausea. And though the smell may fade, off-gassing can continue to occur in our homes.

Fortunately, low-VOC and VOC-free paints are becoming more widely available. If you must use VOC paint, look on the label for information on grams per liter. The safest VOC latex paints contain no more than 250 grams per liter; oil-based paints, no more than 380 grams per liter.

If you want harder-working paint, a company called Aalto Colour has developed a low-emission thermal-insulation paint that reflects radiant heat. The paint is supposed to help reduce energy consumption and cost by essentially acting as another layer of insulation. SR & JF

RESOURCE

Inhabitat
http://inhabitat.com
The blog Inhabitat offers "future-forward design for the world you inhabit," covering sustainable furniture, interior design, and architecture that's both really stylish and truly green. If you want to stay abreast of sustainable style, point your browsers toward Inhabitat first.

Lighting

Good lighting makes a great space. We can customize the atmosphere of a room with any number of different fixtures and bulbs, but most of the products use up about the same amount of energy. If we want to make our home lighting as bright green as possible, we can purchase sustainable products—fixtures made from recycled glass, aluminum, and wood in small-run production; chandeliers made from wine glasses rather than from those tinkling crystal pieces; lamps made from repurposed products, such as recycled chopsticks. We can also be sure to outfit all of our fixtures with compact fluorescent bulbs, which are much more efficient than incandescents because they don't use heat (much of which gets wasted) to produce their light. According to Energy Star, if every U.S. household replaced one lightbulb with a compact fluorescent, the pollution prevention would be equivalent to removing a million cars from the road.

And there's even more we can do. Light-emitting diodes (LEDs) are poised to be the lights of the future, and by investing in them now we can help drive the market for this important technology. Beyond picking

Opposite, left: Aurelle "candles" from Philips use LED lights. **Opposite, right:** The Parans lighting system brings sunlight inside.

between various bulbs, we can now bring the sun inside for even better quality—and lower-impact—illumination. SR & JF

Light-emitting Diodes

Compact fluorescent bulbs are currently the best choice for efficient home lighting (unless we're willing to outfit our homes with office-style big-tube fluorescents), but LEDs may soon outshine them.

Light-emitting diodes give off light when electrons in a semiconductor oscillate. Unlike ordinary bulbs, they don't have filaments (the part that burns out easily), nor do they get hot like incandescents do. Plus, LEDs last far longer than standard bulbs, sometimes up to 100,000 hours. In lab tests, some LEDs have proved to be more efficient than compact fluorescents, though it may be a few years before we see those LEDs on the market.

In the meantime, LEDs are starting to pop up in home-lighting accessories, and even mass-market, low-cost brands are adding LEDs to their product lines. In the next few years, as LEDs get brighter, we'll likely see a major shift away from incandescents and fluorescents.

Designers aren't content to replicate the standard bulbs we've always used. The "Simplicity" LED lights from Philips come in a range of shapes and look so great you wouldn't dare cover them up with a dusty shade. Philips is also working on a line of LED bulbs that change color and brightness with a simple twist or tap.

In purchasing LEDs, we flex our power over the global market, driving down the cost of making them, and thereby increasing the developing world's access to them [see Using Energy Efficiently, p. 167].

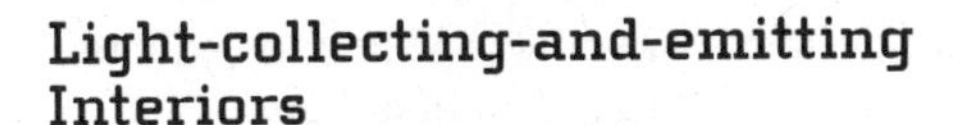

Light-collecting-and-emitting Interiors

Everyone knows that natural light is good for us. Ideally every space would be designed for maximum natural light, but the fact is we have a lot of dismal existing space to deal with, from dreary cubicles and windowless offices to basement apartments.

Full-spectrum lightbulbs are the most common (and affordable) tool we have to mimic the benefits of natural sunlight. But some ingenious designers have taken a more direct approach, letting sunlight flow right into furniture and dark rooms through fiber-optic cables.

Sunlight-transport System

The Swedish company Parans developed a system that collects sunlight in outdoor panels, transmits it through fiber-optic cables, and reemits it indoors through a well-designed overhead fixture. A combination of different types of beams gives the impression of natural light filtered by trees. Placing the outdoor sun-collecting panels at varying angles allows the indoor fixture to maintain a high level of natural light throughout the day, as the sun moves across the sky.

"Energy Curtain"

RE:FORM is one of several research studios under the umbrella of Sweden's Interactive Institute. The studio's current project, "Static!" uses

interactive design to increase awareness of energy consumption. The darling of this project, the "Energy Curtain," is a window shade woven with solar-collecting and light-emitting materials. The shade stores sunlight during the day and emits it at night, providing an ambient light approximately equivalent to the brightness of a lamp.

When we choose how much sunlight we collect (i.e. how long we keep the curtain drawn during the day) and how much collected light we use up at night, we gain insight into our own energy behaviors. It's one thing to wax poetic about local, sustainable energy systems, it's another to understand the trade-offs of such systems (if we don't remember to draw the shade during the day, we won't have its light later on) and to figure out how to adjust our lifestyles accordingly.

"Sunlight Table"

The "Sunlight Table" concept by Random International emits natural light via a grid of fiber optics contained in the table's surface. An input device (which can be placed near any window) collects the sunlight, and a few cables carry the light to the table. Not only do we get the benefit of the light itself, but we also get to watch the light fluctuate as birds or falling leaves brush by, giving us the sense that we aren't quite so disconnected to the movement and the life right outside our windows. Although the table was developed as a concept project and is not available for purchase, it's a sure sign of bright and bold things to come. SR & JF

The Exxon Valdez oil spill alerted the world to the costs of an oil-based economy, Prince William Sound, Alaska, 1989.

Energy

Energy is the lifeblood of advanced civilizations. Just a few centuries ago, our ancestors had only basic forms of energy—the muscle power of humans and animals, the heat and light of a burning fire, and a few simple machines like waterwheels and windmills—with which to work.

Today, in the Global North, cheap power courses through our lives. Flip a switch and electricity from a distant power plant lights up our rooms. Start the car and fossilized plants compressed over millions of years burn in our engines—and off we go to work. Turn the thermostat up and gas piped in from faraway places fires the furnace and makes our homes cozy.

But satisfying today's hunger for energy comes with enormous and severe environmental costs: climate change, air pollution, oil spills, river-killing dams, and nuclear waste. Our addiction to cheap energy is the driving cause behind not only the massive disasters we all fear, but also the longer, slower emergencies that threaten to undermine our entire society, from energy shocks to the climate crisis.

The social impacts of our dirty energy system are no better. Our addiction to cheap oil links us to dictators and repressive regimes, wars and terrorism. It enriches a few, at costs borne by many. And cheap, dirty energy allows us to make many questionable choices—like building sprawling suburbs that require long daily commutes and undermine other values we hold dear (like spending time with our families and friends, and feeling part of a close community of neighbors).

In the Global South, too, the situation demands change. Hundreds of millions of people still struggle to make a living with

only the most basic forms of energy and sustenance (oxen, open fire, hand-pumped well water). They are quite understandably eager to use more energy—to work their fields more efficiently, to light their homes, to refrigerate their food. The kind of energy they adopt will not only determine their quality of life, but will have global impacts. A world of solar-powered villages is a world of less poverty and greater climatic stability.

We do have other options—a clean energy future could be ours. AS

Using Energy Efficiently

The most direct path to a clean energy future starts with simply using less energy to get things done. Most of our homes leak heat; most of our appliances waste energy; and a lot of the power we use is spent doing things (like lighting and heating empty rooms) that we don't actually need to be doing. Green buildings and sustainable products can increasingly deliver us the good life without wrecking the planet, but we can all start being more energy-smart right now.

Sometimes little actions don't mean much. The opposite is true when it comes to powering our lives. Small adjustments, minor improvements, and simple steps can not only slash our electric and heating bills, but can let us take part in a global movement to do more with less. Imagine if this year we each cut the amount of energy we wasted by just 3 percent—say by changing all of our lightbulbs to compact fluorescents and turning off lights when we're not using them—and then, next year, we made the commitment to use 3 percent less again—say by installing a more efficient washing machine or reinsulating our attics. Imagine if we continued like that each year,

Opposite: Increased energy efficiency, whether garnered through better technology or better practices (like insulating homes well) is the most effective way to meet our increasing energy needs.

gradually investing more in making our homes greener, using better technologies as they became available, being more mindful of ways to cut useless power consumption—like the "vampire power" our televisions and cell-phone chargers keep sucking even when they're "off." Imagine if we continued to get smarter about how we use energy, aiming for a constant 3 percent improvement—the "clean three"—year after year. It would obviously grow more challenging as the easier things got checked off our lists, but we would also have more and better green technologies at our disposal as time went on.

What difference could the "clean three" make? A lot. If all we did was deliver our "clean three" each year, and we kept it up for the rest of the century, we could, by 2100, be using less total energy than we do now, even if we had a planet of 10 billion people all as wealthy as the average European. This is what energy experts call "the conservation bomb."

Setting off the conservation bomb would obviously make the world a far better place. But we don't have to hit that "clean three" target every year for a century to have huge impacts. Just getting twice as good—which, given how wasteful we are now, we could do pretty easily for a decade or two, even without a lot of new technology—would mean we'd need less than half as much energy for those 10 billion people to live well. If we do that and keep working hard to develop clean power sources, we could very well have a planet that runs largely on renewable energy by the dawn of the next century. If we combine conservation and clean energy with some of the other tools in this book, the next generation may well live in a solar-powered world.

In the meantime, pushing the efficiency envelope in our own lives not only helps us save money but also changes the world around us. It primes the market for more efficient products. It builds the market for green-power utilities. It creates momentum behind innovations that can benefit the planet's poorest people. It even helps us prepare to disconnect from the grid altogether and start creating that bright green future. AS

Top Five Things You Can Do to Conserve Energy

1. Get a home-energy audit: Many utility companies offer free home-energy audits that evaluate your current cooling and heating systems and insulation, assess air quality and problems with dust or dampness, and hunt out the source of drafts. Auditors provide a range of solutions sensitive to your price bracket. You never know what they'll turn up: something as simple as installing a programmable thermostat could greatly reduce your home's carbon emissions.

2. Use dimmers, automatic timers, and/or motion-detection sensors: All of these will help reduce the amount of energy your light fixtures use. Automatic timers and motion detectors are

perfect for rooms where lights are frequently left on and forgotten about, or for hallways and outside lights.

3. *Lower air-conditioning costs:* Clean your filters regularly; dirty filters can cause energy consumption to skyrocket by restricting airflow and thereby making air-conditioners work twice as hard. If you have central air-conditioning, resist the urge to crank down the thermostat lower than necessary—you won't cool your home any faster by trying to speed up the process. Keeping the thermostat at a steady 70 degrees Fahrenheit (21 degrees Celsius) is actually a more efficient way of cooling your home.

4. *Install proper insulation:* If your home's insulation works properly, the heat it traps does a great job to offset heating needs, which means less consumption and lower bills. Check existing insulation for gaps and wear, which effectively eliminate any savings that insulation offers and allow condensation to accumulate and form mold, a significant health hazard.

5. *Modernize your windows:* Inefficient windows are horrendous energy wasters. Good, triple-glazed windows can save significant amounts on heating and cooling, often counterbalancing a less-than-optimal heating and cooling system. Like proper insulation, good windows also help control condensation and mold, and help keep a consistent air temperature throughout the house. Efficient windows now bear the government-standard Energy Star label, which can help consumers choose wisely. CB

Energy Star and Other Eco-labels for Appliances

The U.S. Environmental Protection Agency (EPA) created the Energy Star label in the 1990s to help consumers make responsible choices—and save on energy bills. The label originally applied to computers and home electronics, but can now also be found on most major appliances, including refrigerators, washers and dryers, and microwaves. The EPA estimates that Energy Star saved enough electricity in 2005 to eliminate 23 million cars' worth of greenhouse-gas emissions, all while saving consumers $12 billion.

The Energy Star label applies to some goods in Europe and Canada as well, but many countries outside the United States have their own eco-labeling systems. Germany's Blue Angel environmental label was in fact the first of its kind, established in 1978. The European Union later created the Euroflower eco-label, which offers retailers and manufacturers voluntary certification to distinguish their commitment to sustainability through corporate transparency and attention to integrated life-cycle approaches to production.

Kill A Watt / Wattson

How much power do we use at home? Our monthly electric bill gives us the total, but how can we figure out which of our various toys and appliances needs to be replaced with something greener?

Two new devices, the Kill A Watt and the Wattson, help make the invisible visible. Plug any appliance into the front of the Kill A Watt, plug the Kill A Watt into a wall outlet, and the system will show you how much power you're using. The kilowatt-hour readout in combination with the time-used readout makes it easy to figure out which appliances should be at the top of the "must replace" list.

The Wattson, currently available only in the United Kingdom, is the next-generation Kill A Watt. A combination energy meter/portable display, the Wattson provides real-time information about household energy consumption, both as a

Opposite: LED lamps designed by the Light Up the World Foundation are distributed in Nepal, Sri Lanka, India, and twenty-three other countries.

text readout of current power demand and as an accumulated "burn rate" of cash. In addition, a colored LED screen indicates your home's overall energy health with an ambient red light that dims in accordance with how much power your electronics are consuming at any given moment. The Wattson can even be hooked up to home computers to archive information and to connect Wattson users to online information and idea exchange. JC

Vampire Power

Much like a leaky faucet—which slowly wastes a valuable resource—an appliance that is turned "off" but still plugged in consumes power through passive energy consumption, often referred to as vampire power. In fact, some devices use almost as much energy off as they do on—certain TVs and stereo systems expend up to 70–80 percent of their power when "off." According to the U.S. Department of Energy, 20 percent of Americans' monthly power bills goes toward vampire power. Many electronics manufacturers don't even use the word *off* anymore, substituting *standby,* which more accurately reflects the state of the appliance: perpetually at the ready.

The U.S. Department of Energy wants to instate manufacturing standards requiring products to be engineered so that they do not draw energy—or at least not as much—when they're not in use. In the private sector, a company called Power Integrations offers electronics manufacturers a technical quick fix in the form of a chip that can be built into products to reduce standby power drainage by 75–90 percent. For now, though, all we need to do to eliminate vampire power is to unplug appliances when we're not using them, or to plug them into power strips that we can switch off when we leave a room. JJF

Power and Light in the Developing World

Across one-third of the planet, nightfall still brings darkness; according to a 2002 World Bank report, nearly 2 billion people live without electricity. When they want to see at night, they burn things: wood, dung, kerosene, candles. Using fire for light is not only incredibly inefficient (only about 0.1 percent of the fuel energy becomes light), it's also dangerous and ecologically disastrous.

A child reading for an evening by the light of a kerosene lamp breathes in fumes equivalent to smoking two packs of cigarettes, according to the World Bank's report. Gathering firewood for fuel is time-consuming for rural people, and buying it is expensive for urban people living in poverty. The smoke from all those fires pollutes the air and contributes to climate change. In every way, being forced to rely on fire to cook, heat, and see hurts both poor communities and the planet.

But better alternatives exist. Better stoves, better lights, better ways of heating homes—these solutions promise to transform life for those almost 2 billion people.

Light Up the World Foundation

How can poor people get the light they need to

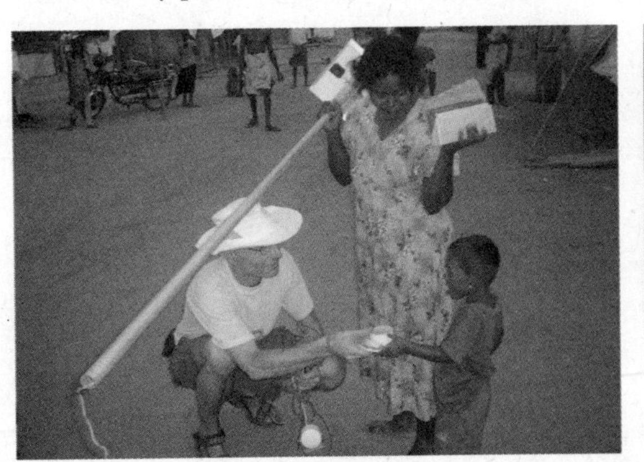

study, work, and socialize at night? The Light Up the World Foundation has an answer. The foundation's breakthrough realization is that a single white light-emitting diode (white LED), run off of a battery powered by a pedal-cranked microgenerator, wind turbine, or solar panel, gives off enough light to read by. Light-emitting diodes can light a hundred homes with the energy it takes to illuminate a single hundred-watt incandescent bulb. They are also cheap, rugged, and long lasting; they use direct current; and the lamps are increasingly easy to manufacture in the developing world. Best of all, the efficiency of LEDs has so far followed Moore's Law, doubling roughly every eighteen months. Based on pilot projects in India, Sri Lanka, Nepal, and twenty-three other countries, there's every reason to hope that tiny diodes will soon be nestled in thatch, attached to corrugated sheet metal, and taped to cinderblocks in billions of homes around the world. JJF & AS

Cooking in the Developing World

More than half of the world's population cooks using biofuels: wood, charcoal, dung, or crop waste. The demand for these wood stoves contributes to deforestation. The smoke they produce leads to respiratory disease, especially in children. Worldwide, cooking-related indoor air pollution kills more people than cigarettes do, according to the British NGO Practical Action's 2004 report "Smoke: The Killer in the Kitchen." Finding better solutions for cooking food is one of the biggest design-for-development challenges. Luckily, better solutions are starting to appear.

Building a better stove reduces the cost of cooking, improves the health of the people who cook, and benefits the environment by using resources more efficiently. But it's harder than you'd think: not only does a stove need to be efficient, but it also needs to be easy to manufacture, repair, and use, and—to a person with limited means—worth the substantial investment. Here are a few success stories:

The Jiko: According to a report by Daniel M. Kammen, director of the Renewable and Appropriate Energy Lab at UC Berkeley, in Kenya, the traditional metal stove delivers only 10–20 percent of a fire's heat to a cooking pot; an open fire is even less efficient. The Jiko ("stove" in Swahili), the latest cooking innovation in the country, has a ceramic lining within a metal casing. Kammen says this structure helps direct 25–40 percent of the fire's heat to the pot, which saves a household 1,300 pounds (590 kilograms) of fuel, or about $65 per year. Many women, the primary purchasers of fuel, have been able to reinvest these savings into schooling for their children or into small businesses, improving their prospects of financial independence. Nearly 1 million Kenyan households now cook with the ceramic Jiko stove.

The Rocket: The elbow-shaped Rocket Stove, used in over twenty countries from Honduras to Zaire, burns fuel and conducts heat to pots more efficiently—and burns only the ends of sticks in the process. A basic Rocket Stove can be built with found materials; the organization behind this innovation, the Aprovecho Research Center, provides do-it-yourselfers with information on how to design the best stove. They've also come up with larger, built-in designs for the home, such as the Estufa Justa, which has two burners and vents all its smoke outdoors.

The Henya: Richard Njagu, a Kenyan inventor and entrepreneur, combined the best of the Jiko and Rocket stoves into the Henya Stove, an attractive, efficient, and inexpensive stove that's designed and manufactured locally. Like the Rocket, the Henya is elbow-shaped—fuel is fed in from the side, and exhaust gases are captured

Opposite, left: Solar cooker in use, Uganda.
Opposite, right: Villagers making Pot-in-Pot refrigerators.

to help produce heat. Like the Jiko, the Henya contains a liner, made from local clay, that contributes to the stove's efficiency by preventing heat from leaking out the sides.

Solar Ovens: Using a solar oven is like cooking with a greenhouse. Typically, solar ovens have several reflective sides, a clear top, and a black pot inside to absorb the sun's heat. By concentrating sunlight, they provide slow simmering, baking, and roasting—on a sunny day, beans, rice, stews, and the like will be fully cooked within two to five hours. Some parabolic designs can generate temperatures high enough to grill and fry food quickly, but these designs require more sensitive engineering, and thus are usually prohibitively expensive; they also require extreme care to prevent user burns and fires. Even an ordinary solar oven is often too expensive for most poor rural users, who also require a wood cookstove for rainy days. However, solar ovens create no toxic smoke and use no fuel but sunlight, saving not only time but money; some households using the technology have cut their fuel costs in half. JJF & EZ

The Pot-in-Pot Refrigerator

How do you keep food from spoiling when you don't have access to ice or electricity for a refrigerator? Millions of people still simply try to eat their food before it goes bad, which keeps them gathering food day-to-day and often makes them sick. But one solution is simple enough and cheap enough for anyone to afford: thermodynamics. When moisture on a surface evaporates, it sucks up heat, causing that surface's temperature to drop. This is how most animals, including people, keep themselves cool.

Mohammed Bah Abba of Nigeria invented a kind of refrigerator that uses this principle, the Pot-in-Pot. As the name suggests, the refrigerator is made up of two clay pots—with water-soaked sand in between. The outer pot is porous, so the water between the pots slowly soaks through, and evaporates on the outside. This keeps the inner pot cool, preserving the food inside. The impact on Nigerians' lives has been substantial. The Pot-in-Pot can keep foods like tomatoes and peppers fresh for three weeks, which means that farmers have been able to store food for later sale. Beyond food freshness, the Pot-in-Pot has had economic and social benefits for the community: Because they have a place to keep inventory fresh, married women have been able to start side businesses selling food from their homes, helping them gain financial independence. Because girls no longer have to spend their days selling food before it spoils, female school enrollment has increased. And because the Pot-in-Pot is locally manufactured, the local pottery industry is being revived, which is crucial in a region with a high unemployment rate. JJF

RESOURCE

Flex Your Power
http://fypower.org
The Flex Your Power Web site is one-stop-shopping for energy efficiency. Although it's designed with California in mind, it's a powerful resource that anyone can learn from. You can access

information on how to buy the most efficient washer, dryer, dishwasher, refrigerator, furnace, air conditioner, or hot-water heater; how to light your home with natural light and compact flourescents; how to insulate and weatherize your home and cool your roof; and how to save water using ultra-low-flow showerheads and toilets. And that's not all. You can also learn how to find and qualify for rebates and tax credits; how to get a home-energy audit; how to get free project design assistance; and how to find demonstration models in your area. There are even tips on how to proceed if you're on a tight budget. Every state, province, and city in the world ought to have its own version of this wonderful site.

Green Power

Most citizens of the Global North rely on the power grid. That is, wires run from our homes to utility poles and power lines that bring us electricity from faraway power plants, which burn coal or oil or use nuclear radiation. We buy this power from a company or from our local government and, usually, we don't think much about what kind of energy we're supporting or what alternatives we have.

But we do have options. Our power utilities can use green energy, and we can use our leverage as consumers to encourage them to do so. What's more, a number of new, innovative power sources are on the way, offering the possibility of clean, diverse, abundant, and inexpensive energy. AS

Buying Green Power

Energy is so ephemeral that it's hard to think of it as a consumer product. But it is, and not all energy products are created equal. Choosing to buy green power generated from renewable, low-impact sources such as solar panels or wind farms is as basic and important to energy conservation as turning off the lights when you leave the room.

You can do your part by buying energy from utilities that generate their power directly from renewable or zero-emissions sources. Bright green utilities are still few and far between, so it's likely that you'll have to choose the green-power option from your standard utility company—meaning you pay slightly higher rates to enable your utility to purchase a certain amount of their power from a renewable source. You don't directly receive that renewable energy, since it all goes into a kind of melting pot. But your utility will use your fees to build new wind

farms, solar arrays, or other systems, which helps transition the electricity infrastructure away from the oil economy.

If there are no renewable sources or green-power programs in your area, you can still do your part by purchasing renewable-energy certificates (RECs, also known as green tags). By doing so, you are funding the advancement of renewable power in other areas that have such programs.

How do I find green power? Organizations like the U.S. Department of Energy and the Environmental Protection Agency maintain online databases of green-power suppliers by state, as well as suppliers who offer green tags. Comparable resources can be found in Europe.

How much does it cost? Usually, green power costs slightly more than nonrenewable power costs. Your utility will either add a cent or two to the standard per-kilowatt-hour rate or will charge an additional flat monthly fee of five to ten dollars. However, in some places, state-sponsored incentives and subsidies mean that buying green is actually cheaper.

How much should I buy? Buy the maximum percentage you can afford. The bigger the demand, the greater the amount of power the utility company will draw from a renewable source.

Are some green-power products better than others? Yes. All renewable resources have some impact on the environment, but to varying degrees. Wind power, for example, has much less of an impact than solar power does. The Power Scorecard (http://www.powerscorecard.org/), developed by a coalition of environmental defense organizations, is an online assessment tool that

The Empire State Building as seen from New Jersey during the 2003 North American blackout.

will help you evaluate the impact of all types of energy available to you.

How do I know that a green-power program or utility is legit? The Center for Resource Solutions in San Francisco has created the Green-e Renewable Energy Certification Program to help consumers determine whether they are doing business with a reliable utility. The certification guidelines were developed by a group of environmental-industry experts.

The Cost of Green Power

A popular myth holds that renewable power is always more expensive than fossil-fuel power. Not so: under the right conditions, good wind farms already generate electricity for three cents per kilowatt-hour, which is almost as cheap as the cheapest energy source around—coal. Energy subsidies, or carbon taxes and credits can make wind power the cheapest energy source available. Solar photovoltaics are expensive compared to the grid, but in remote locations they are often cheaper than running power lines. Ten years from now, the cost of wind and solar power are projected to be one-tenth of what they are now, owing to technology and manufacturing improvements.

Another myth has to do with "intermittency," the idea that renewable energy sources cannot provide power reliably. It's true that photovoltaics are ineffective at night, and wind turbines are nonproductive in still air, but when used together, alternative energy sources can deliver a steady flow of power.

No one source of renewable, clean power is going to single-handedly replace our current energy infrastructure. Instead, we're going to see a mix of technologies, policies, and systems. This will make the resulting energy grid more flexible—but the political path to this future will be challenging. JJF

Solar Power

The sun sends a tremendous amount of power to every square meter of the earth every day. Though clouds and shade can reduce the incoming energy, and though there's no incoming power at night, solar power is a great renewable resource. There are two types of solar-energy systems: solar-thermal systems collect radiant energy to produce heat; photovoltaic-cell systems (those large glossy roof panels) convert direct sunlight into a stream of electrons to produce electricity. Photovoltaic (PV) systems have been around since the 1970s, but they're still fairly expensive, so they're mostly used in off-grid applications. However, as new materials come on line over the next ten to twenty years, prices should drop and make PV systems competitive in grid-connected applications. Flexible thin-film and organic-plastic solar panels are extremely rugged and adaptable to an enormous variety of applications, from building materials and gadget-holding cases to energy-producing backpacks and laptop bags. JJF

Left: A roof-mounted, grid-connected photovoltaic system on the Metcalf Federal Building, Chicago, Illinois.
Opposite: This Dish Stirling solar-power system employs mirrors that focus sunlight onto a thermal receiver to drive an electric generator.

Solar Stirling Engine

The world's largest solar-power facility won't use a single photovoltaic cell. Instead, the 4,500-acre (1,821-hectare) solar farm that Southern California Edison plans to build near Victorville, California, will use technology that's nearly two hundred years old: the Stirling engine.

Stirling engines are "external combustion engines" whose cranks are pushed not by explosions in their chambers, but by the expansion of a gas that's heated from the outside of the engine. The gas itself stays in a closed loop, continually being heated, expanding, using up its energy to push a piston, and condensing again. The external heat source can be anything; the Stirling Energy Systems's product Dish Stirling uses sunlight focused by a parabolic mirror array.

Stirling has been working on solar-power generation for twenty years. Their solar dishes are extremely efficient at converting solar energy into grid-quality power. If all goes well, the 20,000-dish Victorville system will start generating power as soon as the first unit is plugged into the grid and should be fully on line by 2010. Power from this system is expected to run six cents per kilowatt-hour, making it competitive with other sources. JC

China's Solar Boom

According to Worldwatch, China's two largest solar-cell producers are now boosting their production capacities to more than three hundred megawatts worth of photovoltaics every year—that's more than twice as much as the United States makes. At that production rate, by 2020 China could be producing more than double the solar power projected in its renewable energy plan. There's no reason to think that the manufacturing rate won't keep rising.

One reason China is likely to keep making more solar panels is to benefit from the growth of new markets. According to an August 2004 article on SciDevNet.com, Xi Wenhua, the Director of China's Institute of Natural Energy, stated that over the next five years, the country intends to train 10,000 technicians from the developing world on the deployment and use of solar-power technologies. Rather than trying to enter American, Japanese, and European markets, China is apparently seeking to become the business partner of choice for the "leapfrog" [see Leapfrogging, p. 292] nations that need power to support development but can't afford or don't want to buy more fossil fuels.

The value to both China and the developing nations is evident: China gets larger markets for its solar technology and builds relationships with new markets that could last decades; the developing countries get experience with useful technology and the beginnings of a power infrastructure well suited to the diverse and distributed twenty-first-century electricity networks. JC

Workers clean the evacuated glass collector tubes of a solar-thermal system at the Tsinghua Solar Corporation, Beijing, China.

Wind Power

Wind power is the cleanest alternative energy source; harvesting wind via windmills or turbines emits no air pollution. Large wind farms generate the most wind power, but small clusters of megawatt-range utility-scale turbines are popping up. This is especially true in the upper Midwest, where the turbines have received significant public policy support. In addition, farmers are starting to use smaller turbines to power their farms.

The wind's potential to generate electricity is like a free gift waiting to be unwrapped. Wind is plentiful all around the world. It's clean. It'll never run out. So why aren't we letting the wind run our entire planet?

The answer is simple: price. Wind power has historically been more expensive than power from coal, oil, or natural gas. But wind power is growing (from 4,800 megawatts generated in 1995 to 59,322 megawatts in 2005, according to the Worldwatch Institute), and as more people build wind farms, the price of wind power drops. In fact, some recent wind projects are cranking out electricity that is priced competitively with coal and oil, and far cheaper than nuclear power, with none of the radioactivity or greenhouse-gas pollution.

The Battelle Pacific Northwest National Laboratory estimates that as wind power's price drops to a competitive level, wind could quickly supply 20 percent of the nation's electricity. Many other researchers think that, given consistent regulation, equal access to new transmission lines, and more modern power grids (which need to be built anyway), wind power could supply a third of the nation's electricity by 2020—and do so economically. The American Wind Energy Association goes even further, claiming, "North Dakota alone is theoretically capable (if there were enough transmission capacity) of producing enough wind-generated power to meet more than one-third of U.S. electricity demand."

The potential worldwide may be even more impressive. The U.S. Department of Energy has concluded that the world's wind could generate more than fifteen times as much energy as the world is currently using, while a 2002 Danish study sponsored by the European Wind Energy Association, "Wind Force 12," found that with even comparatively modest technological advances and policy support, wind could supply 12 percent of the world's electricity by 2020.

The magnitude of wind power's potential can be seen in two recent projects in the United Kingdom, both of them offshore wind farms positioned in the Thames estuary (offshore wind farms generate more power than land-based ones). Kentish Flats, the largest UK wind farm thus far, began operating in December 2005. Its thirty wind turbines can generate enough power for 100,000 homes while reducing carbon dioxide emissions by 245,815 tons (223,000 metric tons) per year.

But London Array, which is still in the planning stages, will easily steal the spotlight. The project expects to use 270 turbines to power more than 750,000 homes, a quarter of them in greater London. At 1,000 megawatts, London Array would prevent the release of more than 2 million tons (1.8 million metric tons) of carbon dioxide every year. If the project goes through, it will be fully operational by 2011.

Flying Windmills

Wind turbines are constantly getting taller, because the higher they are from the ground, the faster the wind speeds they can catch. With

A wind farm in Kansas.

just 1 percent of its wind power, the jet stream (15,000–35,000 feet [4,572–10,668 meters] in altitude) could supply all of the United States' electrical demands. But building big towers is expensive, especially if we want one 15,000 feet tall. So why not ditch the tower and make the windmill fly?

Several companies and people are trying to do just that. Sky WindPower is the furthest along in its research, with functional prototypes already tested in the field. The corporation's chairman, Bryan Roberts, an Australian professor of mechanical engineering, teamed up with some Americans to commercialize his Flying Electric Generator—a windmill that's tethered to the ground, but that flies like a whirligig in the jet stream.

There are two immediate advantages to these flying wind farms. The first advantage is that the wind is much steadier at altitude—so they would outperform ground-based wind farms, which only operate at their peak capacity 19–35 percent of the time, owing to wind intermittency. The second advantage is ad hoc generation: devices with a reasonably simple tether system do not have to be permanently installed in one place, so they could be trucked out to any location that needed them.

There are a few drawbacks to Sky WindPower's design, however. In an electrical storm, the power-carrying tether could become an enormous lightning rod. The fliers would have to be brought down before a storm worsened, and couldn't be sent back up until the storm passed. Safety is also a concern, though Sky WindPower has plans to address that. The units could be located away from population centers and flown only in restricted airspace so that they would pose no threat to airplanes. According to Sky WindPower, the entirety of the United States' power needs could be generated by units that would fit in one four-hundredths of the nation's airspace.

Another company, Magenn, has created a more modest, more feasible design. The inventor, Fred Ferguson, is a Canadian engineer who envisions a range of devices, from units small enough to be tethered to a boat or RV, to models that can generate power for the average home, to models sufficient for megawatt-range power plants.

Magenn's design is radically different from other windmills on the market: it would not use propeller blades. Instead, it would be a helium blimp with scoops that rotate around a horizontal axis. These would send energy into two generators that would then send the power down a tether. The blimplike design has several advantages: it could operate in winds as low as two miles per hour (conventional tower turbines require winds of at least three miles per hour); it would be safer in a crash, because it would fall slowly and be mostly made of flexible material; it would be safer for airplanes, because it sits below legal airspace; it would be safer for birds, because the moving parts are visible and travel with—not perpendicular to—the wind; and it would be a less risky investment than one of Sky WindPower's devices, because it is smaller, cheaper, and easier to build. There are tradeoffs, though—Sky WindPower's flying windmills, positioned in higher and more constant winds, would be better suited to large applications than the Magenn design would. JJF

Left: Offshore wind turbines at Kentish Flats, United Kingdom.

Right: One of Magenn's flying wind turbines.

Opposite: Wave power represents a vast, untapped energy source.

Tidal Power

Few people are familiar with "hydrokinetic" energy, the use of water motion to generate power. The potential is great, but commercial systems are only now starting to be developed. Questions remain, too, as to the impact of hydrokinetic-power systems on ocean environments. However, hydrokinetic-power systems don't clutter scenic areas like wind-power systems, and they are far less intermittent than either wind- or solar-power systems.

The power of the ocean's constant movement can be harnessed by using the flow of water from tides or the kinetic energy of waves. Generating power from the flow of tides is simple, but less benign than other methods. The power facility involved resembles a hydroelectric dam, built across an estuary. At high tide, water flows freely into the estuary, but when the tide turns, the water can flow out only through a hydroelectric turbine.

In theory, it is also possible to generate power from the ocean's temperature fluctuations, but so far inventors have not figured out how to do it. Oceans absorb a vast amount of thermal energy from the sun—the approximate equivalent of 250 billion barrels of oil per day, according to the National Renewable Energy Laboratory. Only a small fraction of this could be extracted without impacting ecosystems, but with the right engineering, hydrothermal power could be a constant, reliable, minimal-impact power source. Though the idea has been around since the 1930s, ocean-thermal-energy conversion plants have yet to emerge from research labs. Many test projects have demonstrated the concept's effectiveness, but a financially viable system has yet to be produced. JC & JJF

RESOURCE

Wind Power: Renewable Energy for Home, Farm, and Business by Paul Gipe
(Chelsea Green Publishing, 2004)
Wind energy has been growing at double-digit rates for many years, and—as a proven and cost-effective renewable technology—is likely to do so for many more. In *Wind Power,* Paul Gipe provides a realistic assessment of issues involved in building and running a wind-energy installation, including primers on both engineering and practical issues.

According to Gipe, "Today wind energy is a booming worldwide industry. The technology has truly come of age, and with today's heightened concern about our environment, this resurgence of interest is here to stay. Despite wind energy's success, there remains a need for a frank discourse on how to wisely use the technology."

Going Off the Grid

When we're connected to the power grid, we can attempt to influence our local energy utilities only by writing an angry e-mail when they do something irresponsible (like investing in the construction of a new coal-burning power plant), or by voting with our dollars and buying clean energy (if it's offered). Ultimately, though, the power company calls the shots.

Dissatisfied with the status quo, some people have decided to take their business elsewhere. They find a supplier of home-energy systems and get themselves off the grid.

If houses with solar panels on their roofs and wind turbines in their backyards make you think of communes and hippies, your mental picture is out-of-date. Anyone with a bit of a do-it-yourself mindset and a little disposable income can benefit from installing a home-energy system. These setups can save you real money over the long term and provide most or all of your power in clean, homegrown ways.

For some of us, going off the grid isn't a matter of home improvement or rebellion against price-gouging utilities—it's the only option. For people in developing nations who don't have a grid nearby or can't afford to be connected, the advancement of off-grid technologies has been a revelation, creating the opportunity to leapfrog [see Leapfrogging, p. 292] past grid-bound power sources and generate energy independently. AS

Buddhist monks carry solar panels to their monastery in a remote Himalayan village, Ladakh, India.

Home-energy Systems

Of the three main sources of clean, renewable power—solar, wind, and tidal—only the first two are reasonable options for home use. (Ocean power appears to have limited home application, though a wave-powered beach house would be great.) We can add building-integrated solar-photovoltaic shingles, rooftop panels, or wall/window units, and we can install rooftop or wall-mounted wind turbines.

Solar photovoltaics are the most common form of home-generated power, and scores of books describe them. Many homeowners are proud to have solar panels bolted to their roofs; those who want the energy benefits without the bolt-on look can now use building-integrated photovoltaics (BIPVs). Most of these are "solar shingles," which do indeed cover your roof like shingles. The solar roof tiles made by Sharp are full replacements for concrete roofing tiles, providing the same protection against leaks and impact and sized compatibly with conventional roof coverings (one solar roof tile equals five concrete tiles). The modules generate sixty watts apiece, at around 12 percent efficiency. Alternatively, the tiles made by SunPower are not shingle replacements, but they are more than 21 percent efficient, work in low-light conditions, and look nicer.

Wind power is beginning to scale down for home use. Windsave and Renewable Devices are two UK manufacturers who sell small wind turbines designed for the home. The Swift windmill from Renewable Devices is particularly attractive because it is specifically designed to reduce the noise and the potentially building-damaging vibrations that older small wind-turbine designs are known for.

A small turbine is enough to offset a good portion of electricity use in a typical home. But when a microturbine is combined with a building-integrated solar system and high-efficiency consumption—relatively inexpensive home additions—it can turn into a net producer of power. Best of all, these devices can be retro-fitted to existing buildings. JC

Bright Green Meters

Most meters in your home are only there to determine how much money you owe to some company or another. But a new breed of reversible meters could actually make you money. "Net metering" allows surplus energy from your solar-power system to flow back into the grid. You get credit for contributing to the grid, which can greatly offset your electric bill, depending on how much power your system is producing.

Net metering is cheap and simple, and it benefits both you and your utility company. You get maximum value from your system, especially during the day, when you're not likely to be around to use up all the power your system is cranking out. The utility company has to worry less about system overloads when its customers are feeding the grid, particularly during peak times.

Net metering is catching on—thirty-five U.S. states offer it, though their policies vary. Canada, Japan, Germany, and Switzerland also have net metering.

The technology has one drawback: if the power grid goes down in your area, you can't draw power from your home system, even if the afternoon sun is making your solar shingles output power like crazy. Fortunately, a company called GridPoint makes a device that serves as a combined "inverter" (for connecting your solar-power system to the grid) and battery backup. It draws enough power from your solar panels to keep a stash of batteries charged up, so when the grid goes down, you can still run your home. JC

An Off-the-grid Success Story

San Franciscan Brian McConnell retro-fitted his three-bedroom home to generate as much of its power as possible. His goal was to reduce his home's footprint by 80–90 percent, and thanks to a mix of systems and basic energy conservation, he's coming very close to that goal.

As McConnell explains in an O'Reilly Network article, the biggest component of his home power plant is a net-metered solar-electric system, consisting of eighteen solar panels, which were installed on his roof by a contractor. The system generates 65–70 percent of the house's electricity; at midday the system usually generates so much energy that power flows back into the grid. The initial cost of the system was $25,000, but after state rebates the cost dropped to $16,000. McConnell was able to roll the cost of the solar setup into his mortgage. The post-tax payment on a system that adds tremendous value to his property is about $50 per month, less than an average electric bill.

The next big project was to offset heating costs. The house uses a forced-air natural-gas heating system, to which McConnell added a solar-heat system on the roof. Forced-air solar heaters are simple devices, basically insulated boxes. Air ducts move air across a piece of polycarbonate plastic inside the box and pull the warm air into a home's rooms. McConnell was

able to position the three heating units so that each would heat a different zone of the house. This system is much more efficient than the standard heating systems that only send occasional bursts of hot air into a room—with a solar-heat system, which runs constantly as long as there is daylight to power it, every object in the house is heated to 70 degrees Fahrenheit (21 degrees Celsius) and continues to radiate heat after the sun goes down, keeping temperatures stable for several hours.

Once the house's main energy needs were attended to, McConnell tackled his backyard hot tub. To make this luxury greener, he installed a small solar water heater. The unit captures several kilowatt-hours of heat per day, so the hot tub's electric heater doesn't have to turn on as frequently to keep the water warm.

McConnell also reduced his energy consumption using basic home-efficiency techniques: installing motion-sensor switches in high-traffic areas, replacing incandescent bulbs with compact fluorescents, and replacing old appliances with newer, energy-efficient models.

This experiment has both reduced the house's footprint and increased its value (every dollar saved on annual utility costs increases home value by an estimated twenty dollars). Greater energy independence protects Brian McConnell from fluctuations in energy costs (which are likely to rise as fossil fuels become more limited) and protects the neighborhood from unsightly development. San Francisco's zoning laws protect properties with rooftop photovoltaic systems—neighboring property owners cannot build anything that will cast a shadow on solar panels. So it's very unlikely that McConnell will have to worry about a high-rise going up next door. CB

Opposite, left: Using the latest home-energy systems, it is increasingly simple to get our power from the sun. **Opposite, right:** A home wind turbine from Renewable Devices in action.

The Sun Shines for All: Making Solar More Affordable

The World Bank estimates that in Brazil, 20 million citizens still lack power, but Fabio Rosa, a social entrepreneur, has started to bring solar energy to rural communities across the country.

Rosa knew that poor Brazilians could not afford to rig solar-power systems on their own. He felt that, instead, people should be able to pay for solar electricity as they used it. Rosa conducted a market survey and discovered that the average impoverished Brazilian spent about thirteen dollars per month on energy needs, mostly on candles, batteries, and kerosene. So he started renting out solar panels and lighting fixtures at a price of thirteen dollars per month—it's affordable, it's a sound business, and Rosa's initial investment will be repaid in four years.

Rosa's rental business is called the Sun Shines for All. He has also created a nonprofit organization, the Institute for Development of Natural Energy and Sustainability, to deliver electricity and services at subsidized rates to those very poor families who spend less than ten dollars per month on their energy needs. He has built up a network of electricians to install and maintain the solar-electric systems. He's still getting his efforts off the ground, but so far two hundred homes are up and running on solar power.

Micro-hydro

In Nepal, only about 10 percent of the population is on the grid. The rest of the population goes without power, unless they produce their own. However, the mountainous country has plenty of fast-running streams, fantastic sources of energy. Small-scale water-power operations can only provide energy to limited areas, but this type of power is the cleanest, cheapest, and most practical form of energy available in those locations. In most remote Nepalese villages, electricity from water power is even cheaper than buying batteries.

Waterwheels have been used in Nepal for centuries to mill grain and dehull rice, just as the waterwheels of preindustrial Europe were. But now some of these wheels are being retrofitted with generators to provide electricity, too. In some places, entirely new generators are being installed. These systems are called micro-hydro systems, because they are generally capable of powering only a few houses or a small village, and they can be plopped down in a river without significantly affecting its flow, banks, or general ecosystem, as a large hydroelectric dam would. JJF

RESOURCES

Smart Power: An Urban Guide to Renewable Energy and Efficiency by William H. Kemp (Aztext Press, 2004) and ***The Renewable Energy Handbook*** by William H. Kemp (Aztext Press, 2005)
Kemp's pair of how-to books thoroughly outlines our power options, whether we want to live totally off the grid in a rural area, or just make smarter decisions pertaining to our urban dwellings. Despite its subtitle, *Smart Power* offers suggestions for urbanites and suburbanites alike, starting with learning the basics of home efficiency and fanning out to building small-scale solar-power or wind-power systems. *The Renewable Energy Handbook* focuses on larger-scale projects in less dense areas, and discusses both new home designs and retrofits, providing case studies of off-the-grid homes. There's also a chapter on solar-powered pools and hot tubs for good measure. Pictures and diagrams in both books help even those of us who don't know an incandescent from a compact fluorescent make sense of our energy options.

Solar Living Source Book by John Schaeffer (New Society Publishers, 2005)
Real Goods is the godfather of sustainable energy retailers. Since 1978, it's been cranking out ideas for living lightly, with a special emphasis on solar, wind, and other clean energy products. Now partnered with the companies Gaiam and Jade Mountain, Real Goods is the first place you should look if you're considering a home-green energy system, and the store's Web site (www.realgoods.com) is a critical comparative-shopping resource. The *Solar Living Source Book* explores solar how-tos with greater authority and clarity than anything else out there, and if you're thinking seriously about getting off the grid (or feeding green power back into it) this guidebook belongs on your bookshelf.

According to Schaeffer, "In wind energy, size, especially rotor diameter, matters. Nothing tells you more about a wind turbine's potential than its diameter—the shorthand for the area swept by the rotor. The wind turbine with the bigger rotor will intercept more of the wind stream and almost invariably will generate more electricity than a turbine with a smaller rotor, regardless of their generator ratings."

FindSolar
http://www.findsolar.com
Look up your state and county, and FindSolar can return information on your solar-energy potential and on various incentives to install solar technology, and tell you exactly how a solar-power system (photovoltaic or direct heat)

Micro-hydro systems can make sense in remote places like this agricultural cooperative in Cashel Valley, Zimbabwe.

would work for you. A calculator displays the estimated costs of a system, estimated monthly and annual savings, and even how many tons of carbon dioxide your system would save annually. The Web site also provides a database of solar-power professionals, organized by location.

Smart Grids

The "conservation bomb" [see Using Energy Efficiently, p. 164] can make our lives radically more efficient; buying green energy can spur our power companies to invest in a new generation of renewable power technologies; and getting off the grid can turn our homes into little clean power plants. Individually, these are all great steps toward smarter energy use. But what if we put them all together? What if a new kind of power grid could weave together improvements in energy performance, a variety of big green-energy sources, and home-energy systems while maximizing the effectiveness of each?

Meet the smart grid.

The energy grid we grew up with is outdated. A map of the North American power grid shows hundreds of thousands of miles of power lines woven into one another, linking areas larger than most European countries. Power plants at one end produce huge amounts of energy (and greenhouse gases), which get sent down the pipeline to a comparatively small number of substations. These substations mete out this bulk energy to the lines that run into our homes. The slightest disruption in one substation can cause a domino effect that trips outages in multiple states. For instance, in August 2003, an overloaded transmission line in northwestern Ohio took down some power lines and started a cascade of outages that ultimately left 50 million people from Toronto to New York City without power, many for a full day.

Incidents like the 2003 blackout illustrate the limitations of a system that was developed before the age of the microprocessor, yet the power industry's response to such debacles is to call for more and bigger power lines. This approach is

as impractical as it is unsustainable: not only would it fail to shore up the weak links, but it would lead to more greenhouse-gas emissions and the sacrifice of more wildlife habitats to new power-line corridors.

Until recently, the only alternative to supporting this system was to go off the grid, which meant building independent, expensive home-energy systems and dealing with their flaws and fluctuations. But avoiding the grid altogether is no longer necessary—the grid is about to evolve, and those of us who've stayed on it may still get to enjoy a sustainable, and high-tech, shared utility system.

Power poles, emissions-spewing plants, and miles of wires: out. Computers: in. An infusion of digital controls and sensors is making the grid "smart," able to identify surges, downed lines, and outages, and to implement damage control before engineers even realize there's a problem. Better yet, the new digital system is interactive: you can regulate your own utilities, feeding power back into the grid from home-energy systems, and maybe zeroing out your power bills.

Smart grids and distributed energy are central to a bright green future. By decentralizing power generation and adding digital intelligence to the power network, we can build an energy infrastructure that's more flexible, better able to take advantage of renewable energy technologies, and more resilient in times of crisis. PM & AS

Smart Grids: More Reliable and More Sustainable

Our current grid is highly susceptible to natural and human-made disasters. What's more, this fragile grid is currently protected only by a small cadre of operators sitting in control rooms, who during crises can become overwhelmed by the grid's equivalent of the fog of war.

By contrast, a smart energy network is resilient in the face of troubles, capable of rapid recovery from disasters—its ability to adapt parallels nature's. Grid stability in a smart energy network is the task not of a few overworked humans, but of millions of distributed software agents located throughout the system. "Cops on the beat" in grid-watch posts will still monitor the overall system, but they will have the aid of "eyes-on-the-street" software agents on the lookout for trouble and ready to respond. Those agents will automatically check failures that otherwise might cascade through the system.

In Ashland, Oregon, the municipal utility controls power demand peaks, thus eliminating need for overbuilt power infrastructure by sending signals to homes through the cable system. Energy-hungry appliances such as hot-water heaters and pool pumps automatically turn down to consume less power. If that proves inconvenient, users can jump online and override the low-power mode.

With a smart grid in place, distribution-level software agents can respond nimbly when a storm with high winds downs lines and blacks out power. The agents quickly detect the troubles and report the location of broken lines to repair crews, enabling line people to speedily restore

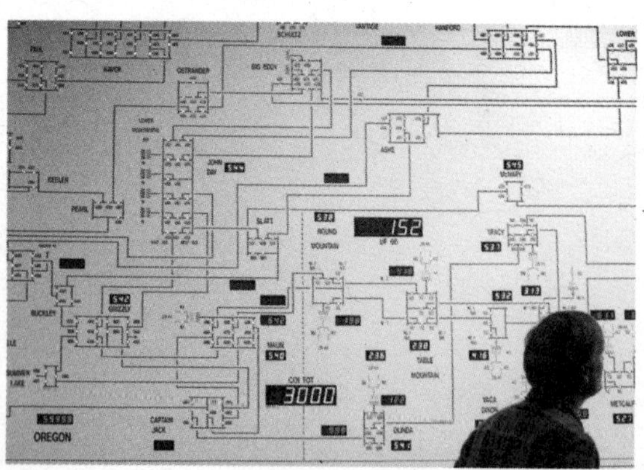

service. As juice begins to flow again, smart appliances turn back on in a staggered fashion rather than all at once, avoiding demand surges that could shut down power all over again.

Smart energy will be cleaner energy. Think of smart grids as "energy webs." Less advanced systems can't handle multiple contributions from outside renewable-energy sources like home-solar outfits or small-scale wind farms, but smart grids can: the smarter our grids get, the greater the range of energy producers they can unite.

Smart grids also make it easier for consumers to buy green energy. With a smart grid in place, power companies could be required to report the emissions coming from each of their plants, as well as their asking price for the energy produced at those plants. Software could then sort and catalog the data and automatically hook consumers up to the cleanest source they can afford. Eventually, entire regions could be giant virtual energy markets, where we buy energy online, getting better prices and cleaner power. PM

Smart Appliances

In your home, you may already have several "smart" appliances—machines that automatically shut off or go into a sleep mode to conserve power. These appliances use built-in features that would react the same way whether they were plugged into the wall outlet or connected to some space-age power supply.

But a new generation of truly intelligent appliances is emerging. These machines actually interact with smart grids to help protect both your wallet and the grid itself.

For instance, the GridWise project, a collaboration between the U.S. Department of Energy's Pacific Northwest National Laboratory, Whirlpool, and IBM, has already developed machines that can detect drops in power from the grid and cycle down just enough to reduce power demands. The shift is virtually imperceptible to consumers using appliances at home, but it eases the burden on the grid enough to prevent a crash.

The GridWise project could be the first concrete sign that a power revolution is at hand. Eventually, interwoven networks of machines, grids, and power sources might pay constant attention to one another, keeping the grid working and safe. In a yearlong experiment in smart power distribution, GridWise connected hundreds of homes in the cities of Yakima, Washington, and Gresham, Oregon, to a new intelligent-power network, combining real-time monitoring of consumption and pricing, Internet-based usage controls, and smart appliances.

GridWise households have access to their energy-use information through both software and Web sites. This kind of transparency is key to the project's success. The grid does a lot to cut energy consumption, but GridWise participants also have to change their energy habits to maximize the benefits of hooking up to an integrated grid. Consumers are most likely to make long-term changes when they can see the direct results of their choices—end-of-month summaries and community consumption averages are simply not enough. Researchers are confident that the success of GridWise will help make the project a standard part of the regional power grid. SR & JC

Opposite, left: Eating lunch at a food stall beneath power pylons, Beijing, China.

Opposite, right: At the California Independent System Operator, which manages the flow of electricity in that state, a map board shows a jump in electricity consumption due to a heat wave.

Water

No matter where we live, there is a water crisis unfolding around us.

For some, this crisis is a matter of life and death. Globally, one in five of us does not have access to clean drinking water. Two in five do not have adequate sanitation facilities—water to wash with and to clean with, or even a safe place to answer nature's call. Hundreds of millions of farmers lack adequate water and/or the proper tools for irrigating their fields. Many more are struggling with the effects of the mass privatization of water.

Those of us in the wealthier parts of the world have our own looming problems. Many aquifers—huge underground lakes, filled up in trickles over centuries or millennia—are being pumped dry so quickly that they will be completely emptied within our lifetimes.

Meanwhile, worldwide, too many people with too many needs using water too wastefully have pummeled natural systems that depend on a steady flow of water. Some of what were the largest lakes in the world, such as the Aral Sea and Lake Chad (once the fourth-largest lake in Africa), have almost completely dried up, and rivers around the world are shrinking as water is pumped from them (the Rio Grande, for example, now sometimes fails to reach the Gulf of Mexico before going dry). Even in the rainy Pacific Northwest, once-abundant salmon runs have been brought to the edge of extinction as huge demand for water has lowered river levels. Add to these problems nearly ubiquitous water pollution from farms, factories, and cities, and nature is in for a hard slog.

Climate change already exacerbates these problems, yet even worse problems are still to come. Rainfall patterns that have held more or less constant for generations are shifting, making some areas drier and deluging others. Depending on where we live, we can expect more droughts, more heavy downpours, or both.

If we are to navigate this crisis, we must start looking at water in a whole new way. We have to save water every chance we get. Just as we need to find more efficient ways of using energy, we need to find more effective water technologies. Many already exist, and many more are on the way—but the speed with which we adopt them will have much to do with our success or failure in meeting this crisis.

We should see all the water coursing through and around our homes and businesses as a valuable resource worth conserving. The rainwater that falls on our roofs, the "gray water" [see EcoHouse Brazil, p. 201] from our showers and sinks, the water used by industry and farming: with the right tools, all this water can be cleaned and used again in a variety of ways.

Much of the world's population, like these children in Uganda, still lacks access to safe drinking and bathing water.

Finally, we need to start thinking of our homes and neighborhoods as part of a natural system, part of the water cycle of plants and streams, rain and runoff. By emulating natural processes in our home systems, we relieve much of the pressure we put on our water resources—and we save money and energy. Neighborhoods that work with natural flows will make us not only healthier and more comfortable, but ecologically friendlier and richer.

Innovations in water use benefit not only the Global North, which has the resources to invent and test them, but also the Global South, which desperately needs better tools. The entire world is in a water crisis—but the search for solutions begins wherever we are. The next time we hear rain falling on the roof, we might consider it an invitation to begin looking. AS

Conserving Water

In the Global North, we enjoy the convenience of turning on a tap and seeing the water flow. Most of us get our water from large water suppliers that pump it from the ground or pipe it in from a river or lake. Everywhere, these large water providers are hard-pressed to deliver all the water customers would like to use.

The most direct way for any of us to confront the water crisis is to use less water. That doesn't mean we should measure out our lives in thimbles, or take showers with stopwatches. Instead, it means making the systems in our homes work for us. When we install the right technologies, a little bit of mindfulness goes a long way.

Low-flow Water Fixtures

Low-flow showerheads are not as wimpy as they sound—by forcing air into the water flow, they reduce the amount of water used, but still produce a strong spray. If we install high-efficiency showerheads along with their cousins, faucet aerators (same concept, same benefit), we stand to cut our water consumption considerably. According to the American Water Works Association, if every U.S. household installed low-flow fixtures and toilets, we would save approximately 5.4 billion gallons of water per day. Installing a low-flow toilet may save us money not only by reducing our water bills, but by entitling us to a rebate from our utility company for conserving water. In early incarnations, low-flow toilets didn't live up to the hype concerning their environmental benefits. They didn't work well, and nothing is less appealing than a semifunctional toilet. Manufacturers who recognized that it takes a great product to make consumers switch from their usual brand accelerated their work

and made significant improvements in low-flow toilet design and functionality. Now consumers have access to a moderate selection of low-flow toilets that work as well as an ordinary toilet while saving water and money. SR

Recycling Shower

To encourage water conservation efforts without depriving us of the pleasure of long showers, design student Peter Brewin, from London's Royal College of Art, developed an ecofriendly shower that recycles water, drastically reducing consumption and saving consumers a pretty penny on their utility bills.

As it runs, the water-saving shower immediately recycles what runs down the drain, cleaning it and sending it back through the showerhead. Once filtered, the water goes to a reheating device that brings it up to a comfortable temperature before spitting it back out. If the recycling shower were widely adopted, it could save municipalities a shocking amount of water—something that draught-prone areas could sorely use. The shower design incorporates a number of appealing features, such as chlorine filtration and a meter for determining water consumption, but the most attractive element is the financial savings that consumers gain by recycling their water as they bathe. SR & JF

reHOUSE/BATH

If low-flow fixtures are good, and recycling showers are better, then reHOUSE/BATH is the next step up the ladder toward the best conceivable level of home water efficiency. Although it does not allow for the occasional indulgence of a long, hot shower, reHOUSE/BATH does make a point: if we could see how much water we really use, we would probably use less—especially if our bathroom were set up to prevent excess consumption.

A group of French designers created reHOUSE/BATH as an exploration of sustainable bathroom design. To use it, a bather sits on a bench over a basin and washes using only a bucket and sponge. Sounds basic, but the basin is just one element of an intricate web of distribution. From the basin, a series of hoses deliver used bathwater to plants in ceramic pots. The size of the plants determines the bather's allotted water supply. As the plants grow, fed only by wastewater, they form an organic screen, to shield the bather. Perforations along the top of the 1.3-gallon (5-liter) bucket set an unforgiving limit on the amount of water the bather can use, by threatening to flood the floor with anything above the maximum fill limit.

Most of us would rather not regulate our water use so mercilessly, but reHOUSE/BATH raises important questions about where our water goes after we wash with it, and what might be possible if we captured and reused it. RD & SR

RESOURCES

Water Use It Wisely
http://wateruseitwisely.com
Want tips on reducing your water usage? This site overflows with them, offering lists tailored to people living in various regions of the United States (though many would apply anywhere). When trying to waste less water, it's the systemic things that produce the biggest impacts, but you can do a lot just by paying attention. This site shows you how.

H_2Ouse: Water Saver Home
http://www.h2ouse.org
This tight little resource offers terrific insight into saving water at home. You can take a virtual tour of a model home, explore the water savings to be found throughout your house, use the water-budget calculator to figure out how much agua you're wasting (and on what), and even find tips on planting a water-conserving garden.

Thinking Differently About Water

Imagine if we woke up one morning to find out that we would be supplied that very day with the entire quantity of fresh water we'd ever be able to use. Given global inequity, our share might fill a wading pool, a swimming pool, or a supertanker, but whatever the amount, that would be it for the rest of our days—not a drop more. Knowing this, and seeing the entire bulk of our water supply in one place, chances are we'd do everything in our power to guard that water and make it last. We'd become compulsive about finding ways to conserve, to reuse, to purify and distill. Remaining aware of how much water we'd used—and how much was left—would constantly be at the forefront of our mind.

Of course, the beginning of this scenario is implausible. We'll never see our whole life's water supply in one place. But the need to measure and plan and remain constantly aware of our water use is real. Although we know that water is an extremely limited resource, most of us in the Global North don't experience a sense of responsibility as we watch it flow into the drain. Any awareness we have of the stream of water running through our lives is likely to be merely a hassle, or at best, just a number on our water bill. Rain falls from the sky, runs down our roofs, and washes into the sewer. We rinse a dusty glass and allow nearly clean water to be piped miles away to a sewage-treatment plant. Then we complain that water rates are too high.

But just as we needn't waste the water coming from our taps, we needn't ignore the water falling on our roofs or running down our drains. A whole slew of innovative tools can capture that rainwater (and recycle the slightly dusty tap water) to serve

other needs. Some of these technologies make even more sense in the Global South, where there are often no pipes or sewers to begin with.

Most of us in the Global North could fairly easily go off the electrical grid if we wanted to; going "off the pipe" is quite a bit harder. But we can save both water and money by making the best use of the water we're given. AS

The Soft Path

The way we manage water is a paradox: we often can't seem to get enough of it, and yet nearly every time it storms, we have too much. We treat water that flows from the tap as a resource (largely because we pay for it), while treating as a nuisance that which falls from the sky and courses through our cities. We have come to rely upon a centralized system that imports water from vast distances and quickly exports that which falls on our cityscapes, overlooking ways to use or detain water locally and on site.

Given the historical context, the centralized system makes sense. In the United States at least, centralized systems for supplying water were first built in response to the problems of urban living: in densely populated places, small, distributed, uncoordinated systems failed to prevent episodes of cholera (and other waterborne diseases) and were of little use in fighting devastating fires. Centralized water systems enable cities to prosper and grow, but they come with costs and weaknesses such as overbuilding the system to match peak demand. Extensive networks of pipes are expensive to maintain and replace, particularly in built-up areas. This centralized approach of traditional "hard path" systems focuses on increasing supply and expediting drainage through pumps, pipes, and storage while discounting the use of behavioral modification and on-site, integrated techniques to manage demand. Perhaps it's time to prime the pump with some new models.

Over the past few decades, distributed "soft-path" approaches have emerged to complement and augment the hard path of centralized systems. Hard-path engineers think mostly in terms of laying pipes and pumps, but soft-path advocates think more broadly, looking to a variety of disciplines (demand management, social marketing, economics, even landscape architecture) to not only deliver the water we need, but to make the system more flexible and resilient, and even to make the places we live more beautiful and comfortable.

Taking the soft path requires thinking about water management in an integrated way: every kind of water—drinking water, rainwater, storm water and wastewater—is a potentially useful resource. Incremental, distributed approaches to managing these various resources add capacity and flexibility to the system. The soft path handles water as a *service,* rather than a *commodity.* Demand is driven not by a need for water itself but by a need for what water can do. We think of water in terms of what it gives us (clean dishes, fresh sheets, houseplants)—not in terms of the number of gallons it takes to have these things. When we approach water as a service rather than as a commodity, we can use pricing, marketing, land-use regulations, and tools from other disciplines to manage demand for water while securing the services we want. We get our dishes, sheets, and plants, but we also get a hardier water system and a healthier environment. PF

Rainwater Harvesting

The tools in the soft-path kit are numerous. Plumbing codes, incentives for highly efficient appliances, social marketing of conservation practices, green roofs, and permeable pavements are just a few examples. Another is rainwater harvesting, a tool that has been around for millennia. Rain that falls on the roofs of homes (and sometimes other structures) runs through downspouts into cisterns that store it for use in a variety of applications. We can use the rainwater—instead of drinkable, or potable, water—to irrigate the garden and flush the toilet. This is one of rainwater harvesting's attractive elements: the quality of the water can be matched to the role that the water will serve. The most ambitious applications provide all of a home's water needs, including drinking water. If properly managed, cisterns are even capable of reducing flooding by capturing storm water and controlling its release. A network of cisterns could potentially serve as a mini–flood pocket, which could be emptied electronically by the local utility in order to distribute runoff slowly.

While simple in concept, rainwater harvesting can get complicated quickly. Rainwater's potential varies dramatically depending on an area's rainfall patterns, a given structure's roof size, and a cistern's capacity. In addition, when a homeowner starts altering plumbing, health officials and utility companies get nervous about the potential cross-contamination of drinking water via rainwater flowing back into the pipelines.

From Germany and Australia to Texas and Seattle, engineers, architects, researchers, and government employees continue to develop rainwater-harvesting technology. Arguably even more work is being done in the developing world. Rainwater harvesting is becoming a legitimate, reliable element of the soft path, and the soft path is gaining support as a tool for a bright, greenish blue future. PF

Australian Regulations

Australia is the driest inhabited continent, albeit one with a highly variable climate. Because of the unpredictable and sometimes infrequent rainfall, and because of a desire to protect coastal waterways, Australians are especially interested in rainwater harvesting. The city of Adelaide recently announced that all homes constructed after July 2006 must have plumbed rainwater tanks. The city may also require that tanks be installed in existing homes when they are resold. What's exciting about this mandate is that developers will have to incorporate rainwater harvesting into their plans for new construction, thereby making rainwater harvesting an embedded component of water use. PF

German Engineering

Germany, which has been tinkering with rainwater harvesting since the late 1970s, is arguably the leader in the developed world in urban rainwater use. According to architect Klaus Koenig, who has lectured on rainwater harvesting at conferences around the world, during the 1990s Germans installed several hundred thousand rainwater-harvesting systems. The prototypical German system is well engineered, with system components that progressively treat the rainwater as it flows from the roof to the cistern. These rainwater-harvesting systems are available in off-the-shelf preassembled filtration-and-treatment packages, which are easy to install. Concerns over bringing rainwater into homes (namely the safety of using it as drinking water) have been addressed by limiting rainwater's use: indoors it is used in toilets and for laundry and cleaning; outdoors, for irrigation. These uses can constitute up to 51 percent of typical demand for water in a German home, according to Koenig, with 33 precent attributed to toilet flushing, 13 precent to clothes washing, and the rest comprising smaller fractions. This means that rainwater can offset the need for that much potable water. In addition, the cisterns help to reduce runoff,

Page 188: The reHOUSE/BATH project offers a unique approach to water conservation.

Right: The Center for Maximum Potential Building Systems in Austin, Texas, has been a pioneer in rainwater harvesting.

which can be a major problem in heavily urbanized areas. The Germans are proving to us all that easy rainwater harvesting is facilitated by good design. PF

Q-Drum

In Africa, a simple plastic cylinder has improved the lives of thousands of women and children. The Q-Drum greatly simplifies the task of fetching water—an activity that usually requires several trips, several hours, and a lot of backbreaking work carrying cumbersome containers. Rather than requiring heavy lifting, the Q-Drum rolls easily along the ground, and can hold up to 13 gallons (50 liters) of water (previous containers held 1.3 gallons [5 liters] of water at best). It's also durable: it's pulled along by a rope—instead of handles or other breakable parts—which can easily be replaced or repaired anywhere on the continent. A screw-on lid greatly improves sanitation by preventing contaminants from entering the water. A similar product is the Hippo Water Roller. Though less sturdy, it has a larger capacity, and comes with a kit for a drip-irrigation attachment. JJF

Worldchanging Water Pumps

It's one thing to improve water-carrying techniques, it's quite another to build water pumps that function without infrastructure—without being connected to a public water system or electricity, both of which are out of reach for many rural poor people. Two new devices do just that.

KickStart's MoneyMaker
KickStart is a nonprofit organization founded by American engineers to create a middle class in Africa. KickStart's Super-MoneyMaker has become their best-selling product; the device allows users to pump water for irrigation using their feet, which facilitates a higher yield for farmers, who can increase tillable area, and sell surplus goods for a small profit. The pumps cost ninety dollars, which is a sizable investment, but the paybacks are huge. One woman was able to radically transform her horticulture business in one year—from a subsistence farm bringing in $93 a season to a five-person operation bringing in $3,200 a season. She was thus able to send her children to school, instead of into the fields.

PlayPump
South Africa's Roundabout PlayPump is powered by a very unusual source: a merry-go-round. As kids play on the merry-go-round, the rotation pushes a reciprocating pump that pulls water from a well up into a small water tower, for storage. Villagers can then get water at their convenience from a faucet. Billboards on the water towers providing public health messages and commercial advertisements pay for the system's maintenance and repair. JJF

Fog Catching

Fog collection is a beautifully low-tech way to supply fresh water in areas with negligible rainfall. Chile, Nepal, and southern Africa have successfully installed fog-catching arrays, fine nets that are stretched vertically between poles, with

a gutter at the bottom. As fog droplets pass through the nets, they impact on the fibers and run down into the gutter; the water is then channeled into reservoirs. Depending on the amount of water collected, the reservoirs can supply homes, irrigation systems, or whole villages with water. The systems require minimal maintenance—they're inexpensive devices with no moving parts and no need for power. The water needs no treatment to be potable; in fact, it is usually much cleaner than all but the best well or river water.

FogQuest, a Canadian nonprofit, is the first organization to successfully deploy fog collectors. According to the organization, its first installation, in the Chilean village of Chungungo, accumulated an average of 3,963–26,417 gallons (15,000–100,000 liters) of water per day throughout ten years of operation. The village was able to stop importing water by truck, and to begin growing gardens and fruit trees; its population subsequently doubled, reversing the migration to cities that had kept it low.

A new and improved fog catcher from UK-based QinetiQ uses biomimicry [see Biomimicry, p. 99] to increase the effectiveness of the nets. A material that has a microtexture like that of a Namibian desert beetle's back composes QinetiQ's nets. The Namib Desert is an incredibly hot, dry environment where occasional morning fog is the only source of water. The beetles that live there have evolved a shell that has a combination of hydrophilic bumps on hydrophobic furrows, which strain moisture from the air and concentrate it. JJF & DD

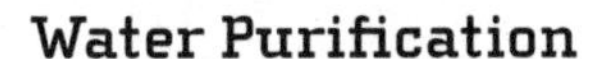

Water Purification

Silver Pot

In Latin America, the group Potters for Peace has designed a low-tech, low-cost water purifier that can be manufactured by local potters; it has been used by the Red Cross and Doctors Without Borders. The Filtrón is a pot made from clay infused with sawdust and colloidal silver. Water poured into the pot filters through its walls into a second, outside pot; in the process most bacteria get trapped, since they're too large to travel through the porous clay; the silver in the clay kills any bacteria that are small enough to pass through. The pots cost about nine dollars apiece and last for three years. Their manufacture also helps boost local economies; a Filtrón factory can be set up for just three thousand dollars.

Solar Watercones

The best-yet solar water purifier is the Watercone, a solar still that uses the sun's heat to evaporate water, which then condenses on the inside of its cone—flip the funnel-like cone over and you can pour the water right into a container. This cheap, rugged system can purify about 1.6 quarts (1.5 liters) of water per day, not only killing all waterborne pathogens but also removing particulates, many chemicals, and heavy metals. Better yet, it can also desalinate seawater—an important function for the world's sizable coastal populations. JJF

Opposite: The Q-Drum (left), a rollable water container, makes it easier to bring water home, while the Super-MoneyMaker Pump (right) makes it easier to irrigate the land at low cost.

Left: Fog catchers strain moisture from the air for collection, Falda Verde, Chile.

TreePeople

TreePeople is an organization pursuing an ambitious effort to retrofit Los Angeles with "green infrastructure" and reconnect the hydrologic cycle. The work for which they are best known involves just what their name implies—planting trees throughout the city where greenery is minimal, air pollution severe, and water poorly managed. Another TreePeople project, T.R.E.E.S., will mimic the "sponge and filter" mechanism of trees and apply it on a large scale by installing a system of cisterns and infiltrators within a small urban watershed. The system is designed to help capture water—which is at times a virtual godsend in Southern California—and recharge the aquifer, instead of letting the water run off and dissipate.

If implemented citywide, TreePeople believes, the project would drastically cut the use of imported water, reduce flooding and toxic runoff, and provide a host of other benefits. Because Southern California, which relies heavily on water piped from the Colorado River, may lose access to some of that water supply, reducing LA's need for imported water looks like a pretty compelling proposition. PF

Permeable Pavement

When it rains in the city, impervious surfaces such as asphalt and concrete repel water and send it directly into storm drains. The flooding that occurs on curbs and corners comes from overloaded drains, and causes problems far more serious than the inadvertent splashing and hydroplaning of passing cars. Among other things, plants and trees in urban environments miss out on the rainwater that should be their most obvious source of moisture. Paving compromises the health of the botanical life that keeps our cities beautiful, shaded, and cool; it also contributes to the wasting of water, since the trees must be irrigated by other means.

Pavement does have its upsides, however: it's stable, even, and firm—ideal for cars, bikes, and pedestrians. A new material called permeable pavement preserves all the functionality of regular pavement but eliminates the downsides. Permeable pavement can be laid anywhere concrete and asphalt usually go; it creates the same hard surface but allows rainwater to filter through into the ground. This prevents street flooding and keeps urban greenery healthier, with less work and less water.

Permeable pavement hasn't found its way into many areas yet, but Vancouver embraced its benefits early. To test the effectiveness of new pavement technologies, the city incorporated the material into its Country Lanes project, a demonstration area in which the city redesigned a series of streets to be only partially asphalt—and partially covered by permeable surfaces, which provide a solid, durable surface, but allow grass and plants to grow through. The permeable surfaces reduced storm runoff and improved air and water quality; they also enhanced the appearance of the streets, adding greenery to stretches of formerly bare concrete.

Another kind of permeable pavement commonly used in parks and other urban public spaces is Biopaver, a system of interlocking concrete blocks that have compost built into the

The Watercone uses sunshine to distill water, Aden, Yemen.

middle. You can lay these blocks in your driveway or on a sidewalk in place of cement. Over time, the compost biodegrades and seeds sprout from within the concrete square; the structure remains intact, but becomes partially obscured by grass. The roots of the plants actually take in pollutants from the ground beneath the pavement and clean up the soil. They also take in contaminants as they absorb storm runoff, both cleaning the water that goes into the ground and controlling flooding. And just like that, the solution to the problem of pavement turns out to make urban areas even more beautiful. SR

Beyond Living Machines

If we want to make a city run like nature, we need to create systems that can handle our waste in natural ways. Back in the 1960s, biologist John Todd began designing systems that used healthy, thriving ecosystems to clean sewage—gracefully coordinated tools he called Living Machines.

Living Machines are hothouses and artificial marshes in which sewage runs through a series of small faux ecosystems—tanks of bacteria, algae, plants, crustaceans, and fish. The plants and animals break down, filter out, or absorb some part of the sewage and turn it into living matter. What comes out the other end, after it is fed on by all these living components, is water (sometimes cleaner than the water in our taps) and biomass (living matter that can be composted for soil, fed to livestock, or otherwise returned to nature). And Living Machines themselves can be quite beautiful: the tanks often resemble well-groomed gardens of water plants.

Living Machines were a great—arguably revolutionary—invention in principle, but in practice they often proved too finicky to build and maintain on a scale that would make them practical for cities. Some of the flora and fauna in the Living Machines could not handle the toxic witches' brew that runs through sewers. As William McDonough and Michael Braungart put it in *Cradle to Cradle* [see Knowing What's Green, p. 115], "In addition to biological wastes, people began to pour all kinds of things down the drain: cans of paint, harsh chemicals used to unclog pipes, bleach, paint thinners, nail-polish removers. And the waste itself now carried antibiotics and even estrogens from birth control pills. Add the various industrial wastes, cleaners, chemicals and other substances that will join household wastes, and you have highly complex mixtures of chemical and biological substances that still go by the name of sewage" (2002).

But some industrial microbes find sewage totally yummy. Researchers are trying to breed microbes and algae that eat sewage and secrete hydrogen [see Neobiological Industry, p. 111]. Other researchers have identified bacterial membranes that can filter most organic matter from inorganic matter. And certain scientists believe we may be able to tame particular kinds of fungi to extract heavy metals in concentrations sufficient to be reclaimed.

Living machines and industrial microbes could, of course, work together, with microbes cooking off hydrogen in vats linked to tanks of marsh plants and hungry snails—producing energy and cleaning water at the same time. (For one

This cistern, developed through the Open Charter Elementary School Stormwater Project, holds 110,000 gallons of rainwater, decreasing LA's need to import water, Los Angeles, California.

vision of how such a system might work, check out Nicola Griffith's excellent science fiction novel *Slow River,* which centers on mysterious doings at a high-tech biomimetic sewage-treatment plant.)

As technology advances and living-machine designers gain experience, these kinds of systems are almost certain to become cheaper and more competitive with conventional sewage treatment. Within a few decades, living organisms might be recycling most of the nutrients in our sewage—if we can stop mixing that sewage with toxins. AS

The "Elevated Wetlands"

One of the best ways to clean large quantities of polluted water is also one of the most beautiful. Bioremediation—the process of restoring contaminated land and water using living organisms—introduces vibrant plant life to formerly lifeless landscapes. In Toronto, marshlands between the Don River and a wide stretch of highway received a revelatory work of ecological art that transformed the space around it. Noel Harding's "Elevated Wetlands" made a noticeable impression far beyond its immediate surroundings, standing as a multilayered example of both bioremediation and transformative public art. Locals refer to the plastic sculptures that compose the piece as a parade of giant molars or as polar bears marching across a green marsh. All agree that even from a distance the "Elevated Wetlands" seems alive. Not long before the artwork was created, the site had been a salt desert: a desiccated piece of land wrecked by too many winters of roadside salt spray.

Although bioremediation can be an astonishingly complex technology, in this case its gesture on the landscape is simple and elegant. A remarkable marriage of functioning public art and scientific invention, the thriving site now attracts wildlife rather than repelling it. Inside the plastic forms lies a key innovation: layers of plastic from old bottles and nonmetal automobile waste form a hydroponic environment for the plants—a soil system made from recycled plastics, with pumps quietly powered by solar panels. As tainted upstream water is pumped through the wetland, the plants absorb polymers and heavy metals, aerating the water and sending it back out clean. "One of the great discoveries was the relationship of solar power and plant intake," says Harding. "Part of the lesson learned is that along all rivers and wetlands, plants take in water by sunlight; you can pump water by sunlight. It's not really quite a work until the water's flowing." DD

RESOURCES

Rainwater Collection for the Mechanically Challenged by Suzy Banks and Richard Heinichen (Tank Town Publishing, 2004)
Don't be put off by the fact that this looks like a children's book—it's full of accessible, amusingly written instructions that will help you set up and troubleshoot your own rainwater-collection system, and enjoy the project, too.

The authors write from their own experience: "We guttered our house and bought our first tank, foolishly believing we were 'settling'

for rainwater. Now, after years of living with 'the gold standard'—water with a hardness of zero, that tastes fresh and leaves our faucets and tile sparkling—we realize no matter what conditions we had faced, choosing rainwater is not 'settling' for less in any sense. Rainwater is, simply, the best."

A word to the wise: the authors acknowledge up front that their experience is very region-specific (Austin, Texas), so you may have to take other factors into account before applying their methods to your home.

The Rainwater Technology Handbook
by Klaus W. Koenig (Wilo-Brain, 2001)
This heavily illustrated book provides snapshots of innovative large-scale rainwater-harvesting programs around the world, from artificial glaciers in India to a thorough description of the German experience. It's a good reference for equipment and specifications, too.

According to Koenig, "Hardship often forces people to settle in areas where access to water is difficult: on mountain slopes and in areas with no water supply or with water that is unfit for human consumption. For them, the use of rainwater is like 'manna from heaven.' 'Why don't you divert the water from the roof of your hut when it rains and collect it?' my African colleague tries to explain to his compatriots. 'In the dry season you have to walk for miles to collect and bring back the same water that has, in the meantime, not only caused erosion, but has also become polluted.' And, he is right!"

A Safe and Sustainable World: The Promise of Ecological Design by Nancy Jack Todd
(Island Press, 2005)
Nancy Jack Todd and her husband had a vision in the late 1960s of building a safer and more sustainable world, a world where they would be directly connected to their food, shelter, and energy. Their vision turned into the New Alchemy Institute, a small nonprofit research organization on Cape Cod, dedicated to experimenting with new ways of leading a sustainable lifestyle. *A Safe and Sustainable World* is the story of their journey: "Fundamental to that journey is the underlying understanding that the instructions as to how human cultures are to live lie encoded in the living systems of our unique and irreplaceable home planet." Todd goes on to note that "the potential is exponential"—the journey is not yet over.

Opposite, left: Permeable pavement allows water to filter through the ground, preventing flooding and keeping urban greenery healthier, Vancouver, Canada.
Opposite, right: Living Machines could provide a sustainable alternative to treating wastewater with chemicals.
Left: Noel Harding's "Elevated Wetlands" at Taylor Creek has transformed its surroundings, Toronto, Canada.

Landscaping

We may not realize it, but we're in dysfunctional relationships with our lawns. They take and take and never give back. Like big green sponges, our lawns suck up water, fertilizer, pesticides, and money, and if we leave them alone for too long, they start to look sad until we give them some more.

The typical American lawn has almost nothing to do with nature. A dense carpet of overbred alien grasses, usually coated with toxins, it keeps competitor plants (like villainous dandelions) withering before they sprout. Almost nothing "natural" can survive there. The average lawn makes an overgrown abandoned lot look like a rain forest.

Maybe we feel compelled to keep up appearances by tending to our needy lawns, keeping them as vibrant and immaculate as our neighbors'. But we will never get ahead of our lawns' constant needs, and they will exhaust not only us, but also the natural resources they require to stay alive.

There are better ways to relate to the spaces outside our houses. Landscaping our property intelligently can mean long-term savings—savings on water, with self-sufficient, climate-appropriate plants; on electricity, with shade trees that decrease the need for AC; and on sundries, with fewer treatment, tool, and nursery purchases. There are many resources available, both from city governments and private organizations, to help us develop relationships with yards that give a little bit back. SR

Kill Your Lawn

Putting a stop to the destructive cycle of all take and no give is as easy as giving the space around our homes some purpose. Lawns are passé. In some places, it's become trendy to pave over lawns, which seems ludicrous for all kinds of reasons (not least of which is that skinning a knee on cement feels a whole lot worse than falling on some nice, forgiving soil). But for all its softness, grass has plenty of drawbacks. If we want to get a little back, the best thing to do is the most obvious: Plant something productive. Create a garden. Fill it with food.

Radical designer Fritz Haeg created Edible Estates as an alternative to the water-guzzling, pesticide-drenched grasslands of American front yards. Edible Estates helps homeowners replace their grassy lawns with productive gardens. Besides the known pesticide-intensive lawn-care issues, the cultural barriers that lawns present are a topic Haeg stresses on Edible Estates' Web site: "The lawn divides and isolates us. It is the buffer of anti-social no-mans-land that we wrap ourselves with, reinforcing the suburban alienation of our sprawling communities. The mono-culture of one plant species covering our neighborhoods from coast to coast

celebrates puritanical homogeneity and mindless conformity."

The first Edible Estates lawn revival took place in Salina, Kansas, where a family offered up their conventional front yard for transformation (like a reality TV show for lawn makeovers), and vowed to maintain the garden as a living, thriving edible installation. Reclaiming front yards through this process not only furnishes families with a hearty supply of nourishing food, it also provides an education in seasonal cycles, organic gardening, and regional biodiversity.

Those of us who think a garden sounds too time intensive are not limited to having an emerald carpet. Even deserts and alpine areas have native grasses, shrubs, and flowers that will thrive on their own with a little water and sunlight. Eliminating the homogenous sod that covers most lawns can be as easy as getting down to the dirt and sprinkling some wild-grass seed. What sprouts will be a beautiful, diverse array of grasses and wildflowers—a little originality in the midst of a green sea of uniformity. SR

Backyard Biodiversity

Garden styles vary almost as much as clothing styles, with trends that change over time. The traditional English rose garden, for example, with pristinely manicured shrubs of identical flowers, contrasts with the complex and strategically chaotic gardens that fill so many beautiful yards today. Besides the aesthetic appeal, a diverse garden that integrates many plant varietals offers a number of ecological benefits. This variety—called biodiversity—actually fortifies gardens, making them more adaptable, more resistant to pests and disease, and more productive.

A growing, though still small, number of Americans have torn up their lawns and planted native ground cover, shrubs, and trees, which not only need far less water and fertilizer than lawns (and often no poisons at all), but also offer homes for passing songbirds, butterflies, even frogs. Add a backyard compost pile or worm bin for your kitchen scraps, and harvest the rainwater that runs off your roof, and your house can soon be a wild oasis in a sea of carefully clipped lawns and asphalt.

The real payoff, though, comes as those oases multiply, forming a mosaic of habitats dotting the city. Indeed, imagine more and more backyard wildlife sanctuaries, woven into a larger urban fabric of green roofs, street trees, and restored streams and wetlands—an urban landscape where nature is at home.

Tips for a Biodiverse Backyard

There are lots of ways to promote biodiversity in our own backyards. Creating curved and irregular perimeters, instead of planting square plots, allows for gradual transitions between various areas and plant varieties. Planting in tiers—or at least being conscious of the gradation in height and size of the shrubs and trees we choose—also works well in biodiverse gardens by mimicking the natural irregularities of plant habitats. To attract bird and insect life, as well as to nourish the members of our households, we can plant edible crops that reach maturity on a rotating cycle through the year, and fruit trees, which have beautiful blossoming cycles before the fruit ripens.

Lawn care can actually be easy: by avoiding invasive species and taking a step back from obsessive maintenance, we can allow naturally occurring plant life to spring up and become part of a pleasant, diverse garden. By planting species that are native to our region, we can integrate our own yards into the larger ecosystem. Plants are better able to take care of themselves in their native habitat, which means we have less work to do.

To attract hummingbirds and butterflies to our backyards, we can plant vibrant colored flowers and masses of plants rather than single

A home before and after its transformation, effected with the help of Edible Estates, Salinas, Kansas.

flowers. Planting flowers in the sun, near a natural windbreak such as rocks or trees, and looking especially for tubular flowers with a high sugar content also works wonders.

Whether we plant a biodiverse garden for its ecological value or simply for the dramatic variety of shape, size, and color it offers, we will find that our yards come alive with fresh vegetables, vibrant flowers, and the constant dance of birds and bees. SR

RESOURCES

The American Society of Landscape Architects
http://www.asla.org
While officially a professional organization with broad interests, the Amercian Society of Landscape Architecture (ASLA) has increasingly moved to the forefront of the discussion about the role of landscape design and construction in building more sustainable communities. Indeed, hitting the Web site or listening to some of the ASLA's leading members, you might imagine you're dealing with an environmental group. So if you're thinking about incorporating bright green landscaping into your professional life, you need to pay close attention to these folks. And even if your interest in landscaping stops at your back fence, you're sure to benefit from the Web site's valuable learning tools.

Space for Nature
http://wildlife-gardening.org.uk
This British group hosts an amazing Web site full of ideas both practical and inspirational for growing a garden that welcomes native birds, butterflies, and other forms of wildlife. While many of the suggestions are particularly relevant to the United Kingdom, there's plenty here for Brits and non-Brits alike—and lots of great leads that will set you on a path toward luring life into your backyard.

The Brooklyn Botanic Garden
http://bbg.org
Botanical gardens, arboretums, and conservatories are great old institutions we tend to overlook, rather like research libraries, but they often (and increasingly) have outstanding programs to inform, educate, and facilitate public action. When it comes to gardening with native plants, local botanists are often your best resource, and the ones who work at public gardens are often eager to help. The Brooklyn Botanic Garden is, of course, just one of many such places; it hosts a fine Web site and offers some great reading on the basics of responsible landscaping.

EcoHouse Brazil

In Brazil's Urca neighborhood, near the base of Sugarloaf Mountain and the shores of Rio de Janeiro, architect Alexandra Lichtenberg tackled a remodeling project that applied green principles to an average urban home. The project illustrated that high-cost, luxury residences and remote, off-grid dwellings aren't the only houses that can go green. The EcoHouse has proven that a good green remodeling job is within reach of average well-intentioned homeowners in ordinary neighborhoods throughout the world.

The EcoHouse Urca project aimed to create an ecofriendly environment for the inhabitants of the house and to serve as a testing site for evaluating comfort levels within ecologically enhanced homes that use highly efficient thermal, water, and lighting systems. The architect's goal was to create amenities comparable to those of or better than a conventional home while improving the structure's ecological impact. The Rio house also served as a case example for ecofriendly houses in hot, humid climates worldwide.

The following features of the EcoHouse remodeling job are available at minimal cost to homeowners, and most are just as easily implemented in an existing building as in a new home:

Alexandra Lichtenberg's EcoHouse is an example of just how green our homes can be, Urca, Brazil.

Rainwater Catchment

In the EcoHouse, a concrete cistern collects rain that flows from the roof and patio through a gravity-driven mechanical filter. The runoff is then pumped to the water-recycling tank, located on the highest green roof, and is distributed by gravity to toilets, the garden irrigation system, and faucets used for nonpotable water. In the first year, the system accounted for 28 percent of the total water use in the EcoHouse.

Sewage Recycling

Even the most hardcore environmentalists sometimes shy away from dealing with sewage and residential "wastewater" (aka gray water, which has been used in sinks, showers, and washing machines, but not in toilets). However, a number of well-designed compact sewage-treatment systems make residential-water reuse easy and clean. The Brazilian company Mizumo supplied a test system for the EcoHouse Urca project. Intended for small urban lots, the unit measures 3.9 x 8.5 x 6.9 feet (1.20 x 2.6 x 2.1 meters). It is meant to provide water for the same nonpotable uses as the rainwater system. Before being pumped to the water tank on the green roof, the water undergoes sand and UV-light filtration to eliminate any impurities.

Passive Cooling

The best means of achieving passive heating and cooling is through well-planned orientation of a house on its site. With existing buildings,

though, there are other ways to make use of passive technologies, such as strategic placement of shade trees, extension of eaves and overhanging roofing, and window glazing. When the walls, windows, and roof of a house are kept cool by deflecting or avoiding direct sunlight, the inside stays cool, as well, without AC or other high-energy systems.

Green Roofs and Facades

At the EcoHouse, the old ceramic-tile roof was replaced with green roofs [see Greening Infrastructure, p. 256], planted mostly with grass and cooking herbs. The house's northwest-facing facade was fitted with an aluminum trellis that supports a vigorous vine. As it grows, the vine forms a shield that absorbs most of the direct sunlight that would otherwise hit the walls.

Natural Ventilation

Natural ventilation is easiest to install during initial construction, when skylights and windows can be designed into a home. For the EcoHouse remodeling job, the architect reconfigured the internal layout to allow for natural ventilation. Air circulation is vitally important, not only for reducing heating and cooling costs, but for the health of a building's inhabitants. Keeping a good inflow of fresh air enhances the interior atmosphere so that it never feels stuffy or stagnant.

Renewable Energy

Two solar systems heat all the hot water in the EcoHouse, both of them working in a passive "thermosiphon" system that takes advantage of gravity and eliminates the need to pump heated liquids around the house. In many systems, solar panels mounted on the roof heat water before it flows to a tank several floors below, and the liquid must be pumped back to the roof for reheating. In a thermosiphon system, the tank is on the roof, above the solar collector. As the temperature of the heat-transfer fluid increases, its density decreases. The fluid rises, causing natural convection, which permits passive circulation in the pipes. In the EcoHouse, one of the two solar systems has an electrical backup source; the other runs completely off solar energy. SR

Refugees

It is one of the worst things that can happen to us: to be uprooted from the place we live and forced by violence or disaster to flee for our survival—to become, in short, a refugee.

The plight of refugees is a tragic and increasingly common part of our world. In 2004, 11.5 million people had been driven from their homes and were living as refugees, according to the World Refugee Survey. (Millions more have become "environmental refugees"— people who find themselves, because of environmental degradation, no longer able to eke out an existence and forced to move in order to survive.) Twenty-one million have been made "internal refugees" by natural disasters or conflict.

Everyone tracking the issue expects many, many more people—perhaps hundreds of millions—to be forced to flee from their homes at some point during this century. Responding to their plight will not only strain our capabilities, but test our character.

Thankfully, a number of brilliant and dedicated people are beginning to revolutionize the field of humanitarian assistance. We are witnessing not only a reinvention of the approach to disaster response, but a reimagining of the refugee camp, and we are learning how to design reconstruction efforts that work. If the twenty-first century is to be, as the slogan claims, the century of refugees, there is at least real reason to hope that we are evolving the right tools and models to meet the challenge.

The EcoHouse's green roofs provide not only great views, but also the perfect platform for a sustainable solar-power system.

Environmental Refugees

Even if we succeed in calming the violence that has displaced so many, human activity is making the planet a more hostile and unforgiving place: deforestation and soil erosion have rendered barren much of the best cropland; overpopulation has forced large numbers of people to live in precarious situations; and, most direly, humankind has been changing the climate.

Climate change may prove to be the ultimate humanitarian disaster. Its immediate effects—droughts, floods, rising seas, worsening storms—directly threaten many of us, of course, but they fall most heavily on those people who are already living on the margins. For a family that survives on subsistence farming, a small shift in rainfall patterns can spell disaster, as happened on the Great Plains in the 1930s, when the Dust Bowl sent farmers streaming toward the cities. When you live on a floodplain, you may be able to rebuild when floods come once a decade. When they come twice a year, you no longer have a home.

Red Cross research indicates that we may already have more environmental refugees than war refugees. By 2010, the United Nations University says, we can expect the displacement of 50 million environmental refugees. If the climate models hold true (and things don't get worse than expected), upwards of 200 million people are expected by the 2080s to have been made refugees by climate instability and rising seas.

The growth of megacities may exacerbate the trend, especially in those places where vulnerability to "natural" disasters collides with local governments' inability to accommodate rapid growth with proper infrastructure. Hurricane Katrina nearly wiped New Orleans off the map

and sent hundreds of thousands into flight; imagine a Katrina-caliber storm directly hitting a city of millions where many people live in shacks. What that storm would have done to Lagos or Dhaka is anybody's guess. Are we prepared to handle the consequences? AS & JC

The Rights of Refugees

In the aftermath of World War II, faced with tens of millions of "displaced persons," the newly formed United Nations worked to answer the question, what rights do refugees have?

The result was a treaty known as the 1951 Refugee Convention. Simply put, the convention guarantees the right to asylum to those who are fleeing violent conflict because their lives or basic human rights are at risk. Countries to which these people flee are obliged to offer them protection until they can safely return to their homes. The convention, though, very specifically does not protect "economic refugees"—those who are fleeing not violence but poverty (no matter how life threatening). And there's the difficulty, because a growing number of people must flee their homes simply because they cannot make a living there anymore. It can be very difficult to differentiate between those refugees who are fleeing in search of better opportunities and those who are fleeing because they have no other choice. Most of us would argue that our moral obligations to the two are quite different. We certainly have the moral obligation to respect everyone's human rights, whether they are illegal migrants or not—but do people who are merely attempting to escape poverty have the same moral claim as do those whose very lives are at stake? With so many environmental refugees expected in the coming years, the question is far from academic. JC & AS

Preceding pages: Providing for refugees, like the two million who fled the war in Rwanda, is a growing challenge, Democratic Republic of Congo, 1994.

Left: Some immigrants, legal and illegal, are economic refugees, fleeing not violence but extreme poverty, Tarifa, Spain.

Reinventing the Refugee Camp

One of the most important tasks at hand is to perfect the art of humanitarian intervention—not only to stop war and genocide, bring criminals to justice, and provide safe havens for refugees, but also to help people in disaster zones and war-torn lands get back on their feet as quickly as possible. We need to save lives, sure, but humanitarian intervention can also help bring victims to a place where they can reimagine their own lives and acquire the skills to forge their own paths.

The first step is to build better tools for helping refugees. Innovate and improve the relief effort, right now, from the start. The demands we put on aid workers are extraordinary. They fly to remote corners of the world, where usually they can expect to find nothing, not even clean water. They create entire cities from scratch, restoring order, pitching tents, digging latrines, finding and filtering water, treating the wounded and diseased, counseling the grieving, and finding ways to bring traumatized people back to emotional engagement with their own lives. This is perhaps the hardest work on earth, and the people who do it—the United Nations' blue hats, the Doctors Without Borders, and the tireless aid workers—are the closest thing we have to true heroes.

We are just starting to invent the necessary tools to help all of these people do their jobs better. We need to spread the best innovations around as quickly as possible, employ better logistics methods, and get aid workers better information about conditions on the ground. AS

A replica refugee camp set up by the Norwegian Refugee Council for World Refugee Day gives Norwegians a glimpse into the everyday lives of refugees, Oslo, Norway, 2003.

Understanding Refugee Camps

In order to comprehend the magnitude of the challenges aid workers face, we have to wrap our minds around a type of situation unlike anything most of us have ever known. One extremely informative resource is the Doctors Without Borders' Refugee Camp traveling exhibit, which has appeared in fifteen countries. The exhibit includes replicas of emergency housing, clinics, and feeding centers, and it displays equipment used by aid workers, such as water pumps.

The Doctors Without Borders Web site (www.dwb.org), another invaluable resource, gives us a sobering portrait of refugee camps: "A poorly planned refugee settlement is one of the most pathogenic environments possible. Overcrowding and poor hygiene are major factors in the transmission of diseases with epidemic potential (measles, meningitis, cholera, etc.). The lack of adequate shelter means that the population is deprived of all privacy and constantly exposed to the elements (rain, cold, wind, etc.)." Imagine 100,000 people staggering to a place of safety, utterly destitute, hungry, sick, traumatized, and grief-stricken. Now imagine trying to feed, clothe, shelter, and care for them all in a place where infrastructure simply doesn't exist (or is quickly overwhelmed), where filth and crowding are almost beyond comprehension, and where every supply must be flown or trucked in and then we begin to see the kind of challenges aid workers face every day. We begin to see what a world full of refugees will feel like, and understand why inventing better refugee camps must be near the top of the planetary to-do list.

LifeStraw

LifeStraw is a drinking straw that contains filters capable of killing *E. coli, Shigella, Salmonella, Enterococcus, Staphylococcus aureus,* and the microorganisms that cause diarrhea, dysentery, typhoid, and cholera. With each sip of water, two textile filters catch large materials and clusters of bacteria; next, a chamber of iodine-filled beads kills smaller bacteria, viruses, and parasites; finally, a chamber containing granulated active carbon catches any remaining parasites and rids the water of the iodine smell. Unlike more complicated water-filtration systems, the LifeStraw costs less than four dollars and can filter about 185 gallons (700 liters) of water (up to one year's worth of water for one person), according to the LifeStraw designers' research and analyses. LifeStraw makes providing safe water to those who need it affordable.

Plumpy'nut

Fighting malnutrition is a top priority for aid workers in refugee camps. This daunting task is complicated by the need to mix standard treatments for malnutrition—powdered milk formulas called F-75 and F-100—with clean water, which is often scarce in areas of extreme poverty and turmoil. Moreover, to ensure that the formulas are mixed properly, workers can only administer them at hospitals and feeding centers; some refugees have to travel miles to reach these centers, and overcrowding nearby can lead to the spread of disease. On top of that, milk-based products are also prone to bacterial growth. F-75 and F-100 are far better than nothing, but this situation clearly calls for a new solution.

Nutriset SAS, a French company specializing in humanitarian nutrition, thinks it has that solution: Plumpy'nut, a peanut-based paste with the nutritional value of the F-100 milk formula. Plumpy'nut requires no preparation or mixing—it can be eaten right from the bag—so it can be distributed directly to affected communities. It's also much more palatable than other formulas (it tastes like a sweeter version of peanut butter), which means it's more likely that people, particularly children, will eat enough of it to recover their health.

Food Force

Because games are active, not passive, forms of entertainment, they have real potential as educational tools. With this in mind, the United Nations has produced a video game about

food aid. *Food Force,* which is designed for eight- to thirteen-year-olds, puts players in the role of the rookie on a food-aid team working in the fictional country of Sheylan. Players do everything from finding, buying, and shipping supplies from around the world to running food convoys over dangerous roads. The final mission is a *SimCity*-like assignment, in which players help rebuild the nation's economy through food aid. JC

Concrete Canvas Tent

Ordinary canvas tents are too flimsy for field offices and hospitals, especially in places where harsh weather conditions wear down fabric within weeks. Concrete structures, on the other hand, are too difficult to transport and too expensive and time-consuming to construct for quick deployment in crisis situations. But it seems that a combination of canvas and concrete might be the next best solution to the problem of getting aid workers and doctors the facilities they require quickly and cheaply.

The Concrete Canvas tent, invented by Peter Brewin and Will Crawford, two grad students at Royal College of Art in London, is a sack of cement-impregnated fabric weighing approximately 500 pounds (227 kilograms). When water is added to the bag and the whole thing is inflated with air (a process that takes about forty minutes), it becomes a sturdy domed structure with 172 square feet (16 square meters) of floor space. It takes twelve hours for the structure to dry out enough for habitation. Such a structure makes an ideal field hospital, enabling doctors to perform surgeries on the first day of a crisis.

Concrete Canvas has won several prestigious awards for its ingenuity, but as of this writing, the engineers are still seeking funding for production. JJF

Compostable Tent City

Today's refugee camps tend to be sprawling, muddy, overcrowded, and septic tent cities where services are rare and opportunities to actually improve one's life are few and far between. They are often nothing more than places to warehouse people most of us don't care enough about to notice.

But future refugee camps could be much, much different. One possibility is the compostable tent city, first proposed by the Rocky Mountain Institute. In this model, the tents themselves would not be tents at all, but treated cardboard shelters that provide basic housing and last for a couple of years. (The shipping containers and packaging for medical goods and food would also be treated cardboard.) When the shelters wore out and the packaging was discarded, they would show their true nature. The panels of treated cardboard, each infused with appropriate local seeds, spores of topsoil fungi, and harmless fertilizing agents, would become very special compost: by tearing the panels up and watering them, refugees could start gardens, complete with mulch, fertilizer, and the microorganisms good soil needs. (Even clothing and blankets can be designed to be compostable when they wear out.) The entire transitional tent city could then be plowed into gardens as refugees settle in to stability. Not only would the soil support food crops, but fast-growing, salt-absorbing hybrid shade trees could go in as windbreaks, helping to check erosion and desalinize the soil. AS

Refugee Camps and Leapfrogging

Refugee camps can themselves become engines of transformation if we can give refugees the opportunity to improve their own lives—creating not just relief but also the possibility of rebirth. Encouraging communities to be active participants in their own rebuilding is key to

Opposite, left: Plumpy'nut, which appeals to kids' taste buds more than traditional therapeutic foods do, helps malnutritioned children survive in refugee situations. **Opposite, right:** The United Nations' video game *Food Force* introduces kids to the challenges of running a large-scale refugee-relief operation.

creating sustainable solutions and reducing the impact of a disaster.

A primary goal of relief and reconstruction work should be to arm communities with the expertise, technology, and capital that will allow them to leapfrog [see Leapfrogging, p. 292] over older, outmoded, costly, and centralized technologies, and start right in on building lives of sustainable prosperity. This process should start the moment boots hit the ground. With that in mind, big international nongovernmental organizations ought to be thinking, whenever possible, about infrastructure's long-term utility to the local communities.

Solar power can save lives today and improve communities tomorrow. Its utility in a crisis is obvious: solar-power systems are portable, can be installed anywhere there's sunlight to feed them, and become much more reliable sources of power than generators (at least for small dwellings and comparable structures in camps). Generators need to be refueled regularly, an otherwise simple task that can become an impossibility when pumping stations are closed or in ruins.

Widespread use of solar energy in disaster-relief efforts can also provide communities with equipment, infrastructure, and expertise around which they can build permanent distributed-energy systems [see Leapfrogging Infrastructure, p. 302]—the kinds of systems that are generally more likely to work for developing world communities in the long run. The solar-energy systems can become a sort of seed stock for new development—smart grids [see Smart Grids, p. 183], light-emitting diodes, computers and communications, and renewable energy—so when relief efforts end, even better shelter can grow from the seeds left behind. AS

Transforming Disaster Relief

It's hard to call disasters "natural" anymore. Climate change, ecological degradation, poverty, poorly designed infrastructure, war—frequently, all of these combine to make what once were serious disasters into catastrophes.

When in 2005, Hurricane Katrina—behaving exactly as we've been told storms will behave in a greenhouse world—nearly wiped New Orleans off the map, who or what was to blame for the havoc it caused? Climate change, which may have made the storm worse? The destruction of local wetlands and alteration of the Mississippi, which made flooding as the storm surged worse? The economic system, which left many people too poor to flee the path of destruction? The New Orleans levy system, which had been allowed to decay from a lack of investment? Or the absence of National Guard troops, who normally would have assisted in disaster response, but who had been deployed to Iraq?

In our day, there's increasingly no way of telling which parts of a weakened system cause it to collapse catastrophically. There's even a term for the feedback loop between an environmental collapse and the failure of the human systems that caused it:

Opposite, left: A Red Cross volunteer comforts a Hurricane Katrina refugee at the Houston Astrodome, Houston, Texas, 2005.

Opposite, right: A survivor of the Kashmir earthquake talks with Pakistani paratroopers, Chautha, Kashmir, 2005.

***Wexelblat disaster.* In such a world, planning for effective disaster relief efforts isn't a luxury: since massive disasters are no longer a question of "what if" but "when," we'd better be ready.**

These "unnatural" disasters—of which the 2004 Indian Ocean tsunami, 2005 Kashmir earthquake, and 2005's Hurricane Wilma are just a foretaste—are beginning to fall more frequently and with more severity than governments alone are capable of responding to. (Though largely ignored in the media following Katrina, Wilma was the strongest hurricane ever recorded.) We therefore need new tools for disaster response, and new methods of allowing citizens to band together quickly in the wake of calamity. And these tools and methods are beginning to emerge. AS

Lifesaving Logistics

Humanitarian relief efforts are logistical nightmares. Aid workers must coordinate supplies and donations, get the permits they need to operate in afflicted areas, find volunteers to help deliver the aid, and figure out some way of moving everything and everyone to the affected site. The information revolution that brought increased efficiency and lower costs to the commercial sector wasn't visible in the humanitarian-relief arena until very recently.

The Fritz Institute, founded in 2001, is dedicated to bringing modern logistics techniques to the world of disaster relief. It provides logistics software and partnering resources to organizations large and small that are engaged in global humanitarian efforts. The Humanitarian Logistics Software, built in coordination with the International Federation of Red Cross and Red Crescent Societies, brings to bear the experience and best practices of the commercial sector on the problems of humanitarian-relief delivery. The institute's annual conferences and "Network of Knowledge" capture the evolving lessons from humanitarian efforts in order to develop and document broad solutions for common problems.

Another tool for the logistics chest is Global MapAid, a recently developed nongovernmental organization (NGO) that aims to provide high-quality geographic information system (GIS) mapping information [see Mapping, p. 518] to rapid-reaction disaster-response teams. In New Orleans, three volunteers made maps of Katrina-affected areas that indicated where returning residents could find food, water, and clothing; they distributed 20,000 of them to residents and Red Cross workers alike. But Global MapAid isn't just about handing out maps to aid workers. One of its goals is to share its mapping expertise with locals to ensure that they can continue mapping projects once the NGOs are gone. After the 2004 tsunami, two MapAid volunteers provided GIS training to a group of students at the University of Syiah Kuala in Bande Aceh, Indonesia. Additionally, the organization provided data-collection equipment that had been retrofitted to withstand the region's high temperatures and humidity. JC & AS

Strong Angel

In May 2000 a refugee camp materialized on a barren lava bed on the Big Island of Hawaii. The refugees weren't victims of a natural disaster or some political action; rather they were actors-volunteers brought together to improve the lives of real refugees thousands of miles away.

The mock camp was created by Strong Angel, an organization led by navy doctor Eric Rasmussen. Strong Angel was created to find new ways for the military, nongovernmental and governmental organizations to employ collaborative systems and cheap technology in order to work better together during disasters and crises. The five-day exercise in Kona was essentially a field test of collaborative tech to determine what works best when it comes to managing a refugee camp. Many different systems and products were tested, from computerized translation systems for doctors in the field (language barriers between medical personnel and refugees are a huge problem) to simple radios for providing one shared communication system. In fact, one of the notable things about the Strong Angel mock camp was that it proved we can get a lot done with simple, cheap, off-the-shelf technology. The Kona refugee camp was managed solely with a Web site and simple wireless-communication gear that workers could carry with them at all times.

In July 2004 Strong Angel II took over the same lava bed. This time the most invaluable tool was the Groove network. Groove is a peer-to-peer application: Information can be transferred wirelessly from laptop to laptop. No central server is necessary, which is important because a server could easily be destroyed during a crisis. In addition, people can work off-line knowing that as soon as they reconnect to the Internet, the Groove software will automatically synchronize all the shared folders, ensuring data is up-to-date at all times.

Unfortunately, few of Strong Angel's discoveries made it into the relief efforts after the 2004 tsunami. The first few days after the disaster, the actions of both military and nongovernmental agencies were so haphazard and poorly coordinated that the problems they caused almost negated the immediacy of their response. The military's standard protocol was effective in gathering information, but inefficient in disseminating that information to the workers actually delivering the aid; in some cases, important data about road conditions, food supplies, and casualties was locked up tight aboard offshore vessels, while nongovernmental organizations (NGOs) sent up helicopter after helicopter trying to gather the exact same data. Some NGOs even duplicated one another's efforts on this front.

Rasmussen actually got a chance to witness the chaos firsthand, and he concluded that one of the biggest problems was that many workers relied exclusively on their cell phones to communicate with one another. There was no central way for everyone to share important information, no database where all of these disparate organizations could compile their findings. A collaborative virtual workspace had been at the top of Strong Angel's list from the start, but what Rasmussen witnessed in Southeast Asia clinched it. The main objective of any future relief effort should be to create such a space, a

Left: Strong Angel is a hands-on laboratory for disaster-relief innovation.

Opposite: SkyBuilt's Mobile Power Station can provide energy in areas where power infrastructure is damaged, lost, or nonexistent.

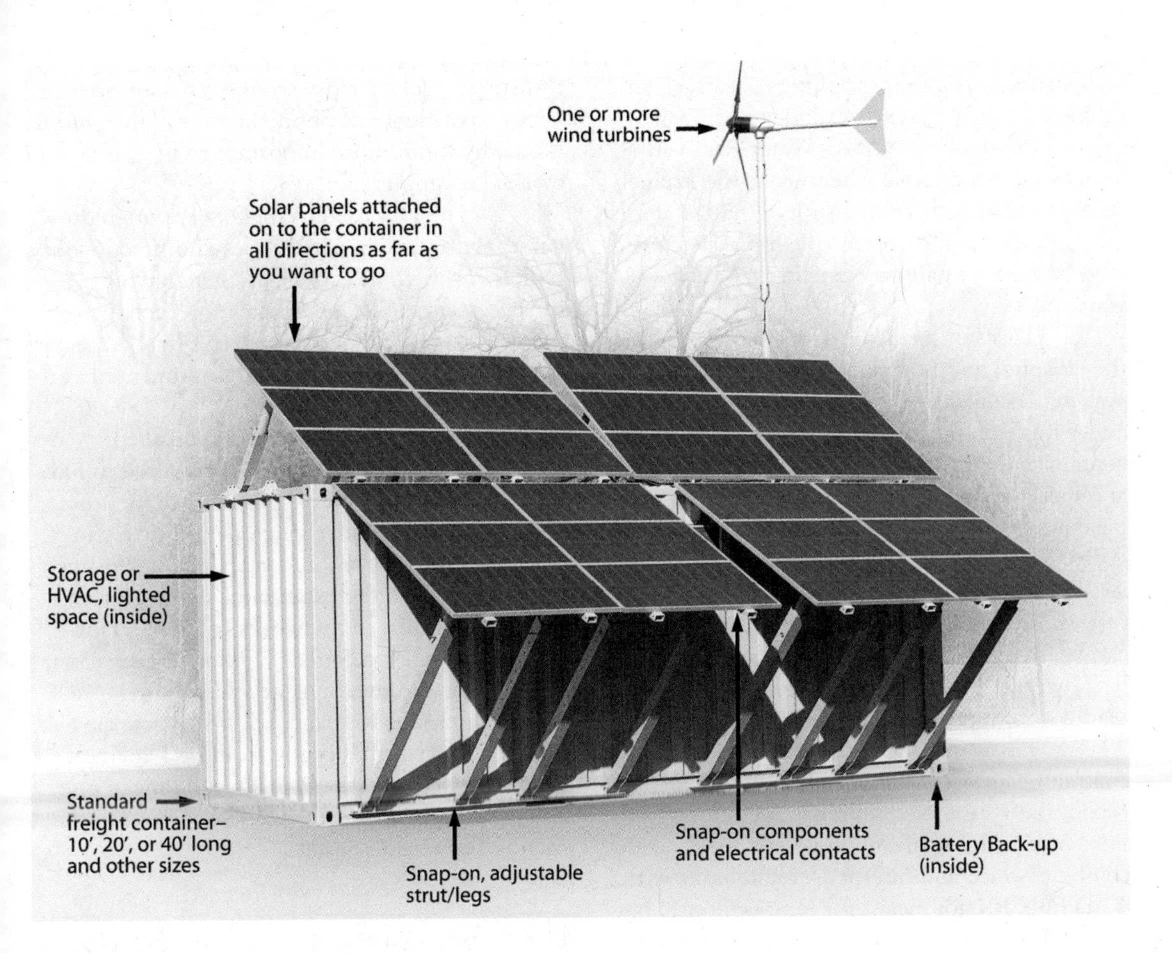

virtual storehouse for all of those field reports and updates—one that is easy to set up, access, and organize.

We don't want to think about the next disaster that may strike, but one inevitably will, and by employing the Strong Angel findings, workers may be able to keep ahead of the chaos. CB

Disaster Response in a Box (or Two)

When it comes to outfitting remote areas, mobility is as important as utility. The best technology out there won't be very useful if it can't be easily transported to the people who need it. With this in mind, several innovators have designed systems for compact medical support, renewable power, water purification, and networking telecommunication.

One of the biggest stumbling blocks for aid workers is finding a consistent, reliable source of power in places where power lines and stations have been destroyed—or places where there was no energy infrastructure to begin with. SkyBuilt Power's Mobile Power Station (MPS) solves this problem beautifully. By using a combination of modular solar panels, wind microturbines, batteries, and plug-ins for fuel cells and biofuel-friendly diesel engines, the MPS can generate a constant 150 kilowatts. It can operate both off-grid and in parallel with grid power, is rugged enough to be dropped via parachute, and requires so little maintenance, SkyBuilt says, that one of their solar/wind units has been operating for a year continuously without being touched. SkyBuilt's open architecture makes it possible for other vendors to build add-on components, confident that the components will work together properly.

A single MPS would provide more than enough power to run a reverse-osmosis water-purification kit now in operation in the Maldives. With only a hundred watts of power, the unit, designed by Solar Energy Systems Infrastructure, can purify 132 gallons (500 liters) of brackish or disease-laden water. Just under a kilowatt of power could help NASA's Water Recovery System—indispensable when water is extremely polluted or in such limited supply that otherwise unusable sources, such as urine, must be considered—create 35 gallons (121 liters) of drinking water per day.

Disaster medicine gets much easier with the Hospital in a Box. This portable medical system, designed by medical technician Alexander Bushell and consultant Dr. Seyi Oyesola, contains a defibrillator, an operating table, an anesthesia system, a burns unit, and plaster-making equipment, allowing a team of up to three doctors to carry out common emergency surgeries. The unit comes with a tent, which essentially turns it into an instant field hospital. The system can be dropped by helicopter into remote areas. It's powered by a truck battery, and can be readily recharged via solar panels.

Lastly, to ensure that aid workers can communicate with one another and the outside world, there's the NetRelief Kit, a communications hub built specifically for relief work, which combines voice and Internet satellite links with a Wi-Fi hub. It's not meant for long-term use, but rather for serving as a ready-to-go communications system for immediate-response workers. Inveneo [see ICT4D, p. 299] provides similar services, but over a wider range and longer period of time. Inveneo is built to link into a local Global System for Mobile (GSM) network and provide Internet and voice communications across an otherwise unserved region—all using free/open-source tools.

If we put all these pieces together, we end up with a system that provides both short-term and long-term support for disaster-struck communities' power, water, health, and communication resources. At this rate, we'll soon have a complete disaster-response center that fits on one flatbed truck. JC

Disaster IT and the Shelter Computer

Access to food, water, and safety is a refugee's primary concern, but the ability to contact loved ones and get news about what has transpired ranks a close second. Disaster shelters try to have plenty of phone lines available, but it is equally if not more important to have networked computers for evacuees.

The PCs used in shelters are often donated, with varying capabilities and functionality. Relief workers can't count on them having all the necessary tools and applications emergency users tend to need, and they aren't likely to have the time to download applications and configure each PC perfectly.

But there's a solution: a so-called liveCD, a bootable CD-ROM configured to have all of the necessary pieces of software. Because it's bootable, the underlying operating system is secure from viruses, and each machine using the CD can have the exact same configuration and applications. In the weeks following Hurricane Katrina, Jon Stokes worked in Louisiana, assisting in the development and maintenance of computer labs for evacuation shelters; the liveCD for Disaster IT sprang from his experience. For future shelters, the liveCD could mean one less obstacle to evacuees' need to connect with friends and loved ones. JC

The South-East Asia Earthquake and Tsunami Blog

The South-East Asia Earthquake and Tsunami Blog (SEA-EAT blog) launched on December 26, 2004. It became the most important repository for news and information about resources, aid, donations, and volunteer efforts around this disaster. Within three days, 100,000 visitors had viewed the blog; within eight days, over a million had. Only three people were contributing on day one; more than fifty people on day three; and more than two hundred at last count. Contributions came from volunteers not only in affected areas like India, Sri Lanka, Thailand, and Malaysia, but also from Europe, the USA, and the Caribbean.

It truly was a global effort that reflected a collective need to overcome feelings of helplessness and do something to make a difference. SEA-EAT became a community, a network, and an open space where anyone could contribute.

The first stone was laid as a spontaneous gesture from Peter Griffin, a blogger who, within hours of the disaster, invited two Worldchanging contributors to blog at SEA-EAT. So began the blog—three people working in real time with real people.

Shortly after the initial impact, people began to respond, transmitting their heartfelt reactions into the most immediate and receptive outlet that they could access—the Internet. Text messages from journalists and volunteers doing relief work promptly found their way onto SEA-EAT and other Weblogs. This became the most basic mode of communication, at a time when cell phone signals were too weak to support spoken messages.

SEA-EAT bloggers aggregated firsthand accounts, reports, and pictures of the devastation from bloggers who happened to be in the affected zone. At the same time, they captured other stories and statistics as they evolved and were published by other new sources, including a page started on Wikipedia by one person, which today has evolved into the best overall record of the disaster. Andy Carvin of the Digital Divide network set up a news aggregator of blogs and sites reporting on the disaster. Groups of bloggers from all over the world set up relief funds and aid channels. Others simply voiced their shock and grief at the event and pitched in by offering useful links to help the victims.

When Hurricane Katrina struck the Gulf Coast, the SEA-EAT team got together again and, by replicating its earlier model, set up one of the most complete repositories of aid and resource information available to the public at the KatrinaHelpBlog and KatrinaHelp Wiki (a related site in which any reader, not just the bloggers, could add or update content). This time, a virtual helpline and phone bank was also set up using Skype technology (a voice-over-Internet system that allows users to make and receive calls that are routed through their computers rather than through traditional phone lines). After the 2005 earthquake in India and Pakistan, the South Asia Quake Help blog and wiki were created, as well as a short message service (SMS) reporter blog.

It is fascinating that these tools that didn't exist a few years ago—blogs, wikis, and photo-sharing Web sites like Flickr—were put to use so quickly in the recent disasters. Today, it's likely that no major crisis will ever be handled again without SMS, blogs, and wikis. The social tools that many of us already take for granted will become a natural extension of rapid adaptation under chaotic conditions. DHM

RESOURCES

Design Like You Give a Damn: Architectural Responses to Humanitarian Crises edited by Architecture for Humanity (Metropolis Books, 2006)

If you care about the future we're building, you ought to own a copy of *Design Like You Give a Damn: Architectural Responses to Humanitarian Crises*. This book should sit on the desk of every designer, architect, and engineer who believes that changing the world is part of his or her job.

Much of the book centers (as one might expect) on housing and shelter, but other fields (sanitation, planning, etc.) are covered as well, with overviews of illustrative design innovations in each field—barefoot solar engineers, land mine-detecting flowers, Hexayurt, Roundabout's PlayPump, the Mine Wolf, Watercone, and Anti-Malarial Bednets. There's plenty of material here we'd never encountered before and, as an overall resource, it's indispensable. *Design Like You Give a Damn* is worldchanging.

Open-Source Humanitarian Design

With humanitarian crises, as with so many of the problems facing our planet, collaboration is not only our best hope for finding solutions: it may become our only hope. The large-scale disasters we face are so profound, their momentum so fierce, that unless we put to use the energy and creativity of every person of good will, we cannot possibly overcome them. Relief and reconstruction need to be approached holistically and openly, or we risk masking a symptom instead of curing an ailment.

Too often, humanitarian and developmental assistance comes from a concept so far removed from the crisis at hand that it simply falls short or fails. It can be especially disastrous when newly formed organizations implement untested ideas on the ground. Historically such ideas have led to economic, environmental, and societal damage in the communities they were meant to help. Many prefabricated structures become useless once a building element fails and needs to be replaced. Implementing culturally and locally inappropriate structures can create resentment and mistrust within a community. Not integrating urban planning strategies of growth and renewal can create permanent refugee camps where villages and towns should evolve. Bringing in large quantities of donated materials and free labor can disrupt an already weakened economy; introducing new high-tech solutions can preclude the potential to hire displaced workers for the rebuilding process, since their skills won't be a good match.

By embracing open-source technologies and design [see Open Source, p. 127], and removing barriers to the improvement, distribution, and implementation of well-designed solutions, we can, more than ever before, ensure that people in crisis receive innovative, sustainable, and, most importantly, dignified shelter. Since the mid-1990s, the sharing of information and technology has steadily gained popularity in the high-tech and arts communities. Why not adopt this approach in the area of humanitarian reconstruction and long-term development? We have an opportunity to provide displaced populations with localized solutions that will last longer and integrate disaster mitigation technologies to protect them in the future. By opening design up to collaboration, we arm ourselves with the greatest number of strategies, skills, perspectives, and tools available, and we guarantee that those tools can be freely distributed and adapted by local teams, on the ground.

When we welcome real, local collaboration into the design and reconstruction process, we have the means not only to build more appropriate housing, but to stimulate local economic development that can help arm communities with new skills and encourage new industry as they rebuild their own lives. As sustainability guru Hunter Lovins replied when asked in a 2006 *Treehugger* interview whether economic development could go hand in hand with sustainable development, "We know how to meet people's needs for energy, for water, for housing, for sanitation, and for transportation, with much more sustainable technologies than are traditionally brought by development agencies. Most of what is called *development* around the world is really donor nation dollars hiring donor nation contractors to deliver last century's technologies, in such a way that the jobs and the economic benefit go right back to the originating donor country." This is a closed system—closed in its process, and closed to those whom it's intended to support. Similarly, much humanitarian design is structured to make use of what donors can provide, not to provide what victims actually need.

Benefits, clearly, should flow toward those in need, not the other way around. To date there is no network of proven design ideas that can be freely distributed to nongovernmental organizations (NGOs) for adaptation, implementation, and use in the field. If we are truly designing for the greater good, why not create a platform that allows for an open exchange of ideas?

In both architectural-design and humanitarian spheres, the open-system concept has yet to take hold. Organizations and individual designers alike fail to collaborate, and end up not only reinventing the wheel, but spending large amounts of limited resources on stopgap approaches that have failed time and again. We need to become more open in sharing both our successes and failures so that we learn and adapt. Open systems of collaboration can help us create better-designed, more workable solutions to the world's humanitarian challenges faster.

What we need is an online tool for sharing and improving upon collaborative designs: a vehicle for distributing, honing, and reinventing existing tools, and mechanisms for innovating new answers to current and emerging problems. The vehicle would ideally take the form of an online resource consisting of a database of thousands of proven designs and best practices, a rendering tool with a built-in simulation of austere environments, localized subsites for regional NGOs, and the capability to facilitate needs-based competitions, project tagging (assigning words that define key aspects of the project to facilitate computerized searching), the integration of local and cultural data, and the protection of designers' intellectual property rights [see Copyfight, p. 336].

Equipped with a searchable database of sustainable and innovative designs, we could better respond to real-world disasters as well as systemic issues that continue to plague our communities. Using the network, an international aid group responding to flooding in China, for example, could identify all the past projects that have dealt with flooding in areas with similar climatic and geographical issues. The group could then connect with an experienced design team to help tackle the situation at hand. Most importantly, they can make decisions about the allocation of funds early in the process, even though rebuilding would not begin for months. This will allow them to begin the reconstruction process earlier and develop an appropriate and sustainable response.

Designers often face a difficult choice between distributing their designs freely so that they can be used for the greater good and protecting them from possible misuse or profit by others. This is where a new system of intellectual property rights, Creative Commons licensing [see Copyfight, p. 337], and the idea of a "developing-nations license" comes into play. We can refine a license for use allowing design and engineering professionals to determine a level of copyright supporting widespread distribution throughout areas in need—while giving them full protection in the developed world. International treaties ensure such agreements are enforceable just about anywhere.

Humanitarian reconstruction must evolve from plastic tarps and Quonset huts to a system of community engagement and sustainable renewal. The reconstruction and design communities need to advocate for innovative ideas and support them with both local and international funding. New avenues for collaboration need to be established between designers, and new systems of intellectual-property protection need to be devised that will let designers protect their own interests while meeting the needs of millions—perhaps billions—for shelter, sanitation, and ultimately, a secure economic future. CS

Land Mines

Land mines are unquestionably some of the most evil devices created in the twentieth century. Deadly and cheap (they cost as little as three dollars to make), they have been scattered extensively throughout war zones around the world. The U.S. Department of State reports that at least 45 million are still out there in the ground, waiting for someone to step on them. Worldwide, as many as 1,500 people a month are killed or maimed by land mines, according to the International Campaign to Ban Landmines.

The carnage land mines cause reflects only a part of the harm they do. Because land mines were generally placed where people were most likely to walk, farmland that might otherwise be feeding hungry people is left unplowed. Roads and trails that might otherwise link villages and allow wealth-building trade go untraveled. Towns that might otherwise shelter people have been abandoned.

Removing buried land mines is incredibly difficult, and can cost up to a thousand dollars per mine when done by trained professionals. In addition, because land mines are mostly made of plastic now, they cannot be found with metal detectors. Some types can be found with trained animals, but the most widely used technique for removing them is for demining experts to crawl along on their bellies poking at the ground ahead of them with a pole. The most widely adopted demining innovation of the last decade? A better pole.

According to the United Nations, at current rates, it will take more than a thousand years and $33 billion to clear all the mines that have already been deployed (hopefully, with widespread ratification of the land-mine treaty [see Connecting with Others, p. 428], existing mines will be the last ones we have to worry about). If ever there was an arena for worldchanging innovation, finding better ways to clear land mines is it. Luckily, breakthrough ideas are beginning to emerge. AS

MineWolf

The MineWolf is, as the United Nations' ReliefWeb puts it, "the first mine-clearing machine that deserves this name." It can clear land nearly ten times as fast as its competitors can, and two hundred times as fast as people can with the aid of dogs and poles.

This breakthrough comes from Heinz Rath, a retired German engineer who decided to approach land-mine removal in the same way as removing a sugar beet—a very cagey sugar beet—from the earth. The MineWolf is essentially a heavily modified tractor, but a nearly million-dollar one. It can clear close to 21,528 square feet (2,000 square meters) per hour.

This is the first mine-clearing machine designed with humanitarian, not military, purposes in mind. Instead of clearing a path for advancing troops, the MineWolf clears land for civilian use after wars are over. But preparing to farm or build on an area laden with land mines involves more than just clearing the mines. The MineWolf is designed to manage all stages of the transformation, from minefield to cornfield. Part tank, part tractor, the MineWolf beats its way across the terrain, detonating antipersonnel mines with its flail, and antitank mines with a tiller attachment. All the while the driver sits protected in a blastproof cabin. After an area has been demined, operators can switch MineWolf's demining tools with a mulching attachment or excavator and begin to improve the land. JJF

The Dragon

What's more powerful than a wolf? A dragon. The MineWolf is a great innovation, but it has one problem: it removes mines by breaking them up or by detonating them. Exploding land mines, even to remove them, can be dangerous for people nearby, and can spray toxic chemicals around the landscape.

By contrast, the Dragon, a device designed by demining specialists Disarmco and explosives experts at Cranfield University in the United Kingdom, is a pyrotechnic torch that destroys land mines by burning them out instead of detonating them. Dragons can be assembled in the field using portable production systems and local materials. They are simple enough to use that civilians can create and employ them without extensive (and expensive) training or the supervision of a specialist.

The first Dragon prototypes were tested in Lebanon in 2004 and used in Cambodia in May 2005. Although the Dragon does not have the broad utility of the MineWolf, it makes up for that by delivering low-cost, easy-use, and safer (for both people and the environment) mine removal. JC

Removing land mines is a costly, slow, and dangerous process. Halo Trust (left) trains de-miners in Kabul, Afghanistan, 2005. Innovations like the MineWolf (right) promise to help eliminate the 45 million land mines still scattered around the world's war zones.

Land Mine–detecting Flowers

Scientists at the Danish company Aresa Biodetection are using nature to help detect land mines. As the explosives used in the mines gradually decay, their chemicals leach nitrogen dioxide into the surrounding soil. Aresa has developed a genetically modified version of a flower (more like a weed, actually) that changes color when its roots come into contact with NO_2. The thalecress is already coded to change color—from green to red or brown—when exposed to adverse conditions such as extreme cold. Aresa simply recoded the plant to react that way only in the presence of nitrogen dioxide. The plant was chosen in part because it can sprout and mature in just six weeks.

The company has begun field tests, and actual implementation could happen within the next couple of years. Though it may seem counterintuitive to try to sow seeds in a minefield, Aresa is confident that any danger can be averted by using crop planes to distribute the seeds or by clearing selected strips of a field using more conventional planting methods. JC

Adopt-A-Minefield

We can all help to eradicate land mines. First, we can urge our countries to sign the landmine treaty—believe it or not, the United States has yet to do so. Second, we can "adopt" a minefield. Through the United Nations Association's Adopt-A-Minefield campaign, corporations or donors with a lot of pocket change can adopt a minefield, funding a demining team for the entire two months it takes to return a swath of land to the civilians that desperately need it for their livelihood. The rest of us can make smaller donations that, pooled with other donations, go toward minefield clearance or toward the organization's survivor-assistance program, which provides people who have survived landmine accidents with medical attention, small loans, job training, and counseling. CS

Rethinking Refugee Reconstruction

Many refugees end up living wherever they land for several years before permanent housing can be established or they can return to their homes. Providing these refugees with housing is a primary and pressing global need.

Most of us don't see the effect of short-term thinking in disaster relief. Often, months after the last of the TV crews have left a disaster zone, children are still being taught under plastic tarps, waterborne diseases have spread throughout camps due to poor sanitation and inadequate infrastructure, and tents have deteriorated to the point that refugees begin deforesting the surrounding land for materials to build and repair housing.

Even where disaster-reconstruction efforts exist, the housing built is often woefully inadequate or wildly inappropriate. There is a crying need not only for better models, but for better methods of working with refugees, to create longer-term housing.

This is a time of accelerated emergence of innovative thinking. From every corner of the globe we've seen proactive and pragmatic ideas for housing, infrastructure, community, and sanitation projects. Some

Right: Thousands of Palestinian refugees, whose families fled the Galilee region of Israel in 1948, live at the Jaramana refugee camp, Damascus, Syria.
Opposite: Teaching children can mobilize a demoralized community in the wake of a disaster, Afghanistan.

of these great designs have already been deployed. But with little funding available for shelter development, only a handful of designs have been implemented, a dozen have been built as prototypes, and most have never made it off of the drawing board. To date, not one of these experimental solutions has been implemented in great enough numbers to make a real difference for a substantial number of people. Without the means to do in-depth research into and evaluation of these ideas, many aid agencies are reluctant to fund and implement the projects, particularly in emergency situations.

If we are serious about facilitating truly sustainable renewal for areas affected by disasters, we need to become proactive in introducing innovative and sustainable solutions. Most importantly, we need to recognize that solutions we provide for (or impose upon) people never work out as well or become as useful as solutions we help people implement for themselves, using their own skills and engaging with their own lives. Innovation in refugee housing must ultimately be innovation in helping others build what they need. Encouraging communities to be active participants in the rebuilding of their lives is key to creating sustainable solutions and reducing the impacts of disasters. CS

Teaching Kids in Refugee Situations

There's been a humanitarian crisis. People are miserable, hungry, sick, perhaps wounded, certainly traumatized. What are the first priorities? Would you believe that the first priority is teaching the kids? While many people still argue that it's best to wait on education until life has returned to normal, research indicates that getting schooling quickly under way after a disaster has far-reaching benefits. In fact, it can create the sense of routine and regularity that falls away in crisis, offering a focal point for distressed kids, and establishing a safe, central location for families to meet up.

As the 2002 issue of *Forced Migration Review* (perhaps the most disturbingly titled social studies publication in the world, if you really think about it) suggests, by creating "safe zones" for kids and providing access to essential knowledge, educators can not only help the community return to a sense of psychological stability, but can also save a generation that might otherwise be lost. Indeed, involving parents and elders in the process of reinstating schooling can help move a whole community out of traumatic shock and into action. AS

Access to Technology

Technology and collaboration are additional tools that can help refugees process emotional trauma and begin mental recuperation. Cheap, discardable video cameras can allow refugees to record their stories in order to begin healing, while serving to inventory their skills. Open-source textbooks [see Education and Literacy, p. 315] rendered in the local language via collaborative translation can help spread literacy and education quickly through the population. *Telecentros* [see Brazil's *Telecentros*, p. 300] and other community technology resources can help bring real opportunity even to extremely impoverished rural people while—in the bigger picture—helping to redistribute the future. AS

The Power of Radio

In the developing world, radio is a critical tool for spreading knowledge, organizing community resources, and pursuing sustainable development. Regions that lack resources and basic services can benefit tremendously when an individual or a group of citizens can build or access a transmitter and organize sufficiently to produce a broadcast or even set up a radio station. Numerous stories have emerged from the developing world that demonstrate the power of disseminating information over the airwaves. When everything else shuts down, radio can keep vital lines of communication open [see ICT4D, p. 296].

Freeplay

Radio can catalyze education for children in dire straits. A solar-powered, hand-cranked radio that is tuned in to an educational station, the Freeplay Lifeline Radio requires no power supply or teaching staff. Placed in a tent or a community gathering space, the radio becomes a tool for learning.

The Freeplay has been distributed throughout the world by international aid agencies, which use it to provide news bulletins during times of disaster, in addition to the regular educational programming. In remote areas, the Freeplay broadcasts school classes and even public health information; in Madagascar, a serial radio drama teaches AIDS education. The Freeplay is considered such a desirable and useful tool that the Nigerian government has successfully implemented a program trading the radios for illicit guns. CS

Darfur Stoves

Fuel-efficient cooking stoves are some of the most important tools being deployed in the developing world [see Using Energy Efficiently, p. 168]. Inefficient stoves cause indoor air pollution, damaging the health of women and children, in particular. The fuel for inefficient

A tsunami survivor makes a much-appreciated satellite phone call to her friends, Banda Aceh, Indonesia, 2005.

stoves is expensive to purchase; alternatively, it is time-consuming for the women—generally—to stock with wood or dung for burning. In refugee camps in Darfur, there's an additional reason that fuel-efficient cookstoves are important: when women leave the camps to gather the materials for fuel, they are vulnerable to attacks from rebel groups and Janjaweed militants. Teaching women how to cook more efficiently allows the women of Darfur to avoid violence as well as to feed their families. Fuel-efficient stoves and techniques help women prepare meals of *assida* (Darfur's staple starch) with less than half the wood they would otherwise use—and to do so free from danger. EZ

CITIES

We live on an urban planet. For the first time in history, a majority of us live in cities. How we grow those cities, how we build neighborhoods, how we provide housing, how we choose to get around, how well we incorporate nature into the places we live—these are the challenges that will largely determine our future.

Those challenges loom large. In the metropolises of the Global North, we face legacies of neglect and pollution, traffic jams, housing shortages, aging infrastructures, and suburban sprawl. Meanwhile, in the booming megacities of the Global South, the problems can seem massive and unsolvable: exploding populations, crippled local governments, poverty, need, and collapsing systems. And with millions and millions of people moving every year from the countryside to the city, all of these difficulties seem even more insurmountable.

Appearances, however, can be deceiving. For, along with the boom in urbanization, we're seeing a boom in urban innovation. Simply put, we're getting better at building cities.

But are we getting better fast enough? Are the problems getting worse quicker than we can imagine solutions? On the results of that race—between urban possibility and urban collapse—hang many of the hopes of humanity. Cities are the key to a better future, and in order to ensure that future, we need to understand them, to consider why they matter, and to try to make them better.

In some ways, urban life feels timeless. In depictions of cities from a thousand years ago, we can recognize the same social patterns we know today, the same economic structures sustaining so many people residing in one place. Human behavior does change very quickly. In a number of different ways, though, the cities we live in today are entirely new creations. Their size, the speed at which they change, the disparity between their richest and poorest residents, the global interconnection they give rise to, and the varieties of cultures they host—these are all unique to the twenty-first century. The magnitude and velocity of change in the world's cities today dwarfs anything we've ever known.

If we could deconstruct a city like we could strip an engine, and lay all the pieces out on a cloth on the lawn (a very big cloth, a very big lawn), we'd be stunned by how many moving parts a city has. Leaving aside the most interesting part of urban life—the people and their relationships to one another (for cities are, above all else, the original social software)—we'd find a mass of large systems: the power lines strung out in a patchwork (in some cities, crisscrossing the entirety; in others, reaching only the richest neighborhoods); the branching pipes that carry water (in some cities, delivering it to home taps; in others, to central pumps from which people carry their water home in buckets); the spun glass of telephone lines and radio waves and satellite signals (in some cities; in others, the weekly mail delivery bicycle). Everywhere, we'd find the overlapping grids and weird rootlike structures of flight paths and train tracks and roads and sidewalks and trails. These physical systems of large cities make for the most complicated machines we have ever built.

When we think of large cities, we're used to thinking of London, New York, Tokyo. But by 2015, there will be dozens of new megacities—many about which most of us in the developed world know little. What, for instance, do you know about Lagos, Nigeria, [see Lagos, p. 279] and the lives of its people? If you answer "nothing," you're hardly alone. And yet, by the year 2015, the United Nations predicts that it will be home to more than 23 million people—making it the third-largest city in the world. As is true in much of the

Global South, many of the people who live in Lagos live in slums. Unfortunately, this common occurrence contributes to why we are used to thinking of cities as problems.

But in reality cities hold out tremendous promise as we try to steer a course toward sustainability. Urban living, especially in compact communities, is a powerful tool for reducing our ecological footprint. Growth that is concentrated rather than sprawling preserves farms and forests outside the city, and helps facilitate the adoption of new technologies and techniques for building green cities, which will have a great impact on the planet's future. If we are serious about sustainability, one of the most consequential things we can do is to vote with our address and live in a city where bright green solutions abound.

Cities also promote prosperity. They are the engines of our global economy, offering greater opportunities for finding a job, educating our kids, starting a better life. But well-designed cities don't just help comfortable people become prosperous, they meet our most basic needs—from clean water and adequate housing to education, health car, and other social services—better than spread-out suburbs do.

And no two cities are alike. Even present-day cities modeled on past ones (like Shanghai's Bund, built to resemble a European city), end up entirely unique as people live in them, use them, and change them.

A few people know just about everything about their cities and can tell you the history, shed light on the character, and reveal the hidden, but those people are extremely rare, and their knowledge takes a lifetime to collect. People like that, however, are the maps and encyclopedias that can help us figure out how to transform our cities—and before the breed vanishes, we need to learn its secrets. We have to know a place in order to make it better, because a given urban-planning tool never works for every city, and because the more we know and love a place, the more we want to participate in determining its evolution. Much of the power to direct our cities' futures rests in the hands of politicians, planners, and powerful interests (the wealthier the place, the truer this is), but it's increasingly the case that we citizens hold the tools, models, and ideas to demand better solutions, and even to begin implementing them ourselves. We are the visionaries and collaborative architects of the cities we inhabit. AS

Nearly three million Greeks, one third of the country's population, now opt for an urban lifestyle in Athens.

The Bright Green City

If you want to live green, live in a city.

Seen from the perspective of the twentieth century, that truth seems counterintuitive: how can cities, with their closely packed homes and tall buildings, be greener than leafy suburbs or small rural towns?

The answer has to do with what is seen and what is hidden. The trees and lawns of our suburbs may look green, but they hide from our eyes all the various ways we consume and pollute. Our backyards in the burbs may have bushes and birds, but our daily commutes are fouling the air and water with smog and oil, and our overly large homes have put a strain on the planet. We notice the birds, but grow a convenient blind spot when it comes to our monthly power bill or to the hour we spend on the freeway.

Studies undertaken by John Holtzclaw of the Sierra Club and Jennifer Henry of the U.S. Green Building Council have shown that people who live in drafty old homes in compact neighborhoods use less energy (and spew less pollution) than even those suburbanites with new green homes and efficient appliances.

How could that be? How could an apartment in New York, where you can see (and taste) the grime, be greener than an Energy Star house on a wooded lot? The answer: Density is efficient.

Spread a neighborhood out—put houses, stores, offices, and schools far from one another—and what choice do we have but to drive? Make neighborhoods more compact and tightly knit, and it becomes more likely that the things we need are close at hand. A trip to the store for milk in suburbia amounts to a five-minute drive to the supermarket; in the city it becomes a quick walk to the corner grocery. This is what planners call access by proximity.

Compact neighborhoods make sustainable living easier. Once your community has about twelve homes per acre (the density of many traditional single-family neighborhoods), public transportation becomes cost-effective, stores can find enough customers to justify locating nearby, and people start walking or biking. Once you reach forty homes per acre (imagine a street of brownstones), not only do the sidewalks come alive, but everything gets cheaper and easier to provide: heating, electricity, sewage lines, and other basic services. And when more people share these services, they all have far less of an impact on the planet.

The denser cities get, the more efficient they get. As David Owen notes in an article in the *New Yorker*, "New York City is more populous than all but 11 states; if it were granted statehood, it would rank 51st in per-capita energy use." Manhattanites

Dense, urban communities use less energy and far fewer resources than do sprawling suburbs.

use fewer resources and less energy than anyone else in America. AS

The Costs of Sprawl

Suburban living is also far less green than urban living because building in natural areas destroys the ecology of those areas, even if they still look pretty. Slapping up even a couple of new homes in a forest can trash local streams and drive off wildlife.

The term *watershed* describes an area of land whose rainfall feeds a particular stream or river. Studies have shown that roofs, driveways, and lawns alone on one-tenth of a watershed's land will biologically undermine the streams and rivers in it. The paradox here, of course, is that our eagerness to live close to nature often contributes to the destruction of the wild areas we moved out of the city to enjoy.

Because homes can be so damaging to nature, the best policy is not to build in untouched areas at all. Conversely, when we build our homes in existing communities—so-called infill housing—our homes have comparatively minimal impact on surrounding ecosystems, since most of the damage has already been done. If we want to live green, we ought to live in a city and leave the woods and meadows alone.

Sprawl costs us in other ways as well.

Unlike crime, sprawl pays. In North America, and increasingly elsewhere as well, powerful interests—big developers, land speculators, construction corporations—get rich off sprawl. The sprawl lobby is strong and ruthless. When an unprincipled developer plops down a new tract on the fringe of a metro area, taxpayers often subsidize the project by paying for the new roads, sewers, and schools to serve that subdivision, which can cost as much as $25,000 per new house, even while older infrastructure is in serious need of repair, and existing urban schools are overcrowded and underfunded.

This is patently unfair, of course, but not at all unusual. Some of our wealthiest suburbs and new subdivisions have a bag of tricks (like requiring new homes to be big and situated on large lots) to keep out everyone but the rich—even as the wealthy new homeowners rely on urban services and amenities from regional hospitals and cultural institutions for which they do not pay. In North America, we've evolved a system in which people who live in central cities not only suffer from the harmful effects of sprawling subdivisions, but often end up paying for the privilege.

There are human costs to sprawl, too. Typically, people living in suburbs are more overweight than their urban counterparts, and suburban kids tend to watch more hours of TV daily than their peers in cities and small towns.

Luckily, we now have a host of tools for understanding and dramatizing the costs of sprawl, from mapping software [see Mapping, p. 518] to growth-management plans. These tools can empower citizens to predict the effects that new kinds of development are likely to have on their communities. Armed with that knowledge, citizens might now be able to put a stop to exploitative land development. AS

RESOURCES

The Next American City
http://americancity.org
This print/online journal glides smoothly along the line between visionary ideas and practical answers, providing a thought-provoking exploration of the ways in which American cities are changing. Recent issues have covered raising kids in cities, contemporary racial segregation, the shifting role of neighborhoods, and new approaches to transportation.

As the Web site states, "All of a sudden, we have left behind the basic assumptions that have shaped debates about urban and suburban economics, policy, and culture for the past half-century. *The Next American City* asks, 'Where do we go from here?' In this rapidly changing landscape, how can businesses and developers thrive? How can cities and suburbs expand their economies? And how can our society successfully address social and environmental challenges?"

Green Urbanism: Learning from European Cities by Timothy Beatley (Island Press, 2000)
If you want a crash course on the recent history of European urban innovation, reading this

book should be your first assignment. Though the writing can veer into jargon at times, the examples are inspiring: you can learn here how the Dutch city of Groningen encouraged its residents to ride their bicycles on 60 percent of their outings; how Denmark has charged ahead on powering its cities with green energy; how Sweden has promoted the construction of "eco-villages." One only hopes that an updated edition is on its way.

Beatley writes, "Perhaps the clearest lesson from European urban practice is that alternative patterns of growth and development do exist, and it is possible to organize space and public investments in ways that create compact, walkable, green communities—places that exhibit many highly attractive qualities. Public transit, walking, and bicycle use, under the right circumstances and with the appropriate long-term public policies, can serve as viable alternatives to reliance on the automobile. Substantially less energy and fewer resources can be consumed by citizens, and at the same time personal choice and mobility can be expanded. In short, these European cities, while clearly not perfect and clearly confronting a host of their own problems, show that different, more sustainable future paths do exist."

Alternative Urban Futures: Planning for Sustainable Development in Cities throughout the World by Raquel Pinderhughes (Rowman & Littlefield Publishers, 2004)
When it comes to explaining what we're capable of doing *now,* Raquel Pinderhughes's *Alternative Urban Futures* does so better than nearly any other book to date. In just over two hundred well-footnoted pages, Pinderhughes covers the basic widely available innovations for dealing with water, energy, transportation, waste, and food. She does a terrific job of rounding up the best established practices in each field and documenting the research and case studies that support them:

"Urban planners and policymakers will have to promote land uses and land use policies principally developed to shape the urban environment in ecologically responsible ways and enhance the livability of human settlements. They will need to support processes and technologies that are explicitly designed to use fewer resources and produce less waste; reduce inefficient patterns of production, distribution, and consumption; use water more efficiently; reduce dependency on nonrenewable resources; and increase reliance on renewable energy sources."

Vancouver

Vancouver, British Columbia, is the crown jewel of North American sustainable urban planning. Blessed with a spectacular setting, a vibrant economy, and rich cultural vitality, Vancouver grew in population by more than half over the last twenty years, yet Vancouverites have largely stopped sprawl, turning growth inward. They built up their downtown, adding tens of thousands of condos and apartments since the mid-1990s, all while making their city one of the most livable in the world.

More than 62 percent of Vancouver's residents now live in compact communities, according to the think tank Sightline, while another 11 percent live in dense neighborhoods of multistory buildings. If Vancouver had grown in the same way as its American neighbor Seattle, it would have sprawled out an additional 251 square miles (650 square kilometers) over the last two decades, Sightline estimates. Instead, Vancouver not only avoided sprawl, but increased its quality of life—improving its air quality, creating vibrant neighborhoods, and becoming the most walkable city on the western coast of North America. AS

Southeast False Creek

For Vancouver, it's not enough to have a few pretty streets and some green buildings. No, the city needs an entire Sustainability Precinct. Vancouver's city planners seem to be trying to jam every good idea they've ever had into the development of Southeast False Creek, an old industrial area near the city's downtown. The first stage is the athletes' village for the 2010 Olympics, but once the games are over, the second stage kicks in: the village will be converted into apartments (for people in a range of incomes), offices, and shops, with a wealth of green spaces: parks, community gardens, and green roofs [see Greening Infrastructure, p. 254]. In keeping with Vancouver's dedication to rescuing the city from cars, planners will design paths and streets to cater to pedestrians first, bicyclists second, public transportation (and there will be plenty of it) third, and cars last. Available parking will be limited, to further discourage car ownership. To cap it all off, all buildings will be required to meet or exceed a silver rating from the U.S. Green Building Council's (USGBC) LEED (Leadership in Energy and Environmental Design) certification program, a set of rankings that indicates the environmental sustainability of a building. If what's built bears any resemblance to the plans, Southeast False Creek will be the greenest neighborhood in one of the greenest cities in the world.

More People, Fewer Cars

Folks in Vancouver are giving up their cars. Though tens of thousands of people have moved downtown in the last twenty years, the number of cars on the road is actually decreasing. This phenomenon proves that when you design a city with pedestrians in mind, it's pedestrians you'll get.

Downtown Vancouver has excellent public transportation options, including its SkyTrain (an elevated light rail system) and a transit mall, but that's only part of the equation. A park rings the entire downtown area, giving walkers and bikers their own green highway.

A walk through the streets of the West End, one of the city's densest neighborhoods, reveals short blocks complemented by plenty of trees and grass. Everything's built on a human scale to make pedestrians feel welcome. Instead of resembling large, unbroken blocks, lower-income complexes are interspersed with luxury buildings and designed in architectural styles similar to the surrounding buildings, making them almost indistinguishable.

Vancouver city council members aren't content with the success they've had so far, and are seeking more ways to discourage car dependence. For example, they want the Insurance

Corporation of British Columbia to adopt a "car insurance by the km" scheme [see Urban Transportation, p. 262]—monthly premiums would be calculated based on kilometers driven, with higher rates for distances driven during peak hours. This pay-as-you-drive plan would be voluntary for drivers, but the council expects that it could cut car use by up to 30 percent—a significant impact on congestion and auto pollution.

Density Done Right

How do we make a city more sustainable and livable? Here are the main planning lessons we can learn from Vancouver:

Make developers pay their way. Hire developers who will treat the opportunity to build a highly profitable large building downtown as a privilege, and insist that the public benefit. Require developers to include public parks or green features in their plans and make financial contributions toward programs such as subsidized housing.

Set specific guidelines. Give developers strict parameters. In some areas of Vancouver, buildings are not allowed to exceed three hundred feet in height, and must be set back from the sidewalk to factor in trees and other green features.

Think big. Look for opportunities to redevelop whole multiblock areas, rebuilding outmoded infrastructure and carving out new public spaces.

Think small. Encourage responsible new building within existing neighborhoods, especially where you can replace empty lots, parking spaces, and old strip malls with high-quality housing.

Demand green buildings. Start with the public, work toward the private. All new public buildings in Vancouver must exceed Canada's LEED Gold standard, and the city is increasingly demanding that private developments meet basic green building standards as well.

Innovate everywhere. Look to the big-scale projects and to the small-scale ones. Though huge developments like Southeast False Creek tend to get the most attention, Vancouver's green initiatives have included projects as small as retrofitting traffic lights with light-emitting diodes (LEDs).

Lay out your city, as former city councillor Gordon Price says, "to be experienced at three miles per hour." Make walking not only easy but pleasurable. Put shopping areas and public transit within a five-minute walk of every home. Discourage car ownership. Invest in making public transportation safe, cheap, and reliable.

Emphasize quality of life. Protect key views. Make streets green and pleasant. Encourage the arts. Invest in public amenities (Vancouver's Central Library boasts not only an outstanding collection, but impressive architecture, a green roof, and free Wi-Fi). Know that the key to a sustainable city is offering a life richer in opportunity and more enjoyable in daily experience than anything that can be found in the suburbs.

Left: Vancouver is one of the most livable cities in the world; its commitment to compact communities, walkable streets, and green building make it an icon of sustainable urban planning.
Opposite: This rendering of the proposed expansion to the Vancouver Convention Centre highlights what will be the largest green roof in Canada.

Three Totems of Vancouver Green Building

The University of British Columbia
All new buildings on the UBC campus have to meet LEED standards, but the C. K. Choi Building, the university's first green building, is particularly impressive. It has continuous dimming light sensors to save energy, composting toilets, and gray-water recycling and rainwater-collection systems that provide irrigation for landscaping. The Liu Centre for the Study of Global Issues is constructed from fly-ash concrete, a by-product of coal-fired power plants and an effective substitute for standard cement. Not only does the use of fly ash in construction put this waste product to use, it helps to reduce emissions by cement-manufacturing operations.

The Vancouver Convention Centre: Canada's Largest Green Roof
The Vancouver Convention Centre is being expanded to accommodate the media during the 2010 Olympics. The roof of the complex will feature a six-acre preserve of native coastal plants. As a living habitat, the green roof will provide a complex system of water catchment, cleaning, and irrigation. British Columbia's rainy climate will allow the roof to capture excess rainwater for landscaping around the building, but when the skies don't supply enough water, a treatment system will clean and reuse water from within the building, so none will be required from municipal sources.

In addition, the greenery atop the building will protect the building from sun damage and help it maintain a more constant indoor temperature, insulating it during winter and keeping it cool in summer. Building operations will enjoy significant reductions in heating, cooling, and water costs. And, of course, the lush landscape will draw birds and insects that can contribute to the city's overall health and biodiversity.

Vancouver Materials Testing Facility
The Vancouver Materials Testing Facility houses an asphalt plant and a lab for testing materials before manufacturing. When it built a new home alongside the Fraser River, the facility made use of the existing warehouse buildings on the site. By salvaging and reusing many parts of the old buildings in constructing the new one, the archi-

tects saved their clients heaps of money, and diverted heaps of debris from area landfills. According to the USGBC's Cascadia regional report, the architects saved a total of C$140,000 (U.S.$125,470) from the project budget by making use of 80 percent recycled materials, from structural materials to furniture.

Additional green-building features at the facility include abundant natural light and views of the river, natural ventilation and operable windows to eliminate air conditioning, and well-planned vent systems in the lab for the protection of workers' health. But it is the highly successful use of junk as construction material that makes this building a great model for future Vancouver projects—and building projects everywhere. SR

RESOURCE

Dream City: Vancouver and the Global Imagination by Lance Berelowitz (Douglas & McIntyre, 2005)
Lance Berelowitz performs what he calls an act of "urban archaeology" as he traces Vancouver's history of growth and its innovative urban planning. Although he writes with great affection about his city and its aspirations (he was, after all, the editor in chief of the city's Olympic winter games "Bid Book"), he understands that archaeology often uncovers tough truths, and his exploration of the city's role in setting a global standard for urban planning goes beyond merely touting its many successes. He takes a critical look at the myths the city has built around itself to give us a balanced picture of Vancouver's model for sustainability: "Vancouver's public space has been increasingly appropriated by commercial and promotional interests, often at the expense of truly ethnic-based, community-based, or class-based activities. Manipulation and dispersal are creating a paradigm that is increasingly hostile to traditional public space. In Vancouver, this paradigm is couched in the seductive terms of 'natural' leisure, making it extremely difficult to express the inherent threat to a populace in danger of pleasuring itself into civic marginalia."

Portland

How we come together as citizens to plan the future of our cities essentially predicts how successful those cities will be at finding ways to grow more livable, more prosperous, and more ecofriendly. Formal plans are of course important for determining what can be built where, how transportation budgets will be spent, and a host of other critical decisions, but what really separates great urban plans from faulty ones is not just legal code, but the degree to which residents act in coalition to create change, work at the proper regional scale to affect the big systems surrounding their cities, and use better tools for visualizing the future of their cities and neighborhoods. Portland, Oregon, in many respects, has the best existing models to offer today, but new tools are also emerging that will shift the balance of planning power away from bureaucracies and back to citizens, so that more Portlands might emerge. If we can create a combination of approaches, so that citizen coalitions work with both the vision of their neighborhoods and the expertise of planners, we can transform the futures that unfold on the streets of our hometowns.

What's the Big Deal with Portland?

If we hang out with planners, we hear a lot about Portland, Oregon. It's small—not a "global city." We might say, "What's the big deal?" The answer is this: Portland is the only U.S. city to have actually created a regional government with the power to control sprawl. The Oregon system was created under the leadership of Republican governor Tom McCall, who lead the effort to pass state Senate Bill 100 in 1969. This bill, motivated largely by the desire to protect agriculture from encroaching urban sprawl, required every local jurisdiction to have a growth-management plan. The system did not emerge all at once but in steps: the Land Conservation and Development Commission formed, the state adopted planning priorities, a land-use board of appeals got to work. Over time, Oregon built a framework for true statewide planning. This system has been challenged in voter initiatives several times. But the basic idea is that state government enacts broad land-use planning goals and requires local governments to comply with state standards. The Land Conservation and Development Commission, which is responsible for overseeing all of this, says that state officials "acknowledge" rather than "approve" local plans, phrasing meant to assuage our very American distrust of centralized government.

In 1978, Portland—Oregon's largest metropolitan area—established a regional government. Metro, as it is called, replaced a conventional council of local governments, which most areas in the U.S. have. Metro was given the job of defining urban-growth boundaries for the region and ensuring that all of the twenty-four cities (and parts of three counties) it served in the Portland metropolitan area complied with the regional plan. If all of this sounds strangely European, it was also given a distinctly American twist: Metro is governed by a directly elected board.

Why does any of this matter? Because planners face a profound problem almost everywhere: the scale of political institutions is smaller than the scale of the actual regions that need to be planned. A metropolitan area includes hundreds of distinct "governments"—cities, counties, transit districts, water districts, port authorities, school districts, and more. Metropolitan areas may include multiple state governments and, in some cases, multiple countries. It is common for metropolitan areas to have an association of local governments, a sort of United Nations for a region. There is almost never an actual regional government with powers that matter. Portland is the exception. Metro shows us that the choice to place wide-scale decisions within the jurisdiction of a centralized governing body can have positive effects, while preserving the voice and participation of small communities within the region. GM

Coalition for a Livable Future

If there's one thing that community advocates and activists know, it's that progress happens through collaboration. Portland's Coalition for a Livable Future draws its strength from the numerous groups who come together to focus on single urban issues. The coalition pools multiple perspectives and voices to build a formidable force for change.

In 1994, Myron Orfield, a state legislator from Minnesota, arrived in Oregon with a warning for Portland residents. Having observed and analyzed patterns of social polarization in Minneapolis and St. Paul over the years, Orfield predicted that even with growth management, urban sprawl in Portland could undermine the social fabric of the city, not to mention the environment, with the added impact of congestion caused by commuting and insufficient public transit. Moved to action by Orfield's prediction, a diverse group of community organizations

Out of all North American cities, Portland, Oregon, has done the best job of managing growth.

banded together into Coalition for a Livable Future, a citizen presence poised to help redirect Portland's course.

Cobbling together expertise in areas ranging from ecology to public transportation, social justice to affordable housing, the coalition created a unified agenda for preserving the integrity of Portland's infrastructure, open spaces, and communities. In 1995, a number of the coalition's member groups, including the organization 1,000 Friends of Oregon, formally presented the Metro 2040 Growth Concept, a fifty-year plan (antedated to 1990) for Portland that put equal emphasis on the protection of urban and rural ecology, support for the local economy, intelligent transportation systems and infrastructure, and accessible, socially just housing options. The 2040 plan, and the savvy way the coalition has subsequently built political support for its execution, serve as an example to cities everywhere that it's possible to accommodate a growing population while improving the urban experience and protecting natural habitats. GM

The Limits of Planning

We have a pretty good idea of what a bright green city would look like. But how do we make it real? If we know that we want to conserve land, enable people to walk to work and school, and promote community, we should build in neighborhood clusters instead of sprawling suburbs. How can we change the process of urbanization to make these things possible when urbanization seems dictated by officials and forces beyond our control?

Cities are not, in fact, centrally planned by anyone. Certain city regulations, such as street layouts or the allowable heights of new buildings, are imposed by the government. But cities are created by the actions of thousands of people over a long period of time. The power to create urban form lies with large numbers of individuals and groups: the farmers who decide whether to keep farming or sell to developers, the families who decide where they will live, the workers who decide how they will get to work, the engineers who decide how they will design streets, the thousands of property owners who decide what they will build on their land. Changing cities is a lot more complicated than laying out a good plan. But the factors that make city planning complex also make the city an extraordinarily open system, capable of being acted upon and improved upon in many different ways.

Northern European countries can adopt national "green plans" that comprehensively redirect land use, transportation, industrial processes, resource flows—just about everything. These plans are worth learning from, but this approach only makes sense in societies with strong governments capable of exercising authority where they need to. Everywhere, governments have lost a degree of economic influence in the face of globalization—the increasing mobility of capital across borders.

In the United States, government is especially weak. Taxes are low, so the government has limited ability to undertake infrastructure projects (say, transit lines) that help to shape urbanization. The United States has also created a set of legal "property rights" that limits the

ability of communities (at all levels) to determine how land will be used.

Simply writing good plans to guide regional growth or to make our society more sustainable has little impact. We must either reform government to give planners more power, or bypass weakened, ineffectual local and regional governments and change urban form ourselves, directly. GM

Tools for Making Better Plans

Plans are part of the DNA of cities: while a million small decisions ultimately determine a city's future, plans often determine what goes where. Today, city plans are very abstract, specialized documents. People pursue advanced degrees simply to learn to read and write them. Unfortunately, this inaccessibility deters city inhabitants both from getting involved in the planning of their own cities and from acting on their own to improve their neighborhoods.

But plans don't have to be so opaque, and residents should have a say in how their cities grow. Learning how to deal with planning and all its problems is critical to learning how to make our cities better places. Environmental historian Richard White writes in *The Organic Machine: The Remaking of the Columbia River*, "Planning is an exercise of power, and in a modern state, much real power is suffused with boredom. The agents of planning are usually boring; the planning process is boring; the implementation of plans is always boring. In a democracy, boredom works for bureaucracies and corporations as smell works for a skunk. It keeps danger away. Power does not have to be exercised behind the scenes. It can be open" (1996).

No natural law requires that planning be boring, but citizens must demand that it not be so. At its root, urban planning is simply the creation of places, and we all have an intuitive understanding of what makes a good place—just look at how kids make imaginary houses and adults play *SimCity*.

With mapping [see Mapping, p. 518] and social software [see Open Source, p. 127], we can essentially turn planning into a game. This doesn't trivialize the process—the military spends billions of dollars a year war-gaming possible future conflicts, and they are dead serious about it. Making planning fun can make it more democratic, which can only make it better. Realistic budget limitations, system approaches, and societal obligations could be built in to keep everyone's feet on the ground.

Imagine having a barbecue with our neighbors and playing a multiparticipant version not of *SimCity* but of *RealCommunity*. This beats listening to a bureaucrat "walk us through" a PowerPoint presentation of how a proposed up-zone would affect projected traffic flows and parking availability, or some such planning jargon. People want to live in a place where their input steers the course of growth and their presence defines the community character.

Indeed, once we start creating such tools, tools that blend the virtual and the physical, other truly useful tools will also become available—tools that help us locate the things we want within the areas of our community to which we can easily walk: our "walkshed." From participating in car-sharing programs [see Urban Transportation, p. 264] to finding the nearest place to get an organic apple, bright green urban living will get easier as walkshed technologies proliferate. AS

RESOURCES

Congress for the New Urbanism
http://www.cnu.org
Though New Urbanism is pretty well established now, the Congress for the New Urbanism (CNU) is still one of the best urban-planning resources

Opposite, left: An aerial view of a Las Vegas suburb illustrates the advantages of compact, planned neighborhoods. **Opposite, right:** Groundbreaking videogames like *SimCity* have increased the public's interest in urban planning.

around. If you're a planning professional or an architect, you'll want to become a member and attend the CNU's fairly legendary conferences, but the rest of us can go to the group's Web site to find a wealth of materials on sustainable design, green-building, and traditional neighborhood planning.

The Death and Life of Great American Cities by Jane Jacobs (Vintage Books, 1961)
No book about urban life has had as great an impact as Jane Jacobs's pioneering study of the qualities that make cities and their streets come alive. Jacobs is passionate, opinionated, discursive—and it doesn't matter a bit that her numbers are decades out-of-date, because she's still right:

"Big cities and countrysides can get along well together. Big cities need real countrysides close by. And countryside—from man's point of view—needs big cities, with all their diverse opportunities and productivity, so human beings can be in a position to appreciate the rest of the natural world instead of to curse it.

Being human is itself difficult, and therefore all kinds of settlements (except in dream cities) have problems. Big cities have difficulties in abundance, because they have people in abundance. But vital cities are not helpless to combat even the most difficult of problems. They are not passive victims of chains of circumstances, any more than they are the malignant opposite of nature."

Retrofitting the Suburbs

We may be living on an urban planet, but the United States is still a suburban place: about half of the country's population lives in the burbs, according to the 2000 U.S. Census. The cookie-cutter, sitcom suburbs aren't confined to the States anymore, either—we can now find identical sprawl around the world. In fact, Beijing has imported its own version of Southern California sprawl—"Orange County, China," a gated community of Mediterranean-style residences that would be right at home in its eponymous California region.

Growth patterns would indeed indicate that the burbs are booming: "first" suburbs, or inner-ring suburbs—the areas closest to major cities—house one-fifth of the U.S. population, according to The Brookings Institution, and they're growing twice as fast as major cities. Newer suburbs—often the site of greenfield development, meaning they encroach on former agricultural land—are growing at twice the speed of first suburbs.

But that's not to say that the suburbs are *thriving*. Inner-ring burbs have

Right: Big-box stores promote auto dependency and can leave asphalt wastelands that are hard on the eyes and the environment.
Opposite: A model of "Orange County, China," built to lure prospective buyers, illustrates that sprawl is no longer purely an American problem.

been in increasing states of disrepair since their initial post–World War II boom. And in both inner-ring and newer suburbs, development patterns defy sustainability. As suburbs develop farther away from urban centers, people rely more on automobiles. The scale of sprawl makes it difficult for residents to walk or bike, practically requiring families to own more than one car; retail development in the form of the big boxes—huge stores constructed inexpensively, housing Wal-Marts, Best Buys, or Home Depots—also contributes to auto dependency. In newer suburbs, developers build housing and retail space on separate plots of land, leaving large gaps in between. Suburbs everywhere seem to contain dead or dying shopping malls, which leave asphalt wastelands in their wake. And, of course, one of the staples of the newer suburbs, the McMansion, with its attendant overmanicured, pesticide-drenched lawn, is one of the least sustainable structures on the planet.

Science fiction writer J. G. Ballard, whose views on suburbia are none too friendly, said in an interview, "I would sum up my fear about the future in one word: boring. And that's my one fear: that everything has happened; nothing exciting or new or interesting is ever going to happen again ... The future is just going to be a vast, conforming suburb of the soul." But Ballard's pessimistic assessment doesn't have to be our assessment. A bold and bright rethinking of the way we approach suburbia can keep this vast conformity from setting upon us, while easing the burden that housing places on the planet.

Already people are reclaiming wasted space and urging smart infill projects, whether they're doing so as part of sprawl watchdog groups or just as concerned communities. Our suburbs and cities are different beasts. But we can make suburbs much better by making them denser—mixing residential, commercial, and open space within prescribed limits, rather than pushing outward. Transportation issues are key: access to rail corridors and public transit is needed, as are designs that support walking and biking. In older suburbs, dead and dying spaces like huge parking lots and,

'60s-era malls should be redeveloped. At the same time, first suburbs should recognize and preserve what's historical and notable about themselves.

To truly rethink suburbia, we need to see it for its social problems. We've got to consider social models that don't just rely on design and redevelopment. How could we *really* mix people of different backgrounds and income levels? How could we shake up the homogeneity? How can we make diverse, sustainable communities? MWS

The Plight of the Inner Ring

As sprawl pushes farther and farther out, the older suburbs it leaves in its wake are falling apart. The inner-ring suburbs of small single-family homes built between the mid-1940s and the 1970s are, in many places, experiencing crises worse than those in the inner cities. The older, blue-collar middle-class families that settled the inner ring are precisely the kind of folks who've taken the biggest economic hit over the last twenty years. The houses they live in have, generally, worn poorly and are neither central enough nor large enough to bring top dollar in today's real-estate market.

Traffic problems in the inner-ring burbs are often out of control, as commuters from outer suburbs crowd roads that were built for much smaller rush hours. The strip-mall culture that marks the inner ring is out of style and the malls can't compete with the kind of upscale retail stores that bring in the highest sales-tax revenues. Major infrastructure systems built on the cheap four or five decades ago are coming due for substantial overhauls, a problem summed up in architecture critic and writer Herbert Muschamp's quip, "plywood has a lifespan of forty years" (from a 1997 article in *The Nation*). Because tax revenues aren't rising fast enough—or even falling rapidly—and so much work is needed, the inner ring's schools are suffering, further depressing prospects.

The plight of the inner ring has been exacerbated not only by the suburban sprawl, but also by an urban revival. As downtown living has become hip again and its benefits recognizable not just to young professionals, but also to empty nesters, gentrification and new development have driven up housing costs in many American central cities. The result is that areas of concentrated poverty are getting squeezed out into the inner ring, where housing is cheaper. The attendant social problems—high public health costs, gang violence, struggling schoolkids—are simply too much for most inner-ring suburbs to cope with. The result has been a growing "doughnut of blight" situated between prosperous urban neighborhoods and newer, wealthier rings of sprawl. AS

Reshaping the Burbs

Just as we can retrofit our homes to be solar powered, we can retrofit our suburbs to be lively and more sustainable.

Addison, Texas, is a first-ring suburb of Dallas that has faced increasing problems with sprawl. The town was at a critical point in the 1990s: it needed to add more businesses and housing, but it was already becoming the victim of haphazard planning. Instead of succumbing to sprawl and adding another ring of single-family homes and strip malls, the city decided to create an official center of town, something it didn't previously have. This center was named Addison Circle, and it succeeded in not only adding the mix of homes, shops, and offices the city needed for steady growth, but also incorporating new parks, pedestrian-friendly areas, and community and cultural attractions, all within close proximity to public transportation. A common saying about the suburbs is "there's no *there* there"; the residents of Addison now have a very distinct *there* to point to.

The Crossings, in Mountain View, California, is an example of how a new development can elevate the convenience, density, and sustainability of an existing suburb. A *greyfield* redevelopment—the term for derelict malls, old airports, abandoned big boxes, and other structures characterized by expanses of empty asphalt parking lots—the Crossings is a higher-density complex than others in the area, featuring three hundred mixed-income homes. It sits very near the Caltrain commuter rail station, as well as other accessible transit systems, making it easy for residents to move between home, shopping areas, recreation spots, and work without using a car. The housing is designed in a vernacular style—that is, it looks like an old-fashioned Main Street. Demonstrating that greyfield and transit-oriented development can be successful on a small scale in an existing suburban community, the Crossings also provides a less expensive alternative to the astronomically expensive housing in the surrounding area.

Suburbs can also look to urban infill projects for guidance. CityPlace, in downtown Long Beach, California, opened in 2002. It is a residential and retail complex with apartments, condos, and over 475,000 square feet (44,129 square meters) of retail space for large anchor tenants and smaller shops. It occupies the site of an unsuccessful 870,000-square-foot (80,825-square-meters) mall that opened in 1982. This infill-greyfield development is unusual in the urban fabric, but it's been a success for downtown Long Beach, reestablishing the traditional street grid and breaking up the old mall's superblock. MWS

Greyfield Redevelopment: From Dead Malls to Vibrant Town Centers

With new shopping complexes going up all over the place, it's hard to believe that one of North America's problems is a dead-mall epidemic. But suburbia is littered with abandoned malls, many of them built in the 1960s and nudged out of use by bigger, hotter malls in the '90s. These run-down strips, along with other greyfields, are some of the biggest eyesores

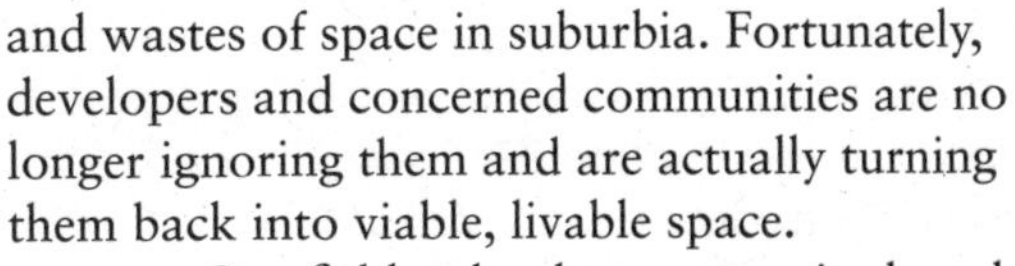

and wastes of space in suburbia. Fortunately, developers and concerned communities are no longer ignoring them and are actually turning them back into viable, livable space.

Greyfield redevelopment varies based on its site and purpose. A mall, for example, might be demolished and cleared outright. Or the inwardly facing stores might be turned outward, to develop a relationship with the street. Then it's a matter of varying the scale of the neighborhood, so that a monolithic, car-scale mall becomes a friendlier neighborhood to navigate.

Smaller streets encourage pedestrian traffic and discourage speeding cars. Public areas like parks, squares, and playgrounds offer places for people to stop as they move through the neighborhood. Mixed-use, multilevel buildings go up where the small cookie-cutter street-level storefronts were. Asphalt parking lots are filled in with buildings or green space—parking can be accommodated by garages that fit in with the street profile.

Greyfield redevelopments are often excellent examples of transit-oriented development [see Urban Transportation, p. 262]. This transition is logical: these sites were chosen by developers years before because they were in central locations. Greyfield developers take advantage of the position of rail corridors, light-rail developments, or rapid buses to provide other options than private automobile use.

Denver, Colorado, is winning national acclaim for its sprawl-fighting greyfield redevelopments. The greyfields are a necessity, because Denver is a booming city: 1.5 million people are expected to move to the region in the next twenty-five years. If Denver's footprint were to

Communities like The Crossings show that better suburban development—with higher density living and easier access to shopping and public transportation—is possible.

follow current development trends, according to the Denver Regional Council of Governments, its metro area would double in size by 2030. But planners from the city and surrounding areas are trying to keep the sprawl in check, containing it in and around a set of defined centers (many in suburban areas). Dealing with the local greyfields is an important part of the plan.

Stapleton was home to Denver's airport from 1929 to 1995; it's now the largest urban redevelopment in the country. In its new incarnation, it will ultimately house 30,000 people in a combination of single and multifamily homes for mixed incomes and age groups. Rather than just functioning as a bedroom community, Stapleton will offer employment for 35,000 workers. A third of its 4,700 acres (1,902 hectares) is green space, which is being allocated to open areas, a community farm, and parks and squares throughout the denser areas. The redevelopment's award list is long and includes recognition from the United Nations for its sustainable approach.

Belmar Urban Center, another acclaimed Denver-area greyfield redevelopment, is the downtown that the suburban Colorado city of Lakewood never had. In 1966, Lakewood built a 1.4-million-square-foot (130,064-square-meter) mall on a 106-acre (43-hectare) site. When it opened, Villa Italia was the largest shopping mall west of Chicago, but by the early 1990s, it began to atrophy when its anchor tenants moved out and competing malls opened in the area. At the mall's lowest point, retail tenants occupied a mere 30 percent of its massive space.

The new plan for Belmar is multimodal—that is, it offers a number of sustainable transportation methods for its residents and visitors. It's positioned on local and express bus lines, with rail corridors only a few miles away. The new street grid creates a number of smaller blocks with public squares and plazas that lead to the larger streets. It aims to be a community with a heavy retail and office base: there will be almost a million square feet (92,903 square meters) of available retail space and 760,000 square feet (70,606 square meters) of office space.

On a related front, some communities are also practicing big-box reuse. Big-box stores go under all the time—even Wal-Mart, which seems poised on the brink of global domination, has left behind a lot of ghost stores around the country. Like shopping malls, if big boxes are naked and unused, they end up just hulking wastes of space. When they're in an actual neighborhood (think large supermarkets) they tend to take the whole neighborhood down with them when they close. Once a big box pulls out, the small businesses around it quickly decline.

These massive and ugly buildings are even more difficult to redeem than dead malls are, but communities are finding clever ways to reincorporate them into town life. The good thing about abandoned big boxes is that they're relatively easy and cost-efficient to move into and adapt. They already have parking lots, fairly new electrical systems, and structural elements that make them reliable and sturdy buildings. In addition, because they're usually at a crossroads of some sort, they are great places for businesses or institutions that need to serve several communities. This perhaps explains why big boxes in the United States have often notably been turned into community centers and churches. Some even become charter schools, like the Snowy Range

Academy in Laramie, Wyoming, which is housed in a renovated Wal-Mart. Perhaps the most wonderfully whimsical reuse of a big box yet is the Spam Museum in Austin, Minnesota—in a former Kmart.

Reusing big-box stores doesn't change the landscape as dramatically as greyfield development does; sometimes the facades change very little, save for a paint job and a new sign. But reuse, in any form, is never a bad thing. MWS & CB

RESOURCES

Crabgrass Frontier: The Suburbanization of the United States by Kenneth T. Jackson (Oxford University Press, 1985)

In what was a prophetic look at the growth of American housing, Kenneth T. Jackson's 1985 classic of urban history was the first full-scale look at the development of the American suburb. Tracing suburbanization from the nineteenth century onward, Jackson argues that the spaces around us condition our behavior, and that "history has a fundamental relevance to contemporary public policy." He goes on to comment, "I would hope that this book indicates that suburbanization has been as much a governmental as a natural process ... For better or for worse, the American suburb is a remarkable and probably lasting achievement."

Once There Were Greenfields: How Urban Sprawl is Undermining America's Environment, Economy and Social Fabric by F. Kaid Benfield, Matthew D. Raimi, and Donald D. T. Chen (National Resources Defense Council, 1999)

If you want the goods on why sprawl is bad, look no further: here are the numbers, the studies, the horror stories, the case made in full. Its emphasis is not on providing tools for change, but as a resource on urban sprawl as it exists, *Once There Were Greenfields* is unmatched.

According to the authors, "The plight of so-called 'soccer moms,' for example, stems from the fact that many women not only hold full-time jobs but also perform more child care and household duties than their spouses. This requires women to make more automobile trips. Because fringe living forces most of these responsibilities to be met by driving automobiles longer and longer distances, the inevitable result is stress, fatigue, and less time with family."

Superbia!: 31 Ways to Create Sustainable Neighborhoods by Dan Chiras and Dave Wann (New Society Publishers, 2003)

Many inhabitants of suburbia would like to see a stronger sense of community and more environmentally sound design in their neighborhoods. Part of the answer is to cooperate within our communities, creating solutions that are hard for any individual to do alone. *Superbia!* offers a range of actionable options—from local newsletters and common houses to community gardens and even physical redesign—for motivated neighborhood builders across the burbs.

"We believe most neighborhoods have the raw ingredients and resources to become sustainable, resilient, and healthy. We want to help make that transition happen because the potential economic, environmental, and social benefits are huge! But it won't happen automatically—it will take cooperation, vision, and strategy."

Opposite: Belmar, Colorado (before and after), demonstrates that greyfields can help reknit sprawling communities.
Left: Even big-box stores can find new lives: this medical center is housed in a former Wal-Mart.

The End of Suburbia: Oil Depletion and the Collapse of the American Dream
Directed by Gregory Greene (2004)
This documentary poses the questions, As energy prices skyrocket in the coming years, how will the populations of suburbia react to the collapse of their dreams? Are today's suburbs destined to become the slums of tomorrow?

Metropolitics: The New Suburban Reality
by Myron Orfield (Brookings Institution Press, 2002)
Though in some ways surpassed now by more recent academic works, Orfield's book remains a classic because it exposes, in extremely frank terms and with concrete examples, how sprawl in the "favored quarter" of our wealthier suburbs creates a reverse-Robin-Hood effect, channeling money from those in central cities and inner-ring suburbs to those who live in affluent bedroom communities. Just as unfair subsidies for oil, coal, and nuclear power make it harder for more recent renewable resources like wind, solar, and hydro to compete, subsidies for sprawl work to undermine the creation of bright green cities. The answer, Orfield shows, is new civic coalitions of older cities, older suburbs, reform groups, religious communities, advocates for working people, and competitiveness-minded business leaders. As former Albuquerque mayor David Rusk explains in the introduction, "The existence of the 'favored quarter' also means the existence of the unfavored three-quarters. The concept of 'metropolitics' is built on uniting the political weight of the unfavored three-quarters."

Suburban Nation: The Rise of Sprawl and the Decline of the American Dream by Andres Duany, Elizabeth Plater-Zyberk, and Jeff Speck (North Point Press, 2001)
If you want to understand the already existing alternatives to sprawl, you should start by reading *Suburban Nation.* Not only does it explain what went wrong with urban planning after World War II, but it lays out the New Urbanist alternative more clearly and accessibly than any other book out there. If you want to know how we might plan neighborhoods to be as walkable, pleasant, and friendly as the ones our great-grandparents enjoyed, you'll find many of the answers here.

As the authors explain, "Recognizing the tremendous cost of auto-dependent lifestyles, the author Philip Langson has proposed a new national holiday: 'Automobile Independence Day.' It would take place on that date each year by which we have earned one-quarter of our salaries, the amount that it takes to support our cars. How appropriate that it's April Fool's Day."

Opposite, left: Architect Norman Foster's Hearst Tower is the first office building in New York to earn a Leadership in Environmental Design (LEED) rating from the U.S. Green Building Council.
Opposite, right: Green skyscrapers, like the proposed Editt Tower in Singapore, provide landmark buildings while shrinking cities' ecological footprints.

Big Green Buildings and Skyscrapers

When we were children, the tall buildings around us seemed wondrous. A city stroll was an adventure simply because of the sheer size and bustle of our public places. Where is that wonder now?

We don't expect much from big buildings anymore. Those of us who work in bland office towers trudge off to work each day knowing that at best the drab interiors will be a bit depressing; at worst, we suspect, the poor ventilation and artificial light will leave us feeling sick. But buildings can be better. We can revolutionize offices by bringing in more daylight, growing indoor plants, improving airflow, and providing access to outdoor space. Productivity in such an atmosphere usually skyrockets, and overhead costs drop. But well-designed buildings can have a positive impact far beyond their footprints.

Truly great buildings can change the entire urban experience, helping to lift depressed cities up to new levels of cultural vitality and transforming the skyline—a symbol of progress to the rest of the world. If we want to create bright green cities, we need buildings that illuminate our course, give us reason to raise our expectations for great architecture, and restore our long-lost sense of wonder.

The new breed of design in contemporary skyscrapers and public buildings aims to affect visitors with dramatic architecture and unconventional use of space while promoting ecological integrity and a sense of well-being. SR

The Editt Tower, Singapore

Imagine walking through a verdant urban park and finding yourself in the atrium of a skyscraper. This will be the experience at Singapore's Editt Tower. Still in the planning stages, under the direction of Dr. Ken Yeang, Editt Tower approaches the fusion of indoors and outdoors by bringing greenery to the whole building—inside and out, top to bottom. Though relatively tall, the twenty-six-story building is designed to minimize the disconnect between upper-floor offices and street-level pedestrian traffic. Visitors will stroll along landscaped ramps and greenways lined with shops that climb to the sixth floor. The indigenous plants that line the walks will be irrigated by means of rainwater harvesting and sewage recycling. When someone in a twentieth-floor office flushes the toilet, that water will run through an on-site cleaning system and into the irrigation lines, forming a closed system for the building's resources. Other green features, such as solar energy and natural ventilation, will keep costs down and spaces comfortable. The Editt Tower design has undergone an evaluation of its strategies for eventual retrofits and the long-term disuse of building components, ensuring that its envi-

ronmental accountability doesn't apply only to its initial construction, but to its entire life cycle.

Dr. Yeang, whose tower design won the Ecological Design in the Tropics 1998 award, made human experience a priority in the development of Editt Tower. In contrast to the cold, disconnected, and hollow feeling of many skyscrapers, this place will be alive with people, commerce, and greenery. Integration of inside with outside and top floors with ground floors will bring the otherwise diffuse energy of a large building into a cohesive whole. Even in its conceptual stages, Editt Tower serves as an inspiring model for what's possible in the revitalization of skyscraper landscapes. Hopefully, by the time it's done, similar concepts will be in the works everywhere. SR

The Reichstag, Berlin

Among all the factors that make a government trustworthy, transparency [see Demanding Transparency, p. 434] sits firmly at the top of the list. When we have access to the official processes that (generally) occur behind closed doors, we can feel at ease with the workings of our social and political systems. In Germany, a country whose history is a case study in the dangers of government secrecy, transparency takes on even greater importance. Nowhere is this clearer than at Germany's parliament building, the Reichstag.

The Reichstag has seen both extreme malfeasance and tremendous civil renewal. A devastating fire in the building in 1933 destabilized Germany and enabled the Nazi Party to take control. Later, after World War II, the Reichstag lay in ruins for decades. It was eventually reconstructed in the 1960s, and in 1990, Germany's reunification ceremony was held there.

At that point, Germany saw an opportunity to create an architectural symbol of their renewed state by rebuilding the Reichstag to reflect the values and vision of German democracy. In 1992, architect Sir Norman Foster was appointed the task of designing a new incarnation of the German parliament building.

The Reichstag was already one of Berlin's most iconic buildings, and Foster's version is a remarkable example of the way cultural, social, and political conditions can be embodied in architecture. The building is full of light; interior meeting rooms are enclosed in glass, allowing visitors to witness parliamentary sessions. This complete visual transparency is a powerful symbol of the openness of German government today.

The features that make the Reichstag transparent also make it green. The building uses a forward-thinking energy strategy, meeting its energy needs with biofuels (refined vegetable oil) [see Cars and Fuel, p. 76]. According to a 2002 report by the Rocky Mountain Institute, this clean electricity means a 94 percent reduction in the building's carbon dioxide emissions.

The Reichstag's cupola, preserved from earlier generations, was restored, and is central to the lighting and ventilation strategies. A light sculpture in the middle reflects sunlight into the parliamentary chamber, with a movable sunshade to control glare. On the project's Web site, Foster calls the cupola "a beacon, signaling the strength and vigor of the German democratic process." SR

The Swiss Re Headquarters, London

The Swiss Re (Reinsurance) headquarters in London, designed by Ken Shuttleworth of Foster and Partners (Sir Norman Foster's firm), has become an icon among modern skyscrapers modeled on biomorphic principles [see Biomorphism, p. 102] and incorporating ecologically sound construction and operations. The concept for the structure was inspired by Buckminster Fuller's theoretical work examining the relationship between nature's patterns and human work environments. Its shape has earned it the nickname the "Gherkin."

The building is encased in a structural skin of aluminum, steel, and glass, which blurs the boundaries between roof and wall and increases energy efficiency by eliminating wasted space. A unique diagonal structure allows interior spaces to be open, and interrupted only by support columns. Separating the offices are open social spaces, placed in different locations on each floor such that they spiral up the building. Ventilation passages in the facade of the building connect these social areas, creating uninterrupted vertical openings that permit light and air to travel through the entire structure. This design reduces the need for forced air and AC, saving energy, reducing operating costs, and contributing to the occupants' well-being. SR

Leadership in Energy and Environmental Design (LEED)

For years, calling a building "green" was nothing more than an assertion, and believing that assertion was a matter of pure trust. A lack of agreed-upon standards meant that anyone could claim that their building was ecologically virtuous. Eventually, green-building groups around the United States began creating points-based systems for rewarding those buildings that conserved water, energy, and resources. As these programs became more sophisticated, the United States Green Building Council (USGBC) developed the Leadership in Energy and Environmental Design (LEED) standards to set a voluntary baseline for green buildings by assigning points to high-performance commercial buildings.

Opposite: The Reichstag in Berlin is a symbol not only of transparency in government, but of environmental responsibility.

Left: The Swiss Re (Reinsurance) headquarters in London, known affectionately as the "Gherkin."

The USGBC uses a public, consensus-based process to evolve and refine the standards—which have grown to encompass interiors, renovations, and new homes—and it is moving toward establishing ratings for neighborhood development and home retrofits. One of the program's greatest successes has been its exportability: Canada uses a version of LEED, and international green-building councils are being set up from Taiwan to Australia, encouraging high-caliber green architecture around the globe.

The Bank of America Tower, New York

The next addition to New York City's skyline will raise the bar for skyscraper design. When New York's Bank of America Tower opens in 2008, it will be the most environmentally friendly skyscraper in the world. Designed by Cook+Fox Architects, the building is intended to reflect both the vitality of midtown Manhattan street life and the movement of the sky above the 945-foot (288-meter) spire. The aluminum, glass, and steel that make up the exterior will act as crystal facets, reflecting the light of the sun and moon.

The designers are aiming for a LEED Platinum certification for the building—the highest possible ranking. Glazed windows will maintain indoor temperatures and provide ample natural light; hot and cold water will flow beneath raised floors for efficient, radiant climate control; and rainwater harvesting and wastewater recycling will conserve massive amounts of water. SR

LEED Platinum at No Extra Cost

If green design techniques make buildings so much more beautiful, cost-efficient, and pleasant to be in, why haven't they been adopted wholesale? Most contractors and architects blame the initial cost of building green. But a new building at Oregon Health and Science University (OHSU) may prove them wrong.

The new OHSU building design complies with the strict qualifications of the Green Building Council's top-level LEED Platinum rating. Plenty of other buildings have attained LEED Platinum status, and most can argue that the long-term payback, achieved through dramatic reductions in electricity, water, and gas consumption, cancels out the higher initial investment, but OHSU can boast an even greater reason for developers to go green. The university's initial building costs came in at 10 percent lower than anticipated for systems and equipment, saving literally millions of dollars up front. Additionally, the university predicts the building will reach a 60 percent energy savings over what's required by Oregon's state building code and by LEED (even with the tight Platinum regulations).

The building will integrate some particularly interesting sustainable-design features, including chilled beams—which offer radiant cooling by running cold water beneath the surface, as an alternative to air conditioning—and an advanced solar-generation system that delivers exceptional energy efficiency and independence from the grid. To encourage wider adoption of green design techniques, the building's engineering firm provides a detailed breakdown of the costs, time frames, and

features of the OHSU design in a freely distributed handbook. SR

RESOURCES

Ten Shades of Green: Architecture and The Natural World by Peter Buchanan
(The Architectural League of New York, 2005)
In *Ten Shades of Green* (based on the traveling exhibit of the same name) curator Peter Buchanan uses ten buildings that combine environmental responsibility and innovative design to argue that sustainability is not only good for the planet, but that it also offers architects new opportunities for creativity and innovation—that there is no such thing as the monolithic "green aesthetic." Chock-full of photographs, diagrams, and sketches, the book truly proves that "the architecture of the emergent new long-term paradigm must be born from an evolutionary and ecological perspective, to be good for both planet and people and grounded in the complex and sensual realities of place and lived experience."

Big & Green: Toward Sustainable Architecture in the 21st Century edited by David Gissen
(Princeton Architectural Press, 2002)
This architectural showcase book is the best single-volume guide to the movement of designing large buildings that protect the planet while improving their communities. *Big & Green* reveals in beautiful illustrations and clear text that even the densest, highest, most modern parts of our city can herald the arrival of a bright green future.

As Gissen puts it, "Most conventional practitioners of modern design and construction find it easier to make buildings as if nature and place did not exist. In Rangoon or Racine, their work is the same. Fossil fuels make buildings in both locales inhabitable, lighting them, cooling them, heating them. An ecologically aware architect would design those buildings different. She would immerse herself in the life of each place, tapping into natural and cultural history, investigating local energy sources, the availability of sunlight, shade, and water, the vernacular architecture of the region, the lives of local birds, trees and grasses. Her intention would be to design a building that creates aesthetic, economic, social and ecological values for the surrounding human and natural communities—more positive effects, not fewer negative ones. This would represent an entirely new approach: Following nature's laws, one might discover that form follows celebration."

The designers of the new Bank of America building in New York (opposite, left) are aiming for LEED Platinum certification, which the Oregon Health and Science University's new campus (opposite, right) has already achieved.

Healing Polluted Land

Development on the rural fringes of cities is a problem. Development in the heart of cities is a solution. Because density is inherently more sustainable than sprawl, and because redeveloping land that has already been developed is more ecologically sound than cutting down forests and paving farms, our goal should be to use every urban acre as effectively as possible. Unfortunately many of the places that have been used most intensely in the past have suffered for it and been left derelict, contaminated by the industries that first brought them to life. In every city there are vacant lots, polluted former factory sites, discarded buildings—places we refer to as *brownfields.* With new and better ways to clean up and restore these places, they are now hotbeds of opportunity for the new frontier of bright green urbanism.

Building on abandoned lots and polluted land, rehabbing forgotten buildings, creating great places from the most wounded spaces: these efforts are on the cutting edge of what it takes to make cities better. The architect Peter Calthorpe said if we want to know how to make a neighborhood better, we should start with the worst places and put the best things we can possibly imagine there. By using discarded lots and brownfields, we not only foster compact development, we also heal the city.

One way to clean up a contaminated piece of land and make way for the regeneration of flora and fauna is to put plants into the polluted ground. Many types of living organisms act as filters for toxins and pollutants in soil and water. Known as *bioremediation*, this organic process involves the absorption of contaminants into the plants, where they either get broken down into nonhazardous components and rereleased, or extracted and concentrated in a way that makes disposal easy. There are many types of bioremediation, each involving a specific type of organism as the cleaning mechanism. Plants have incredibly powerful natural abilities to extract everything from heavy metals to crude oil and turn them into harmless elements.

Many brownfields are now undergoing bioremediation—it's an efficient cleaning method, in terms of both labor and cost. Some of these areas have been prepped for housing and commercial development; others have become spectacular parks. AS

Mycoremediation with Oyster Mushrooms

Can mushrooms clean up our polluted world? Mycologist Paul Stamets says they can. He has coined the term *mycoremediation* to refer to a specific type of bioremediation that uses

Right: Formerly polluted or abandoned lands can be reclaimed, even turned into parks, like this site in Irvington, New York.

Opposite: Certain plants, fungi, and bacteria have extraordinary power to clean horribly polluted sites, like this former industrial waste facility in Houston, Texas.

fungus to break down pollutants. Stamets, the king of this far-out science, has demonstrated that the magic worked by the spores of a few fungi can transform a heap of hazardous waste into a heap of lush greenery—that a vibrant ecosystem can be created anew by letting mushrooms eat up toxins and fertilize soil.

As biological filters, mushrooms work extremely effectively, because they are, by their very nature, decomposing agents. Mycelia—the parts of the mushroom that absorb nutrients and turn them into usable matter and waste matter—are able to eat up some environmental toxins and efficiently break them down into nontoxic components. Mushrooms can remediate soil and sediment contaminated by heavy oils, petroleum products, pesticides, alkaloids, polychlorinated biphenyls (PCBs), and even *E. coli*.

One of Stamets's best-known case studies involved cleaning up a diesel-oil spill using oyster mushrooms. The study demonstrated not only that mycelia were the best tools for the task, but that the mushrooms that flourished in the contaminated soil contained no harmful agents once the pollutants had been cleaned. Many environmental-protection groups are now considering mycoremediation as a viable solution for threats such as marine oil spills, mercury contamination, and mining residues. Stamets's water cleanup work in the Pacific Northwest, where he is based, may clear the way for restoring wild fish populations and protecting marine ecosystems. SR

Duisburg-Nord, Germany

Just because we clean something up doesn't mean we must erase its past. We see the preservation of our cities' industrial histories everywhere now, in the conversion of old warehouses and factories into residential and office spaces, where the facade is preserved intact to evoke the building's gritty past. In the remediation of contaminated land, the relics of destructive forces on human and environmental health can be a profound example of how far we've come, and how far we have to go.

In Duisburg, Germany, the well-known landscape park Duisburg-Nord stands upon a former brownfield. The park was designed to embrace rather than eradicate the industrial history of the site by incorporating artifacts of its past, such as blast furnaces, gas tanks, and storage bunkers. In contrast to many urban parks, such as New York's Central Park or Paris's Parc André Citroën, Duisburg-Nord does not attempt to isolate itself from its city surroundings, nor does it attempt to conceal the remnants upon which it was built. "Landscape is not the opposite of the town," says Peter Latz, one of the park's designers. "Landscape is culture." Walkways weave through the site's blast-furnace buildings; lily pads float in old cooling tanks. A few areas of the site remain too toxic for human use, but most of the formerly unusable, unsightly land has been turned into a vibrant park that serves a city of nearly 100,000 people. SR

Pearl District Development, Portland, Oregon

In many urban centers around the world, run-down industrial neighborhoods have undergone face-lifts that both restored their vitality and boosted their economies. Among many, Portland's Pearl District stands out as a great success.

The Pearl District has come to life in stages. In the mid-1990s, an old rail yard was turned into a residential neighborhood by Hoyt Street Properties; it was an endeavor that required earnest convincing on the part of the developers to keep the City of Portland from building a highway off-ramp through the area's center. By preserving the cohesiveness of the district, the developers were able to create superior alternative- and public-transit options, including pedestrian and bike routes and free bus and light rail, within a designated zone. Today, mixed-use real estate helps to preserve socioeconomic diversity in the area, offering options for people well below the city's median income level.

A subsequent redevelopment near the original Pearl project includes the Brewery Blocks, higher-end residences in converted breweries. The area now includes a number of LEED-certified buildings [see Big Green Buildings and Skyscrapers, p. 247]; the builders diverted some 96 percent of the construction waste away from landfills into recycling plants.

Chosen in 2005 by the Project for Public Spaces as one of the top sixty places in the world to live, the Pearl District stands as a shining example of how cities can invert sprawl and begin rebuilding from their urban heart. Industrial zones frequently offer sturdy old buildings with great structural potential, proximity to downtown, and even waterfront locations; they have also taken on distinctive aesthetic appeal. This is where the next generation of homeowners wants to live. Easy transportation; lively, tight-knit communities; pedestrian-friendly paths and parks; and thriving commercial enterprises make inner-city redevelopment synonymous with a superior quality of life. SR

Art in Remediation Zones: Red Dive

The Gowanus Canal in Brooklyn, New York, is tough to love. For thirty years, it was nothing but a stagnant pool of black goo, and local legend had it that it was a favorite spot for the mob to dump bodies. But as the neighborhoods around the canal shifted from industrial to

residential, local environmental and community groups decided to adopt the canal and turn it into a source of neighborhood pride.

They reactivated a flushing tunnel that had long ago broken down; the oysters that were introduced survived, and soon jellyfish, bluefish, and even one confused harbor seal could be spotted in its waters. Birds like cormorants, ducks, and egrets returned to the water's edge.

In response to the canal's amazing rebirth, a group of artists who call themselves Red Dive created a multimedia celebration of the canal's history, which was performed over two weekends in the summer of 2003.

The performance took place at various points along the canal. As they drifted along in a boat, listening to recorded canal stories told by long-time locals, audience members saw choreographed interpretations of the canal's history.

"I saw the canal as this container for so many forces and needs and drives," said Maureen Brennan, artistic director of Red Dive, in an interview with Diane Cardwell of the *New York Times*. "Here's this place that embodies a history of fear and all the bad things about human waste and pollution and decay, and now it's this container for hope and renewal and reclaiming" (May 19, 2003).

The Red Dive performance was part of the ongoing Peripheral City series, in which artists lead tours of New York's undiscovered, overlooked, or marginalized neighborhoods. Their work reminds us that the first step toward using every acre of our cities well is learning to see the places we like to overlook. CB

Opposite: Brooklyn's Gowanus Canal exemplifies both the pollution we've inherited and the power of imagined change.

RESOURCE

Mycelium Running: How Mushrooms Can Help Save the World by Paul Stamets
(Ten Speed Press, 2005)
Paul Stamets is the indisputable king of mushrooms. Among the more fascinating and potentially worldchanging of his pursuits is mycoremediation—the use of mushrooms to clean up toxic sludge. He's done it a number of times, turning heaps of contaminated waste into verdant hills full of insects and birds. As he puts it: "I see mycelium as the Earth's natural Internet, a consciousness with which we might be able to communicate ... With mycoremediation, brownfields can be reborn as greenfields, turning valueless or even liability-laden wastelands into valuable real estate." Stamets has also been at the forefront of some groundbreaking research into the use of fungus for treating epidemic diseases such as cancer and HIV.

Greening Infrastructure

Living in dense, urban neighborhoods gives us a big leg up on living in a more ecological way. When we live in compact communities, it's easier to drive less, share more, and tread lightly on the planet.

But often our cities themselves are a bit of a mess. Although cities do help us reduce our environmental impact on distant areas (in that we don't have to slash into forests or blast hillsides to create homes), the natural systems closest to cities often degenerate into shattered remnants of what they once were.

This is tragic, because most cities have grown up where nature was particularly bountiful—where the farmland was rich, fishing was good, and water was abundant. So, as cities grow, the best land often ends up paved with asphalt or concrete, the best water polluted with sewage and runoff. The layers of separation that old methods of infrastructure place between urban dwellers and the natural world cloud our ability to make smart, long-term decisions with the interests of the earth in mind.

We need urban infrastructure that allows us to live in harmony with the natural systems that sustain us—we need to start thinking of the entire planet as home. There's no such thing as an entirely human place. When you stand on the sixtieth floor of a Manhattan skyscraper, in one sense you stand as much in nature as if you were in the middle of the woods. The water coming out of the tap came from the hills of upstate New York, and then spiraled down the drain to a sewage-treatment plant on the East River. The walls around you are built of stones pulled from quarries, metals dug from deep mines. The climate that prevents the building from being buried under the snows of a new ice age remains in relative balance through the workings of forests and ocean plants the world over. Tug on any aspect of your life, and you will find it connects back to nature, if you follow the string far enough.

It is time, then, for change. It is time, to paraphrase writer Wallace Stegner, to live on this planet as if we planned to stay. To do that, to build a civilization that is broad in time, we need to begin building cities whose workings resemble as closely as possible the workings of the rest of nature.

That goal, of greening our cities until they are part of the working fabric of nature, seems impossibly far off, surrounded as we are by asphalt and concrete, cars and buildings, clanking machinery and suspended wires, in places where nature (think rats and pigeons) survives off our trash or is shoehorned into tiny parks and the edges of lawns. But the first strides toward greening our cities may be easier than we'd tend to imagine.

How much greener could cities get if we took nature as a model? Well, much of the infrastructure that makes urban life possible is overdue for green replacements. The price tag for rebuilding urban infrastructure is not small, but the cost of maintaining the systems we now use is pretty huge over the long haul. This makes the idea of gradual replacement of these systems entirely realistic, if we approach it as the work of decades. AS

Living walls, like this one at the University of Guelph-Humber in Toronto, Canada, clean the air while providing a beautiful backdrop.

Living Wall

What if instead of plugging in our air fresheners, we watered and weeded them? In Ontario's University of Guelph-Humber's main building, designers built an indoor wall of plants as a living air purifier.

The four-story biofilter holds a thick jungle of ferns, ivy, and other plants that work together to break down harmful airborne contaminants into water and carbon dioxide. Since the plants absorb pollutants and break them down naturally, the wall requires no cleaning. Additionally, the fresh air generated by the mass of greenery reduces the need for ventilation systems, promising big savings for the university.

The "living wall" gives a big boost to the aesthetics of the building, as well. Like a work of living art, the lush greenery towers high into the building's atrium. Scientists have been attempting to quantify the psychological benefits of the biofilter, predicting that the presence of so much greenery will improve attendance. For those of us accustomed to regurgitated oxygen from ventilation ducts, this kind of innovation is a breath of fresh air. SR

Trees for a Green LA

The way we build cities can actually make hot places hotter. Tar, asphalt, concrete, and stone all bake in the sun and radiate heat. The overall temperature in a city can rise by as much as ten degrees Fahrenheit (or by as much as six degrees Celsius) above temperatures in areas surrounding the city, in a phenomenon known as the urban-heat-island effect. The problem is particularly noticeable in Los Angeles, with its streets, parking lots, and freeways. Understandably, Angelenos remedy the situation by cranking their AC, but that costs them an arm and a leg, not to mention releasing copious amounts of greenhouse gas into the air.

A more economical remedy? We can plant more trees around our houses. Shade trees not only cool down our homes, but they also generate oxygen, limit soil erosion, and beautify our neighborhoods. Careful planting strategies, such as arranging the trees to provide maximum shade, can ensure we'll get the most out of a shade tree while putting in the least amount of maintenance. In Los Angeles, residents can even take lessons in tree care.

The Los Angeles Department of Water and Power (LADWP) established Trees for a Green LA to offer free trees and tree-care lessons to local homeowners. Participants need only complete one short workshop (twenty minutes online or one hour in person) on planting and caring for trees before the city will deliver shade trees that they can plant and care for around their homes.

The LA program offers thirty different species for residents to choose from, and each home is entitled to seven trees. This sounds like one of the best deals around: more greenery, fresh air, cool shade, and all for an attractive price: free! Programs like this have caught on in a number of other cities, and for obvious reasons. The benefits speak for themselves: decreased need for AC, decreased cost of electricity, decreased air pollution—not to mention increased property value and beautification of the urban landscape. SR

Green Roofs

While we worry and complain that cities are running out of room, we sit underneath a vast amount of unused space. Remarkably, as real estate prices soar, rooftops never seem to be part of the equation, even though in most urban areas, flat, open roofs offer all the same potential of a yard or deck. Planted out with grass or a garden, a green roof not only cools a building, absorbs

Green roofs, like this one on top of Chicago's City Hall, and green facades offer energy savings and soften sometimes harsh, urban settings.

harmful UV rays that damage exterior surfaces, and generates oxygen, but it adds a beautiful, usable outdoor space for leisure and gardening.

Even with extensive greenery, a green roof can be relatively low maintenance. Unless you live in a desert, rainfall will help water it, especially if you grow climate-appropriate foliage. In extremely dense areas where nobody has the luxury of a yard, green roofs can enhance everyone's quality of life by putting a lawn over their heads. SR

Rain Gardens

In cities where sustainability has begun to influence planning, designers increasingly use artificial habitats full of living plants to cool down roofs, reduce energy consumption, make us more mindful of our water usage, control erosion, and help wildlife move through built-up areas. Now, as it turns out, such habitats can also clean the water running off our streets and parking lots.

When rain falls in the city, it picks up pollutants from the impervious surfaces it falls on; the pollutants then travel along with the rain into storm drains. If that runoff reaches exposed earth, which can absorb water, many of the toxins can be filtered or broken down by plants. Shallow swales and holes called rain gardens can catch runoff and let it trickle slowly through soil and roots, purifying the water before any of it washes into the municipal system.

Of course, the question remains, what do we do with the pollution once the soil filters it out? Adding plants capable of bioremediation [see Healing Polluted Land, p. 250] to the rain gardens—plants that can suck toxins from the soil and lock it in their stems and leaves (or even break it down into less toxic substances)—might provide one solution. But even without further innovation, rain gardens make sense. It's a heck of a lot easier to clean up pollution concentrated in one place than it is to clean up pollution that has been dispersed among rivers, lakes, and seas. AS

Green Facades

Green facades act like sunscreens for the sides of buildings. They cover outside walls in greenery, generally with climbing ivies that can be trained to grow vertically and create thorough coverage. In hot areas, green facades help to keep escalating temperatures at bay by shading absorbent surfaces from the sun's rays. Like many other strategies for natural temperature control, this strategy has the great advantage of making a building's exterior more attractive, and of being largely self-regulating, once a vine is trained in the right direction—and provided that the climate doesn't tend toward serious drought. (In older buildings, or those with painted facades, many people choose to use trellises or other structural devices that hold the vines at a few inches' distance from the building itself, to prevent structural and cosmetic damage.) Like most environmentally motivated design strategies, a green facade also saves a homeowner money on cooling bills. What's better than that? SR

Cool Roofs / White Roofs

Dark surfaces coat our cities. Hot sunshine turns black rooftops into radiant heaters, which contribute to the urban-heat-island effect and bake the inside of buildings below.

The quickest and easiest solution? Turn hot, dark roofs into "cool roofs." Most homes have to be reroofed about every twenty years. Studies by several groups, including the Lawrence Berkeley National Laboratory, have shown that changing from a dark-colored shingle (once traditional because it was more "woodlike") to a light-colored shingle (titanium-based white or terra-cotta red) can shave air conditioning costs by up to 40 percent. JC

Light-colored Concrete

Engineers refer to a surface's capacity to reflect light as *albedo*. The lower the albedo value, the more solar heat a surface absorbs, and the more it contributes to the urban-heat-island effect. We've all seen the heat shimmering off a stretch of black asphalt.

In New York City, the Design Trust for Public Space produced a comprehensive analysis of ways to increase the performance of the city's infrastructure. They found that if New York could raise its citywide albedo by just a fraction, it would make a dramatic positive impact on air quality, energy consumption, and the health of New Yorkers. As one major step, they recommended that the city replace much of its black asphalt with light-colored concrete.

Reducing the heat-island effect would decrease ground-level ozone, which causes eye and lung irritation, not to mention an increased demand for AC. According to the Design Trust for Public Space, if New York City's temperatures came down by just 3 degrees Fahrenheit (or by just 1.7 degrees Celsius), urban air quality would improve as much as if an entire fleet of the city's gasoline-powered buses or service trucks were replaced with electric-powered vehicles (2005). AS & SR

Left: Light roofs save on cooling costs.
Opposite: Turning urban streets into community spaces can make the difference between cities that survive and cities that thrive.

Place-Making

Plans guide development and investment, but people make places and communities. Many small decisions made by many people help make a place feel welcoming, friendly, and healthy. Does the local café have seats outside? Are there street trees? Do people walk instead of drive? These kinds of elements define a well-made place.

Communities are nurtured in much the same way, growing vibrant through the accumulation of small efforts. Whether or not neighbors say hello to each other, whether or not there are farmers' markets and block parties and community traditions, whether or not people get together to improve their communities by picking up litter or planting flowers in the traffic circles: these are decisions that people make together. These practices are the achievements of a strong community. And they have the power to turn even the most disconnected and lifeless streets into well-loved neighborhoods. AS

How to Make a Great Place

The nature of a great place is that it's unique. There's a lot to be learned from going to other people's great neighborhoods, but ultimately the answers we need are going to spring from the particular genius of the places in which we live. If we want to make places that foster community, all we have to do is look around. All around us there are indicators of the potential our neighborhoods hold.

It's sort of like parenting: reading guides to child rearing may give us some good ideas and help us avoid some common problems, but ultimately everyone's parenting style is shaped by their children. Similarly, the right community spaces are the ones that reflect and respect their surroundings. Every good neighborhood is at its best when it is most like itself.

There are some general strategies that tend to work. Meet the neighbors. Promote good, compact development. People naturally gravitate toward certain locations and atmospheres, so allow established spaces to form around organic hubs.

Really good places make us want to hang out. The street, in essence, becomes an outdoor living room. Small parks, bike paths, historic buildings, sidewalk cafés, street trees, benches, and public art can all make being out in public a pleasure; a well-designed neighborhood keeps us from wanting to retreat to the safety of our isolated homes.

Socialize locally. Networks don't live by bits and bytes alone. Networks are made of people, and in order to do truly remarkable things, people need to get together, rub elbows, trade gossip, try out ideas, flirt, schmooze, encourage, and

learn to trust, admire, and love one another. Conferences are great for this. Festivals can galvanize an entire zeitgeist. But movements really rise or fall on the strength of ongoing social events—salons, showcases, the right bar, the right café, the place it's happening. These "third places" are the epicenter of any movement, whether the goal is overthrowing a dictator or getting a stop sign put in on the school's corner.

Want to make a difference? Find the epicenter in your town and look for allies. If you can't find an epicenter, make one. Being barista to the revolution is a nobler calling than most.

Straphangers Campaign

Sometimes the best strategy for changing a city is to focus on a single issue rather than trying to change the overall structure of planning. Single-issue campaigns can be fulcrums for broader social change. These have been very successful for local-scale issues like getting bike lanes on city streets, and regional-scale issues like establishing urban growth boundaries.

One of the most interesting single-issue organizations is the Straphangers Campaign. Founded in 1979 by the New York Public Interest Research Group (NYPIRG), the Straphangers Campaign is widely credited with turning the New York City subway system around, leading to a massive increase in daily ridership, with all of the associated social justice, economic development, and environmental benefits that flow from having a widely used transit system. Straphangers make change through confrontational, grassroots campaigns that mobilize regular New Yorkers to demand reforms to the management and funding of transit.

Some of their campaign slogans have become famous: "With livestock it's called animal cruelty. With people it's called a morning commute." And, on Thanksgiving, "If you want our thanks, fix the subway."

The staff of NYPIRG have come to be highly regarded as "content experts" on transportation policy, but the real source of their power is their ability to turn out large numbers of people and gain media attention. They've marched across the Brooklyn Bridge; they've issued report cards on every transit line; they've put posters up in bus stops; they've met with politicians running for office; they've written letters to the editor. All of these are standard methods of grassroots advocacy, harnessed to a strategically chosen campaign that is changing city planning. GM

RESOURCES

City Comforts: How to Build an Urban Village by David Sucher (City Comforts Press, 2003)
When we find ourselves in love with a particular neighborhood or city, the reasons sometimes escape articulation. It's not necessarily about one feature, it's about the feeling we have when we're there. David Sucher calls this feeling "city comfort"—a sense of ease within dense areas, a pleasant experience in public places, and a regular intermingling of neighbors, all fostered by urban design strategies. *City Comforts* is a visually instructive guide to creating an "urban village"—a contradiction in terms, Sucher admits,

Tenacious urban advocates can create enormous positive change through focused activism; the New York Straphangers Campaign is often credited with cleaning up the city's subway system.

but a useful one for conveying urbanites' "coexisting desires for autonomy and community ... a place of repose as well as a place of activity."

When it comes to creating an urban village, Sucher advises us to take the following three basic design principles to heart:

1. Build to the sidewalk (i.e., the property line).
2. Make building fronts "permeable" (i.e., no blank walls).
3. Prohibit parking in front of buildings.

When we follow these rules, we end up with the kind of urban environment that renews our spirits rather than draining us—a city that is safe, stimulating, and convenient. These kinds of cities are also the ones that use resources most efficiently, have exceptionally healthy citizens, and, perhaps most important, attract migration from understimulating and inconvenient suburbs, which results in urban density and, ultimately, curbs sprawl.

A Pattern Language: Towns, Buildings, Construction by Christopher Alexander et al. (Oxford University Press, 1977)
Few books can last decades and still have the power to transform your thinking. But Christopher Alexander's *A Pattern Language,* though very much a product of the 1970s, is also a remarkably timely exploration of how people live together. Organized into "patterns" that describe the general principles underlying human settlement (the need for holy ground within a community, say, or the value of home workshops), *A Pattern Language* offers tools that can help those who would innovate make sure they are working with and not against the grain of human nature:

"Define all farms as parks, where the public has the right to be; make all regional parks into working farms.

Create stewardship among groups of people, families and cooperatives, with each stewardship responsible for one part of the countryside. The stewards are given a lease for the land, and they are free to tend the land and set ground rules for its use—as a small farm, a forest, marshland, desert, and so forth. The public is free to visit the land, hike there, picnic, explore, boat, so long as they conform to the ground rules. With such a setup, a farm near a city might have picnickers in its fields every day during the summer."

Urban Transportation

When movies and TV shows want to depict a futuristic setting, they often show flashy transit systems: monorails, voice-activated personal-transit pods, maglev people movers. Like robot maids and space hotels, these sleek, pollution-free, hands-off transit alternatives seem to have a hard time escaping from their celluloid home. In reality, the automobile—and the infrastructure that supports it—continues to warp the face of our cities and the structure of our lives.

It may be hard to imagine a city liberated from the automobile, but a lot of people are working toward just that goal, introducing new concepts in urban design and mass transit, and using information technologies to push shared transportation beyond the office carpool. Others seek to *civilize* the automobile, making it a responsible resident of the city rather than the city's master. SJ

Pay as You Go

If carbon dioxide were blood-red, our skies would look ominous indeed.
Bruce Sterling

We like our cars so much that the most obvious problems associated with them—traffic congestion and the need for parking—have utterly changed the urban and suburban landscape. The automobile's less obvious costs—financial, aesthetic, and environmental—are either cunningly hidden in tax bills, borne without complaint, or wafted invisibly away on tainted breezes. Making these costs more visible may convince more of us to leave our cars at home.

The cost to *fuel* a car is highly visible, in flashing liquid-crystal displays on gas pumps. Recent history has shown that severe price spikes do convince some people to carpool or take public transportation. What if the vehicle's other direct costs—insurance and registration fees—were also charged by the mile? Several states are considering "pay at the pump" car insurance strategies, wherein a surcharge would be added to the price of a gallon of gas to pay for a car's insurance. Besides eliminating the problem of uninsured drivers, this approach would make visible the full, actual cost of that solo commute or that extra trip to the mall.

Many states require cars to pass an emissions test every few years, but again this "lump payment" of environmental awareness is easy to dismiss. Roadside zipper signs, traditionally used to warn of accidents, could be used to make *pollution* visible, displaying local carbon monoxide, particulate, and ozone levels. Flashing bar graphs showing the accumulation of contaminants in the air as rush hour progresses could give commuters a daily reminder of their vehicles' contribution to environmental problems. SJ

Charging for Congestion

The cores of many large cities have superb mass-transit systems. It is entirely possible for the residents of Manhattan, for example, to go their entire lives without owning a car—and many do. Tourists and commuters headed to New York are strongly encouraged to leave their cars at home. Still, the streets of New York and other cities are frequently clogged with private traffic,

Opposite, left: London's congestion zone, with its pricing scheme, has proven effective in reducing traffic.
Opposite, right: Curitiba, Brazil's efficient bus-rapid-transit system, is being copied around the world.

making life dangerous for pedestrians and cyclists, and getting in the way of efficient alternatives like buses and taxis. Some cities are fighting back, not by banning private cars, but by charging for the "privilege" of adding to congestion.

One method of discouraging car trips downtown is to make parking expensive and scarce. According to the *Vancouver Sun,* Vancouver [see Vancouver, p. 231] is considering a tax on nonresidential parking stalls totaling C$28 (U.S.$25) a year, with the revenues going to mass transit and transportation infrastructure. Copenhagen reduces parking spaces on its streets by up to 3 percent a year (Makousky 2002); the slow attrition of parking encourages residents to leave their cars at home.

Another tactic involves charging drivers who wish to enter a city center or other congested area. London has had great success with its congestion pricing scheme. Private automobiles (taxis, buses, motorcycles, and bicycles are exempt) entering the city's bustling central zone during its most congested hours are charged an £8 (U.S.$14) fee, payable online or at kiosks, according to the Greater London Authority's Transport for London Web site. A network of cameras around and in the zone feeds images to a central bureau, where an optical recognition system notes license plate numbers. Drivers who haven't paid in advance are "dinged" extra fees. Trying to beat the system altogether results in a fine of up to £150 (U.S.$267).

In the first year after it was implemented in 2003, congestion pricing reduced the number of vehicles entering London's central zone by 30 percent (60,000 vehicles). This success has been noted by other cities tired of congested streets. Cities throughout the United Kingdom are considering a similar scheme, as are Stockholm, Milan, Barcelona, São Paulo, New York, and San Francisco. SJ

Bus Rapid Transit

Passenger buses are one of the most cost-effective methods of mass transit, able to move large numbers of people along existing roads. Unfortunately, in the United States, buses are not used to their full potential. When most Americans think of municipal bus service, they picture lumbering diesel behemoths that get nowhere fast, tracing circuitous routes that emphasize coverage over speed. Given the choice between driving a car and taking a bus that might require long waits and transfers between lines, most of us opt to take the car.

Public buses may never offer every convenience that a private car does, but there are many ways to make them more efficient, more attractive, and faster. Pioneered in Curitiba, Brazil, bus rapid transit (BRT) uses special high-speed express buses in place of commuter rail lines. The buses run on high-occupancy vehicle (HOV) lanes or, ideally, their own dedicated roads, often paralleling existing freeways. Separated from the traffic grid, they are able to run between stops at high speeds, increasing capacity and making the service attractive to harried commuters. Before boarding, BRT passengers pay their fares on the platform (speeding up the process), then they gain entry to special ramps that allow them to embark without climbing stairs.

There's still room for innovation on local bus routes. Smaller, agile vehicles designed to zip through residential neighborhoods could deliver riders to BRT stops or light-rail stations to maximize the system's potential. SJ

Thumbs Up

The layout of our suburbs and cities currently leaves many of us with no choice but to use autos to get to work or go shopping. There's no reason why we have to do these things alone, however—or in our own vehicles. This isn't a new idea: hitchhiking dates back to the 1920s. The more formal practice of carpooling was popularized during World War II to conserve fuel and rubber tires. The inconvenience and social awkwardness of ride sharing have limited its appeal, but new technologies promise to make the practice safer, more appealing, and more efficient.

Slugging

An unusual form of casual carpooling has evolved in some cities. Solo drivers drop by a designated spot (often a mass-transit station), pick up a rider or two, and then hit the freeway, legally able to take advantage of speedy HOV lanes. The practice started decades ago in Washington, DC, where it is called slugging. It has become so integral to DC-area commuting that the developers of express toll lanes hope to encourage slugging by building designated transfer areas.

Dynamic Ride-matching

Craigslist classifieds, discussion sites on company intranets, and dedicated Web sites are already being used to organize daily carpooling, as well as long-distance ride sharing. The next logical step? Combine ad-hoc ride-sharing arrangements like "slugging" with information technologies to implement "dynamic ride-matching."

Early ride-match trials, which used voice mail to coordinate rides, had limited success. New technologies—online mapping systems, GPS units, cell phones, and other information appliances—could revive the notion, turning city streets and willing drivers into a personal-transportation network for passengers. Ideally, no planning ahead would be necessary; getting a ride would require only a little more effort than hailing a taxi. Instead of extending a thumb or waving down a cab, the rider would use a cell phone to enter her location (assuming the phone didn't already know it) and desired destination. The system would alert drivers who were headed in the same direction that a rider was nearby.

Concerns over safety would be the biggest obstacle to such a system. Hitchhiking—once a common, utilitarian practice, and later a rather romantic undertaking—is now considered risky by both riders and drivers alike. (Ask an American to free-associate on the term *hitchhiker* and chances are the short list of words he'd come up with would include *carjacker, molester,* and *shallow grave.*) A reputation-tracking system could help people overcome these fears. Drivers and passengers would be required to register on a Web site, provide a photograph, and agree to be rated by other passengers on their manners, reliability, and punctuality. Registered sex offenders could be blocked from participating. Display-equipped cell phones would act as a security pass, providing not only a photo ID of the person you're catching a ride with but also his reputation as a driver or rider. Additional levels of vetting could be added; for example, women could specify that they'll only accept (or provide) rides with other women.

Texxi

Texxi ("the taxi you text"), a commercial effort along these lines, is being tested in Liverpool, England. Texxi taxis leave from a number of designated points. People who wish to share a ride transmit a text message containing codes for their local point and desired destination. A sophisticated algorithm groups ride requests into bundles that can be efficiently serviced by a single taxi. When a vehicle is on its way, the prospective passengers receive notice to report to their designated departure point, along with a boarding code and the ID of the cab they'll be riding. Result: One taxi picking up a full load of passengers, headed in the same direction and able to share the fare.

With the advent of Global Positioning System (GPS) units, ubiquitous cell phones, online mapping systems (such as Google Maps), and other information appliances, it is conceivable that individuals could set up private systems like Texxi. Drivers and passengers could be required to register, agreeing to participate in a reputation tracking system to ensure reliability and safety by allowing users to rate each other and voice complaints. SJ

Go Ahead, Play in the Street

Any student of urban planning can tell us horror stories about the destruction of once-thriving urban neighborhoods diced up or isolated by new freeways. Paradoxically, totally eliminating cars from a neighborhood or shopping district isn't really healthy either. Pedestrian malls sound good in pitch sessions, and look good in the form of artists' conceptions, but many of these well-intentioned efforts to rebel against Autotopia have resulted in retail ghost towns. Is there a way to lay out a neighborhood or shopping district that is safe and fun to walk around, but that allows access to taxis, delivery trucks, and the occasional private car?

For decades, municipalities have been using "traffic calming" techniques—speed bumps, rotaries, and deliberately inconvenient traffic grids—to keep drivers from turning local streets into ersatz highways. The Dutch have taken this a step further with the *woonerf*. *Woonerf* means "street for living," and the concept includes totally integrating pedestrian and vehicular spaces, eliminating curbs and placing amenities normally associated with sidewalks—planters, benches, and streetlights—in what would normally be the middle of the street. Faced with these obstacles and with the people using them, drivers are forced to not only slow down, but to be aware that they are in a community. The theory is, if we feel like we are driving through someone's yard, where kids play and pedestrians stroll, we'll automatically slow down and widen our awareness, knowing that we have to be on the lookout for the unexpected.

There is also a more scientific theory behind *woonerf* zones. As Linda Baker describes in her 2004 Salon.com article "Why Don't We Do it in the Road?," a pedestrian's chances of surviving a hit from a vehicle going more than twenty miles per hour are slim. At the same time, a driver's ability to make and retain eye contact with things and understand what he or she is seeing above twenty miles per hour is greatly diminished. So if we lower speeds, and add more elements for drivers to interact with, we will undoubtedly produce safer streets. More important, cities often experience less congestion when they lower speed limits; if people are driving more slowly they don't need as many traffic lights and major intersections, which are the real culprits of congestion. Of course, only some streets can be treated in this manner; heavier, faster travel must be handled with ring roads, bus lanes, and other approaches.

But safety and congestion aren't the only reasons to adopt this course. *Woonerf* streets are more attractive. With the exception of a sign alerting motorists that they are entering a *woonerf* zone, there are no signs or traffic signals. The street is all on one grade, and there are no yellow or white lines painted on it. Plus, the elements that make motorists pay attention to their surroundings—small green spaces

A rendering of a recent proposal in London for street space that would be shared by motorists and pedestrians.

with benches; streetlights that are positioned to highlight houses and pedestrians instead of streets and cars; and a smattering of unobtrusive parking spaces—also make streets more pleasant overall. SJ

Transit-oriented Development: An Rx for Sprawl

There's really nothing new about transit-oriented development. We've been doing it for decades. Unfortunately, the transit system we've designed our communities around is the individual automobile, and the results are pollution, congested roads, rambling retail districts, and isolated, unsustainable housing developments.

The fast-growing Georgia capital, Atlanta, could be a poster child for the problems of urban sprawl. It has some of the worst traffic problems in the United States, resulting in long commute times and some of the nation's most polluted air. But less than a decade ago, a local developer started turning what might have been a permanently blighted no-go zone—the contaminated site of a shuttered steel mill—into a model of transit-oriented development. The result? Atlantic Station, a 138-acre (56-hectare) "live-work-play" community that is both sustainable and attractive.

Located at the intersection of two interstates, and convenient to the city's commuter-rail system, the development includes office space, shopping and entertainment centers, and mixed-income housing for 10,000 people. Residents can easily work, play, and do just about anything without leaving the site. Safe walkways, an internal shuttle service, and bike paths (plus other cyclist amenities) make car trips within the complex virtually unnecessary. Residents who work off-site, and employees of the site's businesses, are encouraged to take advantage of carpool, vanpool, and other ride-sharing programs coordinated by the Atlanta Regional Commission. Those who use alternative transportation can take advantage of a "guaranteed ride home" service that will get them home in an emergency.

Atlantic Station isn't an isolated island of development. A nearby bridge was widened and greened as part of the development deal, and a brand-new bridge was built to reconnect sections of Midtown that had been carved in two by interstate highways. The new bridge has two carpool lanes, bike lanes, and sidewalks on each side. All of this frees Atlantic Station from automobile dependence while maintaining its connection to the greater metropolitan area. SJ

Taking Back the Streets, Two Wheels at a Time

Critical Mass is not an organization, it's an unorganized coincidence.

Critical-mass.org

Bicycles are a fast, efficient, and healthful means of transport, but they don't seem particularly welcome in most American cities. Making a municipality bicycle-friendly means laying out bike lanes, providing secure bike stands, and dealing with hazards ranging from dirty gutters to abusive car drivers. But it's hard for cyclists to advocate changes to infrastructure

when so many people view city riding as an aberration undertaken only by bike messengers, adrenaline junkies, or people secretly uncomfortable with city life who should really be living in the country. With this in mind, several cyclist advocacy groups around the world stage defiant, traffic-stopping protests to boost motorist awareness and change the general public perception of biking.

Critical Mass (CM) started in San Francisco in September 1992 as a monthly event to celebrate cycling and assert cyclists' rights to the road. Hundreds of cyclists took to the streets side by side, their sheer numbers forcing motorists to stop and let them pass. These rides quickly spread across the globe. Local branches maintain comprehensive Web sites, letting cyclists know when the next ride is and informing them of policy issues and changes that affect them. The CM chapters also participate in memorial rides for cyclists killed by cars.

Mexico City, one of the least bike-friendly cities in the world, has its own version of Critical Mass: Bicitekas, whose logo is an Aztec warrior on a bicycle. The group organized in 1996, after the mayor's office turned a deaf ear to its ideas about making the streets more bike-friendly in order to alleviate the city's notorious congestion and smog issues. This small group of cyclists holds weekly rides on the city's most heavily trafficked streets. During Bicitekas's biggest protests, members have biked along the city's congested freeways while motorists stuck in Mexico City's infamous traffic jams yelled and swore at them.

The mayor's office has made some concessions in response to the protests: cyclists are finally allowed to bring their bikes on the subway, and the city has built an eleven-mile downtown bike path. But Bicitekas won't be happy until all of Mexico City's streets are safe for cyclists.

While CM and Bicitekas rides assert bicyclists' rights to the streets, another growing movement—Ghost Bikes—focuses more on a single aspect of bicyclists' rights: the right to a safe trek. Seattle-based Ghost Cycle and similar organizations in Pittsburgh, Chicago, and other cities place bicycles painted a ghostly white in spots where cyclists have been killed or injured in accidents. Some stand unmarked; others have signs that tell the unfortunate cyclist's story. MM & SJ

Opposite, left: Transit-oriented developments like Atlantic Station in Atlanta, Georgia, are crucial components of better city development.

Opposite, right: At Critical Mass rallies, motorists learn to share the streets.

Product-Service Systems

Product-service system (PSS) is a new term for an old idea: emphasizing access over ownership. It's simply about sharing products among people, and recognizing that green systems are just as important as green products. We already take part in these systems when we use rental DVDs, Laundromats, libraries, gyms, and taxis; now people are starting to talk about PSS with regard to things that many of us don't usually share, such as cars, tools, appliances, and workspaces. Breaking past cultural assumptions that equate affluence with ownership may still be the greatest challenge to wide implementation of PSS, but what if the alternative is cheaper and more sustainable, doesn't clutter our homes, and connects us with our neighbors? What if we used cars, tools, appliances, and workspaces the way we use Laundromats, libraries, gyms, and taxis? What if, in short, PSS can staggeringly improve our quality of life?

Design researchers Ezio Manzini and François Jégou have spent years researching a growing number of small, bottom-up community solutions for sharing tools, mobility, community spaces, and knowledge. They put forward the hopeful observation that these kinds of systems are emerging organically within communities, and they encourage designers to shepherd them forward, improving their visibility and effectiveness.

In their research, Manzini and Jégou have tracked the emergence of shared "multiservice centers," through which the community shares access to little-used workspaces like professional-grade kitchens and workshops, leases connections to distributed green energy, or uses a mobility network that includes car sharing, buses on demand, assisted hitching, and bicycle networks. Such multiservice centers are part of an "empowered place," a community in which social structures built into everyday life help to create and support sustainable practices.

Design thinker John Thackara [see Designing a Sustainable World, p. 83] observed in his book *In the Bubble* (2005) that while resources may be limited, people are abundant. A fundamental shift from an economy based on stuff to one based on people—designed for systems and services, rather than *things*—is essential in the creation of sustainable communities. Such a change may be easier to adopt where people are most abundant: in the cities of the rapidly growing Global South.

Widespread connectivity is what makes product-service systems a new idea. "These solutions already exist in various forms," Jégou said in an interview in *Dwell*. "[Our project] merely brought them together. There's nothing sci-fi about it. For example,

from Beijing we took the idea of the Lift Club, a sort of safe hitchhiking service organized by way of mobile text messaging. In Milan, we found a scheme among mothers so that kids can walk, rather than ride in cars, to school ... All these things are banal locally, but when we introduce them elsewhere, they are innovations" (2004).

Tool Sharing: The Bay Area, ToolBank

When it comes to possessions, humans are not well known for their resource efficiency: the average power drill, over its entire life span, gets used for about ten minutes (Thackara 2005). It's been estimated that Dutch families are hoarding 9 billion euros worth of unused goods in their attics: all the discarded clothing, unstylish furniture, and forgotten equipment that composes the detritus of an affluent culture (van Hinte 2005).

Thirty years ago, baffled by homeowners' compulsion to own separate lawnmowers, Victor Papanek [see Collaborative Design, p. 124] argued for tool libraries as a way to share resources. It's not as if we lack the systems for sharing: tool lending has been enormously successful in California's Bay Area, where the public libraries in Berkeley, San Francisco, and Oakland have been expanded to lend sawhorses and screwdrivers as well as books. Borrowers tend to take good care of tools that they hope to use again.

One of the best North American tool-lending models comes from Georgia, where the Atlanta Community ToolBank emerged from a program that helped disabled and elderly homeowners with repairs. It now supplies more than 50,000 members with home-repair tools and volunteer support. The organization's Rescue and Reuse program collects unused resources like paint, lumber, and fixtures, tapping community connections to redirect materials away from the landfill.

Community Spaces: The Hub

We sometimes forget that all space doesn't need to be privately owned or rented. One of the challenges for telecommuters, self-employed people, and small businesses is the cost of setting up office space. Isolation is another pitfall: working outside a typical office setting often means there are not enough people around to bounce ideas off of. Shared environments can be cooperative, city funded, or profit driven—but the public places that such workers typically use as workspaces were not designed for that purpose. Copy shops are rarely welcoming, and cafés rarely provide much more than an escape from a home office.

Based in a daylight-filled renovated factory in London, the Hub was designed as a sustainable workspace to provide professionals with access to resources like meeting rooms, reading spaces, and desks; Wi-Fi, fax machines, copiers, and scanners. There's a library nook and a pleasant café for taking breaks. Like the "writing rooms" where writers buy a membership so they have a quiet and respectful place to work, the Hub is a subscription service, rented by the hour, day, or month. For all its well-considered details, the Hub's most important service is conviviality.

Atlanta's ToolBank has tools and wheelbarrows for the borrowing, while London's Hub provides office space on a shared basis.

Attracting a range of small organizations and creative people, it fosters those rarest of things: community and the cross-pollination of ideas. ▫▫

Car Shares: AutoShare, Streetcar

Modern cities are scaled to accommodate cars, which means that even those people who have the privilege of walking to work usually still want access to one. It's an expensive dependency: you're paying to insure and store a machine that loses value as it sits parked in the street. In their travels, Ezio Manzini and François Jégou discovered carsharing businesses and cooperatives all over the world. Car shares are cheaper and less polluting than conventional car ownership, and overall they improve the city experience. Begun in Germany and exported to North America via Quebec City, car sharing has now taken off from Singapore to Århus to Chicago.

Because most of us value car ownership, the services that carsharing provide have to outweigh the inconvenience of not having a car. Besides the cost savings, another advantage is that it can provide an exclusive experience: Toronto's successful Autoshare is one of the only ways that most locals can get access to still-expensive hybrid vehicles. Carsharing also has to provide a seamlessly convenient experience. In their work with Streetcar, a carsharing service for English cities, service designers Live/Work have tried to make the carsharing experience easier, using SmartCards that allow users to access a car in a convenient location, whenever they need it. Live/Work, whose tagline is "You are what you use, not what you own," wants to create what they call "service envy," in which the most desirable product is access, utility, and experience unburdened by ownership. ▫▫

RESOURCES

Bowling Alone: The Collapse and Revival of American Community by Robert D. Putnam (Simon and Schuster, 2000)
If we want to learn to define our lives by the stuff we use, rather than the stuff we own, we need to search out people's rationale for not already sharing more. *Bowling Alone* shines a light on the reasons North Americans have grown distant from one another (the main culprits being sprawl, television, and cars). With detailed arguments and research crowding every page, Putnam's classic is sure to spur new thinking about what's important in life, how we can reconnect with our families, friends, and neighbors, and what that might mean for the future of cities.

As Putnam writes, "Members of a community that follows the principle of generalized reciprocity—raking your leaves before they blow into your neighbors' yard, lending a dime to a stranger for a parking meter, buying a round of drinks the week you earn overtime, keeping an eye on a friend's house, taking turns bringing snacks to Sunday school, caring for the child of the crackhead one flight down—find that their self-interest is served.

In some cases, like neighborhood lawn raking, the return of the favor is immediate and the calculation straightforward, but in some cases the return is long-term and conjectural, like the benefit of living in the kind of community where people care for neglected children. At this extreme, generalized reciprocity becomes hard to distinguish from altruism and difficult to case as self-interest. Nevertheless, this is what Tocqueville, insightfully, meant by 'self-interest rightly understood.'"

Sustainable Everyday: Scenarios of Everyday Life by Ezio Manzini and François Jégou (Edizione Ambiente, 2003)
In the course of traveling the globe in search of community solutions to everyday dilemmas, design researchers Manzini and Jégou uncovered green-roof programs in Japan, hitchhiking services in Beijing, walking teams of mothers and children heading to school on foot—rather than by car—in Italy, and urban gardening initiatives worldwide. They combined these innovations into a clear vision of a sustainable urban existence, by sketching out a picture of an urban community knitted together by a network of services that support the health of the community and the local ecology. The scenarios make daily life more sustainable by creating local systems for sharing knowledge, products, and time; cel-

ebrate the do-it-yourself (DIY) spirit; and extend the home by externalizing some of its functions. Most important, they reveal a vision of what's already possible and easily within our reach.

Chinese Cities of the Future

By the end of this decade, China will have some of the most advanced green urban environments in the world; by the end of the next, China may well have more green communities than the rest of the world put together. At least, that's the plan.

Already home to some of the largest cities in the world, China expects to see roughly 400 million more people—about half of its rural population—move into cities by 2030. The World Bank estimates that between now and 2015 roughly half of the world's new building construction will take place in China. For China to meet these challenges, it will have to change the way it builds cities. Traditional designs and contemporary planning are worse than insufficient, they're dangerous. By some estimates, just manufacturing enough bricks to build the traditional-style buildings needed to house all the people moving into the cities would consume all of China's remaining coal and clay.

Fortunately, Chinese leaders recognize the severity of the challenge, and China has embarked on a bold drive to completely reimagine how cities are created and buildings made. As an initial step, they've put

The influx of domestic migrant workers is putting increased pressure on the infrastructure of Chinese cities, Guangzhou City, Guangdong Province, China, 2004.

forward new rules about building efficiency. They've also brought in some of the world's leading experts in sustainable planning and design and have given them freedom to think in new ways. Finally, China is pursuing its own "eco-city" designs, retrofitting grossly inefficient metropolises with better transit, cleaner water, and an underlying infrastructure of sustainability. No one knows whether these plans will work, but we all have a stake in their success. AS & JC

China's Green Leap Forward

China faces some of the most severe environmental problems on the planet.

China's gross domestic product (GDP) has more than doubled since 1978, and the last decade has seen the Chinese economy growing as much as 10 percent a year. It's no wonder that this period of growth has been called the "Chinese miracle." Unfortunately, this miracle has come at a terrible price. The ecological woes that are the legacy of that rapid growth—achieved with outdated technologies, through governmental mismanagement and corruption—go far beyond quality-of-life issues: in China, pollution kills people, saps economic vitality, and brings into question the future of the entire nation, if not the entire planet.

According to the World Health Organization, China has seven of the ten most polluted cities in the world. Ninety percent of its urban bodies of water are considered polluted, and acid rain falls on nearly a third of the country. Pollution-related diseases—respiratory failure, heart diseases—are the leading cause of death in China. Air pollution is so bad that traffic cops in Beijing only live an average of forty years.

A major cause of that pollution is China's continued reliance on aging coal-fired power plants, which generate 75 percent of the country's electricity, according to the Pew Center on Global Climate Change. Coal is also burnt to heat many private homes. Cars are the other leading cause of air pollution, and the situation may get much worse. According to an article on Chinadaily.com, Chinese officials expect to have 140 million cars on the road by 2020, seven times the number of cars now in China, and more cars than now exist in the United States (September 4, 2004). China is also the second largest producer of greenhouse gases on the planet (China 2004).

China already has such an impact on the earth that it's hard to imagine a bright green future that doesn't depend upon its complete transformation. It can't simply copy the West. A China of SUVs, suburbs, and careless consumption would be a planetary ecological nightmare and would very quickly become a Chinese tragedy. Nowhere is the conclusion of the 2002 Jo'burg Memo (a report compiled by leading thinkers from around the world for the World Summit on Sustainable Development) more true: "There is no escape from the conclusion that the world's growing population cannot attain a Western standard of living by following conventional paths to development. The resources required are too vast, too expensive, and too damaging to local and global ecosystems." Or, as Pan Yue, China's deputy minister of the environment put it in an interview with the German

magazine *Der Spiegel,* "This miracle will end soon because the environment can no longer keep pace" (2005).

China needs to build the first truly twenty-first-century economic superpower, one based not only on a new model of the sustainable city, but on efficient, renewable, sustainable production of goods and energy. China's next miracle needs to be a green leap forward. JC & AS

Green Building Regulations

According to the Energy Information Administration at the U.S. Department of Energy, China wastes enormous amounts of energy—the country uses more energy per dollar of GDP than any country in the West, and more than most other leapfrog nations [see Leapfrogging, p. 292] such as Kenya, India, and Brazil. Nearly all Chinese buildings use two to three times as much energy per square foot as equivalent buildings in other industrialized countries; "energy efficient" buildings account for less than 1 percent of Chinese construction.

But like everything in China, this is changing. Qiu Baoxing, China's minister of construction, has proposed green-building regulations that could give China some of the world's strictest standards. Buildings being built now are required to incorporate technologies to save about 65 percent more energy than previous standards, including natural ventilation and lighting, water recycling, and widespread use of renewable energy systems. By 2010, all Chinese cities must reduce total building energy use by 50 percent—the standard goes up to 65 percent in 2020. If enforced, this standard will trigger the world's largest green-building retrofitting crusade.

Even before 2005, when these new regulations kicked in, Chinese developers began building green. There are now over a dozen LEED-registered [see Big Green Buildings and Skyscrapers, p. 247] buildings in China, and more are being added all the time. In April 2005, the Ministry of Science and Technology, working with the American Natural Resources Defense Council, built China's first LEED Gold standard building—its new 130,000-square-foot (12,077-square-meter) headquarters in Beijing. More such buildings are planned.

Not only do the Chinese recognize that energy inefficiency is expensive, but they also understand that in the future, knowing how to achieve radical efficiency will offer a competitive advantage: construction companies that know how to build green will dominate the market. China is embarking on the world's largest green-building tutorial. JC

Huangbaiyu, Tangye New Town, Guantang Chuangye

The China Housing Industry Association, the China-U.S. Center for Sustainable Development, and famed *Cradle to Cradle* author William McDonough are attempting to build the sustainable dream town—not once, but up to six times, all over China. These cities will serve as models for further green urban development, all with the same goal: "to draw power from the sun, to maintain materials in closed-loop systems of technical and biological nutrition, and to create an intergenerational community of people productively engaged in restorative commerce."

Furthest along is the village of Huangbaiyu, a paragon of sustainable-building processes. The homes are under construction as of 2006, with an ultimate goal of four hundred families in residence—along with a biomass gasification plant for energy, village-wide water recycling, and a mixed-use town center. Next up is the Tangye New Town, which will scale up the conceptual designs pioneered in Huangbaiyu. The Tangye urban district will be home to 180,000 people, and is intended as a working model for replication across China.

Preceding pages: Old City overview, Shanghai, China, 2004.
Opposite, left: A woman collects plastic bottles in a polluted canal in Dongxiang, China.
Opposite, right: Plans for a new sustainable community in Liuzhou, China, include rooftop farms.

All materials used in the model towns must either biodegrade safely or be completely reusable. The walls of Huangbaiyu demonstration homes, 1.6 feet (0.5 meters) thick, are made of pressed-earth blocks lined with straw (a by-product of local agricultural production that would otherwise go to waste). The overall town design is a network of neighborhoods and commercial precincts linked by a series of parks, which also serve to channel storm water.

The Huangbaiyu and Tangye undertakings will soon be joined by the Guantang Chuangye Sustainable Development project; the initial design has only recently been approved. As in the other projects, Guantang Chuangye neighborhoods will be structured around parks, with every workplace and residence within a five-minute walk of the transportation network and a school. Guantang Chuangye will also show how the sustainable-city model can be made to fit with local landscapes and environmental conditions: the city plan will preserve existing streams and wetlands, and maintain the predevelopment water cycle for the local ecosystem.

Building an entirely new model is never easy, of course. These communities aren't meant to be showcases for a handful of wealthy citizens. Originally, in order to keep the Huangbaiyu houses affordable for the growing Chinese middle class, the budget per house was only $3,500; it is proving difficult to keep the price that low. In other cases, sustainable materials have been impossible to source locally. Corners have been cut, say researchers visiting the site. Still, these new communities point the way toward a Chinese future that is both bright and green. AS

Greener Shanghai

Shanghai is the emblematic modern Chinese city. A big, crowded, economic powerhouse, the city has long been held up as the model for today's China. Soon, it may be the model for a green China. British design-consultancy firm Arup is working with the Chinese government to lead the construction of an "eco-city" expansion to Shanghai. Dongtan, the expanded development near Shanghai's airport, will eventually cover about 21,745 acres (8,800 hectares)—roughly the size of Manhattan island—and is intended to be a genuinely ecofriendly city, using recycled water, cogeneration, and biomass for energy, and striving to be as carbon-neutral [see Building a Green Home, p. 147] as possible.

So what does it mean to be a "genuinely ecofriendly city"? There's almost no waste of resources: gray water is captured, purified, and recycled; organic wastes are used as biomass to generate clean energy; and combined heat and power systems provide warmth and electricity for the home in the same process. Buildings use natural light, advanced insulation, and high-efficiency designs to cut energy use significantly. Public transit is commonplace, within an easy walk from almost any location. In many ways, life in the Dongtan eco-city will be very much like life in any modern city: clean, comfortable, exciting. Only behind the scenes, the systems that keep the city alive—that control its flows of energy and people, waste and water—will work with far greater efficiency than they do almost anywhere else in the world.

The first phase, a 1,557-acre (630-hectare) development including a mix of transport facilities, schools, housing, and high-tech industrial spaces, will hold 50,000 people, and is expected to be completed by 2010. But Dongtan is just the beginning. Arup has agreed to work with the Chinese government to extend the green-city model to four more cities around the country. All the cities will be designed to e self-sufficient in energy, water, and most food, and to generate near-zero emissions from transportation. JC

Green Rooftops in Beijing

One of the projects under way to clean up Beijing for the 2008 Olympic Games is an ambitious proposal to increase the amount of green space around the Olympic stadiums by planting more of the rooftop gardens known as green roofs [see Greening Infrastructure, p. 256]. Green roofs have multiple benefits: they help clean the air, limit noise and dust, and greatly reduce the so-called urban-heat-island effect.

By the time the games commence, all buildings around the stadiums and along central city roads will have green roofs, according to the

section chief of the Beijing Municipal Bureau of Parks, Yang Zhihua. The bureau's goal is to add 2.7–3.2 million square feet (250,000–300,000 square meters) in new green roofs every year in Beijing, with an overall project goal of 10.7–21.5 million square feet (1–2 million square meters) of green roofs by 2009.

If that sounds like a lot, consider this: as of late 2005, Beijing had just under 753 million square feet (70 million square meters) of rooftop space. JC

Green Car China

The biggest challenge China faces comes from energy and transportation. Green cities won't matter if they're populated with conventional gasoline automobiles.

Fortunately, there's cause for (cautious) hope. China's auto designers are working on vehicles that could dramatically cut oil consumption—priced to appeal to leapfrog [see Leapfrogging, p. 292] nations around the world. The Wuling SunShine is a small van that runs about five thousand dollars in total cost and gets about forty-three miles per gallon in the city. On sale since 2002, it's among the most popular vehicles in the country, which has helped push it to the number-one spot in the light-vehicles market. It's still a conventional gasoline vehicle, though.

The future looks greener, however, as China pushes to bring in cleaner vehicles. Toyota and Volkswagen are building hybrid-car factories, while GM and DaimlerChrysler hope to introduce fuel-cell-based vehicles in limited numbers over the next few years. Thinking big, Shell plans to help Shanghai, as it has helped Iceland, shift heavily toward hydrogen.

Volkswagen lot, Shanghai, China, 2005.

But while it's good to see Western companies help to bring the green-car future to China, homegrown innovation is even more important—and we're seeing that, too. Geely Motors, China's biggest privately owned carmaker, plans to make hybrid cars available by the 2008 Olympics; Geely joins state-supported automaker Chery, which expects to bring hybrids to market by late 2006. More advanced local designs are on the drawing board, too, such as the Aspire electric car and the Spring Light 3, a fuel-cell hybrid with a five-thousand-dollar target price.

China is also currently the world's number-one market for electric vehicles. According to Beijing's China Bicycle Association, more than 10 million electric bikes and scooters were sold in China in 2005 alone—that's nearly three times more electric bikes sold than autos. Electric bikes are inexpensive, easy to operate, and well suited to the crowded urban Chinese streets. Most are made by a variety of competitive small manufacturers and are used for intracity transit and deliveries. As popular as the bikes are, the Chinese government doesn't like them, since they threaten the "pillar industry" of auto manufacturing. Officials around the country are trying to ban electric bikes from the streets of many large cities. How this plays out will tell us a lot about where China is headed. JC

RESOURCE

The Beijing Consensus by Joshua Cooper Ramo (The Foreign Policy Centre, Spring 2004)
Might all this Chinese work on sustainable urban design and technology signal the emergence of something new? Absolutely, says Joshua Cooper Ramo, a researcher at The Foreign Policy Centre in the United Kingdom. In a report published by the center, Ramo argues that the bold innovations China is pursuing in solar, wind, automotive, telecommunication, and building technologies signal the rise of a new model of development, one based on constant innovation, awareness of ecological limits as a business opportunity, massive investments in education, and a willingness to be honest about the real costs of pollution and corruption. He calls this model the Beijing consensus:

"Without a change to a more sustainable growth model, China's economy is likely to sputter out, choked off by a shortage of resources and hampered by corruption and pollution ... This new view is apparent in the way Chinese thinkers are starting to measure growth. Tsinghua economist Hu Angang, among others, now disdainfully labels GDP [gross domestic product] growth ... 'black GDP growth.' He takes China's impressive black GDP numbers and subtracts off the terrific costs of environmental destruction to measure 'green GDP growth.' Then Hu nets out China's corruption costs to measure 'clean GDP.' This, he says, is how China should measure progress. 'It doesn't matter if the cat is black or white,' Deng Xiaoping famously observed in one of his early speeches on economic reform. 'All that matters is that it catches mice.' But Hu's GDP tools, which I've heard leaders all over the country begin to talk about, reflect the government's new belief: the color of the cat does matter. The goal now is to find a cat that is green, a cat that is transparent."

Lagos, Nigeria—fast becoming one of the world's largest megacities—serves as an example of how pressing the need is for better urban solutions.

Lagos

What do you know about Lagos?

Most of us in the Global North know very little. We may not even be able to find it on a map. And yet this Nigerian megacity is well on its way to becoming one of the planet's biggest: if it continues growing at its current rate, the UN says, by 2015 only Tokyo and Mumbai will be larger, and more than 23 million people will call Lagos home.

Life is not easy in Lagos. Most residents live in informal slums that are growing at astounding rates. Indeed, so much of the city has emerged in ad hoc neighborhoods that there is no way to even be certain of exactly how many people live there. Roads and houses spring up that have never seen a government official; maps are out-of-date before they're even printed. Poverty is endemic (most residents live on less than one dollar a day) and infrastructure is a relative term—only a tiny fraction of homes are connected to sewers; raw waste runs down the middle of the streets; and entire parts of the city are submerged in the rainy season. Traffic is so bad that Lagos's residents have invented their own term for mega–traffic jams: the *go-slow*. A blanket of smog hangs perpetually in the skies. By some estimates, half the population is infected with malaria.

"For the politicians, administrators and community leaders who have to manage this complex, heaving mass of urban humanity," writes Nigerian journalist Paul Okunlola in a paper for the United Nations Human Settlements Programme (UN-HABITAT), "the quantum of decaying infrastructure, widespread urban poverty, massive unemployment, pervasive security inadequacies, emerging slums and overwhelming environmental decay have become the major characteristics that progressively define the city's fortunes."

Most outside observers find Lagos all but incomprehensible; almost all find it terrifying. It is reputedly the most dangerous city in the world, a place where rent-a-cops have submachine guns and half-wild dogs, and where there is less than one official police officer per thousand residents. But more than anything, it is the chaos of the place that puts fear into visitors used to a different kind of city. "Lagos seemed to be a city of burning edges," writes architect Rem Koolhaas of the time he spent studying the city's future prospects. "Hills, entire roads were paralleled with burning embankments. At first sight, the city had an aura of apocalyptic violence; entire sections of it seemed to be smoldering, as if it were one gigantic rubbish heap."

Lagos's residents watch their backs, but say they find their city an exciting place to live. Lagos drives the entire West African regional economy. More than 250

languages are spoken in its streets. Even after the oil boom that started its explosive growth sputtered out, people have continued to pour in, seeking a better life.

"The streets aren't all paved with concrete, let alone gold," writes UNESCO journalist Amy Otchet, "but Lagos appears as an El Dorado in the poverty-stricken countryside, where work can be found and dreams come true." Lagos has a dearth of formal jobs but a thriving underground economy, where at least half of all city residents make their living bartering, hustling, doing odd jobs, and trading on their skills and energy to get by while they keep an eye out for opportunity. In Lagos, being business-savvy isn't a career choice: it's a survival skill.

Even Rem Koolhaas came to admire the spirit of improvisation and enterprise he met there: "Dangerous breakdowns of order and infrastructure in Nigeria are often transformed," he writes, "into productive urban forms: stalled traffic turns into an open-air market, defunct railroad bridges become pedestrian walkways." Lagos, Koolhaas came to believe, is the future—and he means that in a *good* way.

Megacity Opportunities

The problems Lagos faces are uniquely severe, but they are not unique—neither is the energy of its people. Indeed, all across the Global South, emerging megacities like Mumbai, Delhi, Kolkata (Calcutta), Dhaka, Karachi, Jakarta, São Paulo, and Mexico City are poised to become the largest human settlements in history. By 2015, only two of the ten largest cities in the world—Tokyo and New York—will be in what we now consider the developed world.

So many people are moving into cities that we are building an urban area the size of Seattle every four to seven days. By some estimates, two-thirds of the urban areas that will cover the planet in 2030 don't even exist yet; put another way, two-thirds of the planet's future cities have not yet been built.

As Lagos illustrates, poor cities that grow rapidly face potentially disastrous problems. But urbanization is by no means all bad.

In some ways, the rise of megacities offers us incredible opportunities to address challenges that might seem otherwise insurmountable.

Take jobs. Most people leave their farms and villages for the exact same reason that Europeans and Americans did generations ago: because that's where the work isn't. Villages have local economies, but they often burn slowly, like small hearth fires; cities—especially those most connected to the global economy—are, by comparison, blast furnaces. Rural economies may grow slowly, if at all, but some of the fastest growth in the world is in the emerging megacities, as whole new economies sprout up—formal and informal, woven by hand or pulsing with the latest technologies.

On this young planet, more than half the people are under the age of thirty and roughly one-third are under the age of fifteen. Far too many people are unemployed, undereducated, and lacking avenues for working constructively to make their lives better. Frustrated and resentful young men are a breeding ground for radical violence, from gang warfare and terrorism to political paramilitary repression and genocide. Creating better opportunities for young men is one of our best strategies for building a better world, and urbanization can be a tool for doing just that.

But young women can gain even more from city life. While the perils they face—from sweatshop conditions to forced sex work—increase when they arrive in the city, their freedoms and opportunities increase far more. Young women in cities are much more likely to have jobs, to be educated, to avoid childhood marriages, and ultimately, to control their own destinies and family-planning choices.

These ingredients for young women's success have planetary implications as well. Again, because a majority of people on the planet are young, the choices young women make about childbirth have global implications. If the women of the global "youth bulge," as some name it, have many children of their own, we will wind up on an even more crowded planet. If they have fewer, we will find all the problems we face substantially easier to solve. And in general, one rule has held true: urban women have more control over their lives. They marry later, they have children later, and they ultimately choose to have fewer kids. In the long term, this is a very good thing.

Finally, life in megacities offers at least a chance to address many of the social needs that so press us. While local governments often find themselves essentially powerless to channel, much less direct, their cities' growth today, that doesn't mean that innovation is impossible, or that social services and health care can't be provided. Indeed, in many ways it is easier to provide for the public needs of people who live close together—even when they live in a squatter camp—than it is to provide for equally poor people living scattered in the hinterlands.

The very energy and ambition that is driving people to pick up and move to megacities is also fueling an explosion in homegrown urban innovation. Local solutions deriving from ad hoc responses to local problems combine with new models of assistance (like microcredit [see Microfinance, p. 346]) to yield dramatic, rapid improvement in people's lives.

Megacities may still prove to be one of the best levers we have for changing the world. AS

Opposite: São Paulo, Brazil, is an emerging megacity poised to be among the largest on the planet.

Left: Young women, like this one working in a call center in Nairobi, are more likely to find work in expanding cities than in rural economies.

Megacity Innovations

To build megacities that better meet people's basic needs, offer new opportunities for young people, empower women, and foster sustainably is to build an urban way of life unlike anything we've yet seen. This megacity future is going to spawn a whole array of new possibilities, ones we can't anticipate from our armchairs in the developed world. It's already begun. In Malaysia, young architects have come up with a design for a home outfitted with giant solar panels that open like petals as the day warms, shading the home and capturing electricity, and then fold back up as the evening cools, bringing the colder night air into the house and making the surrounding garden a pleasant place to sit, drink tea, and stargaze. In Harare, Zimbabwe, architects have created a biomimetic [see Biomimicry, p. 99] building that resembles an African termite mound. The Eastgate building copies the way termites use earth masses and ventilation tunnels to keep their mounds at a constant temperature. Consequently, the Eastgate needs no air-conditioning system, despite the blistering Harare heat.

Other innovations are being built from simpler parts, but are no less revolutionary. In Curitiba, Brazil, mayor Jaime Lerner led the creation of innovative low-tech social programs, including gardens tended by street kids, payments in food to homeless people who collect litter, converted buses that serve as mobile clinics in the slums, even architectural assistance for the surge of poor immigrants building their own homes. On top of that, Curitiba has built a world-famous transit system, and has expanded its parks and boulevards, giving it the most green space of any Brazilian city.

This is all just the beginning. The new urban future, in full bloom, may be nearly unfathomable to us in the old-fashioned Global North. The future doesn't think like North Americans do: the future is unfolding in places that have mobile phones but still rely on the arrival of the caravans, that sell computer chips in souks and bazaars, that burn sandalwood incense in five-hundred-year-old temples but broadcast video-game championships on TV. A bright green future will smell of curry and plantains, soy sauce and chipotle, and will sound more like Moroccan rap and twangy Mongol pop than Mariah Carey. We in the Global North don't know—we *can't* know—how the next generation of megacity urbanists will respond to the possibilities unfolding in front of them. The best research and development in urban planning won't be done by established professionals in developed-world think tanks, corporate labs, or universities. It'll be done on the streets of developing-world cities, by a younger generation just now coming into its own.

Right: Megacities are booming with fresh designs, like the biomimetic Eastgate shopping and office development in Harare, Zimbabwe.

Opposite: Bogotá, Colombia, has become a hothouse of innovation. A cyclist admires the view of the city (left), which has set aside 186 miles of designated bike lanes. Women celebrate the Night for Women (right), during which female Bogotanos reclaimed the streets in 2001.

They don't need our answers; they need the tools for finding and sharing their own answers. Redistribute the tools for invention and innovation, and the citizens of the megacities will remake the world. AS

Bogotá

Like many emerging megacities, Bogotá, Colombia, is a mess. Unlike many emerging megacities, though, Bogotá has had inspired leadership and a determined citizenry, and is already gaining ground on its problems. Three Bogotá mayors in succession have tackled the city's obstacles with passion, style, and innovation. Bogotá has gone from being almost unlivable to being full of vitality (if still grappling with serious challenges).

Mayor Antanus Mockus gets the credit for getting Bogotanos to think in new ways. Mockus, who was mayor from 1995 to 1997 and again from 2001 to 2003, patrolled the streets clad in red and blue tights, calling himself "Super Citizen" and intervening in places where civility was lacking. During a water shortage, he spoke on a public service announcement to promote water conservation, while naked in the shower. (In just two months, water usage dropped by 14 percent; through continued efforts, Bogotanos now use 40 percent less water per person than they did before the shortage.) He cracked down on corrupt police officers—even shutting down the entire transit police division, which was known for its bribery schemes. He even asked people to pay an extra 10 percent in optional taxes—to everybody's amazement, 63,000 residents did.

Two of Mockus's programs, though, exemplify the creativity he brought to the job. Bogotá can have terrible traffic jams, and even on the best days, its streets can be mayhem. Mockus hired hundreds of mimes to direct traffic—providing not only some needed order, but a hint of levity. More impressively, he launched the Night for Women.

In every city, women are on the receiving end of more antisocial behavior than men are. Great cities find ways to ensure that women are safe on the streets and welcome in public spaces. When Mockus asked residents what they thought the city's biggest problems were, he learned that the danger, harassment, and hassles women faced in Bogotá raised a barrier to women's full involvement in the social life of the city. So in 2001 he called on men to stay home and care for the children, and invited all Bogotá's women to come out for a night on which they would own the streets; 700,000 did, and by all accounts, Bogotá is safer and more welcoming now.

"The distribution of knowledge is the key contemporary task," Mockus said in an interview in the *Harvard University Gazette*. "Knowledge empowers people. If people know the rules, and are sensitized by art, humor, and creativity, they are much more likely to accept change" (March 11, 2004).

Enrique Peñalosa, Mockus's successor, focused an equally innovative program on Bogotá's planning woes. He created the TransMilenio, a rapid-transit bus system, which operates in special lanes and now carries a half million passengers a day. He set aside 186 miles (300 kilometers) of designated bike lanes, built greenways and paths, and launched an aggressive campaign

against motorists who drove or parked on the sidewalks. Realizing that great neighborhoods are central to great cities, he built new parks, libraries, and schools, and started a hundred new nursery schools so working mothers could more easily find good child care. Finally, he raised parking fees and gas taxes, and set up a system that requires people to leave their cars at home on certain days of the week. Overall, Peñalosa's efforts paid off: Bogotá now has 40 percent less traffic than it did before the program went into effect.

The latest mayor, Luis Eduardo Garzón ("Lucho" to his supporters), is tackling Bogotá's social problems head-on. He is expanding school capacity to serve the 100,000 kids who don't currently have access to education. He's creating a network of volunteer doctors who'll regularly visit families in poorer neighborhoods. He's fighting hunger. One way Garzón is paying for these new programs is by attacking corruption and tax evasion, which is a huge drain on the administration's expected revenues; so far his efforts seem to be working. Finally, in an effort to carry on the city's established commitment to its female citizens, Garzón filled all twenty of his submayoral posts with women.

Bogotá is far from perfect, but the progress it has made in just fifteen years is extraordinary: the murder rate has dropped by half; more kids are in school than ever before; traffic (while still bad) has calmed dramatically; and the streets are full of people walking, shopping, and talking. AS

Goa's RUrbanism

Imagine that you and your family live in Panjim (Panaji), the capital of the Indian state of Goa. You wake up in your high-density-yet-very-comfortable, ultra-efficient apartment, open the window, and look out on a capillary of rural farmland that snakes its way in among the other apartment blocks. You wave at the fellow guiding the buffalo past your building. Then you check your bank on the Net and discover that your integrated time-and-money account shows you with very little extra money, but a lot of free time to spend. You decide to spend that time participating in a community discussion about the merits of a new nanomaterials factory planned for a neighboring section of the city.

That's a tiny snapshot of what might be possible in just thirty years' time, according to the creators of RUrbanism, a city growth plan that may turn the state of Goa into a neobiological [see Neobiological Industry, p. 109] mix of human and wild, with high-density neighborhoods that encompass a blend of high-tech superefficient systems and traditional rural living. Rice paddies, fish ponds, and vegetable gardens will interpenetrate urban areas in a "spine-and-filament" pattern like that of a fish's gills, while surrounding forests provide fresh water and clean air. RUrbanism transforms the city from a parasitic consumer of resources into a symbiotic partner with both nature and rural culture.

RUrbanism is the brainchild of a team of Indian experts: Aromar Revi, Rahul Mehrotra, Sanjay Prakash, and G. K. Bhat. Their Goa 2100 project took home a special jury prize at Tokyo's high-profile 2003 International Sustainable Urban Systems Design competition, thanks to both the creativity and quantitative rigor of their hypothetical "transformation plan."

The team chose Goa because it's already a great place to live, with a thriving economy. And Panjim itself reflects many of the challenges commonly faced by India's growing cities, while also having the resources, political culture, and institutional base that make bold moves a clear possibility. The team began by collecting an enormous amount of data—demographic, socioeconomic, economic, institutional—and information related to planning, natural resources, energy, and transportation. They consulted with leading Goan citizens, inspected nature reserves

Opposite, left: Goa, India's RUrbanism growth plan, offers a pioneering blueprint for urban sustainability.

Opposite, right: Subway riders in Mexico City reading free books.

and industrial sites, and then mapped Greater Panjim in detail, using satellite and remotely sensed data, along with Global Positioning System (GPS) technology and Geographical Information System (GIS) maps [see Mapping, p. 518]. The result was a complete land-use and topographical model for the area, probably the first of its kind for the region—and as innovative as anything the world has ever seen.

All that research showed the team that Goa could in the near future implement major land-use changes that would support the regeneration of the surrounding landscape and reduce the city's ecological footprint. Goa could essentially be condensed almost back to its scale during medieval times, without resorting to high-rise, resource-intensive development, by emphasizing compact design and green building.

But the team didn't stop with the structure of the city: they also looked at the pace of life. One of most innovative features of the Goa 2100 project was its analysis of the entire "temporal economy" of the city and region. Comparing studies from around the world of how people use their time, and adapting assumptions to the South Asian context, the team modeled the time use of Greater Panjim and created a "time-use budget" for both the present day's citizens and the citizens of a future, post-transition, sustainable Panjim. This analysis led to a key discovery: that time should be thought of as an additional resource when considering the financing of a transition, since residents can do more with the time they'll save in a better-designed city. The Goa 2100 model—which allows for more than adequate personal, leisure, household, and community time, in addition to the needed time for work, child care, education, and so on—appears to be the first sustainability analysis of the time use of an entire city.

Finally, by combining the time-and-money accounting of the costs of transition, the team firmly established that a full transition to sustainability was both possible and affordable. In their final estimates, they discovered that an investment of only U.S.$60 million per year, coupled with the time investments of citizens from many different sectors (paid and unpaid), could accomplish the transition in just thirty years—much faster than the hundred-year period that the rules of the design competition had stipulated, and at far lower levels of financial investment than most people would expect.

RUrbanism is one visionary, yet feasible, model for how cities throughout Asia could redevelop themselves in coming decades. And the first field test could be Goa itself; the state is seriously considering the results of the Goa 2100 exercise as it ponders its own future development plans. Expect to see more cutting-edge models like this emerge as megacity challenges meet megainnovation. AA

Underground Literature

In a program in Mexico City, free books are being given out in the subway. The stories and poems in the books are short enough to finish on the average subway ride; readers drop the books off as they leave the Metro. Mexican authors were paid about U.S.$300 for use of their works, but their real payment is the increasing in-country interest in their writing. The program gets

books into the hands of those Mexicans who can't normally afford to buy them. And transit officials also hope that crime in the system might decrease if crooks are busy reading rather than picking the pockets of fellow riders. EG

The Hidden Vitality of Slums

In Brazil, they're *favelas*. In India, *johpadpatti*. In Turkey, *gecekonduler*. In Colombia, *callampas*. In Peru, *pueblos jovenes*. In Kenya, *vijiji*. In Indonesia, *permukiman liar* or *kampung liar*. In French-speaking Africa and the Caribbean, *bidonvilles*.

They are the most dynamic parts of the fastest-growing cities in the world—self-built, self-designed, and self-motivated. They are squatter neighborhoods: shantytown communities created when people take over land they don't own.

In English, many call them slums. And, frozen in time, the word fits. Take a snapshot in any of the world's shantytowns and you'll capture degraded and dilapidated communities.

But look at that photo a bit more closely and you'll find health clinics, beauty salons, grocery stores, bars, restaurants, tailors, clothing stores, churches, and schools. In the midst of squalor and open sewage, the streets are lively and business is booming. What's more, study that shack in the center of the photo—the one made of mud and sticks. Is it simply decaying? Or is the family that lives there reconstituting it around themselves like a cocoon, transforming it

Right: Many of the world's children live in squatter settlements—like this one in Nairobi, Kenya—in emerging megacities.

Opposite, left: Slum dwellers in Lagos, Nigeria, improvise a game of table tennis using scrap wood paddles and a makeshift net.

Opposite, right: An entrepreneur sells water to her neighbors in a squatter community, which lacks water for drinking and basic sanitation, Nairobi, Kenya.

from mud to brick, one wall at a time?

And ask yourself how you would cope if your community had no basic services:

- The squatters have no water. They must purchase it at a premium price and haul it to their homes in buckets. But wouldn't you have to do the same if your city hadn't installed the pipes that bring water to your doorstep?
- The squatters have no toilets, only pit latrines, and raw sewage flows down the streets. But what would happen to your waste if your municipality hadn't built a sewer system?
- The squatters have no electricity—or, if they're lucky, they loop wires through the trees and pirate service from far-away poles. But would you have electricity if your local utility hadn't run cables near your home?

Every day, close to 200,000 people leave the world's rural regions and head for the cities. That's 130 people a minute, two every second. They go to cities in search of a job, a way of providing for their families. And by and large, they can find work. But they can't find a place to live. No developer builds for them. No government invests in homes these migrants can afford. So they become squatters, building for themselves on unused or undesirable turf. Squatting, for them, is a family value.

There are 1 billion squatters in the world today, almost one in six people on the planet. If current trends continue, there will be 2 billion squatters by 2030 and 3 billion (more than one-third of humanity) by the midpoint of the twenty-first century.

To keep up with the urban influx, the world must build 96,150 homes a day—roughly 4,000 homes every hour. Generally, only squatters are prepared to make this effort. Their homes start out as mud and cardboard hovels. But once they know they will not be evicted and they can exercise control over their communities, they create permanent, thriving neighborhoods.

This is new urbanism, global style: squatters building the cities of tomorrow. RN

The Difficulty of Squatter Empowerment

The people of Vikas Sagar, in Mumbai, India, still live in one-story huts hacked into the steep hillside above Mahim Bay. They still worry about floods and landslides. They are still concerned about having enough money to make ends meet. But the women who live there have banded together to transform their community. Today, all the homes in Vikas Sagar are permanent, made of concrete instead of mud. The walkways are paved with cement and tile, to prevent erosion. And the people have pooled their resources to create a communal savings plan that functions like a small-scale bank. How did they make these improvements? Instead of agonizing in the face of hardship, they organized.

A tiny squatter community in Vikas Sagar was founded decades ago, but its residents know that no matter how long they have been

living there, the government still considers them illegal. "Unless we take action, nothing will be granted to us," says resident Lali Penday.

Well into the 1990s, the women of Vikas Sagar were traditional housewives, so controlled by their husbands that they seldom left their community. "When we started," remembers Sangita Duby, "we were not able to go out of our houses. We were illiterate and had to sign our names with a thumbprint. Now we are literate and can sign our names in Hindi and English." The women of Vikas Sagar know who the local politicians are. And, even more important, the politicians know who they are, too.

Alone, squatters have little power. Together, they can create great things. "The problems of the urban poor can only be solved by the urban poor, not by anybody else," says Jockin Arputham, head of Slum/Shack Dwellers International, a global squatter-organizing effort. "The urban poor will be the change agents of the city." RN

The Promise of Squatter Politics

A generation ago, the tiny hamlet of Sultanbeyli on the Asian side of Istanbul was just beginning to attract immigrants from the east. These early arrivals lived in hovels, pirated electricity, and survived without water or toilets. But as more people came, the citizens of Sultanbeyli pursued their political rights—and this has made for an amazing transformation.

In Turkey, if squatters build overnight without being caught, they cannot be evicted without being taken to court. This is why Turkey's squatter areas are known as *gecekondu,* meaning, "it happened at night." Further, once a *gecekondu* community has two thousand residents, it can petition the federal government to recognize it as a legal municipality. Sultanbeyli became a municipality in 1989 and a district (a designation with more power and independence than a municipality) in 1992.

Today, Yahya Karakaya, Sultanbeyli's popularly elected mayor, works in an air conditioned office on the top floor of the seven-story squatter city hall building, with a view over the city of 300,000 people who do not fear eviction. Fatih Boulevard, Sultanbeyli's main drag, is lined with multistory buildings full of stores, offices, restaurants, and banks. The city has used its newfound political might to force the Istanbul government to bring water, sewers, and electricity to every home in the district. Sultanbeyli is now a permanent, stable, self-governing, independent squatter metropolis. RN

RESOURCES

Centre on Housing Rights and Evictions
http://www.cohre.org/
An online component of the important global watchdog group that offers reports and action alerts about eviction drives directed against the poor.

Favela Rising
http://favelarising.com/
A Web site and movie about AfroReggae, an inspiring cultural movement that arose in Rio de Janeiro's Vigário Geral *favela* (slum).

One Small Project
http://www.onesmallproject.com/
A site that will, organizers hope, turn into a book—a "small is beautiful" manual for the squatter world, conceived by a caring and committed architecture professor.

Texaco by Patrick Chamoiseau
(Pantheon, 1997)
A richly imagined squatter history of Fort de France, Martinique. As with many works of fiction, it's spiritually true.

Berji Kristin: Tales from the Garbage Hills
by Latife Tekin
(Marion Boyars Publishers, 1996)
A magically evocative novel involves you in the founding of a *gecekondu* community in Istanbul.

Declaração de Guerra (BMG, 2002) and ***Traficando Informação*** (BMG, 1999) by MV Bill
Two recent albums by MV Bill, a hip-hop artist from Cidade de Deus, the Rio de Janeiro housing project/*favela* made famous in the movie *City of God.* Bill (real name Alexandre Barreto), whose

music involves harsh yet poetic social commentary, objected to the movie as a false objectification and glorification of the violence in his community. He recently published a book called *Cabeça de Porco (Pig Head)*, based on two years of interviews with kids from Cidade de Deus, most of whom are now dead.

Global Report on Human Settlements
United Nations Human Settlements Programme
The UN publishes one of these every year. Ignore the mealy-mouthed bureaucratese and search for the disturbing stats, jaw-dropping projections, and hard-nosed facts about the cities of the future.

Shadow Cities: A Billion Squatters, A New Urban World by Robert Neuwirth
(Routledge, 2004)
Robert Neuwirth's book about squatter neighborhoods, based on his experiences during the three years he spent living in squatter communities in Nairobi, Rio, Mumbai, and Istanbul, has become an instant classic.

Planet of Slums by Mike Davis
(Verso, 2006)
A meta-argument that the structural adjustment programs the United States foisted on the developing world caused the explosion of shantytowns and constitute a war on the poor.

The Mystery of Capital by Hernando de Soto
(Basic Books, 2003)
A hypercapitalist view that giving squatters title deeds will liberate billions of dollars in dead capital.

Housing by People by John F. C. Turner
(Marion Boyars Publishers, 1991)
Turner's assertion that "It is what housing does for people that matters more than what it *is*, or how it *looks*" still captures the essence of squatter self-sufficiency.

Preceding pages: The energy and ingenuity of families like this one in Mumbai, India, is the great hope of Southern megacities.

Leapfrogging

What tools will help poor people escape poverty? For decades, the official answer was more or less the same: aid, loans, technology, handouts, and hand-me-downs. Development, it was assumed, was a one-size-fits-all process, and the way for poor countries to prosper was for them to try to catch up with the rich countries. But we have seen, in recent years, another path emerge. Developing nations don't have to play catch-up: they can adopt new technologies and tools—not always from the West—and use them in their own ways, skipping older or outmoded methods and embracing brand-new ones. Surprisingly often, developing countries try out solutions that have yet to take hold in industrialized nations. We call this process *leapfrogging.*

Probably the single best example of leapfrogging is the adoption of mobile phones across the Global South, particularly in Africa. Mobile phones empower both individuals and communities (after all, the more people who have them, the more useful any single phone can be), often acting as a catalyst for economic development and social innovation. According to a study undertaken by Vodafone, one of the world's biggest mobile phone companies, and the Centre for Economic Policy Research in London, mobile-phone use is growing faster in Africa than in anywhere else in the world; presently, more than three times as many Africans have cell phones as have traditional landline phones.

Kenya gives ample evidence of how mobile communications can transform African economy and society. The growth of cellular-phone use in Kenya is startling; out of Kenya's population of 32 million people, nearly 6 million now have mobile phones, up from only 15,000 in 2000. But the story of cellular leapfrogging in Kenya can't be just summed up in numbers. Something new is unfolding there, as more people gain the power to talk to one another and share information.

Kenya faces problems common to much of Africa (and, indeed, to much of the developing world): a shaky democracy, corruption, lack of education and health care, rural poverty, pollution, and explosive population growth in the capital, Nairobi. Most of all, it needs to create an economy that can generate the revenue required to meet its challenges. Better answers, though, may only be a phone call away. JC & AS

Leapfrogging 101

Not only can the Global South leapfrog—skip over outdated modes of development to embrace the cutting edge—but it can itself help redefine that cutting edge.

Leapfrogging means more than simply adopting new gadgets. The red-hot core of the concept is freedom. Being poor or lacking access to established technologies can liberate individuals and communities to embrace the new, because they haven't poured money into the old.

Mobile phones provide a clear example of leapfrogging: if you basically don't have any phone system at all (which is the case in many emerging megacities), there's no reason to spend decades stringing up a grid of copper telephone wires, and *then* start putting up wireless towers—even if that's how the United States, Japan, and Europe did it. If cell phones will fit the bill, why bother with landlines at all? JC & AS

Bringing the Costs Down

You may have gotten a sleek and shiny new phone for free when you signed your cellular service contract, but that doesn't mean the phone itself is cheap—it only means that your service provider figures it will make the money back (and then some) by getting you to sign on as a customer.

But a service provider in a developing megacity can't be so sure, so the phones themselves need to be produced much more cheaply. This is already happening. In 2005, a group of manufacturers launched the Emerging Markets Handset program, aiming to make new phones that cost less than forty dollars. Philips has already blown past that goal with a twenty-dollar handset—and says it's closing in on a fifteen-dollar phone that will be out by 2008. JC

Opposite, left: The rapid spread of mobile phone use in developing Kenya is preempting the implementation of traditional landlines.

Opposite, right: The Grameen Phone helps an entire community stay connected with one phone, Uganda.

Grameen Phone

The community-phone business model is fairly widespread, thanks in part to the efforts of the Grameen Foundation. The Grameen Phone program brings mobile telephones to villages and rural areas in the developing world as a tool for both local empowerment and developmental leapfrogging. Started in 1993, the program has been remarkably successful: with over 3.5 million subscribers, Grameen Phone has distributed more than 115,000 "village phones" throughout Bangladesh and more than 2,000 phones in Uganda. A program in Rwanda is now under way.

The Grameen phones function as owner-operated pay phones, such that a single phone may be used for an entire village. Women are the primary recipients of Grameen phones, for reasons of local relationships (they're better networked) and economic need. The "Grameen phone ladies" offer the community a link to relatives at home and abroad, as well as to services such as hospitals and markets.

Grameen Phone founder Iqbal Quadir is now working with Segway inventor Dean Kamen on village micropower systems fueled by cow manure. Two test systems are currently running in villages in Bangladesh, providing power for twenty businesses. The project combines access to microcredit [see Microfinance, p. 346] with access to low-cost energy to see if microgeneration could work as a village enterprise. JC

Mobile-phone Politics

In 2003, Kenya saw its greatest electoral turnout—and arguably its fairest election—owing in large part to the availability of mobile phones.

The presidential campaigns used cell phones in ways familiar to Western political veterans, but innovative for African communities. Databases of phone numbers allowed campaigners for the major candidates to call or send text messages to potential voters, and facilitated grassroots political networking. The campaigners were also better able to keep tabs on one another, monitor the polls, and keep a lookout for fraud or intimidation. Young people hired to observe the election stations could use cell phones to call in for support in case of any trouble.

Mobile phones also helped make vote counting fairer and more transparent. Rather than shipping ballot boxes to central counting stations (allowing votes to be changed, "lost," or otherwise rigged along the way), each polling station was able to count the votes locally and send the results to officials; the ballots were then shipped off for confirmation with far less likelihood of being tampered with. JC & AS

Markets

In Kenya, the market, or *Jua Kali* as it's known locally, is not just about buying and selling things: it is the very heart of commercial and social life. Mobile communications have transformed Kenyan markets by giving people the power to find out what they need to know—what's a fair price for their crafts or crops, who's selling and who's buying, what's happening in *other* markets. In essence, mobile phones are turning all of Kenya into one large *Jua Kali*. In the process, the country itself is changing.

Purely practical motivations drive this change. A phone call, for instance, can replace a grueling trip to a rural area to look for fresh produce; phones can also help farmers protect themselves from swindlers, by letting them check prices across a range of locations and buyers.

The Kenya Agricultural Commodities Exchange (KACE) sells up-to-the-minute market news via short-message service (SMS) text messaging. Farmers can access daily produce prices from a dozen markets, allowing them to make deals without having to travel around the country. Every month, KACE handles thousands of transactions; for buyers and sellers without mobile phones, kiosks set up near village markets around the country provide cheap access.

Access to these tools helps city dwellers as much as farmers. Small business in Kenya is thriving; it created nearly half a million new jobs in 2004 alone and employed more workers than any other business sector, thanks in part to the spread of mobile phones.

We take for granted in industrialized countries that small businesses rely on phones to gain new customers; until recently, however, few small businesses or independent contractors in Kenya had phone numbers to call. Instead, laborers like plumbers, electricians, and painters would congregate near hardware stores, hoping to drum up work from folks buying supplies. Today, signs bearing the mobile-phone numbers of all kinds of workers can be found throughout Kenya, and business is booming.

Employees, too, benefit. In the past, newspapers and neighbors were the main source of employment information—at best, Kenyans could travel to the nearest cybercafe (sometimes many miles away) to check out the handful of online listings. In 2004, OneWorld International opened a service that posts new job openings (mostly for unskilled and semiskilled labor) and started taking remote applications from job seekers, via text messaging. Instead of pounding the pavement, the unemployed can now work the phones. Thousands have found jobs. JC & AS

Sambaza—A New Kenyan Currency

Author William Gibson once wrote, "The street finds its own uses for things."

Until recently, poor and middle-class citizens of Kenya had few ways to transfer funds in order to pay a bill, make a loan, or simply give a relative a bit of money. Few Kenyans have bank accounts or credit cards; generally, only the wealthiest citizens use the financial services common in the Global North. People needed an option other than carrying around cash, and that pent-up demand served to trigger innovation—and leapfrogging.

In May 2005, Safaricom introduced Sambaza, a service allowing customers to transfer airtime minutes to other subscribers via SMS text messaging. Because the minutes are worth money—and because Safaricom serves the majority of mobile-phone subscribers in Kenya—sending minutes quickly became another way to pay for goods and services.

In essence, Safaricom has become the unofficial national electronic bank. A growing portion of the Kenyan economy now exists purely as bits on a wireless network. And as these electronic transactions are increasingly mediated through a widespread communications network, location becomes far less of a barrier to economic participation than it was before; rural villager and urban entrepreneur alike can send or receive funds, make remote purchases, even provide microcredit services, all from a mobile handset.

Phone-based currencies like Sambaza are not without their risks, however. Since there's a far greater need to secure financial transactions than there is to prevent airtime-minute fraud, emergent, unintended currency systems like Sambaza are likely to become targets of organized crime. Plus, the more people adopt Sambaza as a means of buying and selling, the more the Kenyan government is going to pay attention. Sambaza transactions currently incur no taxes, and are not subject to accounting by anyone other than Safaricom.

Kenya may soon become the world's first real lesson in how electronic currencies work—both how they succeed, and, perhaps, how they fail. JC

Opposite, left: Mobile phones were a key political networking tool in the 2003 presidential campaign in Kenya. **Opposite, right:** Small-scale merchants and farmers rely, in part, on cell phones to price their goods competitively.

ICT4D

For many people in the developed world, it's hard to remember life before the Internet, a time when "mail" involved writing on pieces of paper and handing them to government representatives, who promised to deliver them—maybe in forty-eight hours, maybe in a few weeks. But even in Internet-obsessed nations like the United States, a stark digital divide still exists between wealthy, urban, and predominantly white communities and poor, rural, and predominantly nonwhite communities. This divide turns into a chasm when we compare developed nations to developing ones, on the Internet front. For developing nations faced with more prosaic problems like food security, water safety, malaria, and AIDS, is joining the online community something that should be considered a high priority?

The answer may well be yes. Many developing nations are aggressively pursuing strategies to increase their access to information, motivated by both positive and negative visions of the future. They hope for futures where teachers can supplement meager libraries with online books, where telemedicine—the use of cell phones and other telecommunications devices for diagnosing and treating patients in remote locations—supplements the limited resources of rural hospitals, where local artisans can sell goods to a global audience online, and where the next generation of students writes computer code for international businesses. Developing countries are tapping technology so that computers and the Internet don't make the economic gap between rich and poor nations any broader than it already is.

In response, a movement has sprung up in which activists, techies, social entrepreneurs, and enterprising citizens alike are trying out new approaches to help poor communities use computers, the Internet, and even radio to solve the problems of poverty. The buzzword they use is ICT4D, short for "information and communications technologies for development," but the meaning is far simpler: if we give people, especially people in poor, urban areas, cutting-edge tools to change their lives, we change the entire dynamic. We create space for grassroots innovation to emerge in ways no outsider could ever have predicted or imagined. We open the future up to everyone.

The growth of the Internet in the developed world has been so rapid that it can seem almost inevitable that everyone will be connected soon. But the following are some major obstacles that could make it a long time before we find as many Africans as Americans online:

- Cost: The cost of a PC has dropped from a

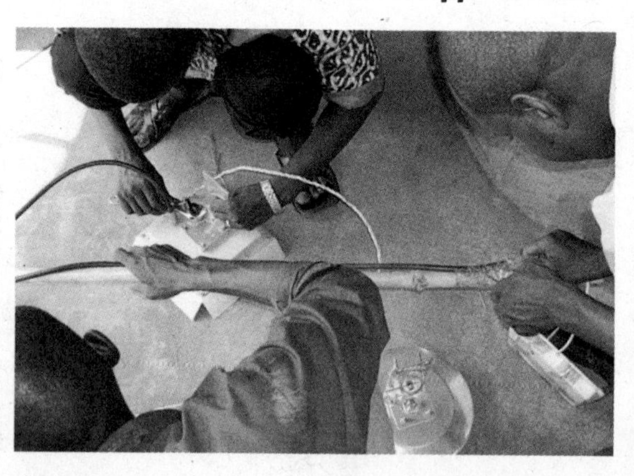

few thousand dollars to several hundred over the past two decades, but computers still cost more than the average annual income in many developing nations.

- Connectivity and power: Phone systems are woefully inadequate in many nations—with fewer than one landline per hundred households—and most households in the developing world lack the basic electricity needed to power their PCs.
- Literacy and language: The Internet is a world of mostly written text, which makes it inaccessible to people who are illiterate (in some developing nations, this can be more than 60 percent of the adult population). In addition, the majority of available content is in English. Speakers of Urdu, Bambara, or Bahasa Indonesian will be severely limited in what they can access and accomplish online.
- Relevancy: While there's lots of content on the Internet germane to the problems of an American student or a European businessman, what's available online to help an African farmer solve the problems he faces every day?

Innovators in the developed and developing worlds are tackling some of these challenges head-on, designing low-cost, low-power computers, wireless networks, and multilingual interfaces. Rapidly spreading cybercafes distribute the costs of computer ownership and network access over hundreds of users. And some of the most promising technologies aren't newly created: FM radio and mobile telephones may well be the key information technologies for the developing world.

Imagine a world where anyone can speak to anyone else, and where everyone can access the knowledge they need to be productive, healthy, and successful. The researchers, entrepreneurs, and innovators who focus on ICT4D are trying to make that dream a reality. EZ

Innovative Networking

How do you connect your computer to the Internet? Connect your modem to a phone line and call your ISP. But what if you don't have and can't get a phone line? Then you've got to get creative.

Wireless Internet has had the less-than-revolutionary effect of letting bored office workers check e-mail while they have a cup in the coffee shop downstairs. But the low cost of Wi-Fi hardware has encouraged experimenters around the world to see if Wi-Fi could be pushed far beyond its ordinary uses and serve as an alternative to wired infrastructure.

Most Wi-Fi access points have omnidirectional antennae: they broadcast a signal in all directions, and transmit data to computers within roughly 325–825 feet (100–250 meters). By attaching directional antennae to conventional access points, wireless hackers have been able to transfer data over tens of miles.

These antennae don't have to be fancy. Wi-Fi hackers in the United States have built long-distance Wi-Fi antennae from Pringles

Opposite, left: Local residents in Bouake, Ivory Coast, use computers at a cybercafe.
Opposite, right: Geekcorps participants are changing their Wi-Fi antenna, which is mounted on a bamboo pole.
Right: A Geekcorps volunteer tests a new Wi-Fi Internet connection along with a young resident, Yanfolila, Mali.

potato chip cans. Using materials commonly available in Bamako, Mali—plastic water bottles, used valve stems from motorbikes, window-screen mesh, television and low-cost coaxial cables—a team from the volunteer organization Geekcorps has started producing long-range antennae designed to connect radio stations to the Internet, for less than a dollar apiece. In Himalayan Nepal, modified television antennae connect rural villages to an Internet access point 22 miles (35 kilometers) away. Using a web of wireless connections, a set of villages now has Net-connected cybercafes, all without wires.

Taxis and minibuses are often the major connection between rural villages and the wider world. They carry passengers, medical supplies, and foodstuffs—why not digital bits as well? That's the logic behind DakNet, a collaboration between the Massachusetts Institute of Technology's (MIT's) Media Lab and villagers in the Indian state of Haryana. Computers in villages store e-mail messages, voice mail, and video recordings, and transfer them wirelessly to a computer-equipped minivan that drives into town once a day. When the van gets to a larger city, it transmits the collected messages to an Internet-connected computer, and from there to the wider world. Early successes with store-and-forward technology have inspired the founders to start a for-profit company, First Mile Solutions, which uses wireless and store-and-forward solutions to bring connectivity to rural areas around the world. EZ

Recycled Computers

As schools and businesses in the Global North replace their obsolete but functional computers, they're looking for ways to "recycle" their machines in the developing world. Groups like Massachusetts nonprofit World Computer Exchange collect discarded computers, refurbish and test them, pack them into shipping containers, and sell them at a low cost—less than fifty dollars per machine—to schools in the developing world. The computers come loaded with Linux and are ready for classroom use.

Yet while recycled computers are attractively cheap, they're also hazardous waste outsourced by the Global North to countries ill equipped to deal with the problem of disposal; most computers contain significant quantities of heavy metals. And many schools and cybercafes are discovering that it's more cost-effective to purchase inexpensive new computers than it is to refurbish old ones.

A Computer Everyone Can Afford

What if computers were as cheap as mobile phones? What if they were so cheap that schools in the developing world could use them in place of books? It's a dream that dozens of groups are pursuing, with different strategies and varying success.

A group of Indian engineers formed the Simputer Trust to design a Linux-based handheld computer with voice recognition and synthesis, so that it would be usable by literate and nonliterate users alike. So far, however, they've had difficulties getting the unit into widespread production.

Microchip company AMD has launched an initiative called "50x15," with the goal of bringing Internet access to half the world's population by 2015. Their tool for achieving this is the Personal Internet Communicator, a rugged, low-end PC running Windows CE and—like the Simputer—priced at about three hundred dollars per unit.

The most ambitious effort, though, is the One Laptop per Child [see Education and Literacy, p. 314] project. EZ

Opposite, left: The goal of the One Laptop per Child program is to make computers a tool every child can use.
Opposite, right: Free Geek turns out-of-date computers into useable desktops for low-income users.

Localized Software

When users in Cambodia sit down at a cybercafe computer, they've got two challenges: not only do they have to figure out how the computer works, but they need to read English to do it, as the programs on the computer aren't written in Khmer.

Frustrated by this digital divide, Javier Solá and the Khmer Software Initiative started the long process of creating a Web browser, an e-mail client, and an office suite in Khmer. They didn't have to start from scratch: open-source office browser [see Open Source, p. 127], e-mail, and office software had already been written in English—it just needed to be "localized" into Khmer.

In the course of the project, Solá met Dwayne Bailey, who heads the Translate.org.za project, which translates open-source software and is translating OpenOffice, Firefox, and Thunderbird—three of the most popular open-source programs—into each of South Africa's eleven official languages. Bailey and Solá are now collaborating on a framework that aids software translation and will help projects around the world localize software. They're also making a great case for the value of open-source software in the developing world: because the source code is open, the software can be translated for free to meet local needs. EZ

Inveneo

Inveneo, a San Francisco–based non-profit organization, is bringing information and communication tools to remote villages in the developing world. These tools are based on open-source software, and rely on Wi-Fi and voice over Internet protocol (VoIP)—which uses computers and the Internet to allow people to make phone calls for free—to connect multiple remote locations to one another and to the broader Internet and telephone networks. Equipment at the stations operates on batteries charged by a combination of solar cells and bicycle generators. In short, Inveneo is using open-source technology and renewable power to improve the lives of poor people around the globe through better access to communication.

This isn't just a fantasy or a "we're hoping to do this soon" sort of project, either. With the assistance of the nongovernmental organization ActionAid, Inveneo has already deployed its system in rural villages in Uganda. EZ

Free Geek

New urban models for bridging the digital divide aren't only found in the Global South, of course. In Portland, Oregon, the Free Geek project ("helping the needy get nerdy since the beginning of the 3rd millennium") takes junked computers, rebuilds them, loads them with free software, and ships the resulting "freekboxes" to low-income people and activist groups in the Portland area and abroad. In the process, they wire poor people, keep toxic computer parts from winding up in landfills, and build skills in local communities. AS

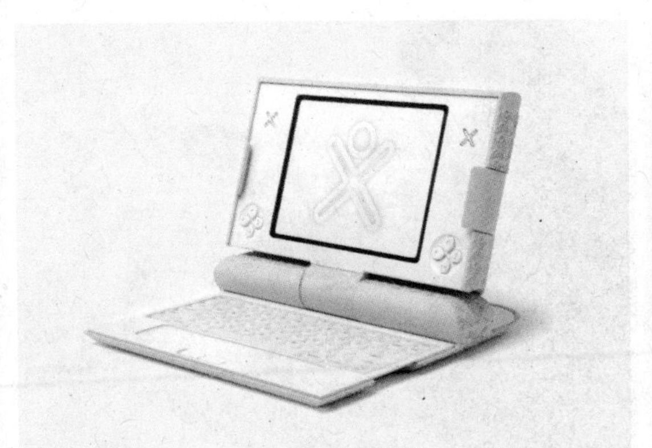

RESOURCE

The Fortune at the Bottom of the Pyramid by C. K. Prahalad (Wharton School Publishing, 2006)
In *The Fortune at the Bottom of the Pyramid,* C. K. Prahalad challenges the notion that the private sector and development are mutually exclusive. Investing in services and schemes that help poor people help themselves out of poverty is not only a good use of profits, but it can be profitable in itself. As Prahalad puts it, the 4 billion people at the bottom of the wealth ladder are the next emerging market. Investments in developing countries are not just a good philanthropic measure, they can also boost a company's bottom line: "If we stop thinking of the poor as victims or as a burden and start recognizing them as resilient and creative entrepreneurs and value-conscious customers, a whole new world of opportunity will open up. Four billion poor can be the engine of the next round of global trade and prosperity. It can be a source of innovations."

Prahalad ties together many important ideas, including the notion that any markets that emerge to serve the bottom of the pyramid must be sustainable ones, and new products must be tailored for people who are first-time users of technologies and can't afford a steep learning curve.

Brazil's *Telecentros*

If we want to understand why open-source software and cheaper computers are important to the future of emerging megacities, we need look no farther than São Paulo, Brazil.

There, in some of the poorest slums in Latin America, an innovative project has created more than a hundred free community computing centers, or *telecentros*, which are connecting locals to the Internet—and changing their lives.

Residents can use the centers for up to an hour a day to learn new skills, hunt for jobs, access community services, and keep in touch with distant relatives and friends. In a city with an official unemployment rate of 20 percent and a growing population of new arrivals from the countryside, these services are critical. Already, at least 250,000 people use the *telecentros* every month (Bacoccina 2003).

All of these community centers/Net cafés use open-source software [see Open Source, p. 127], a revelation for anyone not capable of paying for pricey protected software. It's free. It can also be made to run on simpler computers—which is important for developing nations, where users may

The computers in this *telecentro* in São Paulo, Brazil, run Linux-based open-source software and are free for anyone to use.

be working on older computers recycled from developed countries, or cheaper new models like the hundred-dollar laptop [see Education and Literacy, p. 314]. Finally, open-source software invites experimentation and problem solving, which is vital in places where prepackaged software solutions are not always affordable.

We can't overestimate how important it is for developing nations to have access to open-source software. Sergio Amadeu, the head of Brazil's National Information Technology Institute, says that access to technology is the first, not the last, step in development, that new technology is now fundamental to education, jobs, even democracy and good government all over the Global South.

The cheapest way to obtain technology is to invest in open, collaborative, noncommercial systems. As Marcelo D'Elia Branco, coordinator of Brazil's Free Software Project, told *Wired* magazine, "Every license for Office plus Windows in Brazil—a country in which 22 million people are starving—means we have to export 60 sacks of soybeans. For the right to use one copy of Office plus Windows for one year or a year and a half, until the next upgrade, we have to till the earth, plant, harvest, and export to the international markets that much soy. When I explain this to farmers, they go nuts" (Dibbell 2004).

When a country goes open source (OS), though, it gets something much more important than free code: it gets a native software industry. When people learn how to install, maintain, and modify the software they're using, they aren't just getting better tools, they're learning new, highly marketable job skills and gaining the ability to create still more tools that work even better in the local context.

Lastly, when a country goes OS, it changes that country's relationship to the world economy. The country is no longer just a market for developed-world know-how. It starts becoming a creator of knowledge. Developing countries with an open-source approach also become participants, internationally, in a collaborative culture of problem solving, which has implications far outside the realm of software. Collaborative innovation is smart in the Global North, but it changes the game in countries with lots of smart young people and not much cash.

As science-fiction writer William Gibson says, "The future is here, it's just not evenly distributed yet." Brazil realizes it has limited ability to change the global system to redistribute wealth, but it's in a perfect position to redistribute the future. In very practical terms, that's what open source is doing in the Global South.

Brazil isn't engaged in a science project: it's declaring a revolution. AS

Leapfrogging Infrastructure

Cities are largely the products of their infrastructure. But this bit of common sense is stood on its head in emerging megacities [see Lagos, p. 279], where cities often spring up before there is any infrastructure. How will megacities meet their citizens' needs for the basics of life, like clean water and electricity?

Leapfrogging [see Leapfrogging, p. 292]—skipping over outdated modes of development to embrace the cutting edge—is said to be a great equalizing force, because rich nations have already paid the price of developing technologies, competition keeps the technologies cheap, and the poor people of the world get to benefit from them. The poster child of leapfrogging is the mobile phone, but cheap computers and open-source software [see Open Source, p. 127] are promising to take the leapfrogging revolution digital. In theory, with breakthroughs coming quickly in all sorts of technologies, the Global South should soon be able to leapfrog in any area of development.

To better understand leapfrogging, we need to ask: "What, exactly, did industrialization buy us, here in the Global North?"

To answer that question, all we need to do is to look at our own lives. Let's start with the laptop, an indispensable tool in creating this book: laptops contain dozens of incredibly refined components, requiring tens of billions of dollars of capital to create. They're recharged by being plugged into power outlets, and the wires in those outlets run back to another trillion dollars of capital: the national grid. Our wireless Internet connections go to cable modems, running over incredibly expensive buried copper wires laid to carry yesterday's big thing, cable TV. The houses that contain us and our trusty laptops represent another couple of hundred thousand dollars worth of capital, and rely extensively on the availability of lumber, shingles, and glass—and on centuries of architectural refinement.

This is modernity: a pile of capital, of sunk costs, running into the quadrillions of dollars. It takes this much capital to provide average Americans or Europeans with their average lifestyles. When we speak of developing the world's poorer regions, we generally mean exporting parts of our expensive, interwoven, complex system, which took many, many generations to build up. Our Rome was not built in a day.

But it turns out that consumer goods and global services travel extremely well. You simply put them in a box and ship them where they're needed—leapfrogging achieved. You can put a Game Boy anywhere on the planet and leapfrog the entertainment revolution. The truck that carries the Game Boy from overseas port to consumer is

Old "legacy" infrastructure is often completely overwhelmed in emerging megacities.

also a leapfrogging technology—no Model T, no horse-drawn carriage, but an 18-wheeler likely manufactured elsewhere. In fact, transport technologies, from trains to jets, were the first kinds of leapfrogging. All those Land Rovers in Somalia? Real leapfrogging.

Likewise, services like Global Positioning Systems (GPS) and satellite imaging [see Mapping, p. 518] are already more or less pre-leapfrogged. The hardware already has global reach: you just need to know how to set it up and connect it.

Consumer goods and global services have, for the most part, already leapfrogged into the developing world. There may be some issues with cost and support infrastructure and so on, but basically, this stuff has worked. Television, radio, satellite services, and the like are everywhere.

What's left to leapfrog is, in fact, the dull old industrial-age stuff: reliable electricity, plumbing, and water supply. Most of the fruits of the space age—digital chips and satellite phones—travel easily. It's the backhoe-centric world of civil engineering that's harder to share. We need to find ways to make this kind of technology available to developing nations, or we need to find substitutes.

This is a genuinely hard problem. Industrial-age artifacts are quintessentially different from space-age artifacts. They are raw, big, dug in, enduring. Think about trying to provide rural telephone service in China using precellular technologies. Thousands of miles of cable per village. Trenches, poles, and strung wires. Wire stolen to be sold on the black market (a problem that pops up nearly everywhere you find poor people and exposed cables). It's just never going to happen.

This is the leapfrogging gap: all of the services that our society has delivered—and continues to deliver—via industrial-age means have turned out not to travel very well into the poor world. Satellite dishes are everywhere, but hundreds of millions of households don't even have a source of clean water. We can put satellite dishes on trucks, but the water supply requires extensive infrastructure of the sort built in Victorian England.

Leapfrogging infrastructure is going to require a new generation of tools. But here's the good news: these tools are already here. Solar and wind power, smart-grid technology, small-scale water solutions, light-emitting diodes (LEDs), leapfrogging models—we've got what it takes to leapfrog urban infrastructure. But it's not just the tools, it's what we do with them.

We must reimagine infrastructure in order to make it leapfrog. Instead of thinking in terms of massive government projects, we need to imagine whole cities that can build working systems by linking many small, discrete parts, one household and one neighborhood at a time. Rather than sinking a billion dollars into a power station, and another billion into a grid to distribute the power, we could, with leapfrogged infrastructure, wire together a city from a thousand technologically advanced parts—forming a grid that provides the same basic services without the huge initial investment.

Breaking away from the industrial model, we can find our way back into the "sweet spot" for leapfrogging: easy transport of technologies that can be put on a truck and shipped, that don't depend on a thousand miles of trenches or on Victorian-era engineering. A system for power generation can have the same dynamics as a satellite television system: buy what you need, build as you go.

The billion people in the world without clean water are unlikely to ever get water services from giant water-chlorination factories, with thousands of miles of brick-lined tubes and a spiderlike network of PVC and copper running to shiny chrome taps. But smaller solutions could bring them drinking water. Technologies like solar-powered ultraviolet water purification, for example, have great promise for village-level and even household-level deployment. A box of components, shipped wherever it is needed, could provide developed-world-standard drinking water anywhere there is sunlight.

The solutions described throughout this book are just puzzle pieces. They can be put together in a million different ways to create cities that offer billions of people real hope for healthy, prosperous lives. The new world is modular, interconnected, technologically and economically diverse, and it is shipped in on trucks one piece of the puzzle at a time. It's a curiously future-retro combination of space-age tools in mud-and-stick houses—high density, ultra-connected urban hubs, and villages with solar-electric lighting and drinking water.

This is the leapfrogged world. VG

StarSight

What would a streetlamp look like if it were designed for people who don't have electricity? It might look a lot like the StarSight project. Each StarSight pole combines a streetlight—a tool that itself can bring down crime rates dramatically—with a solar panel, wireless networking equipment, and even hookups for charging small devices like mobile phones.

The designers, the UK-based Kolam Partnership and the Singapore-based Nex-G, describe StarSight as a key element of a "virtual utility," a low-cost, low-maintenance means of providing very useful services such as public lighting and wireless networks. Using LED light to conserve energy, StarSight is expressly meant to be used in the emerging megacities. It is being tested in Cameroon.

None of the technologies used in StarSight are unique in themselves, but the combination is an inspired example of the power of merged leapfrogging technologies. The setup can also incorporate disaster warning systems, pollution monitors, and other location-aware network services. If leapfrogging infrastructure needs to be plug-and-play, versatile, and cost-effective, it may end up looking something very much like StarSight. JC

The StarSight streetlight system would provide a low-cost, low-maintenence means of delivering public lighting and wireless networking to megacities.

RESOURCES

Small is Profitable by Amory B. Lovins et al. (Rocky Mountain Institute, 2002)

Small is Profitable is the Rocky Mountain Institute's book on distributed power generation. A four-hundred-page doorstop that was selected as the *Economist*'s Book of the Year in 2003, *Small is Profitable* outlines the financial and technical case for a new approach to providing high-quality and high-profit-margin electrical services to the world.

This new approach is a "post-industrial" power-generation methodology: fewer industrial-style power stations, lots of analysis and feedback loops, lots of new technology like microturbines, combined solar-and-wind energy, lots of financial modeling to convey the real cost of options. Rather than one-size-fits-all, 120V from every socket, powered by a single continent-spanning grid, highly granular power delivery could direct power from the best available resources to each location: microturbines for some kinds of factories, solar/wind for some parts of a city, and so on.

As the authors explain, "Producing and delivering electricity is extremely capital-intensive—several times as capital-intensive as the average manufacturing industry. Per unit of delivered energy, the electricity system is about 10–100 times as capital-intensive as the traditional oil and gas systems on which modern economies were largely built. Generating electricity by traditional means is also very fuel-intensive. Classical power stations that raise steam to turn turbines that run generators that ultimately deliver electricity through the grid necessarily consume 3–4 units of fuel per unit of electricity delivered, and even the most efficient combined-cycle plants decrease this ratio to only about 1.8. Electricity is therefore a far costlier form of energy than direct fuels: in 2000, for example, the average kilowatt-hour (kWh) of U.S. electricity was delivered at a price of $0.0666—the same price per unit of heat content as oil at $114 per barrel, about 3–6 times the recent world price of crude oil (not yet refined and delivered)."

Infrastructure: A Field Guide to the Industrial Landscape by Brian Hayes
(W. W. Norton and Co. 2005)
When we look at cities, we tend to see the skin—how things look on the surface. Infrastructure is the flesh and bone beneath that skin. It's usually hidden from our eyes, but it is nonetheless vital. Brian Hayes's book is a loving hymn to the big systems that built the modern world: dams, strip mines, steel mills, water towers, feedlots, drilling rigs, freeways, cellular towers, sewage-treatment plants, landfills, bridges, tunnels, and ports. Full of huge, beautiful pictures (sort of engineer porn) and laced with chemistry and physics lessons, history, and amusing trivia, this book will turn you into an infrastructure geek. Even more, your mind will boggle at the ingenuity and scale of the works of the industrial age that's now passing, and the magnitude of the task of replacing all of those systems with ones that better respect people and the planet:

"In the tropical rain forest, ecologists study the communities of plants and animals that live in vertical zones, from the leafy canopy of the tallest trees down through the understory to the rotting litter on the forest floor. Utility poles also have a series of vertically stratified habitats, each with its own characteristic inhabitants. From top to bottom, here are some of the species you might observe in the utility-pole ecosystem: Primary distribution lines for electric power. These are the topmost wires ... Switches, fuses, and surge arresters. These connect to the primary distribution lines ... Transformers ... Secondary distribution lines ... Street-lighting fixtures. They draw their power from the secondary circuits. Traffic signals."

It's fascinating stuff.

COMMUNITY

Many communities these days find themselves under extraordinary stress. Even in affluent communities, taking care of one another—educating our kids, caring for the ill, helping people who need a hand—is proving difficult. In less-than-affluent communities, the stresses of extreme poverty, oppression, environmental injustice, failed educational systems, and diseases like HIV/AIDS are making life a daily challenge.

These communities are not as far away from each other as we sometimes think. As the Mega-Cities Project's report "The Poverty/Environment Nexus in Mega-Cities" reminds us, "Every 'First World' city has in it a 'Third World' city of infant mortality, malnutrition, unemployment, communicable diseases and homelessness. Similarly, every 'Third World' city has in it a 'First World' city with high finance, fashion, and technology" (Mega-Cities Project 1998).

To solve problems like these takes more than individual action—it takes community action. Communities of all kinds need to work together, thinking about the problems they face in holistic ways, and strengthen the fabric that binds them together.

But we also need a global commitment to solving the problems faced by communities everywhere. Problems like HIV/AIDS and other diseases, extreme poverty, illiteracy—these can't be solved without community involvement, but they also can't be solved by our communities alone. They require the application of resources that can only be mustered by governments and international networks.

Our best effort so far comes from the Millennium Development Goals put forward by the United Nations. They are, simply put, the closest thing we have to an international consensus on how to meet the fundamental needs of every person on the planet. They propose programs to help end extreme poverty; feed the hungry; empower women and improve maternal health; fight AIDS, malaria, and other epidemic diseases; protect environmental sustainability; and educate and provide medical care for all children.

In practice, the Millennium Development Goals are far from perfect. Critics rightly point out that they talk a lot about helping poor people and very little about protecting their human rights, building democracy, or righting basic injustices in our economic systems. That said, simply having benchmarks against which to measure our progress in meeting world needs is a huge step forward.

Some say that the goals are too modest. Some argue that new tools for development (which focus more on ensuring people's fundamental economic rights and helping them build livelihoods through small loans to start community-based businesses) could help end absolute poverty altogether. Others point to new collaborative models for scientific research (especially in nations in the Global South) and public health care, and anticipate the day when we'll be able to not only better care for those currently suffering, but stop new epidemics in their tracks.

These are not either/or answers. We need all of them: We need global commitments to fighting poverty and disease. We need new models. We need engaged communities willing to experiment. Together, they compose a tool chest for making all of our communities healthier tomorrow. AS

Holistic Problem Solving

The problems that afflict communities rarely spring from just one source. It's impossible, for example, to isolate one cause of declining test scores in the classroom. Children's success depends not only on the material they study and how they are taught, but also on whether they have the textbooks and computers they need (a problem related to government funding), whether they eat a good breakfast and feel safe at home (a problem related to family welfare and economic security), and whether they feel healthy and energetic (a problem related to health-care access). If we attend to just one of these factors, we may put a Band-Aid on the problem, but we won't solve it.

To tighten the fabric of our communities, we need to pull on all the threads at once: empower the younger generation, support families, promote education, improve health care, and protect neighborhood safety. Around the world, there are gripping examples of communities that have embraced a holistic approach in order to ameliorate challenges. Almost without exception, such an approach touches more people and streamlines more systems than it even sets out to.

Harlem Children's Zone

New York City's Harlem Children's Zone (HCZ) occupies a sixty-block area of central Harlem. Thanks to the visionary activism of one dedicated man, Geoffrey Canada, and the scores of advocates and allies who have flocked in to help over the years, this is the epicenter of one of the most comprehensive, enduring, and successful community-support programs ever created.

Since the mid-twentieth century, Harlem has known the devastating repercussions of poverty, violence, drugs, and a lack of city government support. These conditions stack the deck against kids who have few choices; the poor test scores that come out of schools in inner-city areas like Harlem tend to lead school boards and college-admissions officers to believe that these kids just can't, don't, and won't do as well as their peers from higher-income neighborhoods. But Geoffrey Canada proves them all wrong.

The Harlem Children's Zone begins working with kids literally at birth, and supports them continuously, all the way up through college—with many complementary programs for their parents along the way. It starts with Baby College, which recruits parents—mostly single mothers—of children under three. The program teaches parents about infant developmental stages and guides them in parenting skills. Baby College also provides parents with the opportunity to join a support network of other parents who share similar experiences.

The HCZ leaves no untended cracks for children to fall through. Harlem Gems, which targets four-year-olds, has demonstrated staggering improvement in test scores over a one-year term, according to the HCZ Web site. From there, programs stacked back-to-back help HCZ kids with everything from academics to health and nutrition, conflict resolution to volunteering.

Since 1999, HCZ has also operated an Employment and Technology Center, teaching computer-literacy skills and providing job-placement programs to connect both youth and adults to opportunities and support. It has been a powerful force in revitalizing and strengthening the community.

The HCZ online profiles and reports also confirm that The Renaissance University of Community Education (TRUCE), another HCZ initiative, has been instrumental in preparing junior high and high school students (ages twelve to nineteen) for higher education. Using a multidisciplinary approach, TRUCE helps seniors graduate on time; its average graduation and college-admission rate is better than New York City's general average. Two integral aspects of TRUCE are the student-run production The Real Deal, a cable TV program featuring arts and entertainment by the kids in the program, and the publication of their own newspaper, *Harlem Overheard*.

One of the growing afflictions of inner cities everywhere is respiratory illness. The American Health Association reports that asthma plagues more children in low-income urban areas than it does anywhere else—mostly because city pollution is compounded by smokestacks and other sources of industrial pollution, which are often concentrated in these neighborhoods. In collaboration with the Harlem Hospital, HCZ works to screen all children under twelve for asthma. Not surprisingly, 30 percent of children screen positive, according to research cited in the *American Journal of Public Health*—almost six times the national average. The HCZ prevention and treatment program approaches the problem holistically, incorporating home visits, smoking-cessation support for adults who live with children, resulting in reduced school absences owing to asthma-related illness.

The HCZ programs' developers and administrators look at the situation in their sixty-block district with a wide lens. They know that in order to ensure individual success, there has to be a social infrastructure that allows parents to be present in their children's lives, as well as ample opportunities for employment, affordable housing, health care, and after-school programs for keeping kids constructively occupied. Nothing can substitute for the attention and involvement of adults in the lives of children; knowing that the people around them have real investment in their success drives young people to apply themselves.

But Geoffrey Canada and others at HCZ want more than simply to improve the individual lives of the kids and parents who pass through their programs. They want to prove that it is possible to raise levels of achievement in places like Harlem—that kids who are born into poverty, who live in single-parent homes, who lack the services and advantages that are available to children in affluent areas can excel as well, and in as great numbers, as their more privileged peers. And they've already begun: thanks to HCZ, test scores are soaring and success stories abound. With so many high-achieving kids emerging from Harlem, other neighborhoods and other cities are bound to see the potential for the same success in their communities. SR

Finnish School System

Finland, according to a major international survey, has the best educational system in the world (BBC 2004). This news comes on the heels of several other studies showing that Finland has the highest rate of teen literacy in the world, the highest percentage of "regular readers," and the most "creatively competitive" economy, according to the World Economic Forum.

The Finnish education minister says that heavy investments in education are a matter of economic survival for a small, affluent, high-tech-based nation. Finland spends more per elementary, middle, and high school student than any other nation on earth, and ranks second for higher-education spending. Schools are local, community-based affairs, with extremely low turnover in their teaching staffs and strong expectations of parents. Students all study

Preceding pages: This fishing community in Puri, India, is among the poorest in the region, but also among the most enterprising: with the support of government programs, fishermen have "indigenized" new methods and have adapted innovations in boat design to their traditional vessels.

Left: New York City's Harlem Children's Zone is one of the most comprehensive and successful community-support programs in the country, serving more than 12,500 children and adults.

languages, math, and science (and in Finland, girls now outperform boys on science tests, says the Organization for Economic Cooperation and Development). In short, the Finns go to great lengths, institutionally and culturally, to maintain an exceptionally successful education system.

On the other hand, maybe the secret is what they don't do: Finnish students spend less time in class than students in any other industrialized nation. While some kids in the United States begin preschool by age three, Finnish kids don't even enroll until age six, and their formal schooling begins at age seven. Throughout their schooling, pupils spend the fewest hours in school of any Western country, with longer breaks and holidays. The result of this is a strong emphasis on a family's role in educating kids. Finland scores remarkably high in reading comprehension, which can be attributed largely to an active tradition of reading at home and with family.

Clearly, the Finnish system amounts to a successful employment of quality over quantity. Time and money well spent mean kids benefit more from their schooling without having to spend a disproportionate amount of time in the classroom. And free meals for all students mean kids' health and concentration improve.

On top of all this, Finnish schools do not weed out students who excel and separate them from students who struggle. From age seven to sixteen, all children receive the same education, eliminating the self-fulfilling prophecy prevalent in other countries of not expecting much from certain kids, and ultimately watching their performance decline. The behavioral issues that normally accompany troubled and failing pupils do not seem to plague Finland—parents or teachers give "problem kids" a talking-to, and the issue is usually easily resolved.

Kids generally attend the same school from age seven to sixteen, making for a smoother ride than in other, more segmented school systems, where numerous transitions are the norm. Child-care services are available, and kids walk around barefoot, in an atmosphere that feels homey and intimate. The same freedom infuses the curriculum. Kids hold a great degree of control over their own course selections and schedules, choosing from numerous subjects that complement the core academic program.

At the end of the comprehensive nine-year basic academic period, sixteen-year-olds have a choice between vocational and secondary academic school. Almost all kids choose one or the other; dropout rates remain notably low. Universities in Finland are free, and about 65 percent of Finnish young adults attend.

Another factor contributing to the success of Finnish schools is that teaching as a profession is held in high regard. This is not to be underestimated, as is clear from the United States and other countries where being a teacher often qualifies as an unpleasant job with low pay, low appreciation, and an overabundance of challenges beyond the job description. In Finland, education, literacy, school attendance, and multilingualism comprise some of the most important cultural values. It's no wonder that kids want to be in school—they are born into a system that makes school not only affordable and equitable, but welcoming and fun. SR

Kufunda Learning Village, Zimbabwe

In the midst of the chaos and confusion of modern Zimbabwe, situated on the small red dust roads of Ruwa, outside Harare, sits an extraordinary testimonial to the creativity and resilience of Zimbabweans. Kufunda Learning Village was founded in 2002 to provide the rural populations in Ruwa and beyond with a rich environment where they could learn, and teach, the skills of self-reliance.

Since its beginnings, Kufunda has actively responded to the needs and desires of the local community. Even the security guard, David, who at first simply sat in front of the Kufunda gates, is now busy pursuing his dream of starting an organic mushroom farm. Over the years, Kufunda has regularly run a series of two-week-long residential programs for "community organizers"—generally women—from villages across Zimbabwe. Participants have studied topics as diverse as business fundamentals, soap making, yoga, and Gandhi's philosophies of swaraj ("self-rule"). Because of Kufunda, local villages have committed to building hundreds of composting toilets, which not only reduce water usage but return essential nutrients to village permaculture garden projects.

Kufunda has also established an education fund to support the many Zimbabwe children orphaned by AIDS, and an AIDS education program to create a space for open conversations about the reality of the disease. There is additionally an herbal-medicine program that focuses on immune boosters and medicines for AIDS-related ailments; herbal production gardens at Kufunda and in local villages produce the herbs, and a small lab processes and packages them. In 2006 the scope of the lab will expand into growing and distributing the *Moringa oleifeira* plant (packed with vitamins, minerals, eighteen amino acids, chlorophyll, omega-3 oils, phytonutrients, and antioxidants) and the *Artemisia annua* plant, which is used to treat malaria (a leading killer in the Global South) and boosts the immune system.

Many people in rural Zimbabwe, influenced by the media and by their dependence on outside institutions, perceive their small farms and villages as useless and themselves as without opportunity. They do not see the local assets hidden in their communities. But Kufunda helps them see the value and possibility in their own environment. Kufunda goes beyond models of contemporary education, empowering participants to take on the role of teachers and leaders, as well as learners. Kufunda Village is changing the rules of the game from institutional dependence to self-reliance. In the process, it's creating a living model of sustainability based on a comprehensive understanding of the riches that sit at the heart of rural communities and culture. A particular energy of transformation suffuses Kufunda—and everyone exposed to it is changed. ZH

Education and Literacy

Access to education, or rather the lack of it, is one of the greatest barriers to sustainability. According to Lester Brown, director of the Earth Policy Institute, 115 million children do not attend school and 800 million adults are illiterate—in a world where access to information is key to success, whether one is a subsistence farmer or a factory worker. To be illiterate and unschooled is to be excluded from the possibility of a better future. If we want a safe and sustainable world, we need to provide everyone with the mental tools they need to better their condition. As Nobel Prize–winning economist Amartya Sen warns, "Illiteracy and innumeracy are a greater threat to humanity than terrorism" (Brown 2006).

This networked age has enabled new tools that are revolutionizing teaching and learning, tearing down the walls of libraries and academies, and democratizing access to knowledge in unprecedented ways. Innovations like cheap laptops for children and vans outfitted with on-demand book printers are bridging the last mile of the digital divide in villages across the Global South. Online communities are making the off-line work of teachers and literacy practitioners easier, allowing them to share course materials and lesson plans. And the open-source movement [see Open Source, p. 127] is providing distance-learning tools that are enabling autodidacts anywhere on earth to take the same classes—from Latin to Laser Holography—as students at elite Western universities.

Cliché as it may sound, knowledge is power. If a sustainable global democracy is to emerge in the years to come, sharing that power among vastly greater numbers of

people is not only more essential than ever, it's also more possible than ever. LU

Internet Bookmobile

For six months in 2004, a four-wheel-drive diesel van bounced around the back roads of rural Uganda loaded with a PC, a laser printer, a paper cutter, and a hot-melt-glue binding machine. That van, the Digital Bookmobile, put more than six thousand books into schools, homes, and libraries in an impoverished area of Uganda.

Where did all those books come from? Primarily from the public domain collections on the Internet Archive, a nonprofit online library established to preserve cultural artifacts and data, as media becomes more predominantly digital. The Internet Archive boasts some 30,000 texts, which are free for all users to view, download, and print. Many books that are familiar to those of us who grew up in the West now fill Ugandan school shelves. In addition, the project funded a book-binding station and two scanning stations at the National Library offices in Kampala, where staff could scan in educational materials about AIDS, farming, adult literacy, and other topics for the Bookmobile to print.

While the pilot project was largely a success, it ran into some obstacles along the way. Despite enthusiasm among librarians and many teachers, the current public school curriculum in Uganda places little value on reading. The Bookmobile was most useful at schools where teachers took it upon themselves to initiate special reading periods. Local politics was another challenge. Not all government officials were supportive of providing greater access to public documents; some see such access as a threat. Moreover, technological barriers—especially the cost of printer toner, glue binding strips, and equipment repairs—could hamper the long-term viability of any Bookmobile.

Despite these challenges, the concept is spreading. Since the 2004 pilot, Digital Bookmobiles have appeared in Ghana, India, and Egypt. LU & AS

One Laptop per Child

At the 2005 World Economic Forum in Davos, Switzerland, Nicholas Negroponte was met with stunned disbelief when he announced the One Laptop per Child (OLPC) initiative, an ambitious plan to produce a laptop computer for less than a hundred dollars and distribute it free to schoolchildren around the globe.

Less than a year later, the founder of the Massachusetts Institute of Technology (MIT) Media Lab proved his doubters wrong. The prototype "green machine," unveiled in Tunisia at a United Nation's technology summit, runs on Linux software and sports a color LCD screen, 500-megahertz processor, wireless broadband, DVD drive, and 500 megabytes of flash memory. Mesh networking capability allows the machine to communicate with its nearest neighbors, creating an ad hoc local network. A hand crank allows users to charge the battery far away from the electrical grid. The goal for production models is for one minute of cranking to generate enough battery power for a hundred minutes of use. What it doesn't have is lots of memory for storing large amounts of data.

Using a bookmobile stocked with a PC, a laser printer, a paper cutter, and a hot-melt-glue binding machine, Anywhere Books is engaged in a joint program with the National Library of Uganda to put thousands of books into schools, homes, and libraries.

Why a laptop for every child? Textbooks are notoriously difficult to get into the hands of children in developing countries. Negroponte's idea is to dispense with textbooks altogether. A single laptop provides access to far more texts than any school library can—plus multimedia content, games, and interactive educational software. Moreover, the computers will allow children to create their own media. The main goal, according to the OLPC Web site, is "to provide children around the world with new opportunities to explore, experiment, and express themselves."

The machines may not be available commercially; they'll be distributed by large government programs. The first orders are expected to go to seven culturally diverse countries (China, India, Thailand, Egypt, Nigeria, Brazil, and Argentina). LU

Open-source Textbooks

Textbooks are a huge source of profits for publishing houses, and providing them for each child is a huge expense for any school system. But there are movements in place to provide schools with open-source [see Open Source, p. 127] books—either by placing books under a Creative Commons [see Copyfight, p. 337] license, or by reissuing classic textbooks that have gone out of copyright and entered into the public domain. And the project is by no means limited to the developing world: the California Open Source Textbook Project is a plan to cut in half that state's nearly $400 million textbook budget by developing open-source textbooks. These textbooks will be collaboratively written, like Wikipedia and Wikibooks, and will be designed to meet California's educational standards. If the plan works, providing a developed-world education anywhere in the developing world will be as easy as e-mailing a few text files. PD

MIT's OpenCourseWare

The Massachusetts Institute of Technology has put most of its course catalog online. Well, not just its course catalog, but the courses themselves—the syllabi, readings, assignments, tests, lecture notes, even lecture recordings. Since its launch in late 2003, the OpenCourseWare Web site has grown to include over 1,250 courses, and MIT expects to have its entire course catalog online by 2007. With offerings in fields ranging from architecture to engineering, media arts to materials science, the site provides a pretty good approximation of the academics at a top-notch university education—for free, anywhere, anytime.

Already, students from more than 210 regions and countries are taking these classes. MIT is working to translate the materials into Spanish and Portuguese, with other languages to follow. International online communities have sprung up around some of the courses of study. Best of all, because it is built on an open-license platform, any improvements or expansions that others add to these courses must also be made freely available. OpenCourseWare is arguably the single greatest contribution to the fast-growing open-source education movement. LU

Moodle

Online education is exploding. Universities are granting degrees to hundreds of thousands of students who will never even set foot on a campus, and Martin Dougiamas is partly to thank.

After working for years as a network administrator for an Australian university, the computer programmer grew frustrated with the high cost and low quality of the online teaching tools that were commercially available. So in 1999, he created Moodle, a "free, open source software package designed using sound pedagogical principles, to help educators create effective online learning communities," as the Web site describes. As with so many open-source projects, a swarm of passionate volunteers flocked to the project, collaborating to revise and enhance it into a powerful tool set for creating rich, interactive, online learning environments.

Today, Moodle is in use in all manner of settings: from 40,000 student universities, high schools, and corporate training centers to community education programs. The movement to open up the tools of teaching and learning to us all—whether we're at an American university or in one of the world's million villages—is tremendously worldchanging. Moodle is a step in that direction. LU

WorldCat

The Internet has often been heralded for putting all human knowledge at our fingertips. But even the almighty Google can be frustratingly limited when it comes to finding quality information on certain topics.

Enter WorldCat, an online catalog that lets you search the collections of nearly every major library on earth—nine thousand in all. With more than 60 million records, this massive database covers everything "from stone tablets to electronic books, wax recordings to MP3s, DVDs and Web sites." And searches on WorldCat cover much more than just titles and call numbers. Results often include cover art, reviews, and excerpts. What's more, in any of eighty countries, the system will tell you if the item you want is available in your local library. LU

RESOURCE

National Center for Family Literacy
http://www.famlit.org/
The National Center for Family Literacy is recognized as the global leader in literacy advancement. With programs aimed at increasing educational and economic opportunities among at-risk communities of children, the center provides the most up-to-date tools available for family-literacy training, knowing that the key to educational success is active family involvement. Statistics show that children who are read to three or more times a week are nearly twice as likely as other children to succeed in other areas at school, so there's no denying that universal literacy is truly a noble goal.

Educating Girls and Empowering Women

People in the developing world want many things. They want to reduce poverty and hunger. They want better health care for children, and better early childhood nutrition. They want to curtail domestic violence, fight infectious disease, and protect the environment.

In 2000, the United Nations established the Millennium Development Goals (MDGs) as a framework for achieving these things. Of the eight MDGs, three pertain very directly to women: gender equality, maternal health, and child mortality. But if we regard the goals as a list of articulated objectives for truly achieving a different kind of world, then we see that all of the MDGs are about women. More specifically, they are about the human rights of women.

Not one of the goals stands independent from the need to ensure women's safety and respect unconditionally. Without women's active participation in shaping political, social, and economic progress, the attainment of these goals is, in essence, superficial. We must shift the structure of development to include women, empower them, and secure their human rights. The first step in the process is universal: educate young girls. PD

Educating Girls

If we put a girl in school at age six or seven, ten years later she's more likely to know about contraception, or to be able to find out about it on her own. She's more likely to have fewer children. With increased education, she's able to hold a better job, and perhaps start her own business. She knows about her civil and human rights, and has more self-confidence

to stand up to an abusive husband or sexually harassing boss if she has to. And when she does have children, she knows how to better care for them.

Unfortunately, knowing what to do *for* girls and getting it done are two different beasts. In 1990, at UNESCO's World Conference on Education for All, delegates from 155 countries agreed to work toward making basic schooling available to all children, and to greatly reducing illiteracy by the end of the decade; at a subsequent meeting five years later, the deadline was extended to 2005.

That deadline has come and gone, but the problem of gender disparity in primary and secondary education still exists. In some sub-Saharan African countries, the problem has actually gotten worse since 1990.

In 2000, more than 180 governments committed to addressing this crisis through the establishment of one of the UN's Millennium Development Goals: by 2015, every girl—and boy—must have access to a quality basic education. But achieving that goal involves the mammoth task of shaking up the entrenched bureaucratic and patriarchal structures that keep girls from getting an education. In some cases, it will require the government to open its eyes to child-labor practices it doesn't officially want to see. In others it could mean changing a culture that encourages child marriage and that makes a wife the property of her husband. In many developing countries, taxes do not pay for education, and many parents believe that the extra cost of tuition, textbooks, school uniforms, and school supplies is too great to be "wasted" on a girl who will leave them to join her husband's family when she gets married. On top of that, this goal has been put forward at a time when religious

Educating girls and empowering women is critical to creating strong communities, Jakarta, Indonesia.

fundamentalists are increasingly trying to restrict women's freedom.

Nevertheless, efforts are in motion. The World Food Program provides students in developing nations where chronic hunger is a problem with one meal per day at school—and provides one way to get girls into the classroom. In families for whom food is scarce, the best is usually saved for the primary breadwinner—generally the father—and the sons. Women and girls are often shortchanged, particularly on protein, which contributes to brain development. School feeding is not a permanent solution, but multiple studies have shown that hungry children don't learn or think to the best of their abilities, and one additional meal per day makes a difference. School feeding programs also motivate parents to enroll their children in school and keep them attending regularly. Around the world, feeding programs have been shown to reduce student absenteeism and to increase the number of years children stay in school. Most significantly, cognitive performance increases, and dropout rates and "holding back" rates decrease.

But even when food is distributed at school—*and* even if educational fees are waived and open-source textbooks [see Open Source, p. 127] are provided—girls still face obstacles to their education. Their labor is often an important part of the family income, and cannot be easily replaced. Additionally, male teachers in the developing world frequently set store by the prevailing cultural stereotype that boys are smarter, and female teachers tend to be marginalized by their colleagues and left with the last pick of students, classrooms, and equipment.

As is the case in the inner cities of the developed world, in developing nations, getting parents involved in education is crucial to children's schooling—and that is especially true for ensuring girls' education. In the rural Indian state of Uttar Pradesh, the local government used funds from international charities to establish a parallel school system just for girls. Women in the community walk girls to school, parent-teacher associations and principals reach out to homes where children aren't enrolled, and local groups stress to mothers that schooling is a basic legal right. The program's goals are pretty straightforward: help these girls fulfill their potential, and end the cycle of female illiteracy and impoverishment.

Haydi Kızlar Okula!

The World Bank is not necessarily known as a suicide-prevention program. But that's what it became for Askin Tavuz, a thirteen-year-old girl from the city of Diyarbakir, in southeastern Turkey.

Both out of work, Askin's parents decided that sending their daughter to school had become a luxury the family could no longer afford. But school, and her dreams of becoming a lawyer, meant so much to Askin that she wrote a letter to her principal threatening suicide if she could not attend.

It was not the first suicide note the principal, Oya Senvic, had received from a girl who faced being denied an education. "Parents don't see the point in sending their girls to school. They want the girls to stay home and do housework and get married. The families see schooling as a waste of time," says Senvic.

But Principal Senvic was able to enroll Askin in a special World Bank school-stipend program that provides families with a monthly child-support allowance of ten dollars a month, deposited into an account accessible only by the girl's mother. In return, the child must remain enrolled in school and attend regularly.

The World Bank program in Turkey has been augmented by the UNICEF program *Haydi Kızlar Okula!* ("Let's go to school, girls!"), which goes even further toward addressing the 7 percent gender gap between boys' and girls' school enrollment that leaves nearly 600,000 Turkish girls uneducated (according to the "World Bank Development Report"). The program offers tax credits of 100 percent to any private group or organization that donates to it, and uses that money to give families a 20 percent refund on the cost of their daughters' education. Turkey has been striving to gain membership in the European Union since 1987—one reason its government accepted the UNICEF program. The Copenhagen criteria for membership, which a country must accept if it wants to join the EU, call for each national government to guarantee full human rights plus respect for and protection of minorities. Since the cohorts of today's primary-school-aged girls will be entering the job market at the same time the country is expected to become an EU member, educating girls is in Turkey's best interests in more ways than one. PD

Women's Self-help Groups in India

Women's self-help groups (SHGs) are riding the wave of a silent revolution across rural India. Because they are informal networks, it's hard to determine exactly how many groups exist, but judging by the contact SHGs make with microfinance programs and nongovernmental organizations, it appears that there are thousands, with millions of members. One study from the Global Development Research Center reports that by 2008 a million SHGs will cover at least one-third of India's rural population. While official recognition of self-help groups has generally been limited to their considerable financial activities, the role that women's SHGs are taking in changing the wider socioeconomic conditions of rural India is now being recognized. Studies have demonstrated that such groups have an impact on increasing women's solidarity within villages, resulting in a host of unforeseen benefits (Singh 2003). One study from Kerala reports on the diversity of SHG activities: "The experience of travelling and participating in SHG functions also opened up new ideas and practices to women. Some SHGs held meetings on special topics of interest. For example, one group invited a woman lawyer and human rights specialist to their group to discuss women's rights ... Also, women acquired new skills through various trainings (e.g. umbrella-making, bamboo crafts, etc.)." Across their communities, these groups are driving the establishment of support systems that enable them to deal effectively with social issues ranging from malnutrition to women's rights. "We used to be afraid," comments one member of a women's SHG, "but now we have the group" (Demaret 2004). ZH

Beauty Salons as Tools for Change

Beauty salons are social hubs; for many women they offer the opportunity to socialize and to escape everyday demands for an hour or so. Finding the right hairdresser, then, often has more to do with fostering a friendship and having a confidante than achieving an ideal hairstyle. Though infrequent, visits between clients and hairdressers can also be very intimate—the salon is a safe place to divulge all sorts of secrets, since the hairdresser usually holds a certain distance from the client's personal life.

Because of this, a number of social-service outreach programs have recognized a significant opportunity to target beauty salons for both health education and domestic-violence prevention. Women of all ethnic and socioeconomic backgrounds go to salons, where they are generally relaxed, relatively undistracted, and momentarily focused on themselves. If they are experiencing relationship or health woes that they otherwise might feel ashamed or fearful about, the salon is one place they might let their guard down and talk to someone. Hairstylists are not therapists, of course, and many avoid becoming too involved in their clients' personal lives, but they can nevertheless make resources available, and urge women to take action to protect themselves.

Preceding pages: Girls who get an education are far more likely to succeed in life, Hyderabad, India.
Left: These women are members of the Swadhyaya Parivar, or self-awareness community, a movement founded by the Indian spiritual leader Pandurang Shastri Athavale.

In 2001, hairdressers around San Francisco were offered a training course to teach them how to recognize signs of abuse and how to broach the topic with a client in a comfortable and appropriate way, to ultimately help clients escape unsafe conditions. Whether they dropped subtle information in the context of a conversation or placed pamphlets in waiting areas and restrooms, a number of stylists and salons established a network for safety and prevention that has the power to reach a huge population of women who would otherwise be forced to conceal—and remain in—abusive or life-threatening situations. SR

 RESOURCE

The Global Fund for Women
http://www.globalfundforwomen.org/
Part of a global women's movement dedicated to providing choices and opportunities for women from a wide range of backgrounds, the Global Fund for Women is a grant-giving institution for financing projects outside of the United States that advance women's human rights. Since 1987 the fund has given over $47 million ($7.3 million was given out in 2004 alone) to organizations addressing the specific needs of women in their communities. The fund's Web site features a PDF version of the handbook "Women's Fundraising Help"—an easily digestible initiation into funding women's human-rights-focused organizations or events—in addition to other resources and links pertaining to these issues.

As centers of social activity, beauty salons can be unique vehicles for social change.

Public Health

A bright green future is a healthy future. One of the crowning achievements of our civilization has been the rise of health-care systems capable of keeping us healthy and active to life spans that would have been considered staggering just a few hundred years ago. Moderately affluent young people in today's world have access to the information, support systems, food, and medical care that can allow them to lead healthy, active lives into their eighties, nineties, and beyond. This is an incredible gift—not only to each of us, who can potentially add decades to our time on earth, but to society, which is offered the potential of more elders with the time and wisdom to make a difference (though we have a great deal of work to do on integrating our elders into more worldchanging activities).

We owe that gift not only to the field of medical science, but to the field of public health. Our ancestors died in droves of diseases whose names have almost no meaning to us today, like smallpox, bubonic plague, polio, and typhus. For our great-grandmothers, childbirth was a potentially deadly ordeal, and many of our ancestors' brothers and sisters died before they reached the age of five. One of the biggest reasons we're leading longer lives today is, simply enough, that most of us don't die young.

The same cannot be said all around the planet, though. Largely through the inaction of our own governments, medicine and the public-health system have failed more than one billion people in the Global South. Sicknesses stemming from dirty water, and diseases that can be prevented in the Global North with a simple shot or cured with common medications kill millions of children in the South every year. Despite the fact that drugs exist to treat HIV/AIDS, that epidemic is still decimating the Global South, especially Africa, where it has carried off the better part of an entire generation and left tens of millions of AIDS orphans. Other forms of medical care that we in the North take for granted—from eyeglasses to wheelchairs to simple but life-changing operations—are practically myths to hundreds of millions of our global neighbors, who have never seen them and have no access to them.

Even more ominously, our global public-health system, as it stands, shows signs of unraveling. Having been built up through careful planning, bold investments, and the hard work of tens of thousands of doctors, nurses, and community workers, the system has been all but abandoned recently by shortsighted governments in the North. With new potential epidemics emerging more rapidly as our planet grows more tightly interwoven, this is madness. With billions around the world in need of better care, it is unjust. With all our public-health expertise, medical technology, and wealth, it is patently unnecessary.

We need to bolster our global public-health system, and extend the benefits to everyone. But we can also learn a lot from the efforts of those who have had to fight epidemics and medical injustice with nothing but good ideas and collaboration within communities. If we can combine the powerful tools of the former with the compassion and innovation of the latter, we can build a healthy future for all.

Networking to Eradicate Pandemics

If infectious disease and infant and maternal mortality took as many lives today as they did a century ago, the world population would be dramatically smaller. But while overpopulation contributes to a spate of global problems—health-related and otherwise—none of us would bemoan the advancements in health care that have granted us life and sustained our wellness. We owe our thanks almost entirely to the advent of scientific public-health programs. Focused largely on prevention, education, and access, the public-health system has greatly enhanced the length and quality of lives worldwide during the last century.

One of the greatest examples of a public-health initiative that literally changed the world is the eradication of smallpox. In 1967, it seemed outrageously optimistic to predict that smallpox could be eliminated. The World Health Organization (WHO) motioned to vaccinate virtually everyone in severely endemic regions, which at that time were mostly confined to Southeast Asia and Africa. But massive vaccination was an unrealistic goal for complete eradication—tens of millions of infants would have to be vaccinated each year. Health workers couldn't keep up.

When an outbreak erupted in Nigeria in 1966, aid workers faced a shortage of vaccines. In order to use their limited supplies efficiently, they mobilized to pay house calls to every home in the area, checking on incidences of smallpox and vaccinating only those in close proximity to ailing victims. This method of creating "circles of immunity" around existing cases proved tremendously effective. The strategy was used soon after in India, with equal rates of success. Within a matter of months, smallpox had been eliminated from a number of previously endemic regions in Africa and India. By 1974, smallpox was endemic in only five countries.

Leading the charge in this swift annihilation of smallpox was Larry Brilliant, a young American doctor who had gone to India in the early 1970s. Under the appointment of the WHO, Brilliant led a team of thousands of health workers, who made over a billion house calls throughout India and brought about a victory over smallpox in 1980.

Now polio is disappearing in a similar pattern—it's endemic in only four countries, and might be eliminated soon. Unfortunately, other diseases haven't been as responsive to the method Brilliant instituted. Smallpox was eliminated, but yaws, malaria, and yellow fever were not.

And the nightmare diseases of the future are not the childhood killers that are almost eradicated—they're new pandemics like avian flu or SARS. Brilliant predicts that, should avian flu start transmitting from person to person, people will not get on airplanes, and commerce as we know it will cease for a sustained period, breaking just-in-time supply chains. While he doesn't think avian flu will necessarily become a pandemic, he reports that 90 percent of epidemiologists believe a major pandemic will occur in the next two generations and will make a billion or more people sick.

How do we stop pandemics? The way we stopped smallpox: early detection and early response. We won't have a vaccine for avian flu for at least three years, so we need to isolate cases quickly. Our weapon in this battle is information—systems like the Global Public Health Information Network (GPHIN), which scans news sites from around the world, digesting content in seven languages, to detect illnesses that might be pandemics. We witnessed its efficacy in the speed with which the WHO managed to gain control over SARS. The response took place simultaneously and widely, thanks to networked communication and preparedness. Indeed, technology can in many ways replicate the labor required for all of those billions of house calls, but much more quickly, cheaply, and safely. This is the power of networks.

With more people using the Internet all the time, finding ways to track human travel patterns and the spread of epidemics has become easier. But it's still a challenge. Researchers can't ask people to wear tags and tracking devices, but tracing migration patterns of goods, rather than people, offers a useful and accurate tool.

At the University of California, Santa Barbara, researchers came upon a Web site that tracks dollar bills as they circulate the globe, and

Public-health programs and campaigns, like this Chinese anti-SARS campaign and this Peruvian polio-vaccination program, have been critical for improving and extending lives around the world.

used it to create a breakthrough model of human travel patterns that could greatly boost our ability to respond to emerging epidemics. Where's George? (WheresGeorge.com) allows users to enter their location and the serial numbers of their dollar bills (and other U.S. currency), and to return later to see where those dollars have gone. It's an online curiosity—not really a game, more like an information toy. The research team took this enormous wealth of data and found that human domestic travel patterns, as represented by currency, matched an unexpected, but easily understood (for mathematicians, at least) scaling and diffusion model, which will make it possible to build far more accurate simulations than we currently have of epidemic spread and response.

It's a leap in the right direction, but piggybacking on Internet games can't get us as far as we might hope. Larry Brilliant now has a new plan for preventing pandemics: the International System for Total Early Disease Detection (INSTEDD). The system will assist in promoting and protecting health worldwide by identifying disease early and treating it rapidly. Brilliant wants to build this around GPHIN, expanding the number of sites analyzed, and expanding from seven languages to seventy, so that the information would be accessible (and free) to people around the world, not confined to a few government agencies. He plans to use satellite data and short-message service (SMS) text messages to help confirm media reports. Nowhere is the power of networked technology clearer than in this scheme, where the ability to track, communicate, and analyze information quickly could literally save billions of lives. EZ & SR

Pledge 25 Club, Botswana

Youth education and outreach are crucial to curtailing AIDS around the world. But often it's difficult to get young people to participate in preventative measures—when we're young we feel invincible, and being told something is good for us in the long run can seem too nebulous to make us behave differently.

The Pledge 25 club in Botswana helps prevent young people from contracting AIDS by encouraging them to become blood donors at a young age. They promise to donate at least twenty-five times before they turn twenty-five, thus giving them incentive to stay healthy and HIV-free. The program has had phenomenal success, both in terms of reducing the rate of AIDS among young people and in turning these kids into educators and advocates, providing information and health education to their communities and spreading the Pledge 25 commitment. SR

Community Response to AIDS Orphans

If it takes a village to raise a child, what do you do when you have a whole village full of orphans? This is the case in many African nations, where HIV/AIDS has killed off almost an entire generation of adults. Even a well-established child-welfare infrastructure would buckle under the pressure of so huge a problem, and communities have had to figure out new ways to rally together and pool the limited resources they have to provide these children with a future.

Many villages have responded by forming committees to collect and distribute food and other donations to support the orphans. But these committees are more than just ad hoc nongovernmental organizations; some of the donated funds are put into launching new businesses, the revenue from which is then used to continue orphan-care programs that support schooling fees and other costs.

A Zambian women's group called *Kwasha Mukwenu* ("help your friend") has built on the committee approach in order to ensure that care for orphans is complete and that the children have prospects when they grow up.

Each woman in the group becomes a "caretaker parent," supervising three to five orphans. Though they don't necessarily take the orphans into their own homes, the women make sure the kids get into school and have clothing, shelter, medicine, and adult supervision. They intervene when relatives attempt to throw the orphans out, which, unfortunately, is a common practice for families with very limited means.

The women of *Kwasha Mukwenu* are entrepreneurs, too; they raise the money to provide for the children by cooking for local schools and making clothing and uniforms to sell at local shops. Many of the orphans work as apprentices, ensuring they'll have some skills with which to support themselves. The women have even become de facto outreach workers: they provide AIDS prevention information to the community and have produced a play to educate the public about the plight of AIDS orphans.

Not only do these women pull off an immense feat, made more complex by the illness many of the orphans themselves suffer, but they stand as a powerful testament to the fact that, in the absence of traditional structures, villages still form around the fundamental duty to raise their community's children. CB

PATH.org

If we know anyone who's lost a baby, we know what a devastating loss it is. But in the Global North, infant mortality is fairly rare. That's not true in many developing countries, where the rate of infant mortality is staggering: four million babies die annually, most of them from preventable causes. Vaccines, medical supplies, and early parent education could save a huge percentage of these children. The missing pieces? Funding and outreach.

PATH, an international nonprofit public-health organization, focuses entirely on bringing these missing pieces together and delivering them to underserved populations in the developing world. PATH works toward establishing sustainable solutions that take advantage of technological innovation and tailoring approaches relevant to the locations they serve.

In late 2005, PATH was granted substantial funding from the Gates Foundation to develop several programs for the prevention of infant mortality. Through a combination of midwife care, tetanus vaccinations, education in caring for newborns, and guidance in other basic maternal skills such as breastfeeding and safe contact, PATH expects to improve the mortality rate significantly in eighteen countries, according to its Web site. The solutions will be particularly geared toward cheap-to-use, easily learned

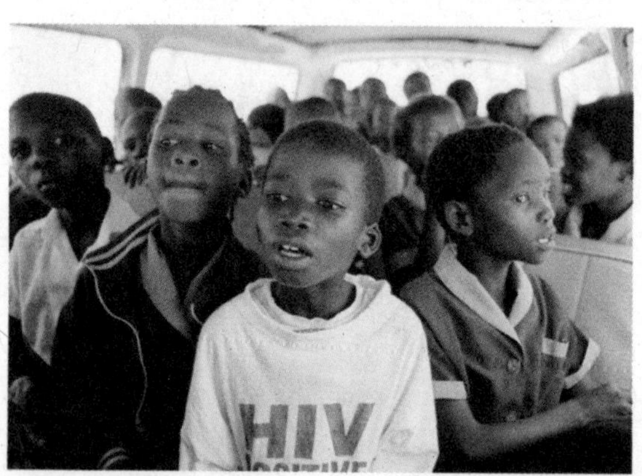

An AIDS orphan in South Africa travels with his classmates to a World AIDS Day event. Tens of millions of children have been orphaned as a result of the AIDS epidemic.

techniques that can be implemented locally without extensive training or preparation. In order to stretch both the funds and the benefits of the interventions, PATH is creating networks of local partners who can spread newborn care and maternal-health education throughout the communities in which they live. SR

Innovative Peer-to-peer Counseling/Education

There's no better way to reach a lot of people quickly than to visit a social hub. In Guayaquil, Ecuador, where HIV is prevalent, HIV-prevention workers have taken that truth to heart, after discovering that the social destinations of at-risk populations are their best venues for outreach. For the transvestite and transgender populations in Guayaquil, those venues are nightclubs, *discotecas*, and bars—and the outreach workers who visit them are not city employees, but members of that community. Fabiana Perez is a transvestite who has been a sex worker and is now a leading participant in outreach efforts along with a team of HIV-prevention educators known as the Amazonas, who work specifically with the city's transvestite and transgender population. The Amazonas' method doesn't involve pamphlets and free condoms; rather, it includes regular nights out on the town, during which HIV-prevention discussions are woven around socializing, eating, drinking, playing bingo, or dancing at clubs. Condom distribution is part of the effort, but the condoms are distributed only where the people that Amazonas targets are most likely to pick them up: one of the project's goals is to install condom vending machines in beauty salons.

In Lusaka, Zambia, condoms and informative leaflets can be found in barbershops and bars, thanks to a group of youth volunteers known as *Bwafwano* ("to help each other"). *Bwafwano* focuses on making HIV/AIDS the business of the whole community, and on offering services to everyone, not just those who are infected with the disease. Much of the volunteers' work happens in the drop-in centers scattered in local shops around town. Peer counselors provide educational materials and offer advice on safe sex. The dedicated group of youth educators is working hard—partnering with business owners and going door-to-door—to establish enough drop-in centers to cater to the needs of the entire city. SR

Siyathemba

In Somkhele, South Africa, a town of 70,000, HIV afflicts half of the girls. Most of those infected will not live to age fifteen. HIV and AIDS are more prevalent in the province of KwaZulu-Natal, where Somkhele is located, than in any other part of Africa, according to the Architecture for Humanity Web site, yet access to medication and treatment evades most of the sick. So how can a community amass those most at risk—youth ages nine through twenty-five—to disseminate crucial information? Easy: Start a football (American soccer) team.

A few things are needed to make this happen: coaches for the team, medical professionals for health education and, most important, a field to play on. If the health educators also coached the team, they could build the players' trust in an atmosphere of camaraderie and recreation, making them more receptive to information about uncomfortable topics like sexual health and AIDS prevention. That's two crucial components—but the team still needs a field.

In 2005, Somkhele's football players, coaches, teachers, and community members reviewed a number of designs for the ideal field. The proposals had been submitted from around the world to Architecture for Humanity's *Siyathemba* design competition. *Siyathemba* (Zulu for "we hope") invited plans for a site that would function as sports field, community hub, and health-education center, using only local, sustainable building materials and staying within a budget of five thousand dollars.

The winning pitch came from designer Swee Hong Ng, who proposed a V-shaped, terraced field, which would create a focal point for conducting health outreach while remaining spatially open. Ng honored the team spirit of the mission by working in collaboration with both young women and men in the community to be sure the design met their desires. The facility will

USING A CONDOM
MEANS THAT
YOU REALLY CARE

be staffed by medical teams, and may eventually house a mobile clinic, but the players will come because it's a place to play sports and to enjoy a sense of connection with their peers. In this comfortable atmosphere, they will find health education much more palatable, and outreach will therefore have a much better chance of success. This is the hope of *Siyathemba*. SR

Swedish Sex Education

No country in the world has put more organized emphasis on the importance of sexual freedom to overall sexual health than Sweden has. The Swedish Association for Sexuality Education, which introduced compulsory sex education in schools in the 1950s, takes a holistic approach, underscoring the significance of personal relationships, socioeconomic stability, and gender equality in forming a healthy self-concept and a positive sexual identity.

As a result of eradicating social taboos around sex and the body, acknowledging that sexual desire is innate and important, and encouraging safe sexual exploration among young people, Sweden has achieved the lowest rate of teen pregnancy and sexually transmitted diseases of any nation in the world. SR

The Ultimate Safety Net

In this age of solar-powered water purification and networked disaster relief, the mosquito net seems laughably simple. But until we develop a vaccine for malaria, mosquito nets are still one of the most reliable ways to prevent the disease. They're also more sustainable than DDT, the chemical usually used to control mosquitoes, which has dire health and environmental consequences.

So the idea isn't to ditch the nets, but to build better ones. In Arusha, Tanzania, A to Z Textiles has created a new local industry by manufacturing the Olyset, a durable net that contains the pesticide permethrin and slowly releases it over a span of five years.

While pesticide-drenched nets aren't new, the Olyset is different from its predecessors: it's woven from durable plastic that actually contains the insecticide instead of just being coated in it. Other nets must be re-treated every four to six months, a process that is expensive, time-consuming, hazardous, and not readily available to people in very remote areas.

Best of all, the venture philanthropy outfit Acumen Fund is funding sustainable bed-net-manufacturing enterprises in the developing world. Their approach simultaneously fights malaria and creates jobs. CR & JC

Public-health education programs are vital for fighting HIV/AIDS and providing family-planning services, Nairobi, Kenya.

Preventing Malaria with Coconuts

Sometimes the best tools are right under our noses. A microbiologist in Peru, which has the highest malaria rates in Latin America, has discovered a way to effectively battle the disease using coconuts.

Dr. Palmira Ventosilla, along with a team of researchers at the Instituto de Medicina Tropical in Lima, has created a community-based program that effectively curbs the spread of the disease—in an ecofriendly way. Coconuts are natural incubators of a strain of bacteria called *Bacillus thuringiensis israelensis* (Bti), which produces a toxin that kills mosquito larvae. Bti is already produced commercially for this purpose, but not in Peru, and the costs of importing large amounts of it are prohibitive for most communities. But by simply tossing a couple of coconuts that have been brewing the strain into a pond or other breeding site, villagers can eradicate mosquito larvae for more than a month.

Ventosilla and her team train villagers to manage the pest-control systems on their own. They are taught how to set up the coconut incubators—which requires cutting a hole in the shell, sticking a swab doused in Bti into the coconut, and sealing the hole with candle wax—how to distribute the fermented coconuts when they're ready, and how to track mosquito breeding cycles and conditions that allow malaria to spread easily. The program also provides villagers with information on other methods of malaria prevention, such as using nets and screens, and planting eucalyptus trees—the branches and leaves of which can be used to fumigate homes. CB

It Takes a Villager

Most often it's not the lack of a solution that allows a medical crisis to escalate: it's the inability to implement that solution in the most affected remote areas. Governments can set up health centers, but if those centers are still miles from people living deep in the forest, they won't do much good. The only way to ensure that medicine and advice are always available is to turn one person in every village into a health-care worker.

In Cambodia, this approach is used to combat malaria, which mainly afflicts rural people. The Cambodia National Malaria Centre has been recruiting villagers to serve as malaria workers for their communities; the workers go house to house to diagnose the disease and administer treatments. The country's Ministry of Health provides the diagnostic tools and medicines; workers get a fresh supply each month when they attend a training seminar at the nearest clinic. Since the program's inception in 2001, 135 villages in the Ratanakiri Province, in northeastern Cambodia, have participated, and malaria rates there have dropped by about 35 percent.

On a related front, traditional healers in Africa are now being called upon to help manage AIDS outbreaks in their communities. South Africa's government, which is struggling to provide health care in a nation where one in nine is infected with HIV/AIDS, has recognized traditional *sangoma* healers as health-care professionals, establishing legal regulations and standards for them. The new legislation removes the stigma attached to *sangomas*—during apartheid they were treated as little more than witch doctors—and facilitates cooperation between the traditional healers and hospitals. This new cooperation is particularly important because an estimated 70 percent of the population visits *sangomas*, who are required by law to refer patients with life-threatening illnesses to medical doctors. People who would otherwise never visit an AIDS clinic on their own end up there at the urging of their *sangoma*, making it possible for them to receive important treatments—and for health workers to track the disease better. CB

HealthStore

HealthStore is a Kenyan franchise model for delivering pharmaceuticals to rural areas; it is owned and operated by nurses, whose salaries are doubled, and gives poor, rural communities access to much-needed medicines. The medicines themselves aren't innovative (malaria meds, antidiarrheals, and so on); the business model is what makes HealthStore revolutionary.

Traditional *sangoma* healers in South Africa are often on the front lines of HIV/AIDS prevention and treatment.

Recognizing that ineffective distribution systems often prevent essential medicines from reaching communities that need them, HealthStore established its network of franchises to reach far-flung locales while simultaneously providing living incomes for the nurse-owners. Combining a nonprofit central franchisor with a for-profit franchise network is part of the innovation; there is no incentive for the HealthStore Foundation to "cheat" the franchisees. Its job is to set the franchisees up for success, and it does this through relationships with the government, a nonprofit drug distributor, regional support offices, and ongoing training programs. Not only do the entrepreneurs get the necessary support to earn a sustainable income, but they also gain the satisfaction of managing their own business without handouts. At the same time, franchisees are closely monitored by HealthStore to ensure compliance with franchise regulations; applying strict franchise standards not unlike those used in Western fast-food chains is another key element of the venture's success.

Ultimately, the local entrepreneurs are empowered to succeed in an environment where they are surrounded by poverty and despair. The communities in which they operate benefit because they are given access to essential, affordable medicine—through a nurse-owner whose very success is a source of inspiration. RK

Dimagi

The developing world faces monumental challenges when it comes to keeping people healthy. We know the will and the ingenuity exist to meet those challenges, and new tools emerge daily, but implementing them in an opportune manner sometimes takes a visionary approach. Vikram Kumar has one: Ca:sh (Community Accessible and Sustainable Health System). Recognizing that doctors already possess the tools to make people better, but that those tools would be more useful in the hands of patients, Kumar created Ca:sh, an electronic handheld gadget that contains medical records and treatment information. Ca:sh is simple enough to be used by children and cheap enough to be carried by traveling nurse-midwives in places like India, where health care is often best provided on patients' home turf.

Ca:sh is now one of several projects under Kumar's start-up Dimagi ("smart guy" in Hindi), which develops open-source software and mobile technologies that protect and empower patients, not only by doling out digital health information, but also by collecting it and creating a medical database that tracks patient records and clues health-care providers in to patterns in disease and treatment.

All over Asia, and progressively elsewhere, information technologies are being used to deliver health care to remote and underserved areas, often in more cutting-edge ways than in hospitals in the developed world. Such approaches are creating a leapfrogging [see Leapfrogging, p. 292] revolution in health care. While the U.S. Centers for Disease Control and Prevention (CDC) still use a paper-based data-collection system for record tracking, the Chinese CDC uses a digital system that tracks daily records from 16,000 hospitals on dozens of diseases. Given concerns over lurking epidemics such as avian flu, this is smart; given that innovations like these are the first steps in bringing first-class health care to all, this is worldchanging. SR

RESOURCE

Institute for OneWorld Health
http://www.oneworldhealth.org/
The only nonprofit pharmaceutical company in the world, the Institute for OneWorld Health works from the premise that "those of us with the ability to provide life-saving vaccines and medications to the world's least fortunate—and to conduct research and develop new medicines—should do everything we can." By securing donated intellectual property and identifying the most promising drug and vaccine candidates, the institute is effectively working to develop safe and affordable medicines for the developing world.

South-South Science

Nations in the Global South have come to realize that their future depends on science. More specifically, they realize that they must build a bright green future from within, that although aid from the North has its place, their future really depends upon the active advancement of local science. Community health, environmental conditions, and the ability to respond effectively to dramatic changes in the global economy are all linked to scientific development. Developing nations can't simply rely on the goodwill of Western scientific communities to solve their problems, especially when that goodwill is negated by actions like applying strict patents to important drugs, putting them out of reach for most of the millions of people suffering from diseases like HIV/AIDS.

These nations have also realized that they're not alone: some of the most exciting breakthroughs are coming out of South-South collaborations. Southern nations can gain by pooling their resources and findings, and troubleshooting their problems and frustrations with those who are also operating with limited resources and within similar confinements. These collaborations have already moved beyond the governmental/institutional phase: they have morphed into formal and informal relationships between individual researchers, businesspeople, students, and tinkerers.

Countries are combining research programs on AIDS, malaria, and other diseases whose impact is most heavily felt in the developing world, creating joint protocols for independently assessing the safety of genetically modified crops; dramatically ramping up basic science education; setting up scientific educational exchanges; signing agreements to increase research funding; and hosting conferences on the scientific priorities of developing nations versus those of the developed world. They have made tremendous strides in bioscience alone. High-quality bioscience doesn't require a big industrial base, just dedicated and well-educated scientists, and many developing nations have plenty of those to go around.

With the emergence of South-South science, the Global South has begun to perceive itself as a generator, not a passive recipient, of ideas, and this reorientation has created an increased sense of national and personal worth, which is having a catalytic effect. What's more, this coalition seeks to do something really revolutionary: tie the fortunes of developing-world science to global collaborative efforts like open-access databases and open-source software. It's a blunt bid to wrestle power over the direction of research away from the "corporate science" of the Global North and to create

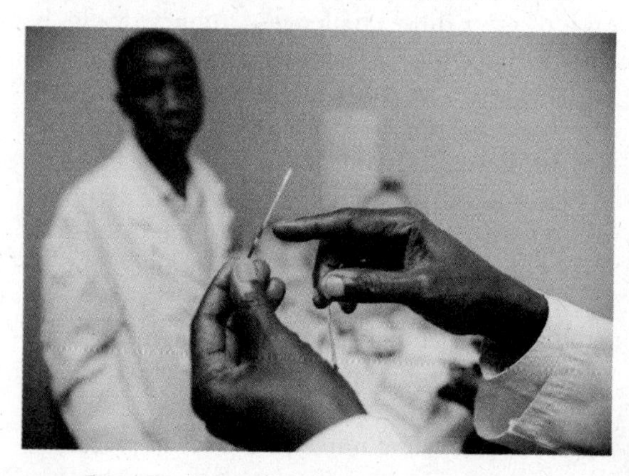

Southern scientists are leading the fight against diseases like malaria, Lagos, Nigeria.

freely available technological innovations that fit the needs of the Global South. If they succeed, we'll soon see how different science can be. AS & EO

The Rise of South-South Science

Many states in the Global South have traditionally sought solutions from the place least likely to provide effective ones: the developed industrialized West. These solutions—when they were offered—usually had a number of drawbacks, such as high maintenance costs. Also, they often didn't take into account the challenges of implementing sophisticated operating systems in countries without the infrastructure to support new industries. In pursuit of "catching up" with the North, states in the Global South ended up focusing on methods they were unable to implement. In many countries, we can still find the rusting hulks of white-elephant projects—huge pieces of technical infrastructure Southern countries don't have resources to maintain. That's all that's left to show for these efforts.

The rise of East Asia, led by Japan and followed by the "tigers" (Hong Kong, Taiwan, Singapore, and South Korea), initiated a period of self-examination in the developing world. (Although there had been similar efforts toward industrial-technological cooperation between China and the USSR in the 1960s and '70s, these initiatives were cloaked in the ideological battles of the Cold War, which ultimately hampered their efficacy.) The addition of Singapore, Thailand, Malaysia, Indonesia, and the Philippines to the cadre of quickly industrializing nations got everyone's attention. These nations had managed to pull themselves out of the vicious spiral of crisis and reinvigorate their economies. In comparison, the sub-Saharan African states seemed unable to keep up with their escalating problems, which ranged from collapsing commodity prices to tottering and disintegrating institutions, and South America had largely become a byword for massive debt defaults. As the gaps between the major Southern Hemisphere blocs increased, a cacophony of voices within the developed world began to question the paths that the less successful nations had taken (even though those paths had been prescribed in most cases by those nations' previous colonial masters or patrons).

The reemergence of China and India as economic powers in the 1990s capped a sense of confidence throughout the Global South, a sense of "if they did it, so can we." This massive shift in perspective is the essential component of the rise of Southern science—nations now have a clear vision of what they're capable of and are dedicating a lot of their limited resources to finding the best methods for managing their own advancement. EO

The Practice

How much do farmers in sub-Saharan Africa know about the protein-rich snail farming in Ghana or the tofu-growing ecosystems of Nigeria? In the desertifying regions of the Sahel, are scientists aware of the successful biogas initiatives at the Kigali Institute of Technology or of the powers of the desert-fighting *Jatropha* plant?

South-South science is rooted in the wide dissemination of information—including new and developing (and often, leapfrogging) ideas, and indigenous and local knowledge. The dissemination can occur casually, in conversations across one country's industries (say, between the tinkerers at Ghana's *Suame Magazine*, a group of small enterprise artisans who operate in an industrial district in Kumasi, and the students at the country's Ashesi University) or more formally (say, India advising Zambia on how to build an entire industry). Breakthrough palm-wine preservation research from the Nigerian Institute for Oil Palm Research does great things for Nigeria, but when that information is shared, it also proves to be very valuable to the sorghum brewers of Zambia.

One of the best early examples of South-South cooperation happened between Malaysia and the nations of West Africa. Malaysia, which is geographically similar to West Africa, adopted West African methods of palm-oil manufacturing, hoping to inject some energy into their economy. Fast-forward thirty years and Malaysia is the world's largest producer of palm oil—and is offering West Africa advice on how to upgrade its palm-oil industry, including how to produce the oil as a biofuel.

One of the best ways to facilitate these collaborations is to progressively introduce them into both traditional and evolving media gateways—from newspapers and television to solar-powered radio stations and cell phones. If there is a widespread sense that science and technology are not elitist, but that they are in many ways relevant to the daily lives of the most underserved people, it will accelerate a change in perception surrounding the importance of South-South exchanges. Stories of Southern innovation should be as accessible to an inhabitant of Kibera, Kenya, as a Nollywood [see Global Culture, p. 368] flick is. Education systems, too, should be progressively rethought and reoriented to encourage creativity and to recognize local resources and their promise. People like Mohammed Bah Abba, who invented the Pot-in-Pot refrigerator, a set of clay pots that keeps food cool without electricity [see Using Energy Efficiently, p. 169], should be transformed into the pop heroes of the sub-Saharan African age. The impact on the daily lives of the disenfranchised is the most evident benefit of these heroes' ingenuity, but the larger, and potentially more important one will be the unlocking of millions of minds. EO

The Potential

The Global South is the greatest source of untapped human potential we have. Too often, policy makers in the North present the rise of South-South science—and its contingent effect on the development of Southern nations—as a threat, not a gain. But alternate viewpoints and methods spawned in the Global South will lead to problem solving that will complement, not subtract from, Northern advancement. The South's need for solutions that are practical, easily implemented, and sustainable will lead to a reexamination of the North's complex and environmentally injurious activities—relics of the age of industrialization. In short, the discoveries of South-South scientists will help us all achieve more robust, less damaging systems that will help to alleviate poverty worldwide. EO

Southern Skies

As of late 2005, forty-three countries had put satellites into orbit. As the cost of the requisite technology continues to drop, a growing number of developing countries are able to design and build their own orbital communication, research, and observation satellites. Some nations—like India and Brazil—have even built their own rocket-launch facilities. Many of these projects are possible because of South-South cooperation.

It's often easier to launch satellites closer to the equator, and Brazil has taken advantage of this fact with its Alcantara launch facility, which it has made available to space programs and commercial outfits around the world. After some solo missteps, Brazil teamed up with China for its own satellite launches, and the two nations have been working together since 1999 on a series of five satellites for monitoring agriculture, water pollution, and the environment.

In late 2005, Iran put its first satellite into orbit, a small observation satellite designed to monitor local agricultural conditions and natural disasters. Sina-1 was launched from Siberia on a Russian-built rocket. Like Iran, South Africa uses foreign launch vehicles for its space efforts, and has put two satellites into orbit. By far the most ambitious developing-world space program outside of China is in India. India operates three launch facilities, and has put dozens of satellites into orbit. The Indian Space Research Organisation (ISRO) has designed and built a wide variety of communication, weather, research, and observation satellites; ISRO has now embarked on the development of a series of satellites offering remote radar sensing, for use in urban and rural planning, resource management, and disaster-relief operations. Beyond that, India has announced plans to put a satellite into orbit around the moon before the decade is out. JC

Biopiracy and Traditional Knowledge Banks

Southern scientists are turning to local knowledge of indigenous plants and traditional treatments to make breakthroughs, particularly medical ones. But this increased focus on reviving and disseminating traditional treatments has brought a novel problem to the developing world: biopiracy—nonlocals patenting treatments based on plants used by indigenous communities.

The best solution so far has been the construction of databases and traditional-knowledge archives, which offer an increasingly popular and effective way to combat biopiracy by establishing "prior art." Prior art disallows patents on anything that has been disclosed to the public in some form before the date the patent was filed. There is, of course, a great debate going on about what exactly constitutes prior art, and everyone seems to have their own definition. Many countries don't recognize oral traditions as establishing prior art, which is problematic for communities that don't have much in the way of a written history. In addition, when indigenous knowledge is primarily passed down orally, the creation of knowledge libraries is more difficult, as remote communities can be unwilling to share that knowledge with outsiders.

But despite these hurdles, several projects have been successful in establishing traditional-knowledge archives, and more are gaining speed. The South Asian Association for Regional Cooperation is building a regional Traditional Knowledge Digital Library (TKDL), covering South Asian traditional medicine, food, architecture, and culture. Participating nations include Bangladesh, Bhutan, India, Maldives, Nepal, Pakistan, and Sri Lanka. The regional TKDL is based on the success of India's own traditional-knowledge library, which was created in 1999 after the country successfully overturned a U.S. patent for medical uses of turmeric; Indians had known about the plant's ability to speed the healing of wounds for centuries.

By working closely with traditional female community leaders, the South African Management of Indigenous Knowledge Systems Project has been able to expand its effort to identify and protect the unique biosystems used by local communities as medicines. The project has led to something even more considerable than a knowledge archive: its leaders are working to improve local economic conditions by establishing community businesses to produce, market, and sell traditional foods and medicines.

Brazil has taken a different path. The Brazilian Microbe Bank is a repository of information about native microorganisms. The bank's researchers have collected detailed information on and examples of nearly a thousand types of microbes, and the facilities are capable of maintaining up to 12,000 microbes. The collection includes soil, water, and plant microorganisms from Brazil's diverse ecosystems, and even microbes isolated from oil fields. In the age of biotechnology, even organisms can be refined into products that can be sold back to the countries from which they originated—for a hefty profit. JC

RESOURCES

The Science and Development Network
http://www.scidev.net/
An invaluable portal for individuals and organizations in the developing world, SciDev.net offers news, views, and analysis of science and development issues, aiming to create a free-access space where decision makers can become

Because of the proximity of the Alcantara launch facility in Brazil to the equator, where the earth's rotation is faster, launches from this station require less fuel and are therefore more cost-effective.

better informed about the effect of their choices on social and economic development within their own communities. The Science and Development Network works to build regional networks of individuals and organizations, and maintains extensive dossiers (available free on the site) on issues ranging from medical ethics and malaria to indigenous knowledge.

The Ingenuity Gap: Facing the Economic, Environmental, and Other Challenges of an Increasingly Complex and Unpredictable World
by Thomas Homer-Dixon (Knopf, 2000)
In a hard-pressed world, what distinguishes people who thrive from people who fail? Thomas Homer-Dixon would answer in a single word: ingenuity. Homer-Dixon argues that success is predicated on both the ability to come up with new technical solutions to the massive problems we're faced with and the willingness to develop new social models for employing those ideas. The spread of new solutions and models, however, has not been keeping pace with the severity of the challenges we face, leading to an "ingenuity gap." If South-South science is about anything, it is about closing the ingenuity gap for some of the people hardest hit by global changes.

According to the author, "Over time, I came to the conclusion that a central feature of societies that adapt well is their ability to produce and deliver sufficient ingenuity to meet the demands placed on them by worsening environmental problems. Basically, I proposed, societies that adapt well are those able to deliver the right kind of ingenuity, at the right time and place, to prevent environmental problems from causing severe hardship and, ultimately, violence."

Copyfight

For five hundred years, alchemists repeatedly learned the hard way that drinking mercury was a bad idea. That's because for five hundred years, alchemists closely guarded their knowledge. We have a name for that period, when learned people kept their knowledge a secret—we call it the Dark Ages.

When an alchemist publishes his results, he stops being a superstitious fool and becomes a scientist. That one, simple step—sharing knowledge—achieves a kind of alchemy more powerful than the conversion of lead into gold: it turns superstition into wisdom.

We live today in a world of unparalleled access to knowledge, and hence, a world of unparalleled potential for human advancement. Raw materials are important. Industrial infrastructure is important. But those aren't the problem: the developing world is rich in materials and industrializing rapidly. What separates a developing nation from a developed one is the right to freely use and reuse knowledge and culture, the infrastructure to spread knowledge far and wide, the ability to use information to bring transparency to governance and to galvanize collective action.

No country knows this better than the United States. For the first hundred years of its post-Revolutionary existence, America was a land of merry piracy. Every invention and artwork of imperial Europe was free for reproduction and acquisition in the USA. Works by domestic authors and inventors were afforded nominal—but critical—protection under a copyright and patent law that explicitly set out to cultivate a post-colonial America in soil enriched by the composted works of foreign powers.

No developing nation today enjoys this privilege. A combination of international copyright, patent, and trademark laws have robbed developing nations of the autonomy that would allow them to embark on a program of self-improvement comparable to that of America's in its first century as a nation. Even when the son of the president of South Africa lies dead from AIDS, South Africa can't afford the economic penalties that would arise from manufacturing domestic dollar-a-dose HIV cocktails.

On the information-technology front, the most pernicious culprit in robbing nations of self-determination is "anticircumvention," first seen in the 1996 treaties from the United Nations' World Intellectual Property Organization (WIPO), a body with the same relationship to wicked copyright law as Mordor has to evil.

Anticircumvention laws make it a crime to tell people how to get around the locks placed on digital works, regardless of whether those locks protect anything guaranteed in law. A digital lock that restricts DVD playback based on region can stop you from watching an American video in India or vice versa. Even though neither country grants filmmakers the right to control where their videos are viewed after they are lawfully acquired, circumventing the technology is still a crime.

Most governments seek to balance the rights given to authors and the rights reserved to the public, but with anticircumvention in place, manufacturers can invent new copyrights for themselves simply by embodying them digitally.

The other major factor threatening universal access to knowledge is the lack of international agreement on exceptions to copyright. Thanks to treaties drafted by Hollywood and Big Pharma—the top-grossing pharmaceutical companies—practically every country offers the same package of minimum rights to every inventor or author. But the rights reserved to the public under each county's copyright are piecemeal. That means that an educator who includes in her course materials an excerpt that is considered lawful in Ghana cannot count on the same excerpt being considered lawful when she shares the course materials with a colleague in Jamaica. This turns international cooperation on humanitarian information projects into a legal minefield—when ten thousand volunteers from around the globe help Project Gutenberg scan, convert, and proofread public-domain books, how are they to know whether various national laws permit the work?

Things are coming to a head. At WIPO, dozens of nongovernmental organizations (NGOs) fight to hold the organization to its charter, under which it is supposed to formulate treaties to advance humanitarian aims; Brazil and other developing nations are coming up strong against copyright and patent laws in their trade negotiations with the United States; software patents were finally axed in Europe. A host of diverse coalitions are banding together in hopes of preserving the freedom to share knowledge and information openly.

The copyfight is fully engaged. CD

Creative Commons

International copyright treaties require countries to place works into copyright's strongbox with all rights reserved from the second they are created. In many instances, these copyrights can last for more than a century—which means that today's napkin doodle won't be in the public domain until the year 2205 or so. Does everyone need that much copyright? Hell no. Lots of authors benefit from having their works freely disseminated, remixed, and reproduced.

Creative Commons (CC) licenses are standard, machine-readable licenses that authors voluntarily apply to their works in order to specify that only some rights are reserved. Creating a CC license on the Creative Commons Web site takes about five minutes; since the project's inception, more than 53 million works have gone under CC license. Projects to convert CC license text into local languages and law are active in more than eighty countries.

Copyfighting
Many of us are not against sharing our work with others, but we would like to be paid for that work when the person using it has the money to do so. On a global scale, that's precisely what the Creative Common Developing Nations License is meant to facilitate. This new license allows creators to make their works available for attributed free distribution (copies can be freely shared, providing the original creator is credited) in the Global South, while still retaining all copyright control in the Global North. Lawrence Lessig, the founder of Creative Commons, notes, "The fact is that most of the world's population is simply priced out of developed nations' publishing output. To authors, that means an untapped readership. To economists, it means 'deadweight loss.' To human rights advocates and educators, it is a tragedy. The Developing Nations License is designed to address all three concerns." In other words, the Developing Nations License is a means of both sharing when it's fair and charging when it's smart. JC

Copyleft

Copyleft, as you might guess, is the opposite of copyright, in that rather than restricting users from duplicating, modifying, and sharing a product, it encourages them to do so. Early movement toward copyleft licensing began in the 1970s among software programmers who believed in free distribution of, and user-driven improvements for, software programs. In the subversive spirit of the time, and in an effort to liberate distribution, programs were often signed with "Copyleft: All wrongs reserved." By definition, any program or work that begins copylefted stays copylefted. If a new user makes changes to a given work, the modified version must continue to be free and open. With this arrangement, copyleft facilitates the development of more free software by requiring participation and compliance from all future users.

What's so worldchanging about copyleft? Besides encouraging freedom—both financial and intellectual—copylefting also fosters cooperation and the formation of strong communities around common goals. When we are encouraged to make systems work better, and are free to implement those improvements, we have great incentive to collaborate with others to create something that serves everyone's needs. SR

Access to Knowledge

The rallying cry of copyfighters on the international stage is "Access to knowledge!" ("A2K" in geek-speak). Since 2004, a coalition of developing nations' governments, health NGOs, telecom wonks, copyright scholars, humanitarian relief agencies, artists' rights organizations, and others has been drafting a new treaty to harmonize the rights of educators, archivists, and those who provide access to people with disabilities.

The drafting process is wide open, taking place on a mailing list and punctuated by regular meetings around the world. You can follow along on the Access to Knowledge Web site (www.access2knowledge.org/cs/). CD

RESOURCES

The Future of Ideas: The Fate of the Commons in a Connected World by Lawrence Lessig (Random House, 2001)
"At the same time that we are being pushed to the world where anyone can 'rip, mix, [and] burn,' a counter-movement is raging all around. To ordinary people, this slogan from Apple seems benign enough; to lawyers in the content industry, it is high treason. To the lawyers who prosecute the laws of copyright, the very idea that music on 'your' CD is 'your music' is absurd. 'Read the license,' they're likely to demand. 'Read the law,' they'll say, piling it on. This culture that you sing to yourself, or that swims all around you, this music that you pay for many times over—when you hear it on a commercial radio, when you buy the CD, when you pay a surplus at a large restaurant so that they can play the same music on their speaker, when you purchase a movie ticket where the song is the theme—this music is not yours."

Consumer Project on Technology
http://www.cptech.org/
Founded in 1995 by former presidential candidate Ralph Nader, the Consumer Project on Technology (CPT) is a nongovernmental organization that deals with issues related to the effects of intellectual property on a slew of fronts, as well as the production of, and access to, knowledge.

As a progressive public interest advocacy group, CPT has tackled issues such as access to medicine and incentive programs to encourage investment in agricultural and medical research.

Urban Community Development

Every city has them: poor neighborhoods blighted by neglect. They may be old inner-city areas hemmed in by industry, 1950s "inner-ring" suburbs [see Retrofitting the Suburbs, p. 240], or squatter slums [see The Hidden Vitality of Slums, p. 286], but they all share one defining characteristic: their social, economic, and environmental problems overlap and magnify one another, making attempts to solve individual problems practically futile. To truly address the legacies of inequality in our communities, we need holistic solutions.

Owing to municipal neglect, decay in these areas often spreads from the infrastructural to the economic and social, rendering communities vulnerable to illness, inadequate school systems, and elevated crime rates. Injustice here is a densely packed crust of environmental, political, economic, and social issues. Finding solutions means examining all the layers at once.

Solution seekers have been emerging at a grassroots level in many cities. A number of truly visionary individuals and organizations have undertaken to heal and reinvigorate communities that urban systems ignore. Cleaning up pollution, establishing

Slums often exist on the outskirts of a city, just enough removed to be imperceptible to city dwellers—a potentially dangerous "out of sight, out of mind" reality.

new recreational areas, increasing access to nutritious food, and facilitating constructive community participation are just some of the goals these groups have achieved. The results have been unequivocally positive, as is evidenced by the attention these inner-city initiatives are finally being paid by governments and the media.

Sustainable South Bronx

For decades, New York City officials have unofficially regarded the South Bronx as a dumping ground. The South Bronx has been home not only to some of the largest dumps and waste-processing plants in the region (particularly Hunt's Point) but also to some of the poorest people in the New York City area.

Majora Carter is working to solve both problems at once. Her organization, Sustainable South Bronx (SSB), tackles environmental, social, and urban-design problems simultaneously, so that the South Bronx can become not only healthier and greener, but more prosperous. We can, as she puts it, "green the ghetto."

To date, SSB has provided ecological-restoration training for area youth, created a Hunt's Point farmers' market, fought against expansion of Hunt's Point solid waste facilities and power plants, and spread awareness of green-building technology throughout the community. Now SSB is getting even more ambitious; it is working to create a recycling/industrial park in the Bronx, which will be a manufacturing-based ecoindustrial park that will create three hundred to five hundred jobs and serve as a cutting-edge model for industrialization. The organization is also pushing for policy changes that would allow for the widespread adoption of green roofs and cool roofs throughout the neighborhood, and is designing a comprehensive rainwater-harvesting and energy-efficiency plan for the community. Working tirelessly to establish new urban-design standards for new development, SSB is demanding that the city move away from big-box commercial development and toward community economic development, and that it move away from an emphasis on freight mobility and big trucks to an emphasis on bike paths and walkable streets.

One of Carter's most powerful levers for changing the minds of government officials and business leaders is illustrating the real costs of getting it wrong, through SSB's cutting-edge research. As Carter said in a phone interview with the author, "having communities this poor actually costs the region more money. And rich communities have an incentive not to truck off their crap to some poor community, both because of the health of the people [there] and for the sustainability of those rich communities." The organization's research is showing that the health costs, lost productivity, and environmental damage caused by the dumping practices in the South Bronx outweigh the economic benefits of running those dumping facilities. That's a message few expected but more and more are ready to hear: "Folks are understanding that there's such a thing as an urban environment. What we need to do now is to educate people on the bigger picture: how sprawl creates social problems as well as ecological problems, and that what happens in our urban communities has a huge

Sustainable South Bronx is an organization that implements sustainable development projects that both cater to the needs of the community and uphold environmental values. Here, a strawberry patch thrives on a roof; the strawberries are sold at a local greenmarket—and help insulate the building.

impact where they live. We need to start seeing regions as a whole." AS

Green Jobs, Not Jails

The connection between social inequality and environmental destruction isn't one made easily by most environmentalists. Sure, they may see a connection between a perceived lack of concern among politicians and corporations about both people and the planet. But that's usually about it.

Van Jones tells another story. For him, the two are inextricably linked: "Both problems are reaching crisis points. We act as if they are separate. But they are linked—economically, politically, and morally. The solutions and strategies for each must, therefore, be one."

The Ella Baker Center in Oakland, California, a nonprofit headed by Jones, sums up the essence of his mission in its slogan: Green Jobs, Not Jails. "In those four words you're talking about the environment, the economy, and the criminal justice system," Jones says. "Those are the three main sources of stress for people of color. And you're saying what you are for and not just what you're against. That, for us, is a real breakthrough. And it will take a while before our capacity and even our understanding will catch up with that vision. One thing I've been saying a lot lately is that Dr. King didn't get famous with a speech called 'I Have a Complaint.' At some point, we have to say what we're for." SR & JM

People's Grocery

In West Oakland, California, one supermarket serves 25,000 residents, and most of those residents don't have the means to get themselves to that grocery. On the other hand, liquor and convenience stores dot nearly every corner, and although the prices there are drastically higher and the quality significantly lower than at the market, most West Oakland residents buy a lot of their food from these corner stores. The other supply line for food in the area comes from government-subsidized assistance programs like WIC (the Special Supplemental Nutrition Program for Women, Infants, and Children), which offer calorically efficient but nutritionally deficient foods, consisting mostly of dairy products, reconstituted juices, and refined starches.

In response, the People's Grocery has arrived on the streets. Initiated by young people, contained in a mobile unit, and armed with hip-hop beats to draw attention, the People's Grocery brings a traveling larder of fresh, organic foods, local produce, and health education seasoned with street cred—to a population long underserved.

People's Grocery wants food justice. In close collaboration with the community they serve, the organizers work to establish a local food system and economy, making healthy food accessible as a basic human right. The organization promotes youth entrepreneurship, social enterprise, and grassroots organizing to help community members achieve a stronger sense of self-reliance and local sustainability. SR

Community Capital

The word *capital* tends to make us think of money. But if we are looking to discover the wealth of a community, we will actually find only a fragment of it by examining economic figures and physical property. There are other forms of capital that are often overlooked, and it's likely that they actually make up the bulk of a community's assets.

In the sustainability framework, human capital—reflected in healthy, skilled, talented, creative, and engaged people—is often valued over economic capital. This acknowledges the need for a human-centered approach to development and the contribution of human capital to broader wealth creation.

Community capital is being applied at vastly different levels—from global institutions to remote developing-world villages—as the driving force in community development. The emphasis is now on a community's *own* resources, rather than on its ability to access *outside* assistance. These resources include money, of course, but they also include people's skills, the community's social bonds, even the local environment's relative health.

It's an incredibly empowering idea, especially for communities across the Global South whom the world often sees—and who often see themselves—as "poor" and therefore in need of help from the outside. Communities discover that they can harness their wealth in the form of human capital and achieve genuine development. They break out of the scarcity mindset and out of dependency. This generates an invaluable sense of freedom and possibility, as well as creativity and self-esteem.

The focus on existing resources also works in terms of dollars and cents: community capital usually leads to growth in economic capital. MB

Soweto Mountain of Hope

At first sight, it doesn't really look like a mountain; after all, the hilltop in Tshiawelo, Soweto, is only a few stories high. But when you meet the people behind the astounding transformation this South African township has seen since 2001, you realize it is indeed a huge mountain—of *hope*.

Anyone passing by the small post office in Tshiawelo for the first time in the past five or six years will notice that things have dramatically changed. The hilltop across the road, today a vibrant community space known as the Soweto Mountain of Hope (SoMoHo), used to be a derelict spot, separating rival tribal groups. It was employed as a garbage dump, and known as a dangerous location where fights, rapes, and murders took place; during the apartheid era

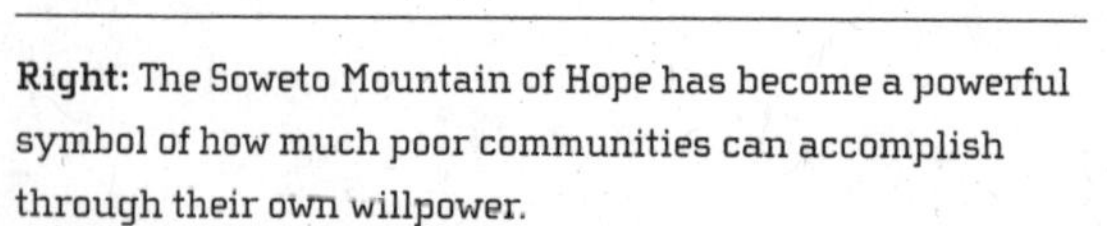

Right: The Soweto Mountain of Hope has become a powerful symbol of how much poor communities can accomplish through their own willpower.
Opposite: The fuel that drives successful development is human capital: capable individuals committed to community betterment.

of racial segregation in South Africa, this was a place for "necklacing"—killing informers by putting a gasoline-soaked tire around their necks and setting them on fire.

Children in their school uniforms used to walk in a wide circle around the hill to get home. Today, they skip right over it every afternoon, perhaps stopping for a climb at the playground, a game in one of the many stone circles, or a glance over Soweto by the old water tower at the summit.

SoMoHo was born out of a small Tshiawelo-based company called Amandla Waste Creations (AWC), which focuses on turning waste into art. Soda cans, plastic bags, paper, wire—you name it: AWC sees all these things as raw material for creative expression. At the time the mountain was established in Soweto, AWC was led by a charismatic social entrepreneur and community builder, Mandla Mentoor, who managed to work with the local people to translate the AWC approach into a broader community vision. What if the area's cultural waste—the squandered natural, social, and human capital—could be turned into social art?

In 2001, Mentoor won an award for his work in the community. At that time his home served as a center for a variety of community groups. The walls were practically bursting, because the space was far too small. He gathered community members, primarily youth, and together they decided to take on the trash-strewn mountain. They spent all the award money he had received on gloves, shovels, and cleaning equipment. After a few months of cleaning, they started landscaping, planting, building, and painting. By the time of the Earth Summit in August 2002, the mountain had become a key meeting spot for the local as well as international activists visiting Johannesburg. The AWC was even profiled at the summit, where it was distinguished as a "best practice."

SoMoHo has now become an overarching title for a variety of projects in the community. It is a living example of how a community can manage to look beyond needs and deficiencies to recognize its own resources and utilize and build four types of capital: ecological, economic, social, and human. SoMoHo fosters ecological capital through community gardening on the mountain, greening and landscaping, recycling waste, and hosting programs that get kids involved with nature. It fosters economic capital by establishing revenue-generating activities—including a shop that sells Amandla Waste Creations, a sewing cooperative, a bakery, a tour-guide service, a recording studio, and a music band. It fosters social capital by creating a shared community space that enables different tribal groups to come together, and by working with people to organize. And it fosters human capital by emphasizing learning by doing and by offering training programs.

The team members at SoMoHo don't hold back if funds don't flow in from the outside. They get started with what they have, discovering in the process that they are surprisingly resourceful. MB

Asset Mapping and Capacity Inventories

Think of a carpenter who has lost one leg in an accident years ago. Clearly, he has a deficiency. However, he also has a skill. If we know he has a missing leg, we cannot build our community with that information. If we know he has a capacity as a wood worker, that information can literally build our community.

Asset-Based Community Development Institute

Why is it that when we want to improve a community, we generally start by focusing on its deficiencies? The "needs survey" is one of the most common tools in the community-development field. It basically identifies all the gaps in a district that need to be filled. One of the key problems with this approach is that it deals with local residents as clients and consumers rather than as citizens and producers. To really make a change, it's more effective to start by mapping out a community's human and social assets—and from there, to work with residents to put these assets to use. The "capacity inventory" or the "community asset map" does exactly this.

The asset map provides a visual representation of the assets a community has to work with. Because it is visually oriented, often colorful, it is accessible even to people who are illiterate. But the real magic of the community asset map goes beyond making these capacities visible: the map generates *new* capacities by drawing people out of a mindset of dependency on outside resource, and into a creative process of mapping their own paths to the future. MB

Udaipur as a Learning City

In the 450-year-old Indian city of Udaipur, Rajasthan, a group of residents are concerned about what is being forgotten. As the city grows and becomes increasingly integrated into the global economy, the residents worry that modernization threatens their cultural wealth and age-old practices. Inspired by Indian leaders such as Mahatma Gandhi and Rabindranath Tagore, the residents are reviving the idea of *swaraj* ("self-rule"), which emphasizes the need to create our own notions of freedom. In 2000, under the leadership of the Shikshantar Institute, the residents launched an initiative called Udaipur as a Learning City (ULC).

An innovative project that involves exploring and connecting the city's diverse learning spaces, ULC is driven by an appreciation for the unique strengths, potential, talents, and skills of the individual. The program recognizes the value of that which is indigenous to Udaipur and rooted in its history. Among a host of activities, it helps local craftspeople continue to pass on their learning to the next generation so that their practices are not lost. Through ULC, residents work together on oral-history projects, revive the local languages, and run intergenerational community reflections and dialogues. They also participate in "unlearning and uplearning" workshops that build the capacity for critical thinking in relation to mainstream media and education (unlearning) and the capacity for creative expression of alternatives (uplearning).

Udaipur as a Learning City has provided a space and an opportunity for people who have a greater vision of their future and that of their city. Both within Udaipur and in concert with cities in India and abroad, residents have been building a network of concerned, motivated people and organizations committed to rethinking and experimenting with urban living. MB

Banco des Palmas

Hundreds of communities in the Global South have decided to end their dependency on external financial resources and their vulnerability to macroeconomic forces by initiating their own local exchange systems and alternative currencies.

While these systems in some ways function like the official monetary systems, they account for community capital in the wider sense. They build the local economy in a way that also develops social networks, revitalizes local cultures, and encourages cooperation and mutual aid. The monetary systems help to build self-reliance and personal skills, and to develop local production for local need, thus reducing the region's environmental footprint. One successful example of such an exchange system is the Banco des Palmas and its currency, the palmas.

The Palm Bank was created in a ghetto of Fortaleza, Brazil, in 1998 by a group of slum dwellers discouraged by the lack of money circulating in their community. The members joined together in an association, determined to fight poverty and improve living conditions through economic development, and to encourage local mobilization and reestablish community spirit. Palmas can be exchanged for the Brazilian national currency, the real, but can be used only within the Fortaleza district where Banco des Palmas operates. The palmas created a strong incentive to keep commerce local, feeding money into the community and thus enabling entrepreneurs to establish businesses that would help the local economy thrive.

For those entrepreneurs, the Banco des Palmas offered several benefits in addition to the palmas currency itself, including interest-free microcredit [see Microfinance, p. 346] loans, and even credit cards. The credit card in particular, called the PalmaCard, helps Fortaleza residents obtain their basic necessities without having to pay for them immediately. Payments can be made one month later without accruing any interest. This has meant that residents buy many more goods locally, rather than turning to vendors outside of the community, as they often had to do before.

According to a 2003 report by Dutch Social Trade Organization Strohalm, virtually all residents of Fortaleza changed their buying patterns to include more locally grown products. The effect of more local purchasing is an increased chance for entrepreneurs to succeed in community-supported businesses. The drastic changes in the community are due partly to the creation of new economic capital introduced by the new currency. However, the system's success depends entirely on social and human capital in the form of organizations such as buyer clubs, local solidarity, relationships of trust, and human ingenuity. MB

RESOURCES

***Development as Freedom* by Amartya Sen**
(Anchor Books, 1999)
Winner of the 1998 Nobel Memorial Prize in Economics, Amartya Sen speaks as an advocate for people faced with persistent poverty, famine, threats to their personal freedoms, and plagues on their local environment. Arguing that freedom and development share a symbiotic bond, Sen contends, "to counter the problems that we face, we have to see individual freedom as a social commitment. The removal of substantial unfreedoms ... is *constitutive* of development."

Sen argues that personal freedom can effectively promote other kinds of freedom, and touts freedom as both an ends and a means to global development: "There are two distinct reasons for the crucial importance of individual freedom in the concept of development, related respectively to *evaluation* and *effectiveness*." The concepts and tools that Sen holds as essential to global development are, in essence, also the most basic: choice, ownership, and personal freedom.

The End of Poverty: Economic Possibilities for Our Time by Jeffrey D. Sachs
(Penguin Press, 2005)
To be born on the earth today is to be the subject of a giant craps shoot. If you're among the lucky small minority, you're born into a family as affluent as middle-class North Americans, Europeans, and Japanese. If you're less lucky, you're born into the global middle class, where life is harder, but where (at least under normal circumstances) your ability to survive is not in question. But if you are unlucky enough to be born into the more

than a billion people living in desperate poverty, your life is likely to be full of suffering, and needlessly short. Jeffrey D. Sachs would like to change that, by funding workable plans to raise every person on the planet out of life-threatening poverty. His ideas are controversial, and his approach resented by some, but if you want to understand the essential nature of global poverty in our times, you simply must read his book.

"Today's situation is a bit like the old Soviet workers' joke: 'We pretend to work, and you pretend to pay us!' Many poor countries today pretend to reform while rich countries pretend to help them, raising the cynicism to a pretty high level. Many low-income countries go through the motions of reform, doing little in practice and expecting even less in return. The aid agencies, on their part, focus on projects at a symbolic rather than national scale, just big enough to make good headlines."

Microfinance

Many of us enjoy middle- or upper-class status not because of the size of our paychecks, but because of things like interest-bearing savings and fixed-deposit accounts, credit cards, mortgages, insurance, mutual funds, and other investment services, which grant us additional financial leverage. If such a complicated mix of financial systems didn't exist, what economic class would each of us really fall into? Would we ever get that leaky roof fixed or be able to turn a hobby into a thriving online business?

These days, financial services are critical for creating wealth. For people to move out of poverty there is an acute need to access financial services that not only provide credit but also help in saving, insuring, and investing. But for the poor people of the world, especially in developing nations, available services are usually limited to pawnbrokers or moneylenders who charge interest rates of up to 1000 percent per year—and even those loans are mainly available to people who already have some assets. On the whole, state-sponsored rural banks in developing countries have also proved to be a disaster.

Microfinance is an invaluable strategy for generating income for people with limited financial resources. It involves minimal risk on the part of the lender with potentially lifechanging results for the borrower.

One solution that has been successful is the offering of small, uncollateralized loans to help entrepreneurs expand existing businesses. This tradition started in the 1970s when several nonprofits experimented with granting "microloans." Grameen Bank [see Social Entrepreneurship, p. 352], started in Bangladesh in 1976, turned this system into a well-oiled machine. Today hundreds of organizations provide microcredit services to untold millions all over the world. The future of microfinance is very bright—though it is largely used by the rural poor, it is now being introduced to the urban poor as well, and has a bigger place in the future of developed nations than we might guess. SA

A Microfinance Scenario

Raghu is a poor laborer in the state of Madhya Pradesh in central India. He lives with his wife and two children. Raghu earns his income by trading his labor for cash. He is paid by the hour, which means on the days he does not work he earns nothing. Also, his ability to grow his income is restricted by the number of working hours in a week and the availability of work.

Raghu's wife, Vimala, is enterprising. She finds out about a microfinance institution called the Self-Employed Women's Association (SEWA), which provides uncollateralized loans to poor women. She discusses the idea with Raghu, and after much deliberation they agree to take out a loan. She borrows a thousand rupees (about twenty-two U.S. dollars) to buy some chickens. She plans to raise the chickens and sell the eggs in her spare time. In due course she succeeds in building a small business out of this, adding a second source of income to the household. However, unlike Raghu's income from manual labor, Vimala's investment continues to grow—to create goods to sell—even when she is not working on it. They have started leveraging the power of credit.

Vimala pays back the loan, and finds that SEWA has begun a new program allowing borrowers to open a micro-savings account through which they can earn some interest. For the first time in their lives, Vimala and Raghu have started levering the power of time and money. This has added a small but guaranteed third income stream, without any effort.

Vimala and Raghu attend a financial education course, where they learn the benefits of insurance and investing. With the help of SEWA employees, they insure their chickens, and also take out a life-insurance policy on Raghu.

In a couple of years, they have built a small amount of savings, which they start investing in a "microfund," an innovative market tool that allows micro-savings customers to invest in a mutual fund and benefit from the growth of bonds and equities.

Vimala and Raghu have changed their lives in a few years—they have conquered poverty and built a stream of steady income, which will ensure that they maintain and increase their standard of living in the years to come. And it all started with one small loan. SA

Armchair Microfinance

While many microcredit institutions have demonstrated great success in assisting poor entrepreneurs, their models are generally set up to foster the creation of more microcredit NGOs. However, several organizations are making it possible for individuals to be microfinance lenders from the comfort of our home computers. The California-based microfinance group, Kiva, built the first peer-to-peer, distributed microloan Web site.

Kiva adapts the idea behind the "sponsor a child" program to allow us to sponsor a business by lending small amounts of money. Founders Matthew and Jessica Flannery, a California couple who have lived in central Africa, argue that a one-to-one process is inherently more transparent than contributing to a charity or NGO, which then redistributes the donations; just as important, making a direct microloan gives the lender a greater sense of engagement than would an indirect donation.

Kiva's Web site (www.kiva.org) lists various businesses that are in need of microfinance aid; individual lenders can choose precisely which business receives the loan, and 100 percent of the contribution goes directly to the selected business. Kiva provides lenders with periodic updates on the start-up's progress and also pays lenders back for their loans. Once repaid, the lender has the option to use the money again to help another entrepreneur, essentially reusing the investment to increase social benefit. Over time, multiple new businesses can receive microcredit support from a single initial loan.

This does mean that lenders do not receive any sort of tax break for the money. Moreover, the Kiva model requires more active participation in the lending process than does a simple donation to Grameen or Opportunity International. But this is Kiva's key strength: lenders aren't just passive sources of money, they're active participants in the microcredit process.

Too often, the charity and NGO system is disempowering for both the recipients of support and those who donate their time and money. The original microfinance model helped to answer part of that complaint, as it explicitly offered support for people who sought to build new businesses. The Kiva peer-to-peer microfinance model may be an answer to the other part, by giving the lenders an active role in the strengthening of global communities. SA & JC

FUPROVI

According to some estimates, a growing percentage of urban dwellers—up to 50 percent in some cities—live in spontaneous settlements that are unplanned and/or illegal, and that lack infrastructure services. To counter that trend, Costa Rica's nongovernmental organization Fundación Promotora de Vivienda (FUPROVI), or the Foundation for Housing Promotion, operates a range of programs to address low-income

Remittances—money sent to poor communities by relatives abroad—are one of the most important sources of capital in the Global South, Morelos, Mexico.

WESTERN UNION
DINERO EN MINUTOS
La manera más rápida de recibir dinero
ESTETICA
UNIS

housing and to promote community development, income generation, sustainable development, and institutional building.

Established in 1987, FUPROVI has developed a self-help-housing program based on the organization and participation of the target population in project design and implementation. The program primarily focuses on low-income households, often headed by women, in implementing its goal. The main idea is to provide essential skills that will help communities face future challenges.

Additionally, FUPROVI has established an innovative revolving-funds model that enables money to be reinvested in new projects so that the organization maintains a continuously growing fund to implement its programs.

Remittances

One of the most important sources of income for people in the developing world are remittances—the money that people living and working abroad send home to family and friends. An estimated $300 billion has been sent from diasporas to developing nations via remittance.

The amount of money diaspora workers send home to Africa—$17 billion per year—is larger than the amount of foreign direct investment in Africa, and rivals official development assistance grants or loans ($25 billion per year). In some African nations, remittance represents as much as 27 percent of the gross domestic product (Nworah 2005).

According to a 2005 press release from the United Nations' Office of the Special Adviser on Africa, the average African migrant living in a developed nation sends two hundred dollars per month home to family. But there are several major problems with the current remittance system. The first is cost—it's expensive to send money overseas. Unless we send money from bank account to bank account, we end up paying substantial fees to services like Western Union. An article by Dilip Ratha, a senior World Bank economist, reports that 13 percent of the average remittance is claimed by transaction fees. Furthermore, sending money via wire services can put our loved ones in harm's way—this is the second major problem with the current remittance system. When our relatives leave the Western Union, money in hand, they're prime targets for muggings. And finally, even if the money does make it home safe, it might not be used for the best purposes—misuse is the third big problem. You may have been sending Aunt Akwe money to pay for your cousins' school fees, but there's no guarantee that Aunt Akwe isn't going to spend it on lottery tickets, or on beer for crazy Uncle Pat.

A further concern is that sending money home may help Aunt Akwe fix up the homestead and send the kids to school, but it won't help pave the road the house is on, or help build a new secondary school nearby. While two hundred dollars a month can have a huge impact on a family's life, only the collective action of many families can make major projects possible.

Today, hundreds of creative efforts are under way throughout the developing world to solve the problems associated with remittances. To address safety issues, MoneyGram is offering delivery service in the Philippines, bringing money to your door instead of making you go collect your funds in town. Alternatively, if Aunt Akwe has an ATM card, MoneyGram will transfer the deposit to her account.

Taking things further, a new remittance strategy—goods and service remittance—addresses the safety, cost, and misuse issues simultaneously. Instead of sending money home, you make a purchase from a Web site or a store in the United States or Europe, and your purchase—powdered milk, cans of corned beef, even a live goat—is delivered to your relatives. Mama Mike's, a pioneer in goods remittance, offers online shoppers the ability to buy supermarket vouchers and mobile-phone airtime for relatives in Kenya and Uganda, as well as more conventional gifts like flowers and cards. SuperPlus, Jamaica's largest supermarket chain, goes a step beyond that: you fill a shopping cart for your relatives and arrange for them to pick up the order at their local SuperPlus store. Goods-remittance services generally don't charge a fee, since they make their profit off the goods themselves.

Cost, safety, and misuse concerns are more easily addressed than is the issue of how to effect positive change from remittances, beyond helping out the immediate family. The Mexican state of Zacatecas has tried a financial incentive: for each dollar a worker sends home to support a local project (paving roads, building schools, digging wells), the state government contributes two dollars. Local governments have increased the match to three to one, and in 2003, $20 million was remitted to support 308 *tres por uno* projects in the state, according to John Authers and Sara Silver in the *Financial Times* (August 2, 2004).

The issues that circulate around remittances remind us that, in a globalized world, helping ourselves locally is an international affair. EZ

RESOURCE

Pathways Out of Poverty: Innovations in Microfinance for the Poorest Families edited by Sam Daley-Harris (Kumarian Press, 2003)
The Millennium Development Goal to cut absolute global poverty in half by 2015 is fast approaching its deadline, and microcredit is becoming an increasingly important and vital part of the solution. In *Pathways Out of Poverty*, Sam Daley-Harris contends that poor people *are* bankable, and moreover, that "microfinance can be a powerful tool in combating poverty," so long as the poorest people of the world are intentionally included in microfinance services. Arguing that to date microfinance has been too cautious and homogenous, Daley-Harris presents a series of case studies from around the world highlighting the ways in which microfinance has been paired with education programs to provide a comprehensive means of empowerment. "We know that microcredit institutions can reach the poorest ... [and that] microcredit institutions reaching the poorest can become financially self-sufficient." Ultimately, microcredit benefits both the borrowers and the financers.

Social Entrepreneurship

David Bornstein has seen the future of the fight against global poverty. Over years of travel and research throughout the Global South, Bornstein has come to know first-hand what social entrepreneurship means. The leading journalist in the field, Bornstein has written two books, *The Price of a Dream: The Story of the Grameen Bank* and, more recently, *How to Change the World: Social Entrepreneurs and the Power of New Ideas*, in which he makes the case for building a better world with innovative ideas and well-distributed microloans. Here he shares his story, and his findings, with the rest of us.

Is it possible for one person to make a significant difference in the world? In 1992 I quit my job and traveled to Bangladesh, where I discovered a surprising answer to this question.

I had been working as a journalist in New York, covering politics, crime, and feature stories, and I was seriously questioning my career choice. One day I was assigned to cover the murder of a Brooklyn grandmother. On the subway ride back to the office, I asked myself, "Why am I doing this? Is this what the world needs?"

Shortly after that, I came across an article about the Grameen Bank, detailing the loans it had made to a million women villagers in Bangladesh. The loans were tiny—sixty dollars a year on average—but with her loan, a woman might purchase a cow, or two goats, or a rickshaw, or bamboo to make furniture. At the end of a year, she would be the owner of the cow or the goats or the rickshaw—an extraordinary development for a village woman. Over time, these women and their families would earn money and build businesses, and move—slowly—from very oppressive poverty, eating one meal a day, to far less oppressive poverty, eating three meals a day, having a corrugated tin roof and a vegetable garden, and being able to send their children to school.

I wanted to see if the article was true. Before I left for Bangladesh, I read everything I could get my hands on about the bank, and I kept coming across references to it as a kind of development "miracle." But when I got there I saw that it was something better than a miracle: it was a system.

I was surprised to discover that the Grameen Bank had actually begun as a tiny experiment in a single village, initiated by an economics professor and his graduate students. The professor, Muhammad Yunus, had been profoundly affected by a famine in 1974, in which thousands of Bangladeshis had starved to death. After that, he began working closely with villagers near his campus trying to find ways to alleviate poverty. He launched several experiments, some successful, some not, before he came to recognize the supreme value of credit, and to found Grameen Bank. Today, the idea that Grameen championed—microcredit—has spread around the world, influencing thousands of antipoverty programs and transforming the field of international development.

It's critical to remember that the Grameen Bank evolved against the backdrop of massive failures in that field. For decades, foreign governments and multilateral agencies had been pouring billions into Bangladesh and other poor countries, only to see most of the money sink into swamps of corruption and waste. What made the Grameen Bank unique was that a wholly different *mechanism* was at work: no consultants or bureaucrats had "designed" the bank and then gone out to hire functionaries to run it in the "field," the way governments or aid operations often do. The bank grew organically, from the bottom up, in an iterative, ongoing process of action and correction, re-action and re-correction.

I spent five years writing a book about the Grameen Bank—*The Price of a Dream*—and I concluded that the one indispensable ingredient in the bank's success was Yunus. To be sure, there were thousands of talented staff members and supportive donors, and millions of competent borrowers who together made the Grameen

Bank a remarkable institution. But I find it hard to imagine that all of the pieces could have come together in the beginning, or that the idea could have taken off without Yunus's energy and vision. Yunus poured himself into the Grameen Bank the way Steve Jobs poured himself into his garage-built computer. He had no other hobbies. He didn't take vacations. And since he launched the bank, he has spent three decades traveling around the world, talking to thousands of journalists, economists, philanthropists, bankers, students—anyone who would listen—to advance the idea of microcredit. In short, Yunus behaved like many successful business entrepreneurs, the distinction being that his goal was not to maximize his wealth, but to minimize others' poverty.

If we could redirect even a fraction of the entrepreneurial capacity in every society toward the creation of social value rather than the creation of purely economic value, what would the world look like? DNB

Childline

India has millions of children living on the streets. But until the mid-1990s, it didn't have any system in place to assist them when they were injured, sick, or abused. Childline—India's national child-protection system—changed that; its story begins with the vision of one social worker.

It was while she was attending the New School for Social Research in New York City that Jeroo Billimoria got involved with the advocacy group Coalition for the Homeless, and saw new possibilities for India. When she returned to her home city of Mumbai, she resolved to apply the experience she had gained at the coalition to India's large population of street children. Jeroo started out small, and her first efforts were extremely modest: she visited shelters in the evenings and gave out her home phone number to children in case of emergencies. Before long, she was being awakened a few times a week with calls—a boy with tuberculosis was rejected by a public hospital because his clothes were too dirty, another boy had been beaten up by a police officer. Jeroo helped the children, then spent weeks looking for follow-up services like long-term shelters and education programs.

Clearly, a better system was needed to combat such a huge problem. So Jeroo decided to launch an official children's helpline, the first such service in India. She paid visits to scores of organizations around Mumbai, urging them to join the "Childline" network (initially more than 80 percent refused to join). She raised funds and developed a program to train street youth as "para-para medics," giving them the rudimentary skills they needed to respond to most emergency calls.

During Childline's first year, in 1996, it fielded 6,618 calls and rescued 858 children. The following year Jeroo approached the Indian government about forming a partnership to make Childline a national organization. India had ratified the Convention on the Rights of the Child, but had done almost nothing to fulfill its commitments to protect children. Jeroo convinced the government that Childline could help honor those pledges: it was low cost, it was effective, and it was popular among street children. Today, Childline operates in more than sixty-five cities

Professor Mohammed Yunus, founder of the Grameen Bank.

across India, has fielded 7 million calls, and has been adopted as a project of the Indian government. Jeroo, meanwhile, has founded Child Helpline International, which has supported and helped to launch similar helplines in more than sixty countries. DNB

Tateni Home Care

South Africa wrestles with three massive problems: poverty, unemployment, and AIDS. While each is its own daunting battle, there is overlap between them that can allow them to be tackled as a whole. That's what the fifty-seven-year-old South African nurse Veronica Khosa realized when she founded Tateni Home Care Services, which trains unemployed youth to become paid home-care attendants so they can assist people with AIDS and other illnesses.

Veronica Khosa grew up in a poor Zulu village. As a child, she had dreamed of becoming a nurse, and she subsequently struggled for years to complete her education and gain accreditation. In the early 1990s, while working in an AIDS center in Pretoria, she got a preview of the catastrophe about to hit South Africa: more than 20 percent of all the pregnant mothers tested at the center were HIV-positive, the public systems were overwhelmed, and patients with full-blown AIDS told stories of being sent away from hospitals without dressings or ointments for open sores, and without so much as aspirin for pain.

Veronica recruited a group of fellow nurses to begin paying home visits evenings and weekends to people in the sprawling former township of Mamelodi. The women found many people suffering at home—sometimes, bedridden patients were left alone all day while the children attended school and the other adults went to work. Veronica quit her job and used her $8,300 retirement savings to establish Tateni Home Care Services.

Given the scope of the problem, she realized she was going to need a lot of help; she called on unemployed South African youth to meet that need. She had no trouble attracting large numbers of young people who were eager to gain marketable skills, as well as to learn how to assist their own family members and friends. By 1998, only three years after Tateni opened, its home-care model was being copied by the government of South Africa's largest province, Gauteng. By 2002, the province was running fifty-seven home-care projects modeled on Tateni and had allocated 40 percent of its health budget to home care and hospice beds. Today, the unstoppable Veronica Khosa, who is almost sixty, is using Tateni's community-based model to develop an effective system for orphan care. DNB

RESOURCE

Banker to the Poor: Micro-Lending and the Battle Against World Poverty by Muhammad Yunus (PublicAffairs, 2003)
Economist Muhammad Yunus is the founder of Grameen Bank, arguably the most successful microlending institution in the world. His book *Banker to the Poor* achieves three things: One, it vividly describes the hardships of the people at the "bottom of the pyramid" and their capability to manage with limited resources. Two, it inspires us to believe that one person can make a difference in the world and that—three—market-based solutions are better in solving the world's problems. It convinces us that a company (in this case a lending institution) with an innovative business model concentrating on the poor can be a billion-dollar success—and economically sustainable.

The central message Yunus is sending is that poor people are no different from rich people in terms of capabilities. Institutions are what matter: "If you go out into the real world, you cannot miss seeing that the poor are poor not because they are untrained or illiterate but because they cannot retain the returns of their labor. They have no control over capital, and it is the ability to control capital that gives people the power to rise out of poverty."

Giving Well

Before most of us learned to talk, we learned to share. It's a lesson we learned from parents at home, teachers in the lunchroom, and friends on the playground. Now that we're all grown up, and we earn our own lunch money, sharing is both easier and more difficult than it was when we were kids. We are in charge of our own assets, and we probably have more of them, yet so many organizations want our money, and it's time-consuming and confusing to figure out who will do the most with what we give.

To a responsible citizen with a desire to change the world, knowing how to give effectively matters. Those of us who are fortunate enough to have spare pennies can spur progress by donating them. We don't need massive wealth to be charitable. A number of great systems exist for stretching modest contributions, building philanthropic networks, and fund-raising successfully from numerous small donations. Remember, it's not the size of the coffer that counts, it's how you use it. SR

Giving money effectively helps your donation have the greatest impact.

New Rules for Global Giving: The Virtuous Circle

While there's no shortage of opportunities to support important causes, there's usually very little opportunity to see our money have measurable effects on the people we wish to help—especially when we only have a small amount to give. But there is a way for us to leverage the least amount of money into the largest measurable effect over time; there is a type of giving that multiplies itself.

Think of this approach as "enabling philanthropy": a virtuous action that enables someone else to take a virtuous action, like giving someone a microloan to start a small business that will eventually provide for all their needs. We don't have to give annual checks to umbrella organizations and hope that our money has actually done some good. We can take a relatively small amount of money and aim it at the precise point where it can do maximum good. We can give this money not as charity, but as an investment in the latent ambitions of poor people in villages and squatter cities, on the condition that the recipients magnify this seed by starting a small business or enlarging an existing one. In addition, we can strongly encourage them to take some small portion of their growing investment to help someone else as well.

This is a virtuous circle that keeps on giving, paying its benefits forward generation after generation. It's a beautiful thing, and it's the only type of love you can dispense with money. There is also an optimistic assumption in this scheme: the 2 billion poorest people in the world are really 2 billion entrepreneurs just waiting for their first seed money. If you give it, they will build upon it.

As you look for opportunities to start your own virtuous circles, keep in mind the following important guidelines:

- Aim your gift at those with the least means, to whom small amounts make a huge difference.
- Give at least two hundred dollars. Though it may seem like a small amount, it's enough to make a real impact on the poorest recipients and to allow them to address their dreams of tomorrow. If you give less than that, the money can only help with immediate needs.
- Ask yourself if the gift will be able to expand itself, gaining amplitude with each cycle.
- Focus your efforts on gifts that have a global range.
- Make sure that the agency that facilitates your donation sends the funds directly to individuals. The more steps between your donation and the recipient, the less impact it will have.

The following three organizations are highly evolved programs that produce amazing results. Giving to these organizations will go far to make you optimistic about the world's future:

Heifer International: For fifty years, the Heifer Project has been providing families in developing countries (and in areas of the United States) with breeding pairs of animals: cows, goats, pigs, rabbits, water buffalo, ducks, and so on. In the world's poorest regions the cost of a cow or goat can exceed a year's income, preventing many families from acquiring animals. When a family receives a breeding pair they get meat, milk, or eggs, but more important, they get a source of income: they can sell the offspring.

Each recipient must agree to give one breeding pair of offspring away to another family, thus paying the gift forward. So a small contribution can multiply as families gain food, a source of income, the means to help someone else—and pride. It's hard to imagine a better gift, or a more practical, proven lever for making a difference in communities of need.

Opportunity International: Microfinancing [see Microfinance, p. 346] is quite the rage in international circles for one truly amazing reason: the payback rate on tiny loans to workers in developing countries is greater than the payback rate on large loans to their home countries. In other words, from an outright profit perspective, you are better off loaning money to a Bolivian peasant than to the Bolivian government. Several nonprofits, starting with the Grameen Bank [see Social Entrepreneurship, p. 352] in Bangladesh, have pioneered microcredit loans on a large scale and for large investors. For a helpful citizen, though, it's easy to contribute funds to a wide variety of microloan programs through Opportunity International. This organization has been providing microloans for thirty years, since even before the term *microcredit* was coined. It works through Trust Banks, groups of twenty to thirty (mostly women) borrowers who meet weekly to cross-guarantee the loans.

Trickle Up: Rather than dispensing loans, Trickle Up issues outright grants (typically two hundred

The Heifer Project is an established and well-regarded program aimed at lifting agricultural communities out of poverty through gifts of livestock.

dollars) as seed capital for microenterprise hopefuls—with strings attached. What the recipients get is some start-up cash and a lot of training. Trickle Up makes grants to those looking to open small businesses like food stalls or repair shops, on the condition that grantees undergo basic business training, commit a minimum of 250 hours in the first three months to their venture, reinvest at least 20 percent back into it, and keep an account ledger. That means that when you contribute to Trickle Up, you are building up a social network of do-gooders to ensure good deeds persist. Last year 10,000 businesses got started via Trickle Up donations, and 30,000 budding entrepreneurs benefited from this global program; about 70 percent of grantees are women. Follow-up expansion grants are offered, too. KK

Smaller Circles, Enormous Impact

The virtues of helping a person in the Global South jump-start a small business are undeniable, but what do we do if we can't afford to make a two-hundred-dollar gift on our own? We can turn our twenty dollars into two hundred dollars by coordinating our donations through giving circles.

Giving circles are easy to set up and easy to manage: we donate a small amount of money and ask our friends and coworkers to match our donation. Pooling our resources and directing the combined donation to smaller, more specific causes is much more effective than writing a small check to an organization that tackles "the environment" or "human rights abuses."

For example, One By One is building an online network to fight obstetric fistula, an injury to mothers during childbirth caused by long, obstructed labor—the kind that would be easily remedied in the Global North with a cesarean section. When left untreated, obstetric fistula can be devastating, often debilitating the mother and rendering her incontinent. This condition is relatively inexpensive to cure (it costs about three hundred dollars), but women in the developing world, particularly in Africa, rarely get the treatment they need.

One By One's network of fistula fighters come together to raise tax-exempt donations through the underfunded United Nations' Campaign to End Fistula. The donation of each giving circle goes directly toward buying one woman the surgery she needs. In other words, when ten people write checks for thirty dollars, one woman's life is drastically improved. AS

Effective Philanthropy

What can $150,000 buy?

If you're a professional musician, it can buy a music video—and it's easy to blow much more than that on staff, union workers, caterers, makeup artists, equipment, and travel.

But if you live in Afghanistan, it can buy clinics and medicine; in Africa, classrooms and books, shelters for refugee camps, ambulances and irrigation and scholarships.

Musician Sarah McLachlan chose to spend the $150,000 allocated to her "World on Fire" music video on services for the world's needy. Instead of buying one ephemeral piece

Sarah McLachlan performed in the G8 concert series seeking, with other performers, to publicize the need for improved trade relations and increased aid to developing nations.

of advertising, the money paid the total running costs for a year of an orphanage in South Africa and a street children's hospital in India; purchased six months' worth of medicine for five thousand people in Nairobi, Kenya; provided counseling and schooling for seventy child soldiers; and much, much more.

And then she went ahead and made a video, for fifteen dollars, that puts faces on the faceless by revealing all that was done with the rest of the money and by making staggering comparisons between what the money could have bought and what it actually did buy. The video reveals, for example, that five thousand dollars can pay for hair and makeup services on a set or it can fund schooling for 145 Afghani girls.

This was more than a mindless publicity stunt. McLachlan and her producer, Sophie Muller, decided to make the video this way after being moved to action by a story they'd read on the Engineers Without Borders Web site. Their hope was that by making the video they would be able to illustrate how easy it is to turn wealth, especially tremendous wealth, into worldchanging actions. JC

The X PRIZE

If you give a kid a hundred dollars and ask him to paint your fence, he'll probably do a decent job, and you'll have a fresh coat of paint. But if you tell all the kids in the neighborhood that you're offering a hundred-dollar prize to the kid with the best idea about how to paint the fence, chances are the competition will work in your favor and you'll end up with the most beautiful fence on the block. Because where there's fanfare, there's motivation. Who doesn't want to shine in the limelight?

This is the idea behind the X Prize Foundation; traditional grants get things done, but prizes provide an incentive that leads to exceptional efforts. Peter Diamandis created the X Prize with a very specific agenda in mind: speeding along the technology for personal space travel—his lifelong dream. After reading a copy of Charles Lindbergh's autobiography, *Spirit of St. Louis*, Diamandis came up with the idea that competitions with large cash prizes could spur progress toward his goal. Lindbergh undertook his historic flight from New York to Paris in 1927 to win the Orteig Prize, which offered $25,000 to anyone able to complete a nonstop flight across the Atlantic.

Not only did Lindbergh's achievement become the stuff of legends, but it jump-started the aviation industry. He proved that the technological barriers cited by industry experts were merely perceived, not actual. Today, commercial space travel faces the same psychological barriers—most people think that it is only the domain of government workers and scientists. But if a prize was able to spur the interest and experimentation to create what is today a $250 billion industry, perhaps the X Prize can catalyze the beginnings of a commercial space-flight industry.

In October 2004, the first private spacecraft made its debut. The inventors, who had essentially executed a very ambitious science project, were awarded $10 million from the X Prize Foundation. Like the Orteig Prize in 1927, the X Prize created a competition that got the world's foremost rocket experts and entrepreneurs working on the idea of making private space travel a reality. Ten million dollars spent on one project would have gotten limited results; ten million dollars in prize money triggered serious efforts by multiple teams.

Whether or not we share Diamandis's lifelong dream of exploring the far reaches of the solar system is secondary to the broader implications that the X Prize raises concerning philanthropy: adding a dose of competition and the promise of notoriety to an opportunity for funding can undeniably yield exceptional results. SR

RESOURCES

Council on Foundations
http://www.cof.org/
The Council on Foundations is a membership organization of more than two thousand grant-making foundations and giving programs. Its Web site is a wealth of information, with resources on everything from "Philanthropy in a Time of War" to a bibliography of Latino American giving patterns. Most important, the council provides training and education for

everyone from newcomers to foundation chief officers—because shared innovation is the key to giving well.

The Center for Effective Philanthropy
http://www.effectivephilanthropy.com/
With a clear mission to "provide management and governance tools to define, assess, and improve overall foundation performance," the Center for Effective Philanthropy is one of the nation's leaders in providing assessment tools, data, and publications to foundations, in the hopes of improving overall performance on a variety of fronts. For the layman, the center's Web site offers some great tools for understanding and assessing foundations and their effectiveness.

One of the most remarkable development programs in the world, the Barefoot College addresses issues of water conservation, women's literacy, sanitation, job training, and more in multiple campuses throughout India.

The Barefoot College

Imagine a university where every activity, every class, every discussion we engage in has a direct and beneficial impact on our community. Imagine a community where values and activities related to sustainability aren't treated as "add-ons" or specializations, but where there is instead a seamless integration between the practical and the sustainable, between community needs and community growth: the elementary school building is a straw-bale construction that the whole community helped to create; the light you read by comes from a solar lamp that your daughter constructed; organic agriculture is the norm, not the pricey exception.

This scenario is not an unrealizable dream. This is the Barefoot College, a remarkable community-development project, which began in 1972 in Tilonia, a village in Rajasthan, one of India's largest, driest, and poorest states. "It was a sleepy looking neglected village of some two thousand people, typical of the 600,000 villages you find anywhere in India," says the college's founder, Sanjit "Bunker" Roy. The only thing that set it apart was an abandoned tuberculosis sanitorium on forty-five acres of

government land. When Roy looked at the vacant buildings, he knew they would make the perfect headquarters for the organization he wanted to start. The abandoned eyesore soon became the thriving home of the Social Work and Research Centre (SWRC), a place that Roy hoped would attract young, urban-educated Indians to work alongside local residents to alleviate rural poverty.

The urban professionals did come, but they were in for a surprise. They expected to bestow the benefits of their higher education upon the impoverished, illiterate villagers. Instead, their own deschooling process was just beginning. They discovered that very often the traditional knowledge—for instance, rainwater-harvesting techniques—that had been passed down orally from one generation to the next surpassed their own expertise. A dynamic partnership unfolded, which respectfully merged modern technology and traditional knowledge. Eventually, the SWRC evolved into an organization that questioned the purpose and process of education itself. They dared to wonder, "What kind of education would nurture individuals and communities who can meet their own needs—and do so sustainably?"

Over time, the SWRC changed its name to the Barefoot College and expanded the campus. The word *barefoot* denotes the grassroots approach that is pivotal for people who have little exposure to formal education. The Barefoot College uses a hands-on education process and has developed programs for every age group, from infants to elders. Students can learn to be solar engineers, hand-pump mechanics, groundwater experts, teachers, midwives, accountants, communicators (in videography, photography, street theater, puppetry), and apprentice in a range of traditional handicrafts. One of the most notable aspects of the program is that women are trained in nontraditional occupations, such as solar engineering and hand-pump mechanics.

People throng to the Barefoot College from all over India to learn about the award-winning program and bring what they've learned back to their own communities. Field centers exist in thirteen Indian states, and other countries have adapted the Barefoot College process to their local contexts.

The Barefoot College stimulates individual and collective creativity. One staff member summarized it by saying, "at the Barefoot College, the only limit is your imagination." The organization's focus on equality, a simple lifestyle, dedication to meaningful work, and openness is as innovative as any creation that comes out of the classroom. The lessons of Barefoot College transcend ethnicity or geography—every community should have such a resource. CO

Barefoot Campus

The Barefoot College campus is every bit as remarkable as its students are. Solar power supplies all of the electricity; a biomass plant provides an additional source of renewable energy. Rainwater is harvested on the roof of every building [see Thinking Differently About Water, p. 190]. Trees have been carefully nurtured through an ingenious drip irrigation system, creating a green environment that gives the impression that the college is a desert oasis. The biggest success? The entire campus was constructed by a dozen Barefoot Architects led by a farmer from Tilonia, Bhanwar jat; a local blacksmith named Rafiq fabricated and installed most of the campus's doors and windows. CO

Barefoot Solar Engineers

Bhagwat Nandan, head of the solar program at the Barefoot College, is a Hindu priest from a nearby village. He grew up in a home that had no electricity, but now he supervises the training of solar engineers, who are responsible for some of the school's greatest successes: the electrification of the entire campus, as well as the lighting of 870 schools and 300 adult-education centers across the country, homes in 28 villages in the Himalayan region of Ladakh (including a Buddhist monastery), and more than 12,000 households in various other villages.

The Barefoot Solar Engineers—usually women and unemployed youth, and usually illiterate—learn to install and maintain photovoltaic systems for individual homes, as well as to create town-sized microgrids. The solar-education model involves community participation and commitment to energy self-reliance. Generally, the village selects someone to train at the college, and that person then returns to install and maintain the solar energy for their community. Women who have very little independence or clout in their village often find that they become impromptu community leaders once they return and demonstrate their new skills. And all of this is achieved without elaborate textbooks.

The developing world is littered with failed solar projects set up by charitable organizations that only had the funds to install them, not to maintain them. These systems either broke after a few years or made villagers dependent on further foreign aid to keep the benefits on which they had come to rely. By contrast, the solar-education model at the Barefoot College is an elegant example of green economic development that supports community resilience. CO

Rainwater Harvesting and Piped Water Systems

In a semidesert climate where droughts can last for years, water is the most precious resource, especially when the few sources of water a village has are brackish or contaminated with high levels of iron and fluoride.

It's no surprise that water initiatives are a big focus at the Barefoot College. Rainwater-harvesting systems designed by students have collected an estimated 7.6 million gallons (29 million liters) of rainwater at 470 schools and community centers. Additionally, communities have been mobilized to build their own piped water systems and learn how to maintain them. Hand pumps at village wells are repaired by mechanics who live in the village, ensuring that there will always be someone on hand to troubleshoot problems.

Prior to the Barefoot College intervention, villages were dependent on government engineering teams to make repairs, which often meant that malfunctioning hand pumps would remain broken for long periods of time until the team arrived. Local control and management of water benefits the entire community and contributes to its economic well-being. Over the past ten years, the Barefoot College has also turned its attention to water conservation and recharging groundwater. CO

Community Health

At a time when public-health efforts worldwide are shifting toward disease prevention and health promotion, the Barefoot College is once again ahead of the game. For decades, the school has been training community-health practitioners, realizing that critical basic health messages—especially when they involve taboo subjects like STDs—are best delivered by midwives and local health-care workers.

Thanks to the college, there are now two hundred village health clinics throughout India that provide basic medicines (including

The Children's Parliament teaches children at the Barefoot College about civic responsibility, a protactive way to shape the values of the next generation.

traditional medicines), administer treatment for minor injuries, and provide transport to government hospitals when ailments are too serious to be dealt with in the clinic. Since clean drinking water is vital to good health, a related program, the Barefoot Chemists, trains young people to test for water quality. CO

The Children's Parliament

The Barefoot College has influenced education throughout India and in other countries. It has been particularly successful with night schools for children.

The night school concept itself has made a huge impact on child education, particularly on the education of girls. Many rural children (especially the girls) are unable to attend school during the day because they are too busy tending cattle or attending to other household chores. Their only opportunity to learn comes at night. Students gather in schools that have been powered by the Barefoot Solar Engineers and pore over textbooks—made from recycled materials. While they do learn basics like reading and math, their curriculum also includes subjects like animal husbandry—practical knowledge that can be applied immediately as the children continue to contribute to their households and communities.

The other component of the college's innovative education program is the Children's Parliament, which teaches kids about civic responsibilities and how elected officials *should* serve their communities. Students elect their own prime minister, who is given a small budget to improve life in his or her village. The program is a timely complement to the current child-friendly movement in which municipalities increasingly involve children and youth in community planning.

Just as Bunker Roy recognized that most development programs simply tell poor people what to do instead of letting them brainstorm their own solutions, the Children's Parliament recognizes that most of the time when people "listen" to children it's in a patronizing way and does nothing to honor their insights on their own communities. But the members of the Children's Parliament are, indeed, very active members of their communities. They oversee and report on the quality of teaching and attendance levels in the night schools. They are responsible for things like monitoring water quality in their villages. They have been able to solve problems that have stumped the adults—one "prime minister" even figured out a way to raise enough money to pay for a piped water system in his village. CO

RESOURCE

Tilonia
http://www.tilonia.org/
The nonprofit retail arm of the Barefoot College, Tilonia.org is based in the United States and is dedicated to providing marketing and business development for the crafts created at the college. If you're looking for a socially responsible way to spend your money, as well to support rural artisans, consider making a donation or buying Tilonia's beautiful block-print fabrics, handcrafted furniture, or one-of-a-kind jewelry.

Travel and Tourism

What's your pleasure—a shopping extravaganza in Paris? A safari in East Africa? Or a weekend ski trip to Tahoe? Whenever we enjoy leisure travel, we are taking part in one of the world's largest industries: tourism. In 2005 alone, there were 808 million international tourism arrivals, according to the World Tourism Organization, a specialized agency of the United Nations.

According to a recent World Travel and Tourism report, more than 9 percent of all jobs are linked, directly or indirectly, to travel and tourism. The World Travel and Tourism Council calculates that the sector accounts for $6.4 trillion of economic activity, over 10 percent of the world total—meaning that a massive number of people the world over count on tourism revenues for their survival. In fact, for many countries, tourism is among the top three sources of foreign exchange.

Though tourism has long been popular, the industry continues to expand at a breakneck pace. In just the past ten years, the distance flown by international passengers each year has increased by 60 percent; this despite temporary setbacks from severe acute respiratory syndrome (SARS), the Iraq war, outbreaks of terrorism, soaring oil prices, and economic slowdowns. The World Tourism Organization projects that by 2010 there will be more than 1 billion international tourist arrivals, reflecting a trend toward more long-haul travel. By 2020, 1.2 billion people are projected to have traveled regionally; 400 million, to have traveled farther afield.

As tourists, we may visit a certain community only once in our lifetimes. But our experiences and memories stay with us forever. The tour operators and hotels in the community contribute to the impression we come away with. Through the growing "Travelers' Philanthropy" movement, both tourists and tour operators are giving back to the communities that host them. For some, giving back is just the right thing to do; for others, it is a way to enhance their image and make it easier to continue working within the community. Whatever the reason, their contributions of time, talent, and treasure are changing the face of tourism. ZC

Don't Our Dollars Help?

When we go on vacation, we spend a lot of money—on transportation, lodging, food, entertainment, and souvenirs, for starters. As anyone who has planned a trip knows, these costs add up. And although we'd like to think that the bulk of our hard-earned money is going to support our destination's local economy, the reality is that much of it never reaches the community we are visiting.

When we travel, we can choose to be part of the problem, or part of the solution.

Let's say we decide to visit Tanzania. First, we look for great airfare deals online. We book a plane ticket, and as we are about to pay, a pop-up asks whether we'd like to add in hotel and car-rental reservations at a great rate. While the price may seem enticing, the package comes with hidden costs.

The vast majority of the dollars we've budgeted for this vacation may never reach Tanzania. Some of our dollars go to an airline company headquartered in the United States or Europe, and the rest to a hotel owned by a multinational chain. This effect is exacerbated if we purchase an all-inclusive resort package, where food and excursions are often arranged through the hotel as well. And though we've just paid for the most expensive parts of our trip (transportation and lodging), we haven't even left home yet.

So what's a better alternative? Selecting a locally owned and operated hotel will ensure that more of our dollars stay with the people who are hosting us. Buying locally produced foods will allow area farmers to benefit from our visit. It will also cut down on the transportation costs associated with bringing food in from elsewhere. The same is true for beverages. Remember that many familiar brands must be imported at a cost to both our hosts and us. Local alternatives are sure to be available.

We can further support the local economy by hiring resident guides, paying the requisite fees when entering protected areas, exploring local transportation options, and buying souvenirs from local shops or crafts cooperatives instead of hotel gift shops or cruise-port shopping complexes. (It's not unheard of to find a "Made in China" sticker on an "authentic" craft we just purchased in Oaxaca, Mexico.) Not only will we inject our dollars directly into the community we are visiting, but we'll go home with richer memories and stories. ZC

Flying Conscientiously

With so many people taking so many trips, it is little wonder that the airline industry is growing. In 2004 alone, 1.9 billion passengers traveled by air.

Choosing to fly, however, is a decision with serious environmental consequences. Airplanes emit carbon dioxide, water vapor, and nitrogen oxides; emissions at high altitudes can contribute to climate change at three times the rate of emissions released closer to the ground.

Efforts are under way to develop planes that can run on biofuels or solar power. Though widespread application of these solutions won't be possible for some time, a Brazilian company has manufactured a crop-duster plane that runs on ethanol, and the University of North Dakota's Energy and Environmental Research Center recently developed a carbon-neutral biofuel that could be suitable for aircraft use—in some respects, it's actually better than kerosene, which jet fuel is traditionally made from. This fuel will potentially cost less than petroleum-based aviation fuel. The fuel is still in its preliminary stages, but the university expects to start testing at a local air force base in the near future.

In the meantime, we can make it a habit to travel by bus, by train, by bicycle, or by car. When flying is our only option, we can offset the harmful emissions released during our flights by purchasing carbon credits [see Ecosystem Services, p. 489]. A variety of organizations offer the credits, which use our donations to support renewable energy projects such as green power plants or clean waste disposal, as a means of contributing something positive to balance out the carbon emissions generated during our flight.

One organization is finding a way to use airlines as an opportunity to effect positive change. Airline Ambassadors International, which was started by airline industry employees, organizes trips to hand-deliver food, clothing, and medical supplies to people in need. Using contributions from four thousand members, and in-kind donations from a variety of airlines, the group has brought $18 million worth of aid to forty-four countries and fifteen U.S. cities since 1996. Volunteers have also escorted a thousand children to new homes or to hospitals in foreign countries for medical treatment. While many of the volunteers are associated with the travel industry, there are special trips that allow children and others to get involved; it's just one more way to make the skies a little friendlier. ZC & JC

Sizing Up Sustainable Tourism

As more and more people take an interest in responsible travel, more and more businesses seem to be throwing around the term *ecotourism*. But many travelers have discovered that companies advertising ecotourism may actually know very little about the concept, other than that it attracts business. What do the terms *ecotourism*, *responsible tourism*, and *sustainable tourism* mean? And how can we find tour operators or hotels that are practicing the real thing?

The International Ecotourism Society states that ecotourism aims to conserve the environment and to improve the well-being of communities, by sponsoring responsible travel to natural areas. We can be responsible travelers by minimizing our impact on a community's environment, building environmental and cultural awareness and respect, contributing financially to conservation and to local people, and supporting human rights and labor agreements.

Whereas ecotourism is restricted to travel to natural areas, sustainable tourism applies these social, environmental, and economic considerations to all areas of tourism. Sustainable tourism could mean reducing water usage at a large hotel. It could be initiating a guide-training program for residents who live near a sightseeing destination. It could inspire tour operators to collect funds for environmental education materials. Sustainable tourism is an approach to tourism, rather than a one-size-fits-all solution.

Distinguishing between sustainable tourism and false advertising, especially when we are booking hotels and tours ahead of time, can be a quagmire of confusion. But there are a variety of certification programs that can help us make wise decisions. Some programs, such as Costa Rica's Certification for Sustainable Tourism, focus on environmental standards. Others, such as Fair Trade in Tourism South Africa, monitor social and economic aspects like wages and working conditions. How do we know which standards a particular program gives priority to? Right now, that's still difficult, but the proposed Sustainable Tourism Stewardship Council would accredit the certification programs, enabling us to expect the same standards from participating certification programs around the world. The most important thing we can do at this point is to ask questions: ask about a hotel's environmental policies; ask tour operators about their involvement with the local community. By being active tourists, we can make businesses understand that we want to be responsible tourists. ZC

Learning Journeys

Travel as education is a time-honored tradition. While it might seem that today we travel more than ever, in many ways our stamina for travel has decreased. We no longer imagine spending decades on the road, as many medieval travelers did. We travel less to learn than we do to get from point A to point B.

The need for travel as learning, however, has not gone away. As John le Carré writes, "The desk is a dangerous place from which to view the world." Since the rise of the "knowledge economy," many of us work in offices, behind desks, and deal with issues with which

An ecotourism sign in the Himalayas outlines proper conduct for hikers and travelers.

we have very little direct, face-to-face experience. This creates a disconnect between our actions and their implications. It makes it difficult for us to understand the true significance and impact of what we do, what our politicians do, and who it all effects. It also means that we have a relatively limited understanding of our planet.

The point of learning journeys is to get out from behind our desks and into the world, to learn about the things that we wish to understand, change, or somehow influence. Regardless of how far we physically travel, the journeys are inevitably about cross-cultural communication. A learning journey is about creating the conditions for honest conversation across barriers.

Mbonise Cultural Concepts

Many tour operators offer visits to "authentic" villages, which turn out to be nothing more than pit stops in rural towns that have been entirely transformed into souvenir shops. Is it possible for us to truly engage in village and rural life in the Global South?

Mbonise Cultural Concepts, based in rural KwaZulu-Natal, South Africa, operates under the principle that we can. *Mbonise* is Zulu for "to show someone." The project's founders, Siphile Mdaka and Roland Vorwerk, shared a desire to offer visitors to KwaZulu-Natal deeper insight into real life in a rural Zulu community. Siphile explains, "This is a homestead, not a lodge or hostel." Visitors stay with families in the community and are guided through nature tours and community dialogues by specially trained hosts, themselves members of the community. The intention is not to simply create another cultural village conforming to tourists' expectations of cultural encounters, but to promote meaningful exchanges that really help us understand the current socioeconomic and cultural reality of a rural community. ZH

VolunTourism

Imagine traveling to a remote part of western Kenya for a safari, during which you assist local villagers, working side-by-side with them, in the construction of a tourist eco-lodge that utilizes gray water and solar power. Future visitors will pay for their accommodations at the lodge, and the villagers can use those funds to invest in a community clinic to improve prenatal care and child health. Or they may build a school to increase access to elementary- and secondary-level education for village children.

VolunTourism is twenty-first-century philanthropy. No longer will we be satisfied to simply mail a check to the headquarters of a multilevel conglomerate nonprofit organization. If we are to change the world and make a significant contribution to the advancement of social capital, we need to experience firsthand the issues and challenges facing the world's communities. VolunTourism gives us a chance to educate ourselves, participate in a travel experience beyond our expectations, and offer our help as volunteers—all in an effort to enhance connections. DC

The Invitation

The invitation is probably the most indispensable tool we have when we visit

nomadic and indigenous populations around the world; stumbling into a community is not a good way to arrive. When no invitation exists, most intercultural encounters serve only to deepen the gap between communities—or go horribly awry when ignorant travelers ignore or disrespect important customs.

The invitation is the first step toward ensuring a mutually beneficial and rewarding engagement. Many communities value their privacy and prefer not to engage with uninvited strangers, who, in the longer view, have included administrators, priests, soldiers, and smugglers—all with their own agendas for the community. If we are interested in a particular community, a first step might be to invite its members into our own community, into our own customs and rituals—even if this is an invitation to share a cup of coffee at a campsite. Suffice it to say, developing sensitivity toward the invitation is critical in building healthy relationships across traditional cultures. ZH

RESOURCE

The Practical Nomad: How to Travel around the World by Edward Hasbrouck (Avalon Travel Publishing, 2004)
One of the best ways to travel and learn about new places is to take a long-haul vacation, including but not limited to around-the-world trips. A veritable bible of world travel, *The Practical Nomad*, now in its third edition, contains invaluable resources on everything ranging from air travel to rail passes to volunteering abroad. Says Hasbrouck, "the biggest mistake you could make, in my opinion, would be not to travel around the world, at least once in your life, if you have the chance." Even if we can't drop everything to troll the globe for six months, *The Practical Nomad* offers tips, resources, and advice sure to benefit the intrepid travelers among us.

We can travel to learn, or we can even volunteer while on holiday, both improving the world and changing ourselves.

Global Culture

Living on a shrinking planet requires, paradoxically enough, broadening our perspectives. If we want to learn to work well with others, and the others with whom we're working come from an increasingly wide array of places and cultures, we need to understand something of their lives in those places and within those cultures. Travel can open new perceptual doors; formal training can equip us with new tools for bridging cultural divides—but we're not all in a position to travel or seek out training. We can, however, broaden our worldviews by watching films, reading books, listening to music, and learning languages from far-off (but rapidly nearing) places.

Global thinking benefits us all, but even more important, opening our mental border can be thrilling. The world is marvelous, complex, and ever changing. To limit ourselves to our own nation's stories, songs, and languages is like going to a huge community potluck and insisting on eating nothing but Jell-O. Life is much more interesting when we sample from all the dishes our neighbors have perfected. AS

Global Film

As the lights dim in the cinema and the curtains draw back, a sense of anonymity and equality is cast over the audience. We lose ourselves in the moment as we connect with the stories on screen—and with one another, regardless of our social differences. Movies are an indispensable tool in this age of rapid globalization: they give us a window into the lives, histories, hopes, and dreams of other cultures. Cheaper tools like digital cameras and consumer editing suites are democratizing the filmmaking process and putting this powerful vehicle in the hands of peoples and nations previously curbed by filmmaking's expense. These bourgeoning film industries are paving the way for a smoother transition into a global economy, while providing a means for individual cultures to retain their identities. MK

Nollywood

As filmmaking grows more democratic, developing nations are entering the global film marketplace with a bold sense of national identity. In just over fifteen years, Nigeria's homegrown film industry, Nollywood, has become one of the largest film/video industries in the world, and it continues to outpace all others in growth. Armed with cheap video cameras and rudimentary scripts, Nigerian filmmakers began committing their long tradition of oral storytelling to video in the early 1990s. The films are made quickly and on the cheap (now on digital video)—usually in less than two weeks and for under $15,000. They're then copied straight to videocassettes and DVDs and sold in open-air markets for about three dollars apiece.

Nollywood films are rooted in traditional folktales and are typically family-based melodramas filled with witchcraft and violence. They represent the first time that African stories have been told onscreen by Africans themselves. Despite Nollywood's trite plots and low production values, Africans can't seem to get enough. The increasing demand across the continent for locally made films, and Nigeria's huge population, have continued to feed the explosive growth of the industry. Nollywood is not only telling new stories, but it's creating good jobs for

Young Buddhist monks browse racks of pirated, imported audio CDs at a music market, Tachilek, Myanmar.

SLIPKNOT
MEGADETH
the second millennium
OzzFest

Nigerians at home and providing African émigrés worldwide with a sense of cultural identity, in the face of rapid globalization.

In recent years, Nollywood has struggled with issues of piracy, political infighting, and the questionable spending of government funding. Still, the young industry is booming and continues to stretch its boundaries. In the vein of the mavericks who first took to the streets with video cameras, some real craftspeople are emerging and taking Nollywood to the next level. Nollywood filmmakers like the famed Tunde Kelani are increasingly touring their films at international festivals. Breaking from the traditional Nollywood mold, the recent *Amazing Grace* (2005), by hot young director Jeta Amata, is a theater-quality 35mm historical epic about the iconic eponymous song—and it's probably the first Nollywood film with North American distribution.

Nollywood films are also becoming increasingly relevant, addressing tough social and political issues. Nigerian filmmakers recently teamed up with artists from the feuding nation Cameroon to produce *Before Sunrise* (2006), which depicts the power of love over hatred and vengeance. Director Fred Amata tells the Nigerian newspaper the *Daily Independent* that the film is his way of calling for an end to the border dispute: "What we have done is to demonstrate that the ordinary people, despite the tension between both countries, are still brothers and [are] collaborating with each other."

Nollywood's success has sparked many other developing African countries to build their own film industries. Kenya's straight-to-video productions, for example, exploded in 2005, only to be slowed down by piracy. Uganda is attempting to jumpstart "Ugandawood."

African filmmakers are proving that what they lack in budget and industry infrastructure they more than make up for in creativity and ingenuity. As these homegrown filmmakers continue to push the boundaries of storytelling and technical quality, more African stories told by Africans will find their way to international audiences. For those of us who want to see Africa through African eyes, that's great news. MK

Atanarjuat ("The Fast Runner")

Director Zacharias Kunuk's *Atanarjuat* (2001)—the first Inuit film ever produced—explores a creation myth from Igloolik ("place of houses") in the eastern Arctic wilderness. Taking place at the dawn of the first millennium, the beautifully produced movie provides insight into the unique world of the Inuits, through the story of Atanarjuat, the Fast Runner. Evil in the form of a mysterious, unknown shaman enters a small nomadic Inuit community and upsets its balance and its spirit of cooperation. The stranger leaves behind a lingering curse of bitterness and discord: after the camp leader Kumaglak is murdered, the new leader Sauri drives his old rival Tulimaq down, through mistreatment and ridicule. Years pass. But when the vengeful Tulimaq has two sons—Amaqjuaq, the Strong One, and Atanarjuat, the Fast Runner—the power begins to shift. As the camp's best hunters, they provoke jealousy and rage in their rival, Oki (Sauri's ill-tempered son). Atanarjuat eventually wins over Oki's promised wife-to-be (the beautiful Atuat), and Oki vows to get even. Egged on by his intimidating father, Oki and his friends plot to murder both brothers while they sleep. They spear Amaqjuaq

and kill him, but Atanarjuat miraculously escapes, running naked for his life across the spring sea ice. Eluding his pursuers with supernatural help, Atanarjuat is hidden and nursed back to health by an old couple—who themselves fled the evil camp years before. After an inner struggle to reclaim his spiritual path, and with the guidance of his elder adviser, Atanarjuat learns to face both natural and supernatural enemies, and heads home to rescue his family. The film pivots around the question of Atanarjuat's choices: will he continue the bloody cycle of revenge, or restore harmony to the community? It's a beautiful film, telling an Inuit story in a traditional Inuit manner, and in the process, opening a new cultural perspective to viewers around the world. The world would be a richer place if every culture produced its own *Fast Runner.* ZH

Global Literature

Reading about places and times distant from our own is one of the great joys of literature. Whether historical, futuristic, or fantastical, a good read transports us, and when we return, we've usually brought a piece of that experience back with us. The first and greatest tool we have for learning about ourselves and the world around us comes from stories. From oral tales to digital graphic novels, stories are the building blocks of culture.

Today, you can find practically any book or author with an online presence by searching under the right keywords on Google, and its "Print" search may even give you a glimpse of the book's inner pages. The "Literature by Country" entry on Wikipedia is a comprehensive list, and a random walk through the "Free Encyclopedia" is always full of surprises.

If you're looking for something specific, try Technorati.com. Using the tag "literature+Africa" as a search, you can find out the interesting novel debuts, and get a good general introduction to what's happening in the Kenyan, Nigerian, Egyptian, or South African literary milieu. You could try this for any region or country in the world. To search the World Wide Web for authors who have been translated into English, try the keywords "in English translation" as a search on Google or Yahoo, and then enter some variations by language or country.

LibraryThing.com

When you visit someone's house, you are sometimes tempted to judge them by what lies on their bookshelves. With friends and people you like, you often find yourself interested in books they are reading, and vice versa. Now you can find people's *bookshelves* online, and upload your own for them to see at LibraryThing.com (there's no charge for the first two hundred titles).

BookCrossing.com

BookCrossing.com is a community of people who stick a paper tag on their books and leave them in public places like parks, bars, cafés, and movie theaters so that strangers can find them and read them, before releasing the book back into the wild. It is possible for a book to travel across the world, and readers can inform one another about its progress online.

Preceding pages: The spread of global culture: people in Yunnan, China, gather around the television.

Opposite: A still from *The Fast Runner*, the world's first Inuit film.

Left: Programs like BookCrossing present a model for new ways of sharing books, films, and music.

Online Audio Books, Readings, and Radio Shows

LibriVox.org is organizing an "acoustical liberation of books in the public domain." You can listen to your heart's content, and you can even volunteer to read a chapter out of your personal favorite. You might lend your voice to Kafka's *The Trial* or to Jules Verne's *A Journey to the Center of the Earth*.

Plenty of other sites now feature online/downloadable audio selections. BBC Radio's features include dramas, readings, and literary conversations that can be streamed on their (largely reliable) player. Garrison Keillor hosts a digital radio show called *Writer's Almanac*. SpaceshipRadio.com podcasts "public domain SciFi radio plays from the '40s and '50s," and also produces its own shows. You can generally subscribe to these shows via Apple iTunes and listen on your iPod or other portable MP3 player.

Online Literary Journals and Anthologies

Arts & Letters Daily is a witty and concise guide to the best essays available online by leading thinkers, critics, and philosophers of the (largely) Western world. Editors Dennis Dutton and Tran Huu Dung post two or three short blurbs daily, in addition to maintaining a directory of notable periodicals and columnists. *The Modern Word* is another impressive Internet literary repository, containing regular columns like Joyous Anarchy by Irish writer Emmet Cole.

There are also a number of bilingual journals online. *The Barcelona Review*, for example, publishes short fiction in Spanish and English, and the French journal *Parachute* features articles in French and English. *New Penguin Parallel Texts* is an anthology of French and Italian short stories with translations in English. And *Sign And Sight* translates ideas and articles from German media into English.

Finally, AlbanianLiterature.com is Robert Elsie's excellent Web site dedicated to Albanian literature that has been translated into English, including modern authors from Kosovo, Macedonia, and Montenegro. Elsie explains in his introductory note why this is important: "The Republic of Albania was kept sealed off for almost half a century, from 1944 to 1990, by its Stalinist leaders, as if it were a planet of its own. Few outsiders were allowed in, and even fewer Albanians were allowed out. For many years, it was politically dangerous for Albanians to learn foreign languages."

Global Music

The term world music was coined in the early 1980s by a small group of influential radio DJs and record-label folks. A lot of great music from all over the world—especially from West Africa—was flooding stores, but they didn't have an established place for it. It didn't fit under "jazz," "rock," "blues," or any other category that the shops used to classify their inventory. So they came up with the classification "world music," and an industry was born.

This term fit for a while, and then it became rather unmanageable. Everything from Indian classical music to Japanese hip-hop was fit into one shelf or category, making "discoverability" very difficult, to say the least. What's more, in the meantime, a new kind of music emerged:

Right: Funky Freddy, lead singer of the Jungle Leaders, records a track criticizing corruption in Sierra Leone.
Opposite, left: Chinese police officers study English in preparation for peacekeeping work abroad.
Opposite, right: English has rapidly become the global language, in many regions changing with and adapting to various cultural circumstances.

modern, transcultural sounds. Indigenous traditional music with deep roots is today being transformed by a new generation raised on electronic music and hip-hop. Megacities crowded with immigrants from a variety of cultures are giving birth to hybrid musical explorations.

Call it modern global music: *modern* because it is often created by youth and incorporates technology in its creation or production; *global* because it mixes elements that might come from anywhere in the world. Whatever you call it, modern global beats and melodies are the soundtrack to which the world is changing. DM

Language

English is the lingua franca of business and scholarship, not to mention the Web and global pop culture. Never before in human history has one tongue been spoken (however imperfectly) by so many. Yet in the coming decades, the linguistic landscape will be increasingly multifaceted. It will be less about learning to speak the Queen's English and much more about being multilingual and mastering a fusion of tongues.

So what will the future speak? By 2050, Mandarin Chinese will hold the top spot. Spanish, Arabic, Hindu/Urdu, and English are likely to be equally ranked. The fastest-growing languages include Bengali, Tamil, and Malay. However, arguably the most interesting development is the growth of new hybrid languages—new argots, pidgin dialects, and creoles. Hundreds of forms of English now exist, each with distinctive grammars and sociocultural contexts. These hybrids, born at the crosshairs of globalization, have emerged either in order to improve our ability to communicate in diverse contexts or to provide a new way to create community.

In Kenya, for instance, Sheng is a Swahili-based patois that borrows from English and the ethnic groups Gky, Luo, and Kamba. It started in the 1970s among the urban youth in Nairobi's ghettos, but has since moved up the social ladder: Kenya's political classes now use it as well. Sheng is also spreading to Uganda and Tanzania, accelerated in part by popular East African hip-hop artists such as Kalamashaka and Nonini.

Hinglish, a mixture of Hindi and English, is arguably even more important. After becoming the preferred language of many Bollywood films (and even used in English hits such as *Bend It Like Beckham*), Hinglish has become hip and respectable, uniting many different groups of Indians through a lingo they feel is their own and not just the elite's. There are already 350 million Indians who speak Hinglish as a second language, exceeding the number of native English speakers in Britain and the United States combined. Some of the Hinglish words in vogue include *airdash* ("travel by air"), *chaddis* ("underpants"), *chai* ("Indian tea"), *dacoit* ("thief"), *desi* ("local"), *dicky* ("boot"), *lumpen* ("thug"), and *would-be* ("fiancé or fiancée").

Some experts argue that Hinglish words are destined to become part of the global vernacular, especially given the many Web-savvy Indians working in high-tech centers around the world. But Hinglish will have some competitors on this front, especially from Globish—a simplified form of English most nonnative speakers use

to communicate with each other when traveling or doing business. "It is not a language, it is a tool," say Jean-Paul Nerrière, a Frenchman and retired IBM executive who is trying to codify and develop Globish as a practical means for global communication. Consider it "English lite" with a limited vocabulary (about 1,500 words), basic structure, and gestures.

Of course, our cultural custodians decry this trend as the debasement of a language's integrity. And their concerns have some merit. However, while linguistic fusion may be a recipe for confusion, these new patterns will also inspire new poetry, art, and creative insight in the world. Overall, multilingualism, like multiculturalism, is good news for the people and the planet. Not only does learning another language lubricate trade and economic growth, but more important, it gives us a richer frame of reference for seeing the world and its cultures. This is essential if we are going to make a diverse society work effectively. Our protean ability to develop new languages is something to celebrate, because it is deeply human—something that enables community, expression, and empowerment during changing times. Many societies are already working this way with some success. In Europe, for instance, an entire generation—the "E-generation"—has grown up being multilingual, thanks to European unification, a model to watch and emulate for the future.

Without question, English will remain the most important language for many years to come, especially in science and on the Web. Yet at the same time, English's preeminence will disguise a more complex picture under the surface of our diversifying and globalizing world. So if you're wavering about taking that Mandarin or Spanish class, don't dilly-dally—or "time-pass," in Hinglish—any longer. If you have the opportunity to learn Urdu from your neighbor, go for it. Or if you're ambivalent about your own multicultural heritage (because it's hard to bridge these worlds), know that this experience is a true asset and not a liability. The future is multilingual—and mighty the mongrel tongues will be. NAB

BUSINESS

Business can be a vehicle for change. Almost nothing we do in our daily lives remains untouched by business; markets and corporations are an inarguable fact of twenty-first-century life. But a lot about the way we currently structure markets and run corporations is not quite right. The companies for whom we work, the things we make, the products we buy—a great many have a role in promoting inequality and injustice, in polluting the planet, in weakening the social fabric. Few of us want this to be the case, but most of us feel powerless to make a difference in the face of large and complex institutions.

Business doesn't have to be destructive. At its core, business is about livelihoods and service: providing for our needs by providing what others need. Increasingly, all sorts of people, from CEOs and economists to consumers and small investors, are realizing that we can remake business to truly serve the public good—and make a lot of money in the process. We can build businesses that embrace sustainability, openness, and fairness not as a sideline ethical consideration, but as the path to profits. Indeed, millions of people are involved in efforts to capture the profit that's available through healing the planet.

The relative success of those green entrepreneurs has a great bearing on our future. As we begin to perceive the scope of our global challenges, we also begin to see that no societal force other than business is capable of delivering the magnitude of economic change we need in the limited time we have. If we want clean power, green buildings, and sustainably designed products, we must have businesses capable of delivering them at a healthy profit. If we want jobs and opportunity for the exploding population of young people around the world, we must have businesses capable of employing their talents. If we want to share ideas, open up innovation, and reform politics to promote transparency and democracy, we must have businesses that see these things as advantages, not impediments. Only when companies begin to operate with the knowledge that creating a better world is a profitable venture can we really move forward.

The measure of a successful business doesn't just boil down to overheads and bottom lines, or even living wages and solar-powered office buildings. Today's successful business is about exploring new ways of making things happen in the world, seeing potential and pursuing it, and recognizing that now—perhaps more than ever—the market is receptive to thinking outside the box.

In some ways, business is the boldest and most exhilarating adventure we could embark on. But it's up to us to keep it fresh, or we risk falling into the groove of business as usual, spreading poverty and pollution, maybe getting a little richer but ultimately feeling sordid and hollow. Money made this way is an ethical cheat, often producing nothing of real value, and sometimes doing real harm. It's also boring: nothing is more tedious than rehashing old business models, paddling around in the backwaters, looking to squeeze out just a little more profit.

If we can avoid that tired groove, we're likely to find the path toward world-changing business, full of men and women who see sustainability, fairness, and openness as prerequisites not just for a better company, but for a better life. These fine minds see a chance to do something extraordinary, to improve the state of the world, and just maybe, in the process, to get reasonably rich. This is the new frontier of business, full of innovative technologies, radically reimagined business models, counterintuitive opportunities—there's room for vision here, and the visionaries see a bright green future. AS

AVEDA
AVEDA
Short for time?

AVEDA
air care
travel light
AVEDA

Your Money

Sooner or later, it happens to the best of us. We've been rolling along in life, minding our own business, and then, one day, we open our bankbook and find it staring us in the face: a surplus. A little bit left over at the end. Capital.

Seeing that extra money, and being reasonably clued-in, we know we should do something with it—put it away somewhere where it'll earn us a little more.

From that moment on, we become involved in a complex web of choices and decisions that will not only define us as people but, in the bigger picture, help determine the kind of future we'll have. We are now investors, and investment is one of the levers that moves the world.

The principle of investment is simple: we lend our money to others; they use it to try to make a profit; if they succeed, they give us back more money than we gave them. That part's pretty straightforward. Gather money, lend it to the right people, let them do their thing with it, sit back and be patient, and we always end up older and, usually, richer.

The problem comes when we stop and wonder, "What are these people doing with my money?"—because the answer may not be very pretty. We can be upstanding members of our community who never miss a chance to help out others—loved by all—and suddenly we discover that our money is out there funding oil dictatorships, sweatshops, clear-cuts, and companies trying to sell tobacco to fourteen-year-olds. Our money has become our evil twin, busily out wrecking the world while we go around trying to save and enjoy it.

Until recently, this was seen as one of those unavoidable trade-offs: you could do well or do good—pick one. But those days are over. Increasingly, with a little smarts you can put your money where it will not only make you a tidy little bundle, but also make the world a better place. GF

Socially Responsible Investment

How do we invest our money so that it helps improve the world and simultaneously grows? That's the question the Socially Responsible Investment (SRI) movement is trying to answer.

At first, the SRI movement employed a simple set of filters to screen out companies whose practices shareholders considered unethical: this investment fund would not buy shares in any company that sold tobacco; that portfolio avoided corporations involved in the arms trade. Eventually those filters grew to overlap, and to embrace the idea that such investments also needed to include reduced risk (less chance that we'd lose our money) and increased returns (more money back on what we'd put in).

As it turns out, increasing our money and being socially responsible are not mutually exclusive. There are investment options that will make us more money than investing responsibly will, but investing responsibly has resulted in better-than-average returns for years now. Geoffrey Heal reported in the *Financial Times* (July 2, 2001), "performance on environmental and human rights criteria is a good predictor of the overall financial performance of companies." Bottom line: When we invest, putting our money into an SRI fund should be our first priority. GF

Using Investments to Change the World

Investing can make us money, but investing responsibly can do that and more: it can help us change the world.

Though SRI began with "negative screens"—excluding companies that engaged in dubious activities—socially responsible capital is now a power in the economic landscape, and investors are demanding that this power be used to not only avoid the bad, but to create the good. Almost a tenth of all the money invested in the United States is being managed with some socially responsible criteria—and that makes for a big lever.

Increasingly, SRI managers and investors are insisting that if companies want their money, they have to stop engaging in destructive activities, and start proactively pursuing business paths that will lead to a better world. To these investors, it's not enough for a company to avoid putting money behind Nigerian oil fields, for instance; the company must also actively invest in wind power. These strategies are already proving to be profitable, as a prizewinning economic study from the UC Berkeley Haas School of Business reveals: "Company managers do not face a tradeoff between eco-efficiency and financial performance ... Investors can use environmental information for investment decisions" (Guenster et al. 2005).

Savvy investors are doing even more than influencing the way companies invest: they are changing the companies' very management, through "shareholder activism." Using their ability to vote for and influence the boards of large companies, shareholders are demanding that management either behave more responsibly or be replaced. This is forcing long-neglected issues—from racial and gender equality to greenhouse-gas policies—onto the agenda of corporate boards.

Finally, SRI funds are beginning to invest in new ways, putting their money to work to strengthen communities that have been left behind. Just as microcredit [see Microfinance, p. 346] has proven to be phenomenally successful both at repaying loan monies and at raising people out of poverty, money invested in intelligent ways in low-income communities tends to pay back *and* make transformative change possible. Skeptical? Consider this: between 2003 and 2005, community-investment assets grew by 40 percent. GF

Transparency in Business

How do we know if a company is behaving responsibly? If it is secretive, holds closed meetings, and carefully guards its books, we can't really know. Which is why one of the hottest trends in investing today is to demand corporate transparency.

And here's the twist: it turns out that a company with transparency, accountability, and social responsibility standards not only makes investors more money, but also makes the company itself more money. This is partly because efficiency, good working conditions, and simple respect, which some consider luxuries that hurt short-term bottom lines, tend to pay back handsomely. Companies with the guts (or the shareholder pressure) to make investments in sounder practices often do better than competitors whose focus on short-term gains clouds their ability to see potential long-range benefits. It's the ones with the long view who can best find and implement real business innovation and responsible practices, which in turn allows them the comfort of being open and honest about their operations.

But it's not all about using compact fluorescents and offering child care. Socially responsible businesses are also run better *as businesses*. It turns out that financial transparency (being honest and open with your books) and corporate democracy (allowing shareholders and even

Preceding pages: Aveda has built its success using business practices that protect the earth, conserve resources, and contribute to a sustainable economy.

Right: The Enron scandal illustrated just how costly secretive and corrupt business practices can be.

employees a strong voice in decision making) are excellent predictors of both corporate profitability and shareholder return. For instance, researchers from Harvard and the Wharton School of Business found in a study published in 2003 that among 1,500 companies studied, those with the most robust shareholder-rights policies delivered 8.5 percent better returns than those with the weakest. Companies make better decisions when those decisions are made in the open and when the decision makers are held accountable. Open books, timely feedback, and employee engagement can actually make a company smarter.

The movement toward transparency is growing rapidly—shareholder pressure is one reason for that; fear is another. Corporate crime and mismanagement (think Enron, WorldCom, Halliburton) are driving many large institutional investors into supporting responsible practices. "Corporate irresponsibility," says analyst Cliff Feigenbaum in *Barron's*, "did for social investing what Watergate did for politics" (Blumenthal 2003).

The World Is Becoming Uninsurable

Reinsurance companies offer insurance for insurers. That is, when we buy insurance policies, the companies we buy from also take out policies of their own, protecting themselves in case of some unforeseen disaster.

There's only one problem: the world is becoming uninsurable. Poverty-fueled political instability, terrorism, piracy, and the threat of pandemic disease all raise the possibility of disaster too great to guard against. But the mother of all reinsurance nightmares is climate change.

Rising seas, catastrophic storms, prolonged droughts, crop failures, floods: the list of global warming–driven catastrophes seems to grow daily, and every time catastrophe hits, some insurance company somewhere is on the hook for a lot of money—more than $200 billion in 2005 alone (Environment News Service, December 7, 2005). Increasingly, the industry (supported by their armies of bean counters) is coming to the conclusion that, never mind scientific certainty, they can't afford even the *risk* of significant climate change.

As a result, huge reinsurance companies like Swiss Re and Munich Re Group are taking bold stands on the issue. Both companies are advocating openly for tighter regulations on greenhouse gases and more investment in clean-energy technologies. More pointedly, in late 2004 Swiss Re served notice to its clients that it might cease writing "directors and officers" insurance for companies that did not have credible programs to reduce greenhouse gases. Guess which gets more attention in corporate boardrooms: regulatory requirements or the prospect of losing the corporate veil against personal liability?

At the same time, some of the world's largest investment banks are finding opportunity where Big Oil offers nothing but denials. Billions of dollars in clean tech investment (the term for renewable energy and environmental technologies) are coming on line—to the tune of 7 percent of all venture funding, by one recent estimate in the *GreenMoney Journal* (de Callejon, Donohue, and Day 2006)—and savvy investors and venture capitalists see the potential to revolutionize an industry. One notable example: Goldman Sachs has pledged to make $1 billion available for investments in renewable energy, as part of a comprehensive "environmental policy framework" that outlines both the business risks of current energy policy and the growth opportunities in clean tech. GF

How to Pick the Winners

So how should you decide where to put *your* money? If you're betting not only on your future security, but on the future of the planet, how do you decide where to put your chips?

We're not money managers or investment advisers, and we're not about to try to pick specific winners in a field that's moving this fast—certainly not in the slow rhythms of a book. But following are some questions you can ask yourself that will lend a worldchanging perspective to your decision-making process:

- Does the industry or company in question fundamentally contribute to or reduce the world's problems? Would the answer change if it did its thing in a more efficient or a very different way? (For example, we might go short on mining and other extractive industries, but long on mining companies that shift their focus from extraction to metals management; or avoid companies with big investments in carbon-based fuels, but pick up a range of clean tech companies.)

- Will the company suffer or benefit from a bright green future? (Energy-intensive industries, for example, face structural challenges as energy prices rise, making their prospects unpredictable.)
- Can the company adapt quickly to unexpected change? Putting aside technology and strategy briefly, what is your take on the company's leadership, management, and culture? Is the leadership trustworthy? Thinking ahead? Open in its decision making?

No company is perfect, so you have to balance the good and the bad, make sure you're comparing apples to apples, read the data, and ultimately trust your gut. As always, be clear on what truly matters: get advice; diversify. (This is also a pretty good strategy when choosing which companies you want to work for. If you wouldn't trust a company with your retirement portfolio, you certainly shouldn't trust it with your career.) GF

RESOURCES

The SRI Advantage: Why Socially Responsible Investing Has Outperformed Financially
by Peter Camejo (New Society Publishers, 2002)
Founder of the first environmentally screened fund on Wall Street, Peter Camejo aims to dispel the myth that investing in socially responsible corporations will yield smaller returns than investing in their less-than-responsible counterparts will. "It will come as quite a surprise that [investors] are sacrificing performance to invest in companies that pollute, violate laws and discriminate." In essence, SRI strategies screen out companies that are in conflict with public opinion, and they tend to provide higher relative returns—something that all investors are interested in. Whether you're managing a large foundation or looking to invest your own money, *The SRI Advantage* will serve your needs: it's full of solid evidence and further resources, driving home the point that your money and your ethics don't have to diverge, and that they can, in fact, support one another in the long run.

Healthy Money, Healthy Planet: Developing Sustainability Through New Money Systems
by Deidre Kent (Craig Potton Publishing, 2005)
"Just as we design houses, and those houses come to shape our lifestyles, so we create money systems, and money systems come to shape our lifestyles." Indicting our current monetary system for failing to provide decent global living standards, Kent advocates for a more balanced financial system—what she calls "healthy money." Given the complete interconnectedness of our social, environmental, and economic problems, changing our current money system is the first step toward creating a more balanced and sustainable world—and it's only through monetary literacy that we can change for the better.

Creating Business Value from Sustainability

A "healthy" company can't last long in an unhealthy environment. True, companies can turn a profit while polluting the air and water, despoiling the landscape, and consuming natural resources with abandon, but they won't be able to go on like that forever. Environmental and social issues are roiling the world of business, causing companies to rethink their products and services, their operations, even their business models. Amid the changes, adaptive companies of all sizes and sectors are finding opportunities to be more competitive and profitable by aligning sustainability goals with their core business strategies. Many companies are already learning the price of operating in places with severe environmental problems: increased regulation, more expensive natural resources, greater difficulty attracting and retaining talent.

The world of commerce relies heavily on a healthy natural environment. Nature's systems provide a wealth of tangible and intangible services to business—some $33 trillion worth of "free" deliverables a year, say experts in the "Millennium Ecosytem Assessment Synthesis Report." Those services include fertile soil, fresh water, breathable air, pollination, species habitat, soil formation, pest control, a livable climate, and a host of other things we tend to take for granted.

None of these services appear on companies' balance sheets, but their availability, or lack thereof, can have a dramatic impact on a company's finances. For example, to prepare for the inevitable regulation of carbon dioxide emissions, many companies are investing in new, more efficient technologies. When nature's services—in this case, the planet's ability to dissipate industrial emissions and keep surface temperatures at a livable level—are overtaxed, this is the cost to business. JM

The Company-level View

The definition of "sustainable business" is rapidly changing. As recently as the early 1990s, being environmentally responsible meant looking beyond what the law required to do a few simple things—say, recycling office paper or increasing the use of recycled materials in products or packaging. By the end of the 1990s, companies were taking a broader view, using energy, water, and materials more efficiently; preventing pollution through improved management controls; and taking other measures that reduced emissions and saved money.

Now, the leading edge of sustainable business practices has moved beyond *saving* money to *making* money. Sustainability is now seen by many leading companies as a means of creating new business value: increasing sales, creating innovative products and services, and expanding markets. Sustainability is also seen

as a way to improve product and service quality, reduce risk, attract and retain employees, and enhance brand value and customer loyalty.

Not every green or sustainably minded effort yields companies the full complement of benefits, of course. But increasingly, the case for sustainable business has become clearer. Both societal and ecological trends suggest that sustainability's benefits to business will only increase.

Concerns about reliable and affordable energy, water, and other natural resources—and the increased likelihood that emission of greenhouse gases will be regulated—are just a few of the realities facing business leaders today. Institutional investors and large insurers [see Your Money, p. 382] are signaling their concerns about the risks and liabilities of companies that don't effectively manage their wastes, resource use, and emissions. And customers of all types—governments, businesses, and individuals—are increasingly buying from companies that show a commitment to sustainability, or are avoiding companies they believe to be shirking their responsibility.

Our expectations of companies will grow as we continue to think greener, and as current events—record heat waves, catastrophic storms, energy disruptions and price spikes, global insecurity—reveal the dangers of ignoring sustainability. Companies that already have started to integrate sustainability into their organizational fabric will be better prepared than their competitors to ride the waves of change and to prosper in tomorrow's markets. JM

Efficiency and clean practices are good for the planet and the bottom line and, sometimes, required by law. Foreseeing increased regulation, the Tennessee Valley Authority installed "scrubbers" on six of its energy plants (opposite). Metalworks, Inc., in Michigan has saved money by cutting 16 million gallons of water use per year (right).

What Can I Do in My Company?

Nearly every choice you and your colleagues make affects your company's environmental performance—and, potentially, its reputation and financial performance. The materials and equipment you purchase, the way you design and deliver products and services, the way you build and maintain facilities, and the relationships you—as a representative of your company—have with stakeholders can all contribute to greater business sustainability. There's no recipe for making a business more sustainable. The ingredients depend on a range of factors, including a company's size, sector, and geography. But in taking steps toward sustainability in your company, you can apply the following general principles of conduct:

Obey the law. Comply with local, state, and federal environmental, health, and safety regulations. This gives beyond-compliance efforts a solid foundation. You may gain additional benefits: some governmental programs reward fully compliant companies with reduced oversight and paperwork.

Understand how your business affects the environment. From the things you buy to your relationships with customers and suppliers, to the full life cycle of your products and services—each step of the way, you can make choices that will go toward aligning environmental responsibility with business success.

Begin to make changes where they can be made profitably. If you can't achieve a profit

immediately, you can at least make changes in a way that will not decrease profits and productivity for more than a short period. It's important to keep in mind that it isn't possible to do everything right; gradual, incremental progress is a worthy goal.

Measure and track your waste. Watch what your company consumes—raw materials, energy, supplies—and what it wastes—raw materials, energy, packaging, emissions. Try to measure and quantify this waste: how much do you spend to purchase, handle, store, and dispose of the wasted material? Your audit may be as simple as counting or weighing the trash bags your company disposes of on a weekly or monthly basis, or checking energy bills regularly.

Draft an environmental-vision statement. It's easier to get behind a vision when all of the players know what the company stands for. Having an established point of departure will show customers, stakeholders, and your community that your business is invested in the environment.

Rally the troops. Bring together a team of employees to promote sustainability in the workplace. Team members can head up the effort to purchase recycled products, educate coworkers on environmental issues, and track environmental accounting for their departments. Consider creating incentives such as rewards and recognition for employees who drive your company's environmental efforts. Name a periodic "green champion" in order to single out employees' environmental contributions. JM

The Natural Step

You don't need an advanced degree to understand that we can't indefinitely extract material from the earth and turn it into waste faster than the earth's systems can naturally reabsorb it. Such an imbalance causes increasing disorder that's manifested in environmental, social, and political discord. Upon this basic principle, and in response to growing concern, a group of Swedish doctors and scientists developed the Natural Step, a framework for making business and organizational operations more sustainable. The Natural Step offers a simple point of reference for prioritizing sustainable practice while also succeeding in business.

Grounded in science, the principles are straightforward and logical, making them effective tools for bringing companies and countries together around the ideas of sustainability. In the Natural Step's framework for addressing environmental issues, consensus is key. Agreement among all players leads to the creation of better benchmarks, comprehensible measurements of progress, and a clearer structure for planning future development.

Time and again, companies (Nike, Interface, and IKEA, to name just a few) have found that embracing the Natural Step puts them ahead of regulation. The benefits include greater efficiency, improved brand image, more effective customer relations, and ultimately, profit.

When companies see sustainable practices as inherently profitable, they can't help but be enticed. The principles of the Natural Step can help to clarify this connection, and can prove to be an indispensable tool in a company's toolbox as it seeks both patent bottom-line profit and the more intangible benefits of a healthier ecosystem and a better-supported community. NA

RESOURCES

Dancing with the Tiger: Learning Sustainability Step by Natural Step by Brian Nattrass and Mary Altomare (New Society Publishers, 2002) Coauthors previously of *The Natural Step for Business,* Nattrass and Altomare have reemerged with this important book revealing the real-life scenarios behind the approach that today's competitive corporations (among them Nike and Starbucks) are taking to establish sustainable business practices. Contending that companies must apply a holistic approach in their efforts toward sustainability, the authors argue, "it is only through people—through the heart, mind, and will of each individual—that the innovations of sustainability will be diffused and adopted within our corporations, our governments, and ultimately our world." If more major corporations followed the models presented in *Dancing with the Tiger*, the business world as we know it would be on a straight path toward sustainability on all fronts.

Capitalism at the Crossroads: The Unlimited Business Opportunities in Solving the World's Most Difficult Problems by Stuart L. Hart (Wharton School Publishing, 2005)
Stuart Hart makes it clear that there is no inherent conflict between creating a better world and achieving economic prosperity. Global capitalism is at a crossroads, he argues, and "corporations are the only entities in the world today with the technology, resources, capacity, and global reach required" to create a more sustainable world. Drawing on his consulting experience with top companies and nongovernmental organizations worldwide, Hart contends that to truly change the world, we need to focus on the bottom of the pyramid, on the four billion people with disposable incomes of less than ten dollars a year—who represent a $4 trillion marketplace.

GREEN MARKETING

How does a company become recognized in the marketplace as an environmental leader? The answer isn't a simple one. Although more than a decade's worth of surveys, focus groups, and studies indicate consumers' willingness to "buy green," consumers' actual purchasing patterns do not reflect their words.

The reasons for this gulf are many and varied. Some companies have been less than forthright about what makes their products green, making it hard for consumers to recognize ecofriendly products when they see them. Most consumers don't adequately understand the environmental impacts of their purchases anyway, and many companies have been reluctant to promote their environmental good deeds or the environmental improvements in their products for fear that their lack of perfection will lead them to be criticized or accused of "greenwashing" **[see Consuming Responsibly, p. 38]****.**

A growing number of companies really do have good environmental stories to tell. But even a pristine ecological track record may not bring in substantial consumer demand. Companies who understand this instead tell a story about quality. The green aspects can be the moral of the story, so to speak. If a company can gain customer loyalty based on a top-quality product, then it has found an entry point to convince consumers of the added value of environmentally sound goods, all while increasing sales.

Electrolux: Efficiency Equals Green

Electrolux is the world's largest manufacturer of "white goods"—refrigerators, washers, dryers, and other large household appliances. The company views environmental sustainability as central to its business strategy and as a source of competitive advantage. Sustainability drives new product design at Electrolux, resulting in some of the most water- and energy-efficient appliances on the market, including a solar-powered lawn mower and a dishwasher that uses only four gallons of water per cycle, versus the ten and a half gallons used by conventional machines.

Despite these efforts, Electrolux doesn't go out of its way to market its products as environmentally sound; instead, it focuses on promoting its products' energy efficiency.

Why not promote the green in the white goods? Companies like Electrolux can get frustrated by consumers' reluctance to embrace their efforts toward aligning environmental sustainability with business success. For example, Electrolux piloted an initiative in Sweden in which consumers were given a washing machine (for a small installation fee), then charged on a per-basis use—ten Swedish kronor (about one U.S. dollar) per load. One objective was that the consumers didn't have to worry about the appliances breaking down or becoming less efficient over time; they could rely on Electrolux to keep the machines in good working order, thereby minimizing their home energy and water usage.

The program was met with a decisive yawn by consumers, who apparently didn't want to change the way they paid for doing the laundry. The company believes the experiment may have been doomed by flawed methodology, and hopes to reintroduce it some day with better success.

Despite that program's disappointing outcome, Electrolux continues to be a leader in sustainable manufacturing. Here are some lessons that companies can learn from Electrolux:

- Demonstrate that environmental initiatives lead to better-quality products.
- Become known as a thought leader in the scientific and technology communities.
- Be ready to experiment publicly, even knowing it won't always lead to success. JM

Philips: Durability Trumps Green

Netherlands-based Philips is considered one of Europe's leading proponents of green marketing and design. Philips believes that ecodesign principles can enhance business, and that ecodesign is not chiefly a technical activity but a concept to be embedded in the business value chain.

Philips' flagship environmental consumer product is the compact fluorescent lightbulb (CFL), which it has marketed since 1978. For years, CFLs, which save more energy and last longer than standard bulbs, languished in the U.S. market—despite their success in Europe, where consumers pay much higher energy costs. (Holland's penetration rate for CFLs is around 50 percent, compared with the United States' rate of less than 10 percent.) Why didn't CFLs take off in the States? Among other reasons, U.S. consumers didn't care for the quality of their light output, and the bulbs didn't fit many existing lighting fixtures.

But things changed as the quality of CFLs' light improved, as they were adapted to fit a wider variety of fixtures, and as they

got cheaper. Equally important was the fact that Philips changed the bulbs' name from Earth Light to Marathon.

Market research at Philips revealed that the environment wasn't U.S. consumers' primary concern—in fact, it was their number-four or -five purchase criterion. Their number-one criterion? Longevity. And CFLs last longer.

Something else came out of the research, and became a key driver behind the company's subsequent marketing efforts—it seemed that consumers were more willing to buy green products when they were bundled with other benefits. Researchers at Philips found that linking such environmental boons as energy reduction, materials' reduction, and toxic-substance reduction with various material, immaterial, and emotive perks (for example, the bulbs' lower lifetime cost; the convenience of changing bulbs less frequently; the good feeling of doing the right thing) raised consumer interest to 60 percent or higher. In other words, 60 percent of the U.S. population would be willing to buy a CFL packaged with all those attributes—and that figure includes consumers who may think negatively of environmentalism.

Since the name change and the bundling of these benefits, U.S. sales growth of Philips CFLs (which started out at essentially nil) has been growing by 12 percent or more a year. Companies inspired by this success can learn the following lessons from Philips:

- Price environmentally sound products comparably to conventional ones.
- Link environmental innovations to other benefits, like quality and durability.
- Brand green product lines with names that emphasize nonenvironmental benefits. JM

Toyota: Green without Compromise

Prius, Toyota's celebrated gas-electric hybrid, may be the first major consumer product to fit practically all of the major criteria for success in the green-consumer marketplace: it comes from a trusted company and can be bought wherever the company's products are sold; it looks and feels like a "conventional" product and doesn't require consumers to change their use habits; it is (almost) comparably priced with similar cars; it can save consumers money to operate; and it is significantly greener than its conventional competition. In short, it represents green without compromise.

When the Prius was launched in the U.S. market in 2000, Toyota did not play up its environmental attributes, but placed emphasis on its ability to save gas and money. Those early marketing efforts were also principally aimed at early adopters—the technology buffs that want the latest, coolest thing. The product's tagline was "Prius/genius," spotlighting the car's smart, hip technology. The first two thousand or so vehicles were sold online—a key medium for early adopters.

But in the years since the Prius's release, Toyota has advanced an increasingly green message—and a key ally in this is the celebrity vote. Cameron Diaz, for example, spoke about her new Prius on *The Tonight Show.* Other celebs have followed suit, deepening the Prius's cool factor, and ultimately helping the car

Opposite: Philips originally marketed its compact fluorescent bulb as the Earth Light, stressing that it was environmentally friendly, but later re-branded it the "Marathon" bulb, emphasizing savings and longevity.

Right: Actor Leonardo DiCaprio is among a number of celebrities who've taken a high-profile stand in favor of hybrid cars, becoming a great advocate of alternative transportation and the environment.

break out of the purely green marketplace—the ultimate success for any green-minded product.

Companies can learn the following lessons from Toyota's success with the Prius:

- Aim to produce green products that are no different than their "regular" counterparts in branding, price, use, or performance.
- Emphasize the cool factor behind environmental innovations whenever possible.
- Seed innovative new products among celebrities—actors, musicians, athletes, and other trendsetters—to give them even more of a cool factor. JM

RESOURCES

Green Marketing: Opportunities for Innovation in the New Marketing Age by Jacquelyn A. Ottman (BookSurge Publishing, 2004)
Ottman, one of the pioneers of green marketing, offers in her book insights and inspiration on environmental issues' impact on the ways consumers choose, buy, and use products and services. She describes how simply "greening" conventional marketing doesn't work and demonstrates how businesses that take the lead now, while industry standards and consumer expectations are still forming, will gain a competitive edge in the market for environmental products and services. The book offers dozens of case histories to show the opportunities—and the slippery slope—behind green marketing. Featured companies include such unlikely "green marketers" as Ethan Allen, John Deere, and Xerox.

According to Ottman, "Finding solutions to environmental degradation involves much more than replacing one supermarket cartful of goods with another. That is because our present modes of production and consumption are simply not sustainable—'sustainability' is defined as meeting the needs of the present without compromising the ability of future generations to meet their own needs—in the face of a global population that is 5.8 billion today and expected to reach 10 billion by 2040. Some experts go so far as to estimate that achieving sustainability over the next few decades requires a radical change in the whole production and consumption in industrial societies—a 'system discontinuity,' characterized by a 90% reduction in the consumption of environmental resources."

Sustainable Marketing: Managerial-Ecological Issues by Donald A. Fuller
(Sage Publications, 1999)
In *Sustainable Marketing,* Fuller uses the "4 Ps" of marketing (product, price, place, and promotion) to show how companies can reinvent strategy and craft "win-win-win" solutions that allow customers to obtain genuine benefits, organizations to achieve financial objectives, and ecosystems to be preserved or enhanced. He advocates the conversion of consumption systems to a sustainable paradigm that represents a circular use of resources, not the linear approach (materials > products > consumption > disposal) that leads to the depletion of ecosystems. Fuller takes green marketing well beyond feel-good sales strategies to show how companies are reinvigorating and reinventing themselves around ecological principles and themes.

BRANDS

Brands are the clothing a company wears—a thoughtfully selected set of signals that tells us what kind of values and style a company has, whose attention it's trying to get, and how it fits into its cultural context. The decision to buy one product over another often happens before a consumer has had a chance to try that product out, which means the most attractive brand wins the most customers.

A well-dressed product stands out. In the early twentieth century, as choices increased in the marketplace, brands became the most powerful mode of communication for companies vying for customer loyalty. As branding evolved, companies realized that the best way to capture a buyer was by aiming not at the head, but at the heart. We'd rather see a cute kid on a cereal box than a corporate logo. Adding an element of emotional connection to a brand usually creates the hook that keeps people loyal to a particular brand. But needless to say, emotional appeal can be deceptive. Attractive brand identities too often veil companies' corrupt or unjust behavior.

If the twentieth century was about corporations learning how to use their brand identity to influence consumer choice—and sometimes snow their audience—the twenty-first century is shaping up to be about consumers learning how brand strategies work, and using their awareness to put a stop to bad corporate behavior—promoting honest business practices and positive change.

Over the last few decades, brand antagonists have warned that culture is fast becoming something that happens to us, rendering us helpless and unable to take control over our own lives. The fruit of a brand is blind desire, and there is more than enough of that to go around. Brands like Nike, McDonald's, Starbucks, Wal-Mart, and their cohorts have built their success on the unfair treatment of domestic and overseas laborers, all the while luring loyal consumers through advertising that strikes an emotional chord.

Anticorporate sentiments are not a recent development, but escalating tensions over globalization and the carefully calculated facades of multinational corporations have spurred activism unique to today's consumer culture. Increased consumer consciousness has exposed many a dishonest corporation, and tactics such as antibranding and culture jamming [see Protest, p. 452] (turning an institution's own mass media into a vehicle for protest) have forced many businesses into transparency.

But here's the thing: brands live on. It's not that the antibranding movement lacks merit; on the contrary, it's a necessary step in the process of recognizing the awesome power of brand identities. Huge corporations like Nike were dragged into transparency and transformation by pressure from consumer boycotts, and now they stand very publicly for social and environmental responsibility. Branding itself changed as corporations recognized that consumers want to feel an authentic and emotional connection to the products that furnish their material world.

But boycotts and renunciation don't themselves create new ways of doing business. Rather than rejecting the tools that build brands, we need to use them to produce different results. It is possible to create a brand that connects to its customers not by turning them into obedient inferiors who seek happiness in products, but by empowering them as co-creators (or at the very least, free agents) in establishing positive identities that fall in line with responsible corporate principles. Branding can be used not only to sell desire, but to create mental models for change. SR

Understanding Branding: Brands That Do It Differently

Plenty of well-intentioned entrepreneurs start businesses that make a point of maintaining a high degree of integrity even as they grow. Some succeed and some—as they scale up their manufacturing—end up sacrificing responsibility in the interest of profit. The brands that manage to stand by their founding message are usually those that build their identity around their commitment to subverting the paradigm and doing business differently.

Muji

The Japanese company Mujirushi Ryohin, meaning "No brand, good product," is better known as Muji. Often described to Westerners as a merging of IKEA and Target, Muji sells products for all aspects of life, out of large retail spaces in urban areas. But unlike so many other companies that fit this profile, Muji makes products that feature no distinguishing markings, logos, or trademarks. Muji is a brandless brand.

Of course, simply by virtue of labeling itself as an "unbrand," Muji is a highly unique and recognizable company. But it makes a point and stands by it: customers don't need bells and whistles; they don't need to form personal identities based on the things they own; they simply need useful, well-made, efficient products—and that is just what Muji gives them. The company also makes a point of utilizing recycled materials whenever possible, and reducing packaging to what's absolutely necessary. Muji thus satisfies customers' increasing desire to avoid aligning themselves with questionable brands, while meeting the growing demand for environmental consideration in manufacturing. SR

Blackspot

Shoes symbolize a lot about what is wrong with consumer culture these days, what with Nike's notoriously questionable treatment of overseas laborers in the early 1990s, the resulting consumer boycotts, and the scads of other companies that have done their best to distinguish themselves as "better." Of course, as a result of the dramatic response from shoe dissidents, Nike may now have one of the cleanest acts around. But it's hard to know what's real from such a mammoth brand.

Amid all of this footwear ambiguity, one shoe brand came along and spelled things out in black and white: Blackspot. A product created by antibranding advocates at *Adbusters* magazine, Blackspot is a plain black canvas shoe with a white spot where the logo would otherwise go. The creators of the minimalist shoes—manufactured in a family-run Portuguese factory—claim an ethical foundation superior to that of other shoemakers, aiming to "create a cool that is more powerful, and [to] cut into the market share of Nike." It might be more of a paper cut than anything, but the undertaking is indicative of growing consumer sentiment that is at odds with the omnipotence of brands like Nike. Blackspot has found, as Muji has, that simply creating a "logo-free" brand can itself be a powerful branding tactic.

If you're going to make a product, you need to sell it, so you better have a plan for attracting customers. But Blackspot and Muji drive home the point that as consumers, we don't have to be walking advertisements for the corporations that clothe us, feed us, and furnish our homes. Brands can tone down their tightly constructed facades and still retain a loyal consumer following, one based in trust and in loyalty to a high-quality product. SR

Brand-new Justice: Is Branding the Key to Wealth in the Developing World?

In 2004, the marketing consultancy Interbrand estimated the total value of the top one hundred global brands at $988 billion. That is to say, treating brands like Diet Coke and Pentium as tangible assets, Interbrand approximated that the top hundred global brands are worth almost $1 trillion—more than the combined gross domestic product of the world's sixty-three poorest nations combined.

One possible interpretation of that data is that brands are tricking us into paying billions of dollars more for branded goods than we would pay for their generic equivalents. Simon Anholt is interested in another interpretation. Author of the book *Brand New Justice* and one of the founders of Placebrands—a consultancy that advises national governments on brand strategy—Anholt believes that understanding how brands work might well be a key to helping poor nations become wealthy.

To fully participate in a global economy, nations must be willing to brand the goods they produce. While this is an uphill battle, some brave companies have succeeded, and we're starting to see Slovenian-branded skis, Indian-branded software, and Brazilian-branded aircraft.

Some of Anholt's most interesting ideas are about "national brand," the idea that nations, like products, have brand associations. Anholt believes that a nation's brand can affect everything from that nation's currency exchange rates to allegations of cronyism and corruption in the country.

Anholt argues that consumers want brands "to come from somewhere," and that, increasingly, they don't want that somewhere to be the United States. More and more, people are interested in discovering unfamiliar types of food and music, and Anholt believes these trends are indicative of an increasing enthusiasm for exoticism and authenticity—in other words, the best advantage a brand from Mongolia may have is that it's Mongolian, something a product from a U.S. firm can never be.

According to Anholt, developing-world exporters can take advantage of an unprecedented opportunity to increase the value of their exports, by moving away from the charitable marketing strategies used to create products like Fair Trade Coffee and banking on the intrinsic value of authenticity. By moving goods out of the

By offering an antibrand alternative, Blackspot, a "logo-free" shoe, illustrates the influence branded footwear has come to have on our culture.

category of "things we buy because it's the right thing to do," and into the category of "things we buy because they're exciting, new, and we really want them," developing nations can tap into a major new market for their products. EZ

Cotton: "The Fabric of Our Lives" (Manipulative Branding)

Even the staunchest corporate denouncers among us cannot have been immune to the ubiquitous cotton campaign, which etched the emotional slogan "the fabric of our lives" into our collective consciousness. The commercials and ads from this campaign took a family-values, tearjerker approach, leading viewers to believe that cotton had to be the purest, most natural fabric a person could buy if they really loved their family.

But there's something suspicious about so sappy a campaign, right? Suspicion warranted. The conventional cotton industry is the number-one user of pesticides of any agricultural-crop industry in the world. This heavy-duty toxic enterprise is not only polluting water, depleting soil, and contaminating neighboring farms, it's also leaching toxins into its by-products (like cottonseed oil, which we eat) and contributing to illness and disease among farmers who cultivate and harvest the fields.

"The fabric of our lives" campaign is one example of how branding can paint a wash over the true behaviors of an industry. Many a well-intentioned parent, thoroughly convinced, bought cotton bedding and clothes for their newborn. The cotton campaign makes it all too clear that a misused tool is a weapon. But forward-thinking brands that are bold enough to respond to consumer pressure for transparency and reveal their corporate practices can ultimately embody the image they present. SR

Brand Sensitivity to Climate (Future-conscious Behavioral Shifts)

Progressives, citizens who are concerned about the course of globalization, and supporters of diverse local economies alike lament the power that national and global brands have over our economic choices. But this emphasis on brand is arguably a lever for change: as people's attitudes toward a certain issue evolve, brands that become associated with that issue can emerge as positive or negative forces, depending on their stance. Specifically, as global warming and climate disruption become real public concerns, brands that are linked to dubious environmental practices will see their value plummet. Companies wishing to avoid being labeled greenhouse villains a decade hence need to start paying attention *now* to their carbon footprints. JC

RESOURCES

Commodify Your Dissent edited by Thomas Frank and Matt Weiland (W. W. Norton and Co, 1997)
This classic from the heyday of the dot.com era remains one of the most articulate and viciously funny attacks in print on branding, advertising, public relations, celebrity, and "alternative" culture. To read it is to laugh mostly at ourselves, since none of us are immune to the come-ons of companies that would have us believe that by buying their product or service we will become somehow more special or unique. To understand how we're manipulated into thinking this way is the first, large step on the road to pivoting the power of brands so as to serve free thinking and social justice:

"Turn on the TV and there it is instantly: the unending drama of consumers unbound and in search of an ever-heightened good time, the inescapable rock 'n' roll soundtrack, dreadlocks and ponytails bounding into Taco Bells, a drunken, swinging-camera epiphany of tennis shoes, outlaw soda pops, and mind-bending dandruff shampoos."

Pattern Recognition by William Gibson (G. P. Putnam's Sons, 2003)
What does it mean to be swimming in a sea of brands, advertisements, and rapid cultural evolution? Gibson's novel, a sort of science fiction set in the present, offers us a new lens through which to glimpse the answers to that question. The story itself concerns Cayce Pollard, combination cool-hunter and brand psychic,

and her hunt to find the creator of a mysterious series of wildly popular video clips that have surfaced on the Net. But that story (though a page-turner) is really just a structure on which Gibson hangs insight after insight:

"'Of course ... we have no idea, now, of who or what the inhabitants of our future might be. In that sense we have no future. Not in the sense that our grandparents had a future, or thought they did. Fully imagined cultural futures were the luxury of another day, one in which 'now' was of some greater duration. For us, of course, things can change so abruptly, so violently, so profoundly, that futures like our grandparents' have insufficient 'now' to stand on. We have no future because our present is too volatile ... We have only risk management. The spinning of the given moment's scenarios. Pattern recognition."

Thriving in a Bright Green Economy

All of the changes discussed in this book have dramatic implications for businesses. Many of the bedrock assumptions that form the foundation of our economy are shifting. The ability to look ahead and anticipate the ways business can harness new innovation is becoming a fundamental skill for businesses that want to accelerate through this shift and come out ahead. Smart businesses that regard change as an opportunity will not only adapt, they'll thrive.

What defines the companies that will excel in this new world? Four values are essential: anticipation; transparency and collaboration; long-term planning; and vision. AS

Anticipation

With radical gains in energy efficiency, cascading waves of better designs, new technologies like nano- and biotechnology, and faster, cheaper computers, delivering better products with less becomes easier every day. Businesses that embrace steady advances in efficiency will gain a gigantic competitive advantage, while those that don't will find themselves wasting money and increasingly struggling with regulations.

How much better can businesses' environmental performance get? As Jesse Ausubel puts it in "The Environment for Future Business," it's practically impossible to measure:

"High environmental performance forms an integral part of the modern paradigm of total quality ... In general, efficiency in energy favors efficiency in materials; efficiency in materials favors efficiency in land; efficiency in land favors efficiency in water; and efficiency in water favors efficiency in energy ... Nowhere do averages appear near the frontier of current best practice. Simply diffusing what we know can bring gains for several decades. Overall, society hardly glimpses the theoretical limits of performance" (1998).

To recap: efficiency equals profits; we're getting better and better at being more and more efficient; efficient companies will eat their inefficient competitors' lunches over time. AS

Transparency and Collaboration

Transparency is making corporate behavior visible as never before.

Corporations can now build intimate relationships with customers. We all know that corporations are honing their marketing research down to us as individuals, amassing data on our preferences and buying patterns, trying to anticipate our needs. But we don't often think about how this truth extends two ways: we're increasingly well informed about the goods we buy and the companies that make them. Intimacy goes both ways.

When shopping for a computer, for instance, we can use the scads of online product-evaluation sites to get a startling level of insight into not only models' prices, but the origin and performance of those models and their constituent parts. And more and more of us want to know what's inside. Fanatical user groups dedicated to everything from Prius hybrids to soap operas don't just bring the kind of publicity money literally can't buy, but they're also the first to cry havoc over a product's faults. The result is that it's growing harder for companies to get away with stuff—corporate crime and shoddy quality quickly find their way to the Net. Even in remote corners of the world, blunders and bad behavior are ever more likely to be captured on video and phoned in over a satellite. William Gibson, in the 2003 *New York Times* article "The Road to Oceania," issues a strong warning to that effect:

"It is becoming unprecedentedly difficult for anyone, anyone at all, to keep a secret. In the age of the leak and the blog, of evidence extraction and link discovery, truths will either out or be outed, later if not sooner. This is something I would bring to the attention of every diplomat, politician and corporate leader: the future, eventually, will find you out. The future ... will have its way with you. In the end, you will be seen to have done that which you did."

In this context, it's simply stupid strategy to depend on secrecy or distance to cover your tracks. Anything less than full disclosure—open, willing, eager disclosure—is likely to backfire. Even the appearance of opacity will set off alarm bells. And if transparency rewards good behavior with more loyal customers, it also allows sharper, more decisive consumer activism, quite above and beyond the scope of government regulations. Hell hath no fury like a pissed-off consumer, and the avenging spirit of the people is now empowered as never before; actions are much more easily coordinated, damage to reputations much more easily spread [see Protest, p. 450].

Anyone who thinks corporations don't care about public dissatisfaction should talk to an executive who's found herself at the business end of a serious international boycott (and in today's connected world, such boycotts are easier than ever to coordinate). Although it's unlikely that any large corporation to date has been

driven out of business through citizen action alone, it's just a matter of time until it happens.

In a transparent world, businesses must at the very least *appear* to be good, or appear to be making a concerted effort to change. "Today, social responsibility is no longer a matter of corporate discretion, due in large part to the NGO [nongovernmental organization] community's growing influence," says Bennett Freeman, managing director for corporate responsibility at Burson-Marsteller. "NGOs and other stakeholders are more likely to acknowledge progress and success if companies are candid about problems and even mistakes."

And once one business has proclaimed a desire to change, or admitted its failure to do so, pressure on other businesses begins to mount, driving even reluctant companies to do better or suffer dire consequences. As Don Tapscott and David Ticoll argue in their 2003 book *The Naked Corporation*, this is simply the new face of business.

"If you have to be naked, you had better be buff. We are entering an extraordinary age of transparency, where businesses must for the first time make themselves clearly visible to shareholders, customers, employees, partners, and society. Financial data, employee grievances, internal memos, environmental disasters, product weaknesses, international protests, scandals and policies, good news and bad; all can be seen by anyone who knows where to look. Welcome to the world of the naked corporation. Transparency is revolutionizing every aspect of our economy and its industries and forcing firms to rethink their fundamental values."

Opposite, left: The United Farm Workers boycott of Gallo persuaded the winery to change its policies toward its workforce, San Francisco, 2005.

Opposite, right: NGOs use the media to demand transparency, helping this union leader appeal for the release of arrested South Korean protesters.

In a transparent world, companies can't beat 'em with secrecy, and if they're smart, they won't try.

On the other hand, companies that are willing to embrace transparency find themselves with allies who'll see them through their efforts to change—investors, collaborators, journalists, users, and employees. Corporate innovation today is all about having deep, open, honest two-way relationships with long-term investors, NGOs, government regulators, collaborative networks, and consumer groups. Fostering and nurturing those relationships is increasingly a major part of business operations, because they are the founts from which slow and deliberate capital, new innovation, and customer loyalty all spring. And those relationships can only be fostered and nurtured by folks who consider themselves active forces for positive change.

This, in business terms, is the lesson of the open-source movement. These networks first emerged in software, where they've built operating systems like Linux at dizzying speed, with unmatched complexity and quality. But the paradigm is rapidly spreading to enfold all sorts of technical projects, from sustainable-energy systems to medical technologies for the developing world. It's even sweeping through nontechnical arenas. Nets of people are now translating schoolbooks into foreign languages for underserved communities, building online collaborative encyclopedias, and creating democratic news services.

Whatever the goal, though, all open-source projects share these defining characteristics: they break huge goals into small problems, encourage innovative answers, recognize quality contributions, and allow anyone to look for (and fix) problems. As *Wired* magazine editor Thomas Goetz writes, "Open-source projects succeed when a broad group of contributors recognize the same need and agree on how to meet it." They use distributed methods of working, so large numbers of people can work together across distances, in what O'Reilly Media founder Tim O'Reilly coined an "architecture of participation." And they mandate that the results be freely available to all. "Think of it," Goetz says in the November 2003 issue of *Wired*, "as the triumph of participation by the many over ownership by the few."

This kind of collaboration is redefining our economic landscape, and it's barely just begun. Distributed collaboration produces better results—any company that lines itself up against it is toast. But collaboration requires those corporations that would benefit from it to act as collaborators themselves—not as owners. In the new model companies don't own the building blocks of a new innovation, even if they helped design them. Instead, they sell conveniently prepackaged assemblies of bricks, using their expertise as mortar for putting them together, and a handful of new kinds of bricks as add-ons. Apple gets this. In 1999 it opened up the code for its Mac OS X software while keeping proprietary certain key surface innovations (like its user interface).

Trying to cop a free ride much less attempting to establish wide-ranging proprietary domains, practically guarantees the hostility of these collaborative networks, whereas allying with them maximizes the opportunity to work with swarms of innovators around the world. The choice for businesses is simple: ride the rising tide of collaboration, or waste more and more effort piling up leaky dikes of legal protections around your intellectual property. AS

Long-term Planning

DuPont CFO Gary Pfeiffer once observed that sustainability advocates often say, "Wall Street doesn't get it about sustainability."

"They actually 'get it' just fine," Pfeiffer asserted. Wall Street's legendary focus on the short term, Pfeiffer explained, is more precisely a focus on cumulative discounted future cash flows. In simple terms, Wall Street is happy if a company will (a) make more money in the future, (b) make money sooner rather than later, and (c) face less risk. In a market changing as rapidly as today's is, all three of these criteria can form an argument for investing in sustainable innovation.

Working backward through the criteria, making more responsible business choices often reduces risks, particularly risks from lawsuits and changing regulations. Spending on efficiency and pollution reduction and on the company's social capital can also save money, which makes it an investment, not a cost: that saved money goes straight to the bottom line, contributing to profit just as surely as increased revenues—even more surely, in fact. Finally, investing in change will undoubtedly pay off handsomely in the long run so long as the business environment itself is changing.

The modern culture of investment has been a key barrier to harnessing efficiency. When investors demand more profits *this quarter*—and most do—it's pretty hard to invest in technologies and processes that will return even greater profits in five (or even in just two) years. But with patient, principled capital—or better still, with investments that are both smart and responsible in the long term and profitable in the near term—the dynamic changes. This is increasingly available in the form of SRI (socially responsible investment) funds [see Your Money, p. 382].

More and more, knowing how to think long term, connect long term to near term, deliver results in both time frames, and explain your thinking to the Street is proving to be a critical survival skill in business. AS & GF

Vision

In his classic 1975 *Harvard Business Review* article "Marketing Myopia," Theodore Levitt skewered industries that lost market share—and critical opportunities—because they didn't understand what business they were really in. Railroad companies identified themselves with "rail and locomotives" rather than transportation, and stagnated in the face of the rise of the automobile and the interstate highway system. Hollywood considered itself part of the "motion picture" business, rather than the entertainment business, and was battered by television.

The same key question—"what business are we really in?"—applies to companies facing the new world of bright green business. Consider this: business visionary Joe Romm reminds us that just four industries—pulp and paper; petroleum; petrochemicals; and primary metals— generate 85 percent (by weight) of U.S. solid waste, use 80 percent of industrial energy, and release more than 70 percent of toxic emissions. There is not a business or home that isn't a customer of these four industries.

But all four are also sectors for which greater attention to environmental and social concerns is bound to compel change. How might each industry reimagine itself?

Take the pulp and paper industry: is it in the business of turning trees into paper products, or is it in the "forests and fibers" business, managing forests for the long term and recycling paper for the things we need? Paper or forests?

Or take the petroleum industry: is it in the business of pumping and refining fossil fuels from the earth's crust, or is it really in the business of providing energy, which might increasingly take the form of renewables or biofuels? Oil or energy?

How about the petrochemical industry: is it in the business of solving problems through the invention of clever new (and often toxic) petrochemicals, or of providing the best solutions to technical problems in the best way possible, using biology and the rethinking of processes to minimize the need for chemicals in the first place? Chemicals or solutions?

And the metals industry: is it in the mining and smelting business, or is it in the materials business, providing high-quality ingredients for the modern economy, whether those are recycled metals, new substances (like carbon nanotubes) or better design? Mining or materials?

We could ask the question of any industry. Is the automobile industry in the business of selling cars, or of moving people? Is agriculture in the business of raising massive amounts of food, or of providing nutrition and healthy soils?

The point is that companies that can't take a second look at the nature of their business will increasingly find themselves fighting rear-guard actions against tougher legislation, shifting consumer tastes, and changing technologies. That's a formula for bankruptcy. On the other hand, in sectors where the biggest players are resistant to change, huge market opportunities will open up for smaller companies that see the profit to be had in change. If, as many believe, the shift to sustainability represents the biggest transformation of the business environment in a century, empires will fall and fortunes made based on the speed—and effectiveness—with which leaders embrace these new realities. GF

RESOURCE

The Natural Advantage of Nations: Business Opportunities, Innovation and Governance in the 21st Century by Karlson Hargroves and Michael H. Smith (Earthscan Publications, 2005)

Anyone interested in the business of building a bright green future needs to read this book. It's an absolutely critical overview of our progress thus far toward sustainability, with 500-plus information-packed pages on what's working best; pragmatic examinations of best practices in business and government; issue-wrangling essays on profitable climate solutions, greening the built environment, and sustainable transportation; probing inquiries into institutional responsibility for major planetary problems; and studies evaluating new industrial and regulatory models. It's not beach reading, and indeed, unless you plan to make green business or governance your calling, it's probably not a book you need on your shelf. But if you are serious about organizational change, it's a treasure chest of hard evidence that better ideas can work.

According to the authors, "Government eco-efficiency programmes can be especially helpful for small to medium sized business, which make up a significant percentage of the business sector in any country. Fields like energy efficiency are moving so fast that if firms have not checked what is best practice within six months they will probably be out of date. Most small businesses do not have the time or resources to source the best information, let alone the funds. It makes sense then for governments to address these information and market failures to help them implement resource productivity programmes wisely."

Seeing the Big Picture

Most of us dwell in the blissful ignorance of partial knowledge. If we knew everything about the inner workings of the systems, governments, and corporations that run our world, we might be in a perpetual state of anxiety. Or outrage. On the other hand, allowing these things to stay comfortably invisible implicates us, to some extent, in continued corruption, inequity, and environmental degradation. So how do we drag our heads out of the media matrix and survey the big picture, the true numbers, the deeper drivers behind the brief strobe-flashes of information that we call news?

Indicators. Indicators measure critical information and contextualize it. They make the invisible visible. They reveal the past and predict the future. They reduce enormous and unfathomable systems into useful information. We're talking about trend data here—but trend data reinterpreted into something that we can all understand and respond to. Think tachometer (an indicator of how fast a car's engine is running). Think gas gauge. Think blinking red light on the dashboard.

Data alone doesn't do the job, because too few people can read tables of data. We need that up-or-down arrow, the simple graph, a colored map, even a smiley face.

It turns out that smiley faces and other visual simplification strategies are critical to helping people understand what's doing well and what's doing poorly in the enormously complicated systems we must now manage. Why? People can immediately react to smiley faces (or blinking lights), while they may not even read, much less understand, charts and graphs or tables of data. Power to elicit action in response to a trend—even when that trend concerns big systems and seemingly slow changes—is the mark of a good indicator.

Take air quality. Technicians measure the particulates and exhaust coming out of our tailpipes and smokestacks in ways that only technicians and scientists understand. Their numbers get crunched into a set of graphs—indicators—that show a range of specific pollutants getting better or getting worse. Even those graphs are too much for many of us to take in, so the graphs get crunched together into a single "Air Quality Index"—a cumulative, numerical indicator. To make it even easier to understand, the various numbers are then assigned a color: green, yellow, or red. Then, when we want to know how good the air is where we live, we don't need to try to remember our college statistics, we just need to know what a traffic light is.

Indicators make it possible to change the world by clearly communicating which parts of the world need changing. AA

Systems Models

If mere indicators of the status quo are not enough for you—if you want to understand cause and effect, see correlations in past trends, play with future scenarios—then what you want is a "systems model." A systems model takes data and makes it dynamic, using the magic of differential equations and microprocessors. Fortunately, you don't need to know about either form of magic to use a systems model successfully. But it does help to be a little geeky.

To see what a dynamic global model can do, try the online version of International Futures (ifsmodel.org). Built originally for the CIA, it is now free to all on the Web, and includes an astonishing range of data and a robustness that impresses even the most learned geeks in the business. You can dial in the countries, the indicators, the starting conditions, the interventions you might like to try—and see what happens to the world.

It wouldn't be a good idea to try to *run* the world using indicators and systems models, but these tools do help us make the links between, say, oil, water, industrial agriculture, war, and climate change—and other clusters of trends that are often presented as having nothing to do with one another, but that are actually intertwined in ways that probably determine the fate of civilization. It helps to know these things. Knowing these things makes it possible to know what needs doing. But no indicator or model, no matter how clever or persuasive, will ever change the world: only people can do that. AA

RESOURCES

EarthTrends Environmental Information Portal (World Resources Institute)
http://earthtrends.wri.org/
On the EarthTrends Web site you can pull up readable maps, information on a world of issues, plus specific data on individual countries, in a heartbeat. Don't let the word *earth* fool you into thinking this is only environmental data; this site gives you the whole sustainability enchilada, including economic flows and social trends. Need to resolve that ongoing argument with Uncle Phil about climate change, fisheries' collapse, whatever? Go here first.

Gapminder
http://www.gapminder.org/
If you're looking for brilliant visualizations of vast seas of global data, take a peek at Gapminder. These Swedes can help you grasp an entire century in just a few minutes. With their time-lapse graphics, you can watch the world (largely) get richer and healthier over the course of the twentieth century (with notable exceptions like Russia—which is now getting less healthy, albeit a bit richer—and Cuba—which is among the world's healthiest nations, despite being desperately poor). Then watch China's great big blob of data explode into a lot of little regional blobs of data, some very rich (the city of Shanghai), some very poor (the province of Sichuan). These images stick in your mind and help you make sense of the deeper news behind the headlines.

Start-Up 101

Ideas at rest tend to stay at rest. Change only happens when someone stands up and kicks an idea into motion. One powerful way to do this is by starting a business or an organization focused around an idea. All powerful organizations, from the U.S. government and the Girl Scouts to the Democratic Party and your local food co-op, began when one person decided to stop following along and start up something better. Start-ups of all kinds, for profit or not, create more positive influence and potential for change than any individual could ever create alone. Start-ups are tools for changing the world. SEB

How Your Business Will Change the World

Anything can be improved. When we look closely at how manufacturers make clothes, restaurants make food, architects make homes, or programmers make software, we always find things that could be better. And that's where the opportunities lie. What should be there, but isn't? We can start with what we know or are passionate about. What bothers us about the urban planning of our street? What's missing on the shelves at the supermarket? We can probably come up with a pretty long list of all the things we'd like to change, and the things that we have the skill, or the passion, to do something about. A surprising number of businesses are born from these simple lists. Jeff Bezos, founder of Amazon.com, sat down and listed all the things he thought could be sold easily over the Internet: books came out on top—and the rest is history. He didn't have a magic formula or special powers. He just found an opportunity to make something work better, and jumped in. The same principles apply whether we want to sell fair-trade products, generate clean energy, or launch a nonprofit to create social change.

That so much is wrong with the world today actually means there are more opportunities to launch enterprises that aim to make a business of improving it—especially since so many tools for improving it now exist. One of the simplest ways to find a new business idea is to look for the innovations and social trends larger companies are ignoring. Tremendous opportunity exists in building a business around doing good things better at a time when slower, more conservative competitors are stuck in their old ways. SEB

How Business Works

There may be copious MBAs and business texts in the world, but the fundamentals of what makes a start-up work are dead simple: bring in more money than you pay out. End of story. If we have another job (or a trust fund), the equation changes, but in the long run our organization runs only as long as we can afford to put energy into it. There is no law that says we must earn millions of dollars, must have scores of employees, or even that we must have a groundbreaking idea. As long as we can bring in enough income to cover our costs, we can stay in business for another day. Nonprofits rely on grants, donations, and volunteers to make ends meet, but even for-profit start-ups often use donations from friends and family to get things going (or to stay alive). The raw economics of business are simple, and most successful businesses work on quite ordinary terms. In short, you don't need a big bank account to start a company. Instead you need commitment, passion, discipline, business fundamentals, and a support system—all things obtainable without cash. Start-ups can begin as small, part-time efforts or as big career- and life-changing events: it's up to you to decide how much change you're willing to take on at once.

Do you have something to offer that's of value to people? Something that distinguishes itself just enough that people will seek it out, or pay a little more for it? That's the pivotal question of any business. Whole Foods is a great example of the sensational success of one individual who was sure he had that slightly better, slightly different idea. Whole Foods was not the first health-food store in the world, nor the first gourmet supermarket. When John Mackay opened his first store in 1974, he had a small staff and modest ambitions. He wanted to create a store that sold healthy food, rather than a place that sold mostly vitamins and supplements, as so many "health-food" stores of the day did. That first store succeeded just enough—selling a sufficient volume of more expensive, but higher-quality foods—that in a few years he expanded to a second store and then a third. The business simply brought in more income than it cost to run, and generated enough profit to eventually expand into larger and larger stores. Consumers saw the value of the store's offerings, were willing to pay for it, and we can see the rest of the story in the gleaming Whole Foods palaces around the country today. SEB

Left: Whole Foods Market has become synonymous with the booming organic-food industry.

Opposite: Amazon.com founder Jeff Bezos identified a way to sell books a little differently and has, since 1999, been on the *Forbes* list of the World's Wealthiest People.

How to Start: Finding Support and Making Decisions

The most important action entrepreneurs and organization founders must take is to build a network of support for their new venture. Friends, significant others, and family all need to be involved in helping to manage the new stresses and challenges that come with a start-up; emotional and social assets can be just as powerful as economic ones. Even a simple conversation asking for understanding and additional encouragement dramatically improves our chances of success: we know from the beginning whose support we can count on. Start-ups by definition force change into the lives of everyone involved, and it takes time to restore balance in our lives. Managing the emotional and psychological challenges is just as important as managing the business and logistical ones.

The first decisions we must make are metadecisions: How will we allocate our time? Which factors are most important to our success, and how can we devote enough time to them? What sacrifices are we willing to make concerning time, money, family, and hobbies?

The next decisions will help us start off favorably: What product should we make or what service should we sell first? What are the goals of the organization? Who should we invite in as partners? Make a list of such decisions—it will be a long one. Take some time to prioritize, and consider the best use of time. Which decisions need to be made this week? This month? This year? Some small decisions (like the brand name) will probably need to be made before bigger ones come into play. Don't get stuck on making big decisions if they are not immediately relevant; be willing to go after low-hanging fruit, make steady progress, and build momentum. With a rough outline of the decisions to be made, and their order of priority, we can bring other people into the conversation and use them to help refine, add, and trim the decisions into actual plans of action to get from this stage into actual operation. SEB

Being a Good Manager

Ideas aside, organizations depend on relationships among people to make things happen.

If we're the leader of the organization, everyone looks to us to set the tone for behavior and interpersonal dynamics, whether among employees or among companies we're doing business with. And it's not what we say, but what we do that people will respond to. If we want to work with honest, hard-working people who listen to good ideas from others, we have to be the first to demonstrate that behavior. People will expect us to define our expectations of them. If we're clear and open about how hard we expect people to work, what we want them to be responsible for, and what the rewards will be, they'll be clear and open with us about delivering on those goals.

Transparency, openness, collaboration, integrity: these are the operating principles of businesses that aim to change things, and they always start with the people running the show. If we want our businesses to change the world, we've got to manage them in worldchanging ways. SEB

RESOURCES

The Art of the Start by Guy Kawasaki (Portfolio, 2004)
Former Apple marketing maven and current entrepreneurial coach Kawasaki packs decades of advice on starting companies into this fun and concise read. Drawing on his experience funding and advising start-up companies, he describes the ins and outs of getting a company off the ground, with plenty of sound advice on how to avoid common mistakes:

"Find a few soulmates. History loves the notion of the sole innovator: Thomas Edison (light bulb), Steve Jobs (Macintosh) ... Anita Roddick (The Body Shop). History is wrong. Successful companies are started, and made successful by at least two, and usually more, soulmates. After the fact, one person may come to be recognized as 'the innovator,' but it always takes a team of good people to make any venture work."

Start Your Own Business by Rieva Lesonsky (McGraw-Hill, 2001)
This comprehensive reference covers every topic imaginable, from business planning and administration to finance and marketing. You won't need everything here on day one, but this book will be a well-worn companion on your bookshelf as the days and weeks spent running your new business fly by—packed as it is with Lesonsky's excellent advice:

"A business plan is your road map to success, guiding the growth of your business at every stage along the way. We'll show how you how to craft a business plan that puts you on the fast track."

KaosPilot A–Z by Uffe Elbaek (KaosCommunication, 2003)
By most accounts, Denmark's KaosPilots (an "international school of new business design and social innovation") offers the coolest entrepreneurial education on the globe. With the business world in upheaval and values like sustainability, transparency, collaboration, and vision at a premium, the KaosPilots don't so much offer a standard business course as an immersion in what just might be the future. Even reading their materials is slightly mind-altering, full as they are of discussions of strange things like streetwise systems thinking, value-based entrepreneurship, grassroots aristocrats, and competence environments. One would almost suspect an overabundence of hand-waving, if it weren't for the raves the school and its graduates have won from business leaders and nongovernmental-organization advocates alike. Even if you aren't in the market for a graduate program, their lavishly designed signature book gives you a glimpse into the world they're preparing to build:

"This is why the KaosPilot's skills and attitudes are so important. In today's highly volatile environment we need to learn how to react quickly to events in the marketplace; how to learn from our experiences—especially the bad ones—and above all we need to become world experts in collaboration and coordination."

POLITICS

Building a better world is a tough business. Many of the steps demand long-term vision, and the real payoffs for each of us as individuals (prosperity, health, safety, clear consciences) come gradually and diffusely. Moving toward a bright green future takes a lot of thinking and debate, careful innovation, and good government policies.

In the meantime, however, people willing to act in dangerous and immoral ways often gain simply, immediately, and directly. Protecting a grove of old-growth trees can benefit us all a little in the long run, but clear-cutting that forest can yield a company massive profits—instantly. Educating every child will benefit us all over time; putting those children to work in a sweatshop will benefit a factory owner right away. Greedy, corrupt, violent, power-hungry people have every incentive to be fierce in the pursuit of their own interests. And the rest of us sometimes have a hard time knowing how to stop them.

But stopping them from their nefarious practices and defending the public good are vital. We can't change the world if we can't keep dangerous people from destroying it. Politics is our means to do just that.

Making politics work isn't simple. We need to show how the future could be better, and band together with others to create the conditions for that better future. We need to bring forward new ideas, and connect people who are passionate about seeing them implemented. We need to demand transparency in government, elect the best politicians we can find, and hold accountable leaders who do wrong. We need, in short, to spread, grow, and protect democracies.

It won't be easy. Power, as the great abolitionist orator Frederick Douglass taught us, concedes nothing without a struggle. The small number of powerful men who benefit the most from oppression and exploitation will fight with every means at their disposal. Dictators will murder and jail and torture those who speak against them. Wealthy interests will attempt to corrupt existing political systems and use the power of propaganda to divert attention from their actions. Demagogues will use hatred and fear to frighten people away. In many ways, the odds are stacked against us.

But in many other ways, the tide is moving in our direction. Never have there been so many tools for allowing citizens to reclaim their nations. Information technologies make it easier for people (with access) to not only speak up against injustice, but to find others with a common cause and unite with them. We have better models than ever before for working effectively together in civic groups and for mobilizing others. We know how to create better civic conversations, pursue better public visions, and establish better public policies. We know new ways to fight corruption and open the workings of governments to everyone. We're even learning how to work together to protect human rights advocates, prevent wars, confront dictators, and build peace.

Best of all, we're not alone. All over the world, a new generation of millions of worldchangers is emerging, connecting, sharing information, and developing tools together. The strongest force on earth is not an army or a police force or a government or a corporation—it is we ourselves, awakened to the dangers we face and the possibilities we are creating. We are everywhere. We have powerful tools. And though we come from all nations, races, and creeds, we're increasingly fighting for the same kind of future—where democracy, transparency, human rights, and peace prevail. If we work together well enough, we just might get there. AS

Movement Building

Optimism is a political act.

Entrenched interests promote despair, confusion, and apathy to prevent change. They encourage us to think that problems can't be solved, that nothing we do can matter, that the issues are too complex to allow even the possibility of change. It is a long-standing political art to sow the seeds of mistrust among those you would rule: as Machiavelli taught, tyrants do not care if they are hated, so long as those under them do not love one another. Cynicism is often seen as a rebellious attitude in Western popular culture, but in reality, our cynicism advances the desires of the powerful: cynicism is obedience.

Optimism, by contrast, when it's neither foolish nor silent, can be revolutionary. When no one believes in a better future, despair is a logical choice—and people in despair almost never change anything. When no one believes there might be a better solution, those who benefit from the status quo are safe. When no one believes in the possibility of action, apathy becomes an insurmountable obstacle to reform. But when people have some intelligent reasons to believe that a better future can be built, that better solutions are available, and that action is possible, their power to act out of their highest principles is unleashed. Shared belief in a better future is the strongest glue there is.

Great movements for social change always begin with statements of great optimism. Facing as we do today so many interlocking challenges, one of our biggest tasks is simply this: to be willing to look so many looming catastrophes in the face and courageously point out that radical changes for the better are possible. History attests that if we can show people a better future, we can build movements that will change the world. AS

The Abolition Movement

Can a small group of people change the world? The answer is unquestionably yes—a minority of dedicated and savvy people can trigger big things. Social values don't always change at glacial speed.

Sometimes they change astonishingly fast. A powerful case in point is Britain's massive and swift shift in attitudes and policy concerning slavery in the early nineteenth century. As the popular historian Niall Ferguson puts it in *Empire: The Rise and Demise of the British Empire and the Lessons for Global Power,* "It used to be argued that slavery was abolished simply because it had ceased to be profitable, but all the evidence points the other way: in fact, it was abolished despite the fact that it was still profitable. What we need to understand, then, is a collective change of heart. Like all such great changes, it had small beginnings" (2003).

Preceding pages: After months of pro-democracy demonstrations and two days of brutal assaults by the Chinese government's army, "Tank Man" stepped out in front of a line of moving tanks in Tiananmen Square. Although his identity remains unknown, he has become an enduring symbol of the power of individual defiance, Beijing, June 5, 1989. **Left:** Red and blue political bracelets demonstrated partisan affiliations during the 2004 U.S. presidential elections. **Opposite:** Dr. Martin Luther King, Jr.'s "I Have a Dream" speech electrified its audience with an inspiring vision for the future and is a compelling demonstration of the power of words to change the course of history.

Antislavery activists had to start small, because overturning an ancient, almost universal practice was a seemingly impossible challenge. While considered barbaric today, slavery was accepted for most of human history as a necessary, if unsavory, part of the natural order of things. Neither the Bible nor Christian tradition explicitly opposed it (although religious antislavery activists would later use the teachings of Jesus to support their cause), and the dominant thinking of the period easily rationalized its existence: oppression could be justified and explained by a sort of proto-social Darwinism. Given the widespread belief in European superiority, and the powerful, entrenched economic interests supporting the slave trade, the quick and decisive defeat of slavery was amazing; that citizen action was able to end it all was more amazing still.

Yet end it the abolitionists did. The British antislavery movement was officially born in a printshop in London in 1787; in August 1838, Parliament granted emancipation to 800,000 slaves. In just over fifty years, a fundamental and previously unquestioned institution of the British Empire was challenged, made odious to the public, politically discredited, and abolished outright.

How was slavery ended so quickly? Through the efforts of a small but committed group of antislavery activists. Remarkably diverse, the champions of abolitionism included religious leaders across dominations—Quakers, Evangelicals, Unitarians—as well as some of the day's leading thinkers, including ex-slaver John Newton, politician Edmund Burke, poet Samuel Taylor Coleridge, and industrialist and pottery giant Josiah Wedgwood.

The abolitionist movement is especially relevant today because it spawned the modern nongovernmental organization (NGO). Abolitionists invented a new kind of politics, the politics of the pressure group, and marshaled an impressive groundswell of support that profoundly influenced legislators in Britain. The movement also pioneered such NGO tools as direct mail, the newsletter, the boycott, and the media campaign.

A letter-writing campaign essentially marked the start of the movement. Former slave Olaudah Equiano presented himself to the English scholar Granville Sharp, desperate to get someone, anyone, to pay attention to an obscure court case. The case involved a ship captain facing insurance-fraud charges for throwing 133 sick slaves overboard to their deaths (he wouldn't be able to collect insurance on them if they died of natural causes). Equiano was outraged that no one seemed to think the man should be tried for murder. When Sharp heard the tale, he was equally outraged and wrote to everyone he knew, decrying this act in particular and slavery in general. One of the people to whom Sharp wrote was the vice-chancellor of Cambridge University, who was so moved by the letter that he decided to make slavery's morality the topic of a prestigious essay-writing contest. The winner of this contest, Thomas Clarkson, became obsessed with the issue in the course of writing his entry; he would become the abolition movement's unstoppable leader.

In a time when many people did not have the right to vote, petitions to Parliament were some of the most powerful tools the abolitionists had—it was only after receiving petition after petition, signed by tens of thousands of people, that the government decided to hold hearings to consider abolishing slavery. (Though nothing ultimately came of these hearings, they were notable: antislavery petitions had never been seen before, and here they dominated the debate.)

Adam Hochschild, in his book *Bury the Chains*, describes what may have been the first large-scale boycott: "Within a few years, another tactic arose from the grassroots. Throughout the length and breadth of the British Isles, people stopped eating the major product harvested by British slaves: sugar. Clarkson was delighted to find a 'remedy, which the people were ... taking into their own hands ... By the best computation I was able to make from notes taken down in my journey, no fewer than three hundred thousand persons had abandoned the use of sugar.'"

As quickly as fair-trade labels today have popped up on food products, advertisements flooded the press: "BENJAMIN TRAVERS, Sugar-Refiner, acquaints the Publick that he has now an assortment of Loaves, Lumps, Powder Sugar, and Syrup, ready for sale ... produced by the labour of FREEMEN." Then, as now, the full workings of a globalized economy were largely invisible. The boycott brought these hidden connections to light; poet Robert Southey even labeled tea "the blood-sweetened beverage."

The abolitionists used the media with particular skill, employing powerful images and icons to dramatize the horrors and inhumanities of slavery. At an inquiry into the slave trade by the Privy Council in 1788, for instance, abolitionists revealed a diagram of a slave ship, the *Brookes*, showing slaves tightly packed and chained in rows; this started shifting people's perceptions.

Today's activists know that getting people to change their minds often depends on getting them to see what they don't want to see—something the abolitionists became masters of. As Hochschild explains, "When the famous one-legged pottery entrepreneur Josiah Wedgwood joined the committee, he had one of his craftsmen make a bas-relief of a kneeling slave, in chains, encircled by the legend 'Am I Not a Man and a Brother?' American antislavery sympathizer Benjamin Franklin, impressed, declared that the image had an impact 'equal to that of the best written Pamphlet.' Clarkson gave out five hundred of these medallions on his organizing trips. 'Of the ladies, several wore them in bracelets, and others had them fitted up in an ornamental manner as pins for their hair.' The equivalent of the lapel buttons we wear for an electoral campaign, this was probably the first widespread use of a logo designed for a political cause. It was the eighteenth century's 'new media.'"

The lesson for us today is that entrenched beliefs can shift quite dramatically, seemingly overnight. In the last several decades, we've seen seismic shifts in social values regarding gender, race, religion, and sexuality. We've seen the end of apartheid in South Africa—barely imaginable just a few years before it happened (indeed, not long before apartheid ended, Margaret Thatcher had said, "Anyone who thinks the ANC [the African National Congress] will be running South Africa is living in cloud cuckoo land"). Of course, sometimes even discredited ideas can hold on with a nearly unshakable grip: it took the bloodiest war in American history to end slavery in the United States, and another century of struggle to overthrow the legally enshrined racism that sprang up in its wake. Some of the global slave trade's effects are still with us today.

The story of the antislavery movement raises some questions that require us to take an imaginative leap. Which of today's practices and beliefs might be considered barbaric and inconceivable just decades from now? Perhaps the intergenerational injustice of destroying the biosphere to satisfy fleeting desires. Perhaps our tolerance of absolute poverty. Perhaps, even, the notion that national boundaries and accidents of birth, rather than merit and hard work, determine one's opportunities in life.

If the history of the abolition movement teaches us anything, it's this: when a small group of people commit themselves utterly to righting an injustice, they redefine the possible. The very act of commitment has the power to change the world. NAB

Mont Fleur: Scenarios as a Tool for Social Change

Stories are tools for knowing and judging. Change the stories and you change how people live.
Brenda Laurel, social entrepreneur

It's easy now to underestimate the improbability of the fall of apartheid and the rise of democracy in South Africa. Of course we now know that Nelson Mandela was elected president of South Africa, and that democracy has taken root there, and that—while racial prejudice is still powerfully entrenched in some parts of society—by and large, South Africans are learning to live with one another.

But before all that transpired—and not so very long ago—a peaceful outcome was not at all a given. During the transition, a popular joke went that there were two ways of creating a better future for South Africa—one practical, and the other miraculous. The practical approach was to get down on your knees and pray a violent revolution wouldn't tear the country apart. A negotiated peaceful solution would *truly* take a miracle.

How, then, did South Africa make such a quick and peaceful transition out of apartheid and into a rising democracy? One part of the solution came from a unique project in 1992 that brought together a diverse group of South

Africans to imagine the country's future. This included not only concerned citizens and businesspeople, but also embittered enemies—leaders from the paramilitary white right wing, and left-leaning blacks in exile. Together, they created the following four plausible schemata (known as the Mont Fleur Scenarios) imagining South Africa's evolution over the upcoming decade:

Ostrich was a future in which the white government stuck their heads in the sand, their inaction eventually provoking a violent backlash and civil war.

Lame Duck imagined a prolonged transition with a weak government that tried to be all things to all people, and thus ended up doing very little.

Icarus was a future in which a black government came to power, but eventually—in reference to the Greek mythological character—crashed, along with the economy, as a result of unsustainable public spending programs.

Flight of the Flamingos imagined what it would take to create a positive transition and future. Alluding to the fact that it takes some time for a flock of flamingos to fly off together, this was a "slowly but surely" scenario.

When the Mont Fleur Scenarios were released, few people could see one path forward for their nation: suddenly, they had four possible paths to think about and debate. Because the schemata were memorable—simple with strong imagery—they spread quickly, shaped policy options, and accelerated negotiations between opposing sides. No one wanted a civil war. Even F. W. de Klerk, the last apartheid-era president (who agreed to end the system), said years later on a radio program that he was "no ostrich" (Beery, Eidinow, and Murphy). Lame Duck and Icarus helped everyone to agree to avoid mistakes like overspending, difficult given that many of the emerging black leaders had come from strong Socialist or Populist backgrounds and there was strong (yet economically unrealistic) pressure to right all of South Africa's wrongs immediately. Significantly, the Mont Fleur process gave one-time enemies practical experience in working together. Not only did this ease the negotiation of the transition into democracy, it paid dividends when it came to the hard work of rebuilding the country.

Since Mont Fleur, imagining shared futures has become an important tool for effecting social change around the world. Happiness, as the Buddha is said to have taught, demands giving up all hope of a better past. Building a vision of the future to which all sides in a conflict can agree helps people let go of the past and begin working together. This constructive engagement in turn breeds hope, and hope energizes and empowers people. Hope makes even bitter enemies behave more responsibly and strategically, because they start to see how certain actions might jeopardize their stake in their future—and that of their children. Mont Fleur helped all South Africans see a proud role for themselves in an important story—building the modern world's most successful example of peaceful systemic change. NAB

RESOURCES

Solving Tough Problems: An Open Way of Talking, Listening, and Creating New Realities by Adam Kahane (Berrett-Koehler Publishers, 2004)
Any book that has a blurb from Nelson Mandela on the front cover must have something valuable to say about resolving conflict and political stalemates reasonably, peaceably, and compassionately. Kahane has worked with corporate, government, and civil leaders all over the world to help them understand how the hardest problems we face can be solved collaboratively if we are willing to revise the way we've been trained to communicate: "Most conventional approaches to problem-solving emphasize talking, especially the authoritarian, boss or expert, way of talking: telling. In a debate, each party prepares their position and speech in advance and then delivers it to a panel, which chooses the most convincing speech ... This approach works for deciding between already created alternatives, but it does not create anything new."

There are many books about creating better dialogues—it's a booming industry—but few can connect their methods to successes in systemic afflictions on the scale of apartheid. Kahane's combination of elegant prose, solutions from the small to the large, and stirring success stories makes this one required reading.

Stir It Up: Lessons in Community Organizing and Advocacy by Rinku Sen (John Wiley and Sons, 2003)
Sponsored by the Ms. Foundation for Women, *Stir It Up* offers a complete set of tools for effective community organizing, based on case studies of fourteen community organizations. According to Sen, "Today's movements for social and economic justice need people who are clear about the problems with the current systems, who rely on solid evidence for their critique, and who are able to reach large numbers of other people with both analysis and proposal." With step-by-step advice on how to build and mobilize a constituency, and on how to focus internal movement-building so that it leads to external social change, Sen's work is an essential manual for anyone looking to alter public policy.

Opposite, left: South African President Nelson Mandela waves to a crowd in London near the monument erected there in his honor, 1996.
Opposite, right: Two women kiss after being legally married during the first week of state-sanctioned gay marriage in Massachusetts in 2004.

Networking Politics

The Internet is changing what is politically possible. It has evolved as a kind of operating system, a platform both technical and social, which helps people connect and helps networks of people talk through issues and approaches. Social networks are, in fact, "shaped" like the Internet, with nodes, connections, and hubs (nodes with numerous connections). The Internet is also a path to the digital convergence of many kinds of media in a variety of electronic devices. With emerging and new media, far-reaching tools for communication are no longer restricted to media companies who produce content for mass consumption. Internet-based forms of media such as blogs and podcasts possess low barriers to entry, so consumers can be the producers, and any Internet user can publish content and make it accessible to any other Internet user in the world.

As access to the Internet becomes more commonplace and broadband connectivity supports affordable worldwide multimedia distribution, our sense of who we are and how we relate is changing. Though mindshare is inherently limited, and new media outlets don't tend to gain mass audiences, a "long tail" of bloggers and multimedia users—as *Wired* magazine editor Chris Anderson calls it—is developing smaller, loyal audiences and becoming part of a global conversation. Media circulates through social networks with many dynamic nodes and connections, repeatedly shared rather than wholly consumed. This world conversation has broad implications for the social and political consciousness of the next generations—our children and grandchildren. They will see a world of diversity, humanity, and cultural creativity as these media trends progress that we can barely imagine today.

This is no techno-utopian vision. A system that is more democratic, that brings more voices into the mix, is not an inherently "better" system—we don't want democracy because it's a more effective form of governance; we want democracy because it's morally right. As Winston Churchill famously said, "Many forms of Government have been tried, and will be tried in this world of sin and woe. No one pretends that democracy is perfect or all-wise. Indeed, it has been said that democracy is the worst form of Government except all those others that have been tried from time to time."

Democracies are inefficient, especially as they scale up, and the democratic process can slow decision making to a crawl. Social technology brings more voices into the mix, and larger public conversations bring with them significant challenges. There are interruptions, misunderstandings, outright lies, and arguments. In networked democracies, transnational networks of Web-enabled citizens will use technology to facilitate civil discussion and debate, civic engagement, and participation in an ongoing public conversation across borders and cultures—and misunderstandings will be even more likely.

One helpful trend: we are beginning to see the development of a class of interactive technologies designed to mitigate social and political problems and prevent disharmony. A classic example of this technology is the "bozo filter," which is used in many forum-based online communities. If we find another member of the community particularly obnoxious, we can employ the bozo filter to hide his or her remarks. This prevents virtual collision and conflict. On systems like Slashdot.org, an online technology news source, robust systems of moderation rate users' comments. Each member of the Slashdot community can select a tolerance level on the ratings scale—if another user's comments are found by the community to be worthless, a tolerance setting allows us to avoid seeing those comments altogether.

New social conventions and practices are also emerging. The best online communities have moderators or facilitators who nurture conversation. Online

facilitation has evolved into a sophisticated professional practice. JL

Architecture of Participation

A key concept in the latest thinking about the Web, categorized under the general term Web 2.0, is the architecture of participation. Web systems are increasingly social, interactive, and participatory. Though support for interaction and group forming has always been an important aspect of the Internet, the emphasis on conversation and collaboration increased with the appearance of blogs (online "journals") and wikis (online workspaces for collaborative content development) in the late 1990s. The social-software movement followed, based on the insight that blogs can be more or less public conversations. This movement influenced political volunteers for several 2004 U.S. presidential campaigns, especially Howard Dean's. Since then blogs have become important components of political and advocacy campaigns.

In developed nations, the Web is becoming a platform for conversations about the challenges and decisions that define social structures. It is also a source of accessible, effective opportunities for civic engagement. How do we foster this kind of participation in emerging democracies? From the 1960s on, the "whole earth" perspective has helped us to "think globally, act locally." In the '90s, as Internet access became more widespread, we began to "think and act globally and locally" online, developing casual acquaintances with people from around the globe and learning that the aspirations and dreams of people in diverse nations and cultures are not so different from one another.

Now many of us have intermixed the local and the global, and our hearts and minds are in a "glocal" space. Using interactive tools and technologies that have emerged from the information age and the network society, we've formed friendships and alliances across borders.

Networked democracies can work because they are enriched by social capital [see Community Capital, p. 342]. Reed's law, formulated by David P. Reed, states that the usefulness and value of networks, particularly social networks, scales exponentially with the size of the network. The increase in value is an increase in the potential connectivity for transactions. A social network is rich with connections that increase the range of possible (social, political, economic) transactions. Will networks replace existing hierarchical systems of social organization? Almost inevitably: as we study networks, we learn that they are fundamental structures—our bodies are networks of cells; each cell itself is also a network.

Network models can be effective in pushing power to the edges. Rather than depending on a central authority to make decisions, we can use networks to make decisions at the nodal level, closer to the point of impact. The Electronic Frontier Foundation (EFF), an online activist group founded in 1990, was originally conceived as a community-based organization with chapters in various states. But the EFF met with the leadership of potential chapters in January 1992 and announced that it was no longer interested in forming chapters. The EFF had decided it would make more sense to operate as a single member of a network of activist groups—each of which could operate autonomously—rather than as a central authority. For several years such a loose network did exist, and its members worked effectively and cooperatively in sharing information and educating constituents about "cyber liberties" issues. JL

Networks and Politics

Network models are already shifting some of the power away from our politicians. When in 2004, the Howard Dean presidential campaign found itself without the funds it needed to directly pursue certain campaign activities, the organization worked at leveraging networks of supporters and making the best use of Web-based tools—Web sites like meetup.com (which allows like-minded people to locate one another and form interest-based groups), a custom social network called Deanlink (which allowed Dean supporters to connect to one another), blogs, and online donations. Howard Dean recognized the need to relinquish some control over the messages and supporters behind his campaign in order to make effective use of networks; as a result, he built a viable, well-funded presidential campaign.

In his 2004 paper "Movement as Network," Gideon Rosenblatt of ONE/Northwest suggests that the environmental movement be reorganized so as to shift "the focus away from individual people and organizations working on environmental issues and toward the connections that link them together." By networking many organizations, movements can leverage synergies and network effects. Redundancies can be eliminated, and each organization in the network can focus on what it does best. "Movement as Network" may have growing influence within the environmental movement, and perhaps within other advocacy movements as well. But these technologies will become most influential when we no longer notice them, when we take them for granted as everyday tools.

We're already beginning to take blogs for granted. According to David Sifry of the blog-ranking service Technorati, 70,000 new blogs are started every day, and 50,000 new blog posts are created per hour. Blogs can be the product of one writer or several, and can range from personal creations to slick corporate-PR-team productions. Wikis, which allow for collaborative writing and editing, are also increasingly popular, as are more robust open-source content-management systems like Drupal and Zope, which can be used as platforms for highly functional and interactive, yet relatively inexpensive, Web sites.

An increasing number of online networking systems support "tags"—categorizations that can be attached to various content elements (blog items, bookmarks, images). Originally, tags allowed individual users to create and sort categories on the fly; then came "social tagging," which allowed users in a single network system to see all of the content, theirs and others', associated with specific tags. Now some systems are sharing and aggregating tags across networks; tags have become an important way to find and share content. For example, David Isenberg encouraged everyone attending his Freedom to Connect conferences to use the tag "f2c" on posts and photos. If you search under that tag on Technorati, you get an aggregate of links to photos from the conference and blog posts about it. Content is also syndicated and aggregated via a couple of technologies, RSS and Atom, which are formats for encoding content so that it can be parsed and displayed in other environments.

In all these manifestations of "social technology," the "social" is just as prominent as the "technology." A social network may be supported by a Web system, but in essence it's a bunch of people who are connected to one another. Web software may mediate the connections, make them visible and perhaps more usable, but they are human connections. And that, ultimately, is the real nature of the networked democracy: people, connected. JL

RESOURCE

The Revolution Will Not Be Televised: Democracy, the Internet, and the Overthrow of Everything by Joe Trippi (Regan Books, 2004) Howard Dean's campaign manager, Joe Trippi, reveals how he used the Internet to

Presidential campaigner Howard Dean reaches into a crowd during his 2003 tour. Dean's campaign was known for taking advantage of online networking tools as a means of leveraging support.

make Dean a front-runner, if only temporarily, claiming that the Dean campaign was open source—meaning that it was a relatively transparent political campaign. This is a good read for anyone who wants insight into the Dean campaign, though it's important to remember that it's only one man's perspective.

According to Trippi, "One problem with running an open-source campaign is that you also open it up to every crank with a computer who despises your candidate. We had a policy that we wouldn't remove critical posts from the blog. If people wanted to take an honest shot at Howard or at the campaign, we welcomed it. Disagree with his stance on the war? Fine. Want to argue abortion? Great. It sparked some of the most interesting debates on the blog to have differing opinions."

Amplifying Your Voice

Change begins when people speak up. Wrongs observed in silence are rarely righted, and perhaps more importantly, the problems we face today are so complex that we need everyone's ideas to fix them. We have never been in greater need of a lively, informed, passionate public debate.

Each of us has a role to play in building that debate. We have more tools than ever to make our voices heard: not only the familiar methods of attending public forums, writing to our newspapers, and calling our radio stations, but a whole array of new means powered by the Internet. We have online forums on which we can debate, Weblogs on which we can publish, wikis on which we can collaborate, video blogs and podcasts through which we can broadcast images and sound. Indeed, new tools roll out every day, most of them extremely cheap, if not free. Never before has the average person had greater ability to make him- or herself heard.

In some places, such speech can come at a high price. Despotic regimes around the world are cracking down on bloggers and online communities, trying to shut down sites that have become centers of dissent. These cyber dissidents deserve our support.

Chinese blogger and high-tech investor Mao Xianghui cofounded China's first blog, setting off an explosion in the use of blogs for expressing sentiments that, in other media, would likely be censored by the government.

But for most of us, the challenge is less a matter of being forbidden to speak than of learning how to advance the dialogue in a meaningful way. But rest assured, resources abound, and with a little hard work, anyone can help put good ideas into the "blogosphere" and help others find new answers and ways of making change. Journalism is too important to be left to the professionals. AS

Bridge Blogging

Mahmood Al-Yousif, a Bahraini entrepreneur, is one of millions of people around the world who are taking political matters into their own hands, creating their own online media, and talking directly to global audiences. In June 2003, he started a blog. On his "About" page he wrote, "Now I try to dispel the image that Muslims and Arabs suffer from—mostly by our own doing I have to say—in the rest of the world. I am no missionary and don't want to be. I run several Internet websites that are geared to do just that, create a better understanding that we're not all nuts hell-bent on world destruction. I hope that I will be judged that I made a small difference."

The desire to make a small difference is the essence of "bridge blogging": writing online (or broadcasting audio or images onto the Internet) for an audience beyond your immediate community, while making an effort to explain, interpret, and contextualize the concerns and conversations of your community for a global audience.

Alliances of "bridge bloggers" are now forming, in an effort to bring critical mass and attention to their voices. Global Voices Online is an international citizens' media project focused specifically on bridge blogs from outside the United States and Western Europe. A global team of blogger-editors and volunteers provide links and summaries of what they think are the most interesting and globally relevant conversations taking place in the "blogospheres" of their countries and regions. RM

OhMyNews

Started on a shoestring budget by a group of politically progressive freelance journalists, the South Korean online newspaper OhMyNews.com mixes writing by trained professionals and volunteer "citizen reporters" to produce timely, in-depth coverage of current events. The goal is to include a variety of opinions and viewpoints previously ignored by South Korea's conservative mainstream press. With 27,000 citizen reporters and about 1 million daily readers, *OhMyNews* has become perhaps the most influential news source in Korea.

Japan Media Review recently interviewed *OhMyNews* founder Oh Yeon-Ho, who said, "Journalism is changing. The form of 20th century journalism and the form of 21st century journalism will be fundamentally different. For 21st century journalism, if a reader wants to, he can convert himself into a reporter and this is realized through the Internet ... Where professional reporters once exercised their influence exclusively, now they compete with citizens, so professional journalists could be in trouble if they still try to confront general readers with their authority and arrogance ... Thus, it is necessary that the reporters quickly figure out how the world is changing and that they change themselves along with it."

The English-language "international" version of *OhMyNews* has become a vital source of global perspectives, features, and opinion. Now *OhMyNews* is opening up its "citizen reporter" system, making it possible for anyone to contribute stories. Contributions are read, edited, and fact-checked by the site's staff. There's even a system for compensating contributors for good work. *OhMyNews* is the future of journalism. JL

Citizen Media

One impact of the availability of low-cost, powerful personal computers has been the appearance and evolution of citizen media, beginning with the desktop publishing revolution and the appearance of many zines (small, generally noncommercial magazines) in the late 1980s. As Internet use grew throughout the

1990s, zine culture drifted onto the World Wide Web, and by the end of the decade, start-ups and established businesses were experimenting with large-form Web portals and destinations; smaller Web publications had appeared; and many people had mastered Web technology sufficiently well to build personal Web sites. By 2000, content-management software for Weblogs had appeared, spurring the creation of millions of blogs over the next several years.

Though the traditional publishing world was slow to comprehend the blog world, the journalists who were early adopters saw in blogs the potential to democratize and transform media. Dan Gillmor of the *San Jose Mercury News* was one of the first journalists to create a blog and regularly post new entries. Gillmor began to see the Internet as a foundation for a new kind of citizen media, produced by individual bloggers operating independently and creating blogs with RSS feeds. What RSS amounts to is a format for syndicating blog content; it is available with most, if not all, blog applications. With RSS, blog content can be more broadly distributed; readers can then subscribe to content from many blogs using a news aggregator, an application that facilitates management of blog subscriptions. JL

Citizen Journalism

In war, good journalism is almost always independent journalism, and the Internet is making independent journalism easier than ever, even for reporters in war zones. Not only can journalists say things that mainstream media would never publish or broadcast, they can also strike out completely on their own, free from minders and PR flacks and the passive standards of most "embedded" journalists.

Take Kevin Sites: a freelance multimedia journalist, Sites often works as a one-man unit, using portable digital technology to report, write, edit, and transmit his stories from conflict areas around the world. Sites blogged from the front lines of the Iraqi war zone beginning in 2003, while covering the war for CNN. His personal eyewitness accounts of the war provided an authentic, behind-the-scenes perspective. After he videotaped a U.S. soldier shooting and killing an unarmed Iraqi, Sites used his blog to give an eloquent and fair account of the shooting.

This new breed of journalist is changing the rules of engagement for reporters, bringing truthful, independent reporting to the most dangerous places on earth.

Radio Any Which Way You Can

In 1999, Daoud Kuttab, a Palestinian entrepreneur, journalist, and professor, wanted to start a radio station in Amman, Jordan. The government wouldn't issue him a license, so he started his station online.

It wasn't the first time Kuttab had turned to the Internet to overcome local barriers—in 1995, he founded the Arabic Media Internet Network, which allowed Arabic speakers in various countries to find out how newspapers throughout the region reported their news and politics. For his new project, he armed smart young reporters with digital

During the Democratic National Convention in Boston in 2004, a "Bloggers' Boulevard" was established to allow the burgeoning new media a platform for reporting on events.

video recorders and sent them into the streets, the markets, and Parliament to produce some of the highest-quality journalism in Jordan.

The audience for AmmanNet online was far smaller than it would have been on FM radio, but it was influential throughout the region, and was rebroadcast on FM by Palestinian radio stations. After five years of operation, AmmanNet was granted a license by the Jordanian government, and it is now on the air, providing critical journalism and a community voice for Amman's million residents. EZ

Machinima

Movies can have transformative effects, but they can also be difficult and expensive to make. Machinima attempts to turn video games into cinema. Players can record video games and review their own adventures or pass them along to friends to review; when players combine edited versions of these game recordings with amusing voice-overs or music, they've got a simple digital movie.

By and large, the first machinima movies were done for laughs or to tell action stories closely related to a game's source material, and they appealed primarily to people familiar with the games in question.

But machinima may finally have had its breakout moment in a fascinating short film called "The French Democracy" (2006). Using a game called *The Movies,* French machinima maker Koulamata tells the story of three young men in Paris who end up taking part in the race riots that rocked France in 2005. All three suffer different kinds of indignities at the hands of French society, triggering their decisions to fight back; the movie is very clearly on the side of the rioters. Whether or not one accepts the filmmaker's political perspective, he has done something truly remarkable: he has taken computer-game characters and told a story with clear social relevance, demonstrating that machinima has the potential to be much more than a medium for artistically exploding jeeps.

Koulamata is aided, in part, by his chosen software. *The Movies,* ostensibly a linear role-playing game, set in Hollywood, allows players to create their own stories using the game's characters and sets. This explains some of the oddities of "The French Democracy"—the "electric power station" looks like an Old West shack (complete with horse); the Parisian skyline includes the Empire State Building. Koulamata is able to tell a good story despite these limitations. Undoubtedly, future tools will allow for greater manipulation of settings and characters, much as the online game *Second Life* allows players to create and modify objects and settings within the 3-D world.

In the mainstream movie industry, computer graphics are generally used to show things that can't easily be filmed. A mainstream version of "The French Democracy" could doubtlessly have been made with human actors and location shots, but it would have required at least a half dozen performers and crew. Machinima movies can be made with much smaller teams.

Machinima has the potential to open up politically and socially meaningful filmmaking to individuals. As the tools get more powerful, and the characters and sets more expressive, we will certainly see more movies like "The French Democracy." JC

RESOURCES

We the Media: Grassroots Journalism by the People, for the People by Dan Gillmor (O'Reilly Media, 2004)

"What industry is among the least transparent? Journalism."

If you doubt that citizen media is overtaking big media—or that this is a good thing—you won't after reading nationally known journalist Gillmor's treatise on the rise of amateur reportage. Gillmor uses case studies and personal anecdotes to trace the emergence of and implications of citizen media; he offers a thorough primer on the tools available to citizen journalists, from blogs to wikis.

Moreover, Gillmor urges corporate media and corporations to embrace and collaborate with citizen media outlets—not fight against them. He offers advice to corporate media participants and lays out new rules of conduct for everyone from PR flacks to CEOs: "Make sure your Web site has everything a journalist might need ... Post or link to what your people say publicly, and to what is said about you. When your CEO or other top official gives an interview, transcribe it and post in on the Web site. If it's an interview being broadcast, put the audio or video online as well ... Find out which micro-publishers are talking about your product or service. (Use Google, Technorati, Blogdex, and Feedster, not just Nexis and clipping services.) ... Then make sure you keep these people well-informed. Treat them like professional journalists who are trying to get things right, and they'll be more likely to treat you with a similar respect."

Handbook for Bloggers and Cyber-Dissidents
(Reporters Without Borders, 2005)
Bloggers are the front line of contemporary journalism. But making waves also makes enemies. With increasing frequency, bloggers are finding themselves persecuted by repressive governments in exactly the same way investigative reporters were in decades past.

Journalists under threat need tools, and Reporters Without Borders has produced an excellent toolbox. Their *Handbook for Bloggers and Cyber-Dissidents* provides basic information that will benefit those new to blogging—explaining blogging terminology, how blogs differ from other kinds of Web sites, how to select a blogging tool and Web host, and how to get started—and goes on to offer tips that even veteran bloggers will find useful. Journalist, blogger, and author of *We the Media* Dan Gillmor tackles the issue of journalistic credibility and standards. Journalist Mark Glaser offers tips on how to "make your blog shine." French Internet consultant Olivier Andrieu contributes a discussion on search-engine visibility.

The most inspiring aspect of the *Handbook* are the "Personal Accounts," short essays from bloggers around the world about why they blog and why blogging matters. German blogger Markus Beckedahl of Netzpolitik.org describes his blog defending civil and human rights online; anonymous Bahraini blogger "Chan'ad Bahraini" explains how Bahraini bloggers have "broken the government's news monopoly"; Hong Kong's Yan Sham-Shackleton, aka Glutter, discusses blogging for Chinese human rights and free speech; Iran's Arash Sigarchi, who did prison time for his blogging, reflects on the importance of blogs in Iran as an outlet for nongovernment-approved speech; and American Jay Rosen points out that even in the United States, blogs are a revolutionary new way for writers to circumvent various powerful "gatekeepers."

Importantly, the book takes a hardcore technical look at how bloggers like Chan'ad might avoid being "outed," how bloggers like Arash might avoid arrest, and how bloggers and blog readers in countries like China, where the Internet is heavily censored, can get around the political "firewalls." Other topics include anonymous blogging, overcoming Internet censorship "firewalls," and ensuring e-mail privacy. If you want to know how to blog for impact, the *Handbook* should be your first stop. JL

Connecting With Others

"To work at this work alone is to fail," says the poet Wendell Barry (1990). Almost no one changes the world single-handedly. To make real change, we have to learn how to find and work well with others. We need to search out allies, build networks of people who share our values, support good causes, and trade ideas. To change the world, we need to connect. AS

Backing the Right NGOs

There's a change coming to the world of advocacy, one that is as fundamental as anything we've seen since the abolition movement [see Movement Building, p. 412] gave birth to the modern civic group. That change? The move from centralized, mass-market nongovernmental organizations (NGOs) to decentralized advocacy networks driven by their members.

Right now, many large advocacy NGOs consider their members, like you and me, mostly a source of small donations. By and large, it doesn't matter what we think, how we act, who we know, or how strongly we're committed, as long as we keep writing those thirty-five-dollar checks for our "memberships," which usually yield nothing more than a tote bag or a coffee mug or some lame newsletter. It's a dysfunctional model, all the way around. Mass marketing, direct mail, subsidiary income tracks (like selling T-shirts), and the rest of the modern NGO racket degrade everyone involved. The model turns passionate advocates into carnies and charitable citizens into consumers of change-related program activities and products.

One of the biggest problems with mass-market NGOs is that they limit and control information. With a few excellent exceptions, they make no effort to educate their members about the broader field of activism in which they are involved, instead regarding their communications with members as marketing opportunities for ensuring continued financial support. When they do disseminate information it is impersonal and largely irrelevant to our real concerns.

Furthermore, members have few ways in which to interact with one another. Most NGOs, even those with sophisticated online presences, restrict the flow of information between members. Criticism of the NGO, dissent, endorsement of other efforts, even the sharing of outside information on the issue at hand—these just aren't welcome on most NGOs' Web sites and e-mail lists. Being a member of a modern mass-market NGO often provides the member with fewer opportunities for acting. In most NGOs, independent efforts, personal (rather than personalized) messages, creative approaches, and new ideas are actively discouraged in favor of "action alerts" and "calls to action," in which members are asked to send prepackaged messages to politicians or corporate leaders.

How might online advocacy networks begin to change this sad state of affairs? By putting the member in the driver's seat: we choose the flows of information we receive, the people we are allied with, the establishments we give money to, our affiliations.

Advocacy networks can encourage the exchange of information. By making available news feeds on the issue at hand, member discussion areas and listings, and other discussion tools, an advocacy network can allow us to choose the best mix of information sources for our concerns. Better still, it can let us contribute to the debate. Got a blog? We can add it to the

Opposite, left: Health-education workers from a nongovernmental organization (NGO) hold up a poster depicting an AIDS patient before and after receiving the benefits of antiretroviral drug treatment, Nairobi, Kenya.
Opposite, right: A woman displaced by the 2005 Kashmir earthquake gives birth, aided by doctors from Citizens' Foundation, a Pakistani NGO.

list of feeds from which members can choose to aggregate their news. Got a great idea for a new campaign or a beef with an existing NGO? Start a discussion topic. Because we choose the information we receive, the information we get is by its very nature more relevant to our concerns.

Advocacy networks can encourage relationships; after all, they're a form of social software, like Friendster or the Omidyar Network, where we can make connections and foster our social and professional circles online. When our working relationships are not subject to the control of any third-party organization, all manner of cooperation becomes possible. We can identify allies online and create informal networks and groups among ourselves, and we can use reputation systems to help evaluate the worth of causes and the truth of information. This allows us to then lend support to the causes or ideas most relevant to our concerns ("Hmmm ... eight of my friends identified this article as important. Maybe I'll take a look.").

Advocacy networks can treat our money as ours. Making online contributions securely is easily done now, so why should Friends of the Mud-sump Salamander own our personal information or restrict our choices? Why shouldn't we be able to view a whole array of opportunities to give, and choose between them ourselves? Perhaps in the end we'll trust in a "name brand" NGO to use the money wisely, but perhaps we'll discuss with our fellow members (some of whom may have expert knowledge) who's doing the best work most effectively, and give money to some great outfits with whom we were previously unfamiliar.

Some NGO leaders claim that advocacy networks will hurt organizations. There is no doubt that such a shift will drive some NGOs out of business. This is a good thing. NGOs were never intended to be perpetual. They should exist as long as they meet the world's need to change—not stumble on, zombie-like, until the heat death of the universe.

For groups that excel at including members in their activities, advocacy networks will be an asset, making them bigger, leaner, faster, stronger. For organizations with an extremely specific focus and the humility to take the time to explain why that focus is important, advocacy networks will be an incredible boon, providing an effective tool for finding focused allies—a "long tail" [see Networking Politics, p. 418] effect, as it's called. For groups willing to learn how to collaborate on the fly and work from a campaign-centric model, advocacy networks will be transformative. For all of us, networks are the future. AS

NGO in a Box

Nongovernmental organizations, especially those operating outside of the industrialized world, rarely have access to sophisticated technology infrastructure in the field. Unless an NGO focuses on information technology, chances are its computers and networks are a hodgepodge of donated hardware and off-the-shelf commercial software (which may or may not be legally acquired)—and far too much time is spent on technology hassles. That's where the NGO in a Box program from

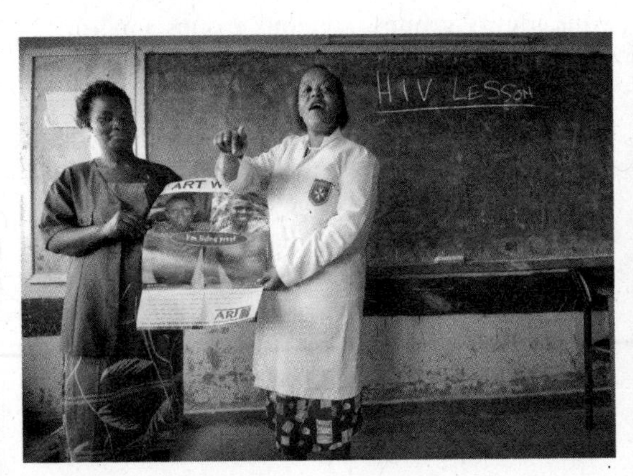

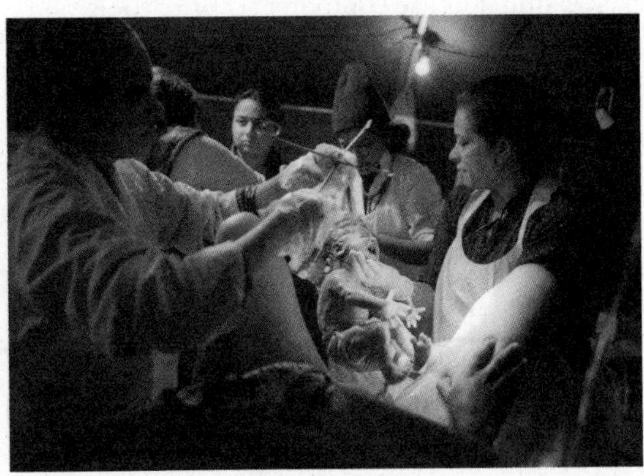

the Tactical Technology Collective comes in.

NGO-in-a-Box is a set of specially selected, high-quality free/open-source software chosen to meet the needs of NGOs. The included applications range from the familiar (Firefox and Thunderbird) to the highly specialized (VNC, a tool for managing and controlling remote computers). The first NGO-in-a-Box kits, which came out in 2004, were localized for specific regions, but the subsequent Phase II version of the program seeks to provide tools for specific categories of applications. First up is the Security Edition, which helps NGOs find and use computer security tools.

Due out soon are the Base Box (core tools for NGO offices) and the Advocacy Edition, intended for groups creating audio and video media for political and social activism. Tactical Technology plans to develop an Open Publishing Edition, for print media activism as well. Tool kits like these mean that a greater number of small NGOs can be both more effective and better networked—and therefore more capable of creating change. JC

International Campaign to Ban Landmines

If we want to see exactly how much a small, loosely connected network of dedicated people can accomplish, we need look no further than the International Campaign to Ban Landmines (ICBL).

The ICBL is a network representing more than 1,400 groups in over ninety countries—human rights' groups, demining groups, humanitarian groups, children's groups, veterans' groups, medical groups, development groups, arms control groups, religious groups, environmental groups, and women's groups. Together, they work locally, regionally, nationally, and internationally to ban antipersonnel land mines [see Land Mines, p. 218]. The network was first organized in 1992, with the coordination of six separate NGOs: Handicap International, Human Rights Watch, Medico International, Mines Advisory Group, Physicians for Human Rights, and Vietnam Veterans of America Foundation. All six had encountered the devastation caused by land mines while trying to implement projects in the developing world, and had concluded that the only solution was a comprehensive international ban on land-mine use. The organizations were able to pool considerable regional knowledge and experience, as well as worldwide networks and contacts. From this group sprang dozens of national campaigns.

From the start, the ICBL, with its tiny staff of four, rejected a rigid, bureaucratic structure. Today it has overall strategies, but each member NGO is allowed to implement solutions in ways that make sense for the group and the regions it serves. The campaign itself, which resulted in the international Mine Ban Treaty, is also an example of network advocacy [see Networking Politics, p. 418] through community. Faxes and (later) e-mail were among the campaign's most powerful and effective vehicles. The ICBL Web site includes a collection of online tools that allow any organization to create campaigns to advocate the ban and to replicate the ICBL's success. JL

Fahamu: Pan-African Text Messaging for Social Justice

The nonprofit Fahamu uses information-communication technologies to conduct pan-African campaigns for human rights and social justice. Founded in 1997 by Firoze Manji, a former director of Amnesty International's Africa program, Fahamu ("consciousness" or "comprehension" in Kiswahili) fields digital publishing and communications—CD-ROMs, the Web, e-newsletters, and now, mobile-phone text messaging—in diverse ways. The group, Manji told Worldchanging in an interview, can "pack a punch larger than our weight" to support progressive social change in Africa, empower Africans to control their own economies and political systems, and stave off a repeat of the tragedies that hang over the heads of African social-justice activists.

Fahamu's distance-learning courses for African human-rights organizations are distributed on CD-ROM and coordinated via e-mail networks that also give participants the opportunity to forge relationships and alliances across borders. Through these courses, Fahamu has trained about 300 people in 160 organizations throughout Africa. Subjects range from activist intensives on topics such as the basics of human-rights activism, or campaigning for access to information, to organizational capacity building in areas such as financial management, leadership skills, and fundraising, to understanding the role of the media in the Rwanda genocide.

"What we do at Fahamu—the distance learning courses [and the] production of an activist newsletter which gives a platform to challenging dominant views," says Manji, "is absolutely critical to contribute to creating a movement that prevents Rwanda from becoming a mirror of our future."

Fahamu's team grew intrigued about the possibilities of text messaging as a tool for social justice because of the phenomenal growth of its flagship publication, the electronic *Pambazuka News*. *Pambazuka* (meaning "dawn" or "daybreak") combines news, commentary, analysis, and resource sharing on social justice, serving as a primary outlet for progressive views from across Africa. Key focus areas include women's rights and debt cancellation. A reliable and unique outlet for conveying African progressive thought, ideas, and analysis—and for dispelling myths about Africa—it has a huge, pan-African audience. The publication grew exponentially since its inception in 2000, quickly becoming the leading social-justice newsletter on the continent.

In the staff's analysis, much of *Pambazuka*'s power comes from its dual role as both a public space for idea sharing and a protected space, one that activists can count on to support them in their struggles for social change at home.

A technician from the Mines Advisory Group prepares to detonate recently discovered unexploded land mines.

In 2004, Fahamu joined a coalition of human-rights groups supporting ratification of an African Union protocol to protect a wide array of women's human rights, in some ways breaking new ground in international law.

To promote ratification, Fahamu staff took advantage of the huge growth of mobile phones in Africa, all of which feature short-message service (SMS) functionality—the ability to deliver a brief text message. They decided to test whether this mass of "texters" could be mobilized for a social-justice campaign; they set up a Web-based petition in support of the protocol with an SMS service that would let mobile users text in their signatures for entry onto the online petition page. The campaign was a success and played a part in the ratification of the protocol. EG

Text Messaging for Debt Relief

Many poorer countries are staggering under the weight of massive debt to developed nations. Much of the debt was accrued as development loans under despotic and corrupt local governments, and is now being paid for by the people they robbed. The payments on those debts often represent large portions of a debtor nation's budget. Even if a country makes large budget-busting payments, it may still be unable to dig out of debt. Between 1970 and 2002, debt-cancellation advocates point out, the lowest-income countries in Africa paid $298 billion on their original debts of $294 billion, and still owed over $200 billion as a result of extremely high interest rates, according to the Web site MakePovertyHistory.ca.

These funds could otherwise suppport education and poverty alleviation. Fahamu launched a text-messaging campaign—Thumbs Down 2 Poverty—in support of the larger Global Call to Action Against Poverty (GCAP), a loose alliance of more than seventy global groups calling for cancellation of Africa's debt to the developed world.

The GCAP SMS campaign (www.gcapsms.org) asks texters to send the message "No to debt" to their number, along with a call to end poverty expressed in their own words—making use of a new Fahamu-developed technology that allows users to send longer political messages. The GCAP SMS Web site fields Fahamu's technology strengths neatly; it includes fresh content from *Pambazuka News*, regularly updated SMSs from all over the globe (after review and approval by a site administrator), and a community-building feature: an RSS feed of these messages.

"We didn't have a clue what would happen, or what the reception [to it] would be," Fahamu founder Firoze Manji says. "It was just such a crazy idea, and even if it didn't work, out of failures—you learn. Some of the best stuff we've done has come out of stuff that's gone badly." EG

RESOURCE

The Power of Many: How the Living Web Is Transforming Politics, Business, and Everyday Life by Christian Crumlish (Sybex, 2004)
Crumlish uses interviews and case studies to explore how social software is creating ways to form and sustain political, social, and community organizations. This book was written during the 2004 election season and shows how Howard Dean's and other presidential campaigns used social software to organize and build support:

"The changes now affecting how political campaigns are conducted include a trend toward greater transparency and lowered barriers to entry for folks who want to participate, as well as more control and a greater sense of satisfaction for their efforts. Internet technologies, according to Mary Hodder of the University of California at Berkeley, collapse the time and space between like-minded people, and individual targeting 'can lead to finding people who are the most passionate supporters of a particular cause or candidate.' This can ideally provide the core of support online that can translate into a successful campaign 'in the real world.'" EG

Tools For Talking

All movements begin in conversation.
Vicki Robin

Café Slavia, on the banks of the Vltava River, is one of the most famous cafés in Prague. The food and coffee are only OK; the waitstaff, a bit slow. Its fame derives rather from its history as a spot where dissidents met during the Communist era. Here they would gather to talk, perhaps scribbling on napkins, envisioning freedom for their society and strategizing how to bring it about. Around the world, historical markers and local legend commemorate public spaces that have been venues for conversations that shaped history.

Where are these conversations today? What form do they take?

The new social movements that have arisen in the recent past are innovative in their use of digital technologies and online conversations. A parallel surge has provided new tools and approaches that enable face-to-face conversations to be richer and to have more impact. These tools for talking draw inspiration from the revolutionary cafés, from the African marketplace, and from human ways of organizing. They also recognize that the most groundbreaking ideas and relationships are often created not at conference sessions and formal meetings, but in the hallways and during coffee breaks.

The new tools and processes facilitate more democratic ways of organizing at all levels—from the family unit to the international nongovernmental organization. With them, citizens become a source of innovation and take responsibility for shaping their own reality. They are not sitting back, waiting for institutions and governments to act on their behalf. Power structures are leveled out and rebuilt in more democratic forms. Collective intelligence arises.

Political issues are becoming increasingly polarized: "Are you for us or against us?" One of the paradoxes of the information age is that more and more information channels are available to us, but we increasingly have the freedom to choose to hear only those we already agree with. In the context of the new conversation movement, however, exposure to a diversity of views becomes crucial. Approaches that emphasize learning from diversity create spaces where participants can listen to different points of view that challenge their assumptions and judgments, and step beyond their comfort zones.

In times of paralyzing complexity, conversation becomes extremely relevant. Party politics is no longer as predictable as it used to be. Casting a vote every few years really doesn't seem sufficient anymore. It is not easy to find solutions to major issues such as climate change, global inequality, terrorism, and migration, or to discover what our roles as individuals can be. We turn to conversation as a way of making sense of what is happening around us. It enables us to generate new ideas—to think collectively—about how to live together and mend torn social fabrics. Far from being idle talk, innovative conversations help change the world. MB

Spreading the Power of Café Conversation

Undoubtedly, right now there are thousands of café conversations going on—one of them may indeed change the world. Several groups have merged the power of the café debate with structures that make even the quietest members feel safe and respected, that create new ways to help people come together to exchange views and solve community problems.

Conversation Cafés are small meetings in coffee shops or public spaces throughout the United States, Canada, and Australia. Through the Conversation Café Web site (www.conversationcafe.org), diverse strangers find their way to a scheduled café, where a facilitator helps to guide the conversation. Participants decide together what topic to discuss, but sometimes a "talking piece" is employed. The process is simple: an introductory round, where people speak but don't comment on each others' statements, is followed by an open conversation and a closing round of reflections.

World Cafés (www.theworldcafe.com) are meetings that range in size from 12 to 1,200 members, split up into smaller groups of four for deeply participative, problem-solving conversations. After a conversation has heated up, a facilitator asks people to switch tables, so members network and cross-pollinate the conversations, allowing the group as a whole to access collective intelligence—usually the best way to make a tough decision. Since the inception of the World Café in 1995, tens of thousands of people on six continents have participated in these dialogues. A number of them are featured in the book *The World Café: Shaping Our Futures through Conversations That Matter.* MB

AmericaSpeaks

How do you engage five thousand citizens actively in one town meeting, and enable each of them to give substantive input to public decision makers? This is what happens in the 21st Century Town Meetings held by the nonprofit organization AmericaSpeaks. Updating the traditional New England town meeting to accommodate today's democracy, AmericaSpeaks restores the citizen's voice.

Since the organization's founding in 1995, AmericaSpeaks has engaged over 65,000 people in more than fifty large-scale forums in every U.S. state. Meetings have addressed local, state, and national decisions on issues ranging from the development of municipal budgets and regional plans to Social Security reform.

At the gatherings, tables of ten to twelve participants engage in facilitated deliberations. These discussions are then electronically processed into synthesized recommendations: each table submits its ideas through wireless computers, and the entire group votes on final recommendations. Results are compiled into a report in real time for participants to take home at the end of the meeting. MB

The Israeli-Palestinian School for Peace

The Israeli-Palestinian conflict is one of the most intractable problems of our era. From a distance it appears completely stuck. But if we zoom in on the School for Peace, located in the village of Neve Shalom/Wahat al Salam, Israel, where Arabs and Jews have chosen to live together since 1972, we see something very different. Here, Arabs and Jews are in conversation, not just polite and empathetic conversation, but honest and genuine conversation. Relationships are evolving. The facilitators at the school believe that what happens here reflects the path the overall society is on, and the journey Israel and Palestine need to go through.

The School for Peace runs "encounter programs" in which equal numbers of Arabs

and Jews come together to create understanding and bring about a more equitable relationship. Encounter programs are relatively common around the world. What happens in many of them, though, is an exchange along the lines of "you can be my friend, you're not like the rest of them." The School for Peace doesn't think this is enough to shift politics and xenophobia. Sure, participants build friendships, but they go beyond that. One of their key assumptions is that the solution to the Israeli-Palestinian conflict depends upon encounters between groups, not just between individuals.

Tali Latowicki, a Jewish participant, wrote a compelling account of her experience. She says she arrived prepared to express guilt, shame, and empathy. She left with a profound realization that empathy is not enough. At first she was hurt at the Palestinian participants' inability to distinguish between Jewish supporters and opponents (like herself) of the occupation. Through conversation, she came to realize that their words weren't personal—they were trying to explain to her the general process of Palestinian society relative to Israeli society. She also realized that if she was supportive of their cause in principle but was not doing anything about it, in their eyes, she might just as well be supporting the occupation.

So far, 35,000 people from different walks of life—from attorneys to activists, schoolchildren to schoolteachers—have attended the programs. The aim is to create a safe space that allows participants to examine their feelings and thoughts in a group, to pose new possibilities, and to challenge the existing reality.

Through its many encounter programs, the School for Peace has not only affected the individual participants but also their friends, colleagues, and families. This is truly courageous work. MB

RESOURCES

Appreciative Inquiry Handbook: The First in a Series of AI Workbooks for Leaders of Change by David L. Cooperrider, Diana Whitney, and Jacqueline M. Stravos (Berrett-Koehler, 2004)
We grow in the direction of our inquiry. By asking positive questions, we can generate new, positive images of the future. These powerful visions of ourselves, our organizations, and the world can in turn inspire action and innovation. *Appreciative Inquiry* asks us to pay special attention to what works—the best of the past and the present—and to allow this to ignite our collective imagination of what might be. This is the core inspiration for the global "Imagine" movement that started in Chicago and has spread to other cities around the world, where people are gathering to reimagine their communities and enact a new future.

Open Space Technology: A User's Guide by Owen Harrison (Berrett-Koehler, 1997)
The idea of inviting people to a conference with no preset agenda can seem risky and maybe even foolish. Perhaps that's why one of the central mottos of Open Space Technology (OST) is "be prepared to be surprised." When OST is at its best, it's an incredibly appealing idea. The basic concept, which Harrison outlines in

Opposite, left: AmericaSpeaks holds large-scale forums around the country where citizens can get together to discuss local and national issues.
Opposite, right: In the Israeli-Palestinian School for Peace, Arabs and Jews have lived side by side in peace for several decades, amid violence and unrest.
Left: An Arab girl and a Jewish girl who grew up together in Neve Shalom/Wahat al Salam remain close friends.

his book, is that participants are empowered to create and manage their own agenda, built up around parallel working sessions related to a central question or theme. Participants take responsibility for their own learning and action by volunteering to host sessions on topics they care about or by choosing to attend sessions hosted by others. The meetings can take place among very small groups or very large groups of a thousand or more. And as Harrison suggests, they can be energetic, creative, productive, efficient, and inspiring:

"Open Space Technology has now successfully been utilized in India, South America, Africa, Europe, China, the United States and Canada with groups of from five to one thousand members. The purposes have ranged from corporate redesign in the face of intense competition to national redesign in the face of massive transformational forces, as in South Africa. The technology is not magic, nor does it solve all problems. However, in those situations where highly complex and conflicting issues must be dealt with, and solved, by diverse groups of people, OST can make a major contribution."

Right: Andrus Ansip, the prime minister of Estonia, demonstrates the ease of voting online, in 2005. Estonia was one of the first countries to enable nationwide Internet voting. **Opposite:** The Corruption Perceptions Index is one of Transparency International's best-known tools, ranking more than 150 countries in terms of perceived corruption, as determined by expert assessments and opinion surveys. The lower the number, the higher the perceived level of corruption. This one is from 2005.

DEMANDING TRANSPARENCY

Corruption is a global problem. Those of us in the Global North face electoral systems that essentially allow open bribery, in one form or another. Those of us in the Global South often face systems that skip over the elections altogether and go straight to the bribes. And corruption tends to reinforce itself across borders, as dirty deals know no boundaries. From oil to diamonds, the commodities that make life rich in the North support corruption and oppression in the South. The corrupt are their own global network.

Corruption breeds in dark corners. It also makes good government—the intelligent policy making that underpins most solutions—virtually impossible. Schools and hospitals suffer when the bureaucrats overseeing them are skimming money. Businesses trying to make money without polluting suffer when compromised regulators let their competitors get away with breaking environmental laws. Elected officials who owe their offices to the campaign contributions of powerful interests find voting on the merit of laws a threat to their jobs. If we want governments that work, we need to demand transparency.

What is transparency? Openness, in all things: transparent societies demand that

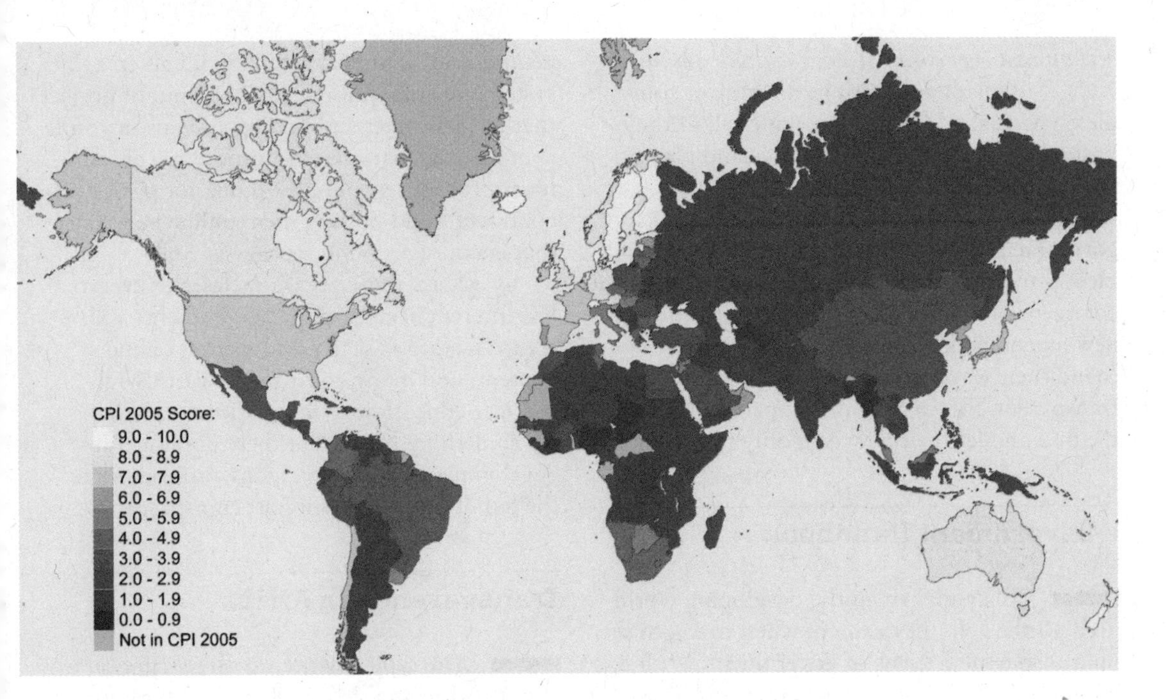

their leaders conduct public business *in public*—they demand to see the books, to inspect the records, and to be kept abreast of potential conflicts of interest. They demand that those in power be accountable to independent legal authorities when they break the rules—that leaders be public servants in more than name alone.

Transparency is the battle cry of twenty-first-century politics. AS

Estonia's Tiger Leap

After the fall of the Soviet Union, an international aid assessment summed up Estonia as being "bankrupt, polluted and decaying." Today, Estonia is climbing up through the economic ranks of the "upper-middle-income" countries (like Mexico, Brazil, and the Czech Republic), and many of its social indicators are better than those in certain countries in the "upper-income" bracket. The reasons are complex—cohesive national identity, a well-educated population, traditional ties with Scandinavian countries—but one of them stands out: Estonia has made the world's strongest commitment to providing technology to all its citizens. In fact, in Estonia, Internet access is a human right.

The groundbreaking *Tiigrihüppe* ("Tiger Leap") program has wired 98 percent of Estonia's schools, provided an average of more than one computer for every twenty students (running about forty new software packages in Estonian), created a national learning network for teachers, and trained 40 percent of them in advanced computer skills, according to a program executive summary.

The Estonians now use the Net at a higher rate than the French or Italians do. To accommodate the demand, the Estonian government has opened up hundreds of free Internet-access centers, from downtown Tallinn to remote islands in the Baltic. In addition, there are already over 280 Wi-Fi hotspots in the tiny country, according to a BBC News report, and over two-thirds of them are free (Boyd 2004).

But this is much more than a story of leapfrogging or redistributing the future through technology: even more amazing than Estonians' appetite for Internet access is the Baltic nation's absolute commitment to digital transparency. Estonia's former president Lennart Meri (1992–2001) supposedly answered his own e-mail. The state IT adviser attends cabinet meetings. Almost all state services are online, as are tax filings and the state budget. Almost everything the government does online is open to public scrutiny, *and almost everything it does is done online.*

Parliament meetings are almost completely virtual—legislators conduct all of their business, including reviewing bills and placing votes, on networked computers. They sign important documents with digital signatures. A government Web page contains all bills, amendments, and proposals, and citizens are allowed to comment on the drafts and make suggestions for new amendments, some of which have already found their way into law. By making their society transparent, Estonians have nipped corruption in the bud and let their new economy flourish. AS

e-Government Handbook

Policy makers in the developing world need all the help they can get when trying to set up transparent systems of government. With this in mind, the Center for Democracy and Technology and infoDev have created an e-government handbook to help policy makers reduce or eliminate corruption. The section on transparency suggests that it should be embedded in the design of policies and systems, especially in ways that will help citizens see how decisions are made.

Transparency becomes much easier with technology. For example, customs' services are notoriously corrupt—goods come in the door and lots of money changes hands under the table. If we put that system online, it becomes much harder to subvert. When the whole thing is on paper, it's easy for corrupt officials to charge money for an official stamp or refuse to process an invoice unless they get a bribe. When it's all online, it's much easier to say, "Here's my money, here's my form, where's my shipment?" The Center for Democracy and Technology points to the government of Seoul, South Korea, as an illustration of this. As part of a comprehensive campaign against corruption initiated in 1998, the government streamlined the regulatory rules of systems like licensing and permit approval that are most subject to corruption through bribes. An online tracking system now helps citizens monitor the status of government applications.

It's unlikely that technology will turn nondemocracies into democracies, but it may just make the difference for young and fragile democracies. Empowering individuals to avoid systematic corruption is just the kind of project that has leverage, and finding systems in young democracies that aren't transparent—procurement procedures, public bidding for government contracts—and putting them online in a way that encourages public review is vital.

Sure, there are obstacles: not everyone has Internet access; many governments still try to operate in secret and restrict online content; and many constituents worldwide lack the education or inspiration to work for something better. But there are hopeful developments, including many projects—like the handbook—that support transparency. EZ

Transparency in Africa

The typical write-up on corruption in Africa is dire and despondent. And for good reason: according to the African Union, as much as $146 billion, or a fourth of the continent's average gross national product, is lost every year to corruption. In Africa, corruption limits economic growth and keeps people in poverty, especially since poor people are most vulnerable to bribes in exchange for basic services, and because corruption diverts public resources away from sectors like education, health, and housing. To say that this has undermined people's trust in government and diminished their expectations for the future would be an understatement. In fact, being an honest public servant in Africa today is considered an act of heroism.

One of the problems with the fight against corruption in Africa has to do with the way corruption is often oversimplified and branded as a "uniquely" African problem. What is often glossed over is the global nature of African corruption. It is not just about the corrupt "Big Man" leaders like Mobutu, Abacha, Moi, and their coterie of politicians who will do anything for a price, or the lazy civil servant who has to be given "something small" before doing his job, or the waste of aid dollars. It is also about campaign finance, lobbying, white-collar crime, multinational companies with a take-no-prisoners approach, shady single-source contracts, greedy individuals, complex networks, and apathetic electorates. In fact, change some people's names and you could be talking about the goings-on in Washington, DC, or the difficulties of doing business in China. Indeed, that $146 billion is going somewhere: often into the pockets of transnational corporations and Swiss bank accounts.

Africans, though, are fighting back. Over the last few years, a number of African international institutions working for better governance have sprung up and partnered with international donors, newly elected multiparty governments, and civil society groups like Transparency International to build a focused, growing movement to clean up governance—a movement with lessons for the whole world.

Corruption in Kenya

Perhaps no country better illustrates both the challenges and the alternative avenues for tackling corruption and governance issues than Kenya. The republic has a deservedly poor reputation where corruption is concerned. It has remained one of the lowest-ranked countries in Transparency International's annual "Corruption Perceptions Indexes" since the mid-1990s. Kenya is particularly notorious for corruption on a grand scale, such as the 1990s Goldenberg scandal (in which government officials granting shady subsidies to gold exporters may have stolen as much as 10 percent of the nation's gross domestic product) and the more recent Anglo-Leasing scandal (involving gigantic overpayment on a government contract), exposed by the former anticorruption czar John Githongo, who now lives in exile in the United Kingdom.

After the election of the new National Rainbow Coalition (NARC) government in December 2002 and the government's first few months in office, NARC seemed poised to become a textbook example of how to fight corruption in Africa. The government was elected on a policy of zero tolerance toward corruption, and within the first few months had passed several new laws and created new institutions to address the problem.

Four years later, little of that promise has been fulfilled. In fact, the running joke in Kenya is that NARC stands for Nothing Actually Really Changed.

Part of the problem is that in Kenya (as in many places), it's hard to clean house politically. Politicians who've worked for decades to get power are reluctant to turn their backs on their old friends. Entrenched, corrupt bureaucrats are hard to dislodge.

What's more, fighting corruption takes government institutions that are strong enough to take on the people involved and demand transparency. Kenya doesn't have them yet. Of the thirty-five cases forwarded post-NARC to the attorney general by the Kenya Anti-Corruption Commission, not one has yet gone forward (as of mid-2006). Corruption laws without prosecutions are of little value.

Finally, while governments have changed, the way politics is conducted in Kenya and indeed in most of Africa remains fundamentally the same. A long history of political patronage, combined with the lack of any formal campaign-finance mechanisms, forces politicians to rely on corruption to ensure their survival. Indeed, the now disgraced former minister Kiraitu Murungi has admitted that Anglo Leasing was essentially

In Nairobi, Kenya, minibus drivers are often stopped by police demanding bribes. This kind of corruption has posed great challenges to the Kenyan government.

a vehicle for financing campaigns for the upcoming 2007 elections (see the *East African Business Week* March 6, 2006, article "Anglo Leasing Storm").

Combine systemic campaign corruption with low expectations on the part of many Kenyan voters and you have a recipe for disaster. As Jaindi Kisero, a columnist with the *Daily Nation* astutely observes, "When we elect a leader in Africa, we enter into unwritten contracts committing him to reward our political loyalty with appointment to parastatal jobs, to raise money to build rural schools, and to award the elite of our tribes with contracts at inflated prices. As we learn from Goldenberg, the Ndung'u Report and Anglo Leasing, an elected leader must—to remain in power—hand out benefits such as lucrative supplies contracts and public land to their relations and political allies ... We must start thinking about how to address the root cause of political corruption in Kenya. If we don't do so, then we must be prepared to go through this boring ritual of replacing one corrupt administration with another after every five years."

The Power of the Press and Rebranding Corruption

But despite all these problems, things are changing in Kenya, and they're beginning to change quickly. The press has fought for, and gotten, more freedom. The impact of being able to find out about scandals via the press, of being able to discuss corruption openly over the airwaves via call-in shows, of being able to write about it on Kenyan blogs, and of being able to circulate "confidential" documents electronically cannot be estimated. While the Freedom of Information Bill is still pending in Parliament, Kenyans have generated their own de facto version of freedom of information: they are relishing their right to know, and as a result have forced the government and those involved in scandals to respond to demands for information. Things can no longer be swept under the rug.

The Anglo-Leasing scandal in Kenya reveals more than just the insidious nature of corruption in government: it also demonstrates the impact that investigative journalism and freedom of the press can have in fighting corruption.

First, Kenyans learned about the scandal as it was ongoing and have stayed riveted to the news as the complex story has slowly unfolded. The government had no time to cover up the story and divert attention to other issues, as they undoubtedly would have done if they had the opportunity. In fact, the government's bumbling attempts to try and conceal their actions once the story broke just made things worse. In addition, the increased vigilance of the press and the impact it is having is encouraging journalists to "break" other scandals. Perhaps Anglo Leasing is Kenya's Watergate.

Second, the Anglo-Leasing story refused to die. The story broke in May 2004 and is still dominating headlines as of early 2006. The press and other actors—including, most recently, former government anticorruption czar John Githongo—kept it in the headlines once it became apparent that no one was being held accountable. The relentless manner in which the Kenyan press has pursued this story has been important, not just in getting heads to roll but also in fighting the culture of political amnesia that has afflicted

Kenyan protesters call for the resignation of corrupt leaders in Nairobi in early 2006.

Kenyans. The fight against corruption requires persistence—keeping scandals in the headlines as the Kenyan press has done shows real tenacity.

Third, that the government felt compelled to engage in some (ineffective) damage control when the story broke demonstrates change. That ministers linked to the Anglo-Leasing scandal have recently resigned in response to the public's clamor for action demonstrates the empowering effect of access to information.

Corruption-fighting tactics have also gotten more innovative. Civil society groups like Transparency International Kenya, the Operation *Firimbi* ("Whistleblowers") Network, and We Can Do It have begun to focus on addressing corruption at a local level, with the goal of increasing grassroots awareness about the pernicious nature of corruption. The groups have also begun to "rebrand" the fight against corruption by emphasizing the tangible impact corruption has on Kenyans' daily lives.

This may seem like a no-brainer—of course corruption has a negative impact on citizens, how can anyone not see that? Well, if you are a Kenyan living below the poverty line, struggling to put food on the table, and not paying taxes, the Anglo-Leasing scandal is very remote to you. It's just politicians doing what politicians always do: scamming people. So why should you get all riled up about it? In fact, you aspire to get into a position where you too can eat one day without having to work too hard.

The Kenya National Commission for Human Rights (KNCHR) is an independent national institution established by an act of Parliament, and is one of the organizations at the forefront of rebranding corruption. The commission's campaign demonstrates the link between corruption and the lack of access to basic rights like food and shelter, and highlights the diversion of public funds from government coffers by focusing on wasteful government expenditure. In conjunction with Transparency International Kenya, the commission recently released the first in a series of "Living Large" reports that highlight government waste in a dramatic fashion. For instance, the KES 878 million (about $12 million) spent on luxury vehicles for government officials from 2002 to 2006 could have provided antiretroviral treatment for 147,000 Kenyans for a whole year, or seen 25,000 children through eight years of primary schooling (*New Zealand Herald,* January 31, 2006). The report also highlights countries, like Rwanda, that have more prudent policies. In response to the public outcry caused by the report, the government recently announced that it will cut spending on cars and review its policies.

While it remains to be seen whether the culture of corrupt, conspicuous consumption will change any time soon, the work being done by KNCHR is having a tremendous impact as far as making the corruption's true ramifications tangible to the average Kenyan. Kenyans now want to do something about corruption not because of some massive civic education effort, lofty convention, or donor-sponsored program, but because they recognize that they ultimately pay the price for corruption. It is their money, and they want to see how it is being spent. ▯▯

RESOURCES

Transparency International
http://www.transparency.org/
Transparency International (TI) makes the struggle against corruption a global endeavor. It has more than eighty-five independent national chapters, and works on both national and international levels to combat corruption through a coalition of civil society, business, and government. The organization is focused on reforming systems and preventing corruption, and publishes the *Corruption Fighters' Toolkit.* An invaluable resource, the kit "highlights the potential of civil society to create mechanisms for monitoring public institutions and to demand and promote accountable and responsive public administration." By underlining the variety of tools used by TI around the world—from street theater to access to information law—the guide serves as an invaluable "compendium of practical civil society anti-corruption experiences, described in concrete and accessible language."

MoveOn's 50 Ways to Love Your Country: How to Find Your Political Voice and Become a Catalyst for Change by MoveOn.org (Inner Ocean Publishing, 2004)
You gotta start somewhere, and MoveOn's book will help the political-action newbie get inspired and get started. The fifty short personal essays are written by MoveOn's members, and are woven together with the two cents' worth of some impressive leaders, including Al Gore, Nancy Pelosi, and Gail Sheehy. Topics range from creating a successful online petition to starting your own radio show. Although the book focuses more on "traditional" voting-based actions, such as petitioning and mobilizing underserved voters, than on conquering new media, it does provide additional tips and resources to tie on-the-ground action to online action. There are also a few pithy reminders of things we too often forget, like this statement from Al Gore: "Voting is how we come together, as Americans and as believers in self-rule. There is no greater or more profound right of citizenship. Take it from this veteran of a close and controversial election: The process matters. Sometimes you win, sometimes you lose. And then there's that little-known third category. But democracy wins when all of us get in gear and participate with enthusiasm and passion and heartfelt commitment."

Africa Unchained: The Blueprint for Africa's Future by George B. N. Ayittey (Palgrave Macmillan, 2005)
Controversial and confrontational, Ayittey doesn't pull punches when describing his theories of what has gone wrong with Africa and what might be done to fix its problems. The blame, he says, lies with an older "hippo generation" that, on the pretext of fighting colonialism and Northern exploitation, has itself grown fat off corruption and complacent about the challenges the continent faces. On the other hand, Ayittey places much hope in the emerging "cheetah generation," who are more interested in seeding an African renaissance through transparent government and indigenous models of entrepreneurship than in rehashing the battles of the past. Whether he is right or wrong is a subject of great debate, but there's no doubt that anyone who reads him will learn a lot about Africa, its future, and the value of transparent leadership.

"The cheetahs know that many of their current leaders are hopelessly corrupt, and that their governments are ridiculously rotten and commit flagitious human rights violations. They brook no nonsense about corruption, inefficiency, ineptitude, incompetence, or buffoonery. They understand and stress transparency, accountability, human rights, and good governance. They do not have the stomach for colonial-era politics. In fact, they were not even born in that era. As such, they do not make excuses for or seek to explain away government failures in terms of colonialism and the slave trade. Unencumbered by the old shibboleths over colonialism, imperialism, and other external adversaries, they can analyze issues with remarkable clarity and objectivity."

Opposite, left: During a 2006 press conference in Mexico City, the executive director of Human Rights Watch holds up a report calling attention to violations against rape victims' human rights.
Opposite, right: Former Liberian president Charles Taylor arrives in Monrovia en route to his war crimes trial in Sierra Leone, 2006.

Demanding Human Rights

Change demands people who are willing to point out abuses of power, challenge authority, and hold wrongdoers accountable. The world needs courageous watchdogs, because it's their vigilance that creates and preserves the transparency that holds power in check.

This is precisely why being a watchdog tends to shorten your life expectancy. Speak up against a dictator, and you are likely to find yourself arrested, jailed, tortured, or shot. To the outside world, you may just simply "disappear." Even in more democratic countries, blowing the whistle is often dangerous.

If we're serious about reforming politics, defending human rights is job number one. Thankfully, we are beginning to get better at doing just that. We are becoming more effective at holding human rights abusers internationally accountable. We're spreading tools for defending the lives and work of human rights workers themselves, mostly by making sure the eyes of the world are upon them, so they will not simply disappear.

Those of us in wealthier democracies have tremendous power in the watchdog arena. Not only can we pressure our own governments to improve their records on human rights and to promote rights abroad, we can get directly involved in efforts to watch over the watchdogs. AS

Universal Jurisdiction

General Charles Taylor has a lot to answer for. During his insurgency, his term as president of Liberia, and the civil war that followed from 1989 to 1996, at least 150,000 Liberians were murdered, tens of thousands were raped or mutilated, and Liberia became one of the poorest, most devastated countries on the planet. In Liberia today, hunger and disease are still killing; clean water and electricity are no longer available; and schools, hospitals, housing, and roads have been utterly destroyed. To paraphrase a UN official, it'll take two decades for Liberians to get back to where they were in 1988, and the psychological scars may never heal.

Taylor bears much of the responsibility, having unleashed his army—drugged-out child-soldiers in wedding gowns, combat boots, and voodoo regalia—on his own people. His era was marked by unimaginable extreme violence, including mass rape, ethnic cleansing, ritual killings, and cannibalism. He is, by nearly all accounts, guilty of crimes against humanity.

Unfortunately, Taylor is not unique. The twentieth century was putrid with dictators, strongmen, criminal bosses, and tribal leaders who went far beyond any norms of human behavior—even human behavior during times of war—in ordering the grossest atrocities.

How do we stop the tyrants? Certainly, the global community needs to be more responsive.

Peacekeeping after trouble has broken out is noble, but proactive and early intervention (entirely possible in military terms) would save millions from blighted futures and death.

There is, however, a tool other than military intervention: universal jurisdiction. Universal jurisdiction makes certain heinous crimes matters of universal law, subject to prosecution in any duly constituted court in the world. At the end of March 2006, Charles Taylor was arrested on charges of murder, sexual slavery, rape, the exploitation of child soldiers, and other war crimes. He faces trial by the Sierra Leonean Special Court, under authorization of the UN Security Council, which will take place at the facilities of the International Criminal Court in The Hague. If he is convicted, the international community is expected to cooperate in finding prison facilities for his incarceration.

Taylor's trial stands to demonstrate the strength and effectiveness of universal jurisdiction. If no country is exempt, if no bank account is sacrosanct, if no murderous rampage is sufficient to cover their tracks, dictators (and their cronies) will be held responsible for their actions. There will be no escape, no comfortable retirement. It may not save humanity from the bullies and lunatics, but it will weaken their grip and make them think twice: for them, somewhere, the gavel will always hang in midair. AS

Ending Torture

Governments torture people to get information. Repressive governments also brutalize their citizens as punishment, and in order to control people through fear. In supposedly legitimate regimes, torture (including "soft torture," such as forcing victims to stay awake for days on end or humiliating them, as the U.S. military did to detainees in Abu Ghraib) is meant to function as a tool for finding out things that wouldn't be revealed in normal interrogations.

Torture, of course, is unethical no matter who's attaching the electrodes. Torture is, plain and simple, a gross violation of the victim's human rights. In addition, it nearly always leaves lasting damage, even (sometimes especially) when no physical scars can be seen. As the United Nations puts it, "Rape, blows to the soles of the feet, suffocation in water, burns, electric shocks, sleep deprivation, shaking and beating are commonly used by torturers to break down an individual's personality. As terrible as the physical wounds are, the psychological and emotional scars are usually the most devastating and the most difficult to repair. Many torture survivors suffer recurring nightmares and flashbacks. They withdraw from family, school and work and feel a loss of trust." Torture can ruin the lives of the victims, while degrading the torturers.

Those who defend the use of torture often acknowledge the horror involved, but claim torture is necessary for national security or other important aims, like preventing terrorist attacks. There's one problem with their arguments: torture doesn't work.

Professor Darius Rejali knows more about torture than nearly anyone else in the world, and his research shows conclusively that nearly all the arguments made in favor of torture are bogus. Modern advocates for the use of torture often claim that democracies use torture in different ways than dictatorships do, that torture can be, in Rejali's summary "safe, legal, professionally administered, rare and more effective than other forms of interrogation" (Rejali 2002).

That's bunk. Once soldiers or police begin torturing, the historical record clearly indicates, they always slide into more extreme and systematic violence. Nor does "professionalism" prevent excessive torture. There is no known way to use torture safely. Indeed, the emerging medical field of "torture treatment" has found unequivocally that torture victims and torturers alike almost never avoid suffering lifelong consequences.

Furthermore, physical pain and humiliation neither secure the compliance of the victim nor yield secret truths. In fact, torture may even produce less reliable information than traditional interrogations do. In the 1940s, the U.S. Supreme Court prohibited police from applying physical pressure on suspects. Police departments have since taught a range of new psychological techniques that are equally (or more) effective. People will confess to anything to escape the pain of torture, and false confessions and coerced false intelligence are worse than useless. After all, if a cop is chasing the wrong guy, the bad guy is running free.

Indeed, if we want more effective law enforcement and more protection over human rights, we ought to be simultaneously spreading better and more legitimate policing techniques and more aggressively holding accountable those who torture. If what we want is a safe and democratic world, torture chambers can have no place in it. AS

Witness

Founded in 1992 by musician Peter Gabriel, the human rights organization Witness has partnered with more than two hundred human rights groups in fifty countries, supplying video cameras and communication gear to allow people on the scene to document abuses of human rights. Witness attempts to create pressure for change by shining a light on injustice around the world. The participating volunteers are incredibly brave; they have only their cameras and the truth for protection.

Videotaping abuses is only the first step, though, and getting that video out to those who can bring it to the world has traditionally been difficult. That's where the Witness Web portal comes in, allowing people to send in images and recordings from digital cameras and camera phones. Things change when you can send your exposé over the Internet; speed and breadth of access are the best allies for transparency, and the Internet has both in abundance. Once damning photos or video have been released onto the World Wide Web, there's no bringing them back—efforts to do so are only more likely to draw attention.

The execution room at Abu Ghraib Prison near Baghdad was once used to hang opponents of Saddam Hussein and is now a notorious U.S. military prison.

More people bought camera phones than any other kind of camera or any other kind of mobile phone in 2005; with the Witness Web project, every single one of those phones, and every digital camera, can be a tool for the protection of human rights. JC

Forensics and Human Rights

Dictators kill people to silence them, to literally bury their voices. But forensics experts are helping the dead speak. Thanks to the work of one dedicated organization, the use of forensic anthropology to document and validate war crimes and human rights violations is starting to go mainstream, even in countries where the carnage has only recently ended. Forensic experts are doing more than granting closure to traumatized families—they are creating lasting public records pertaining to some of the worst criminals in history, and the evidence is being used to prosecute those criminals whenever it's still possible to do so.

The Argentine Forensic Anthropology Team (EAAF) was founded in 1984, shortly after democracy was restored to the country. The military government that ruled from 1976 to 1983 left behind a terrible legacy of more than 10,000 people who had "disappeared," the majority of whom were kidnapped, tortured, murdered, and then dumped in anonymous mass graves in local cemeteries. The first efforts by local officials and doctors to locate and identify these victims were often clumsy and inefficient—for example, they used large cranes for excavations that destroyed bones and other evidence—so victims' rights groups asked the American Association for the Advancement of Science for help. The association sent a delegation, including some of the world's experts on forensic anthropology to Argentina, and the EAAF was born.

Over the next five years, the delegation exhumed remains using traditional archaeological and anthropological techniques, all the while training a local team that could continue the work after the delegation was gone. It didn't take long for the newly formed EAAF to extend its efforts beyond Argentina. The EAAF has provided evidence to the United Nations War Crimes Tribunals on the former Yugoslavia, and has made ongoing missions to South Africa to investigate apartheid-era disappearances in conjunction with that country's Truth and Reconciliation Commission [see Ending Violence, p. 470].

In every case, the EAAF's objective is manifold: locate the victims; exhume the remains; determine the causes of death and patterns of torture and human rights abuses; return the remains to the victims' families; and finally, present the findings to the press and judicial bodies to ensure that word of these abuses is disseminated to the public, and that the criminals involved are brought to justice.

An important part of EAAF's work is forming local forensic teams; in countries where these sciences are not developed and oral testimony is still given precedence over physical evidence, the teams are often the first of their kind. Not only can local teams reach their communities better, as they're already familiar with language and customs, they are an absolute necessity: the work can often take decades. The EAAF is still searching for the disappeared in Argentina, more than twenty years

after the first delegation came and went. Without a local infrastructure to support continued research, the last case would have been closed when the last foreigner flew home.

Outreach is just as important as investigation. The EAAF holds regular presentations for key decision makers, and reaches out to victims' families by involving them in every step of the process, ensuring they have access and information throughout the recovery. This level of involvement and transparency is crucial to families' healing.

Finally, the EAAF is pushing to get its forensic teams much better access to DNA information. Genetic testing is indispensable to these investigations, but it's incredibly costly. To remedy this problem, the EAAF has started to create DNA blood banks—designed to hold multiple samples from victims' families and to cross-check data quickly and efficiently.

Each stage of the EAAF process is thoroughly documented with written records, photography, and videography. The organization is creating an audiovisual center in Argentina to further educate the public about its findings. Working with the human rights organization Witness, EAAF made its first documentary about its work, *Following Antigone*, which featured footage recorded by EAAF members in Argentina, El Salvador, Ethiopia, Haiti, and East Timor. CB

Benetech

It takes human rights organizations and forensic teams years of painstaking documentation to build their cases. All of that work can be lost in a flash: human rights workers are increasingly the targets of abductions and harassment, during which they suffer personal harm and their computers are stolen and offices ransacked.

Benetech, a nonprofit that develops technology for humanitarian purposes, has created an open-source database program designed specifically to protect the work of human rights observers. Their Martus software is pure genius: not only does it provide an easy way to log and organize all the countless details of each case (and share that information with other groups), more importantly it backs up all information on remote servers, so the case files survive even if the laptop they were created on doesn't. With this remote backup system in place, workers also have the option of deleting sensitive files from their hard drives completely if that level of security is needed. The Martus program also comes with a simple-to-use but nearly impossible-to-crack encryption feature for files and e-mail alike.

A story from Colombia illustrates why Benetech's software is so invaluable to human rights workers. EQUITAS, the Colombian Interdisciplinary Team for Forensic Work and Psychosocial Services, has been recovering the remains of the many people who disappeared during Colombia's decades of complex and violent internal conflicts. The team's involvement in an excavation at a sensitive site—a former ranch that might hold the remains of hundreds of people—drew the worst kind of attention to the group. A female staff member was hijacked, assaulted, and robbed of her computer and cell phone. Fortunately, EQUITAS was already using Martus—all the files on the worker's computer were encrypted and saved on the remote server, and her attackers got none of EQUITAS's precious information.

Martus is available for download (www.martus.org) in English, Spanish, Arabic, French, Russian, Thai, Nepali, and Persian versions, and so far people from over seventy countries have installed the program. The Thai version is being used to help Southeast Asian groups document human rights abuses in Myanmar.

Benetech's role in aiding human rights work doesn't end with this amazing software. The company's Human Rights Data Analysis

Opposite, left: A forensics expert examines remains found in a mass grave near Sarajevo, Bosnia, 2005.

Opposite, right: Witness trains villagers, like this man, on the Thai-Burmese border to record human rights abuses.

Group (HRDAG) program conducts and publishes statistical analyses of large-scale human rights abuses. Most recently HRDAG compiled a report that definitively quantified the magnitude of human rights violations in East Timor's long struggle for independence. The team established that at least 102,800 Timorese died as a result of the conflict, approximately 18,600 of whom were murdered, while the remainder died of hunger and illness directly related to conditions caused by the conflict. CB

RESOURCES

Amnesty International
http://www.amnesty.org/
Held as the gold standard of human rights groups, Amnesty International works to fulfill and protect the rights detailed in the Universal Declaration of Human Rights, focusing on state oppression. The organization's mission "is to undertake research and action focused on preventing and ending grave abuses of the rights to physical and mental integrity, freedom of conscience and expression, and freedom from discrimination, within the context of its work to promote all human rights." Not affiliated with any nation, religion, political doctrine, or economic interest, Amnesty International has consistently fought for real social change, from ending the arms trade to campaigning against domestic violence.

Human Rights Watch
http://www.hrw.org/
The largest human rights organization based in the United States, Human Rights Watch (HRW) has continually been at the forefront of international human-rights-violations inquiries and campaigns. The organization conducts fact-finding missions around the world and publishes annual reports, which garner extensive media coverage—and embarrass both the offending nations and the world. HRW was instrumental in the global response to human rights atrocities in Rwanda and Kosovo, and in moments of crisis puts hard pressure on global governments for immediate action, including but not limited to withdrawal of military and economic support. The extensive array of reports available on the HRW Web site is organized by both country and global issue.

Why Societies Need Dissent by Cass R. Sunstein (Harvard University Press, 2003)
We value free expression not because free expression makes us feel good: we value it because dissent, differences of opinion, and new ideas are the rudders of democracy—without them, democracies inevitably sail straight ahead into tragedy. This lesson should be tattooed onto the brain of every student on the planet, but unfortunately, it's one that defenders of democracy must teach their grown-up peers again and again. Sunstein argues this point clearly, and goes further to explore how any group can become irrational when the tendency to mentally herd together is not balanced with encouragement to think for oneself:

"The problem is that widespread conformity deprives the public of information that it needs to have. Conformists follow others and silence themselves, without disclosing knowledge from which others would benefit ... Conformists are often thought to be protective of social interests, keeping quiet for the sake of the group. By contrast, dissenters tend to be selfish individualists, embarking on projects of their own. But in an important sense, the opposite is closer to the truth. Much of the time, dissenters benefit others, while conformists benefit themselves."

Rights of Man by Thomas Paine (1791) (Penguin Books, 1984)
Thomas Paine is in many ways the archetypal worldchanger: besides being the writer most important to the American Revolutionary cause (his pamphlet "Common Sense" helped stir popular support for independence and still bears reading today), he was also the first to announce in clear, passionate terms what human rights are and why they're worth defending. His language is sharply barbed—he spared no one's feelings—but his arguments still thunder, perhaps even more loudly in an age where many attack human rights in the name of national security or religion:

"The inquisition in Spain does not proceed from the religion originally professed, but from this mule-animal, engendered between the church and the state. The burnings in Smithfield proceeded from the same heterogeneous production; and it was the regeneration of this strange animal in England afterwards, that renewed rancour and irreligion among the inhabitants, and that drove the people called the Quakers and Dissenters to America. Persecution is not an original feature in *any* religion; but it is always the strongly marked feature of all law-religions, or religions established by law. Take away the law-establishment, and every religion reassumes its original benignity. In America, a Catholic Priest is a good citizen, a good character, and a good neighbour; an Episcopalian Minister is of the same description: and this proceeds, independently of the men, from their being no law-establishment in America."

Watching the Watchers

More and more, we live in a world where what we see, hear, and experience is recorded wherever we go. As our recording technology gets more portable and easier to hide, few statements or scenes will go unnoticed or unremembered. It's possible that, in the not too distant future, our day-to-day lives will be archived, saved, and even available over the Internet for recollection, analysis, or sharing.

This isn't happening because a Big Brother government is looking over our shoulders, or powerful corporations are watching us with security cameras and RFID (radio frequency identification) tags. That kind of surveillance is far outstripped by the millions of cameras and video recorders in the hands of millions of Little Brothers and Little Sisters—us. In our camera phones, we carry with us the tools of our own transparency, and we do so willingly, even happily.

Let's call this world the Participatory Panopticon. "Panopticon," originally a prison building design that allowed guards to see all inmates at all times, has more recently come to mean a *society* under constant observation. And it's already developing.

Camera phones and similar devices are constantly getting better at gathering and sharing information about the world. As it becomes cheaper and easier to just leave them running, we may find that it's very useful to have a "backup memory." We could use them as "TiVos"—the service that allows us to watch TV shows whenever we want—for our own lives, recording everything around us, including what people say and do, all for easy review. Microsoft and Hewlett-Packard are among the companies already writing software to help manage this kind of abundant information. Done right, such tools would make it possible to never

again forget a face, a name, or an important bit of information.

Many of us would see this as a serious loss of privacy, with enormous legal implications regarding liability, self-incrimination, even intellectual property. But it could also serve the cause of truth. A police officer lying about hitting a protestor, a despot lying about human rights abuses, an executive lying about dumping toxic waste—these are easier to catch in a world where everything can be on the record. The results might surprise us: corporations could be forced to become more transparent to stakeholders, and certain officials could be required to wear a recorder while on duty (already, some police cars automatically videotape police stops). Ironically, it could be a world where trust would be easier, because it would be harder to get away with lying.

Is all of this really possible? Not yet, but the pieces are coming together faster than you might think. The Participatory Panopticon is not something people are building intentionally; it is the accidental result of technologies we've come to think of as the basic tools of our lives. JC

Wearable Cameras

Think all of this is far off in the future? Think again. At least two companies already sell wearable cameras.

Deja View offers a small hat-mounted camera and recorder that constantly buffers the last thirty seconds of whatever you're looking at, saving the buffer to permanent storage at the press of a button. It's designed to hold four hours of video, transferable to your PC for later viewing. The company is marketing the cameras to anxious parents who don't want to miss out on capturing baby's first steps or words.

DoubleVision, from Second Sight Surveillance, is a head-mounted camera hooked to a portable hard-disk system, with an optional LCD display for on-the-spot review. Looking a bit like a small gun stuck to the side of the head, it's similar to the Deja View but aimed at a very different audience: military, police, and security outfits.

Anyone who dismisses these devices because of their obvious limitations (bulky camera and cable, clumsy belt-pack storage, limited battery life, inability to wirelessly send signals) hasn't been paying attention. *This* version is ugly, ungainly, and far too limited—but it's just a hint of things to come. JC

Sousveillance

Who watches the watchmen? Maybe we all should. This is known as "inverse surveillance," or "sousveillance," meaning "watching from below." When citizens use cameras to record the actions of officials, either political or corporate, that's what they're doing. As the New York City government discovered recently, sousveillance can be a surprising equalizer.

In 2004, New York City police arrested nearly two thousand people during demonstrations around the Republican National Convention. The mayor and the chief of police condemned protestors for "rioting"

Protesters on the streets of New York City hold up video cameras, demonstrating the power of civilian surveillance during the 2004 Republican National Convention.

and "resisting arrest," and provided the press and the courts with videotapes taken by police officers, showing protestors out of control.

But it turned out that the police weren't the only ones armed with video cameras. Cheap digital video cameras employed by free-speech groups showed people being swept up without cause and without resistance. It turned out that prosecutors selectively edited the official video record to prove their cases, and police officers repeatedly misrepresented the protest events at trial. According to the *New York Times,* 91 percent of the nearly 1,700 cases ended with the charges being dropped or with not guilty verdicts. A startlingly large number of these cases included citizen videos that clearly showed that the police and prosecutors were lying. So the next time around, don't expect the police to politely ignore citizens with video cameras.

Unfortunately, people carrying video cameras, even small ones, are pretty obvious. But people carrying mobile phones are not. Videophones and higher-bandwidth networks will transform activism. So the next time around, we'll see the transmission of dozens, hundreds, thousands of views from marches and protests live over the Web.

As easy-to-use, portable cameras become commonplace, we're seeing more examples of sousveillance in action, from student camera-phone recordings of teachers harming classmates to the Video Vote Vigil, an online clearinghouse of recordings of voter obstruction and harassment. In the United Kingdom, the delightfully named Blair Watch Project was an effort, coordinated by the *Guardian,* to keep tabs on Prime Minister Tony Blair as he campaigned around the country. The project was sparked by the Labour Party's attempt to limit Blair's media exposure on the trail; instead, Blair had more cameras on him than ever. JC

Earth Phone

Imagine a world of networked environmentalists, something like the Witness project [see Demanding Human Rights, p. 443], but focused on environmental wrongs rather than human rights.

Imagine building a Web portal, collecting recordings and evidence of what's happening to the planet, and putting news and data at all kinds of people's fingertips, from researchers and activists to business and political figures. It would highlight the changes to the planet, and would give voice to people who are willing to work to see a new world, a better world, take shape. It would give everyday citizens a chance to play a role in the protection of the planet.

This Web site would document the ecological problems, human-caused or otherwise, but would also showcase the good ideas, successful projects, and efforts to make a difference that deserve much more visibility. It would show us two worlds: the world we are leaving behind, and the world we are building for generations to come.

The key components are already here. The site itself would encompass ways to send photos and trade messages over the Web, combining a photo-sharing service and social-networking platform system. All of this data could be tagged with geographic information and combined and synched with online maps for easy viewing and analysis. The online component would be created by the users, working together and working openly.

For the participants, camera phones would be fundamental tools. We may not remember to bring our digital cameras with us in our day-to-day travels, but we rarely forget our phones. Camera phones are incredibly popular, outselling both other types of mobile phones and other types of digital cameras, according to statistics on WirelessMoment.com.

But imagine making a phone that's dedicated to this purpose—an "Earth Phone." Besides the camera, it would have small, cheap environmental sensor attachments. Such add-ons for phones aren't out there yet, but students and engineers around the world have attached atmospheric sensors to bicycles, handheld computers, and cheap robots. They've even stuck sensors on the backs of pigeons to measure smog-forming pollution. It would be easy to put the same sensors on a phone.

The idea of a phone with a sensor chip is not new. Tech companies around the world offer specialized phones that can sniff for bad breath or process blood tests in the field, and an enormous variety of tiny, inexpensive sensors are already on the market. We could easily build phones with environmental add-ons that measure temperature, CO_2 and methane levels, or even the presence of biotoxins. Such a phone might even be built by its users, spinning off from current open-source mobile-phone design projects.

The Earth Phone Web site would mix what you witness with what thousands or millions of other people around the world witness. It would give us far better knowledge of what's happening on our planet environmentally than could be gathered with satellites and the handful of government and academic sensor networks alone. As we work to figure out ways to mitigate the worst effects of climate disruption, every little bit of information matters.

The Earth Phone would be a collaborative, bottom-up approach to environmental awareness and protection. Moreover, considering how today's global youth rely on mobiles, this system could put the next generation at the front lines of gathering environmental data. A system like the Earth Phone would be a tool for us all to improve our awareness of the environment and, ultimately, improve the planet itself. JC

Protest

Things have changed a lot since the protests of the 1960s. These days, protests aren't about gas masks, bail money, and picket signs.

For one thing, scales and targets have changed dramatically. Your parents (or grandparents) may have made a righteous stand against Nixon's White House, but as a post-millennial protester, your targets may include not only your own government, but also other countries' governments, multinational corporations, and even abstract alliances and trade agreements like the Group of Eight (G8) and the World Trade Organization (WTO). The situation grows more complex if you believe, like the hippest factions of the protest scene, that there is no difference between corporations and countries anymore.

It's a complex web of tangled alliances out there; you may show up and discover that the corporation you're picketing is also the corporation that, directly or indirectly, made your shoes or printed your last paycheck. Not only has the line between friends and foes become blurred, but it has become fractal in its complexity—if it even exists at all.

In fact, you may not even have to carry a picket sign, or even show up

Rosa Parks's refusal to give up her seat on a bus in Montgomery, Alabama, and her subsequent arrest signaled a turning point in the civil rights movement, 1956.

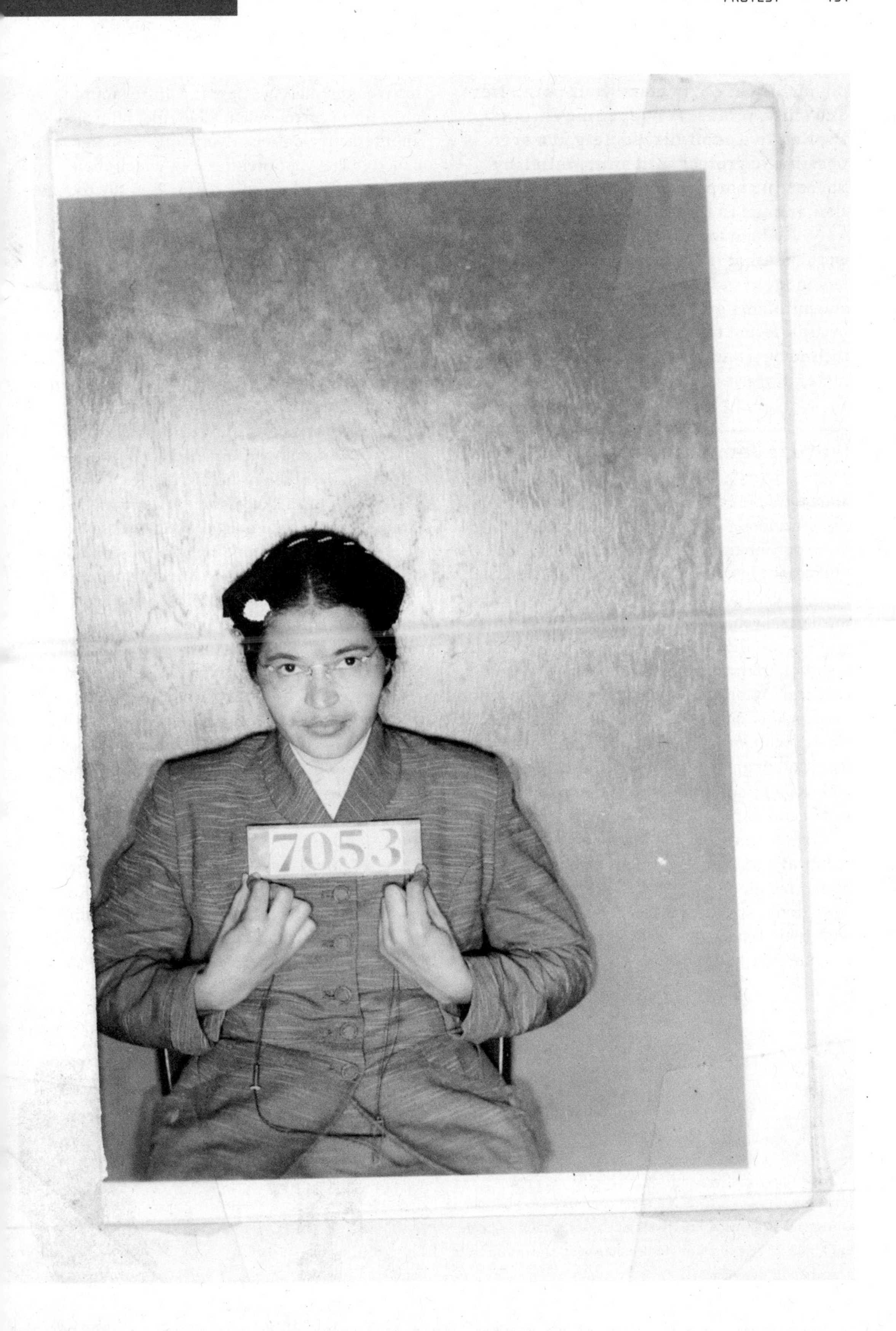
7053

physically at all, to make your point. More and more, protest is happening through the Net. In a capitalist society, it's even possible to protest with your wallet, by boycotting corporations and nations that engage in unethical practices.

Along with everything else in the world, dissent is changing. Political action the postmodern way requires no massive time commitments and no membership in dodgy groups. It just takes you, your personal technology, and a desire to make the world a better place. JE

Culture Jamming

If the boycott is the resolute parent of the consumer-activism family, culture jamming is the defiant and subversive teenager: brash, but often easily ignored. Culture jamming is the act of using existing advertisements and other public media as a vehicle for protesting the institutions that created them. It's a symbolic protest, drawing attention to the realities behind many deceptive corporate facades. Culture jamming could only exist as a recent phenomenon: it's meta-activism—a billboard altered in protest of the proliferation of giant ads on our landscapes; cons and slogans modified to make damning political statements about the companies they represent.

Advertising is a numbers game, where the more often a message can be hammered home, the more successful it is. Culture jammers don't have the budget to saturate society with their images like commercial advertisers do, but their one-time acts have proven effective nonetheless. A great culture jamming moment occurred when Chevrolet launched an interactive online advertising campaign called "Chevy Apprentice," in which the company asked visitors to the Web site to help create commercials for their giant SUV, the Tahoe, by uploading their own video clips, music, and text. Of course, Chevy didn't specify that suggestions had to promote the Tahoe, which gave culture jammers a prime opportunity to use the company's own platform to soapbox about global warming and the other environmental woes SUVs contribute to.

Culture jamming in general, and online efforts in particular, catalyze outreach and solidarity among consumers, two necessary (though not sufficient in and of themselves) ingredients for a successful movement. Culture jammers often scheme and execute their ideas in groups, thereby building strong communities of activists and pooling creative strategies for conveying a message. Those who view the culture jammers' modified ads and logos often feel a sense of solidarity and inclusion as the recipients of a particular and unexpected statement. Oftentimes we don't feel motivated or empowered to act upon cultural trends that bother us until we realize how many others are sharing our experience. When culture jammers strike, many otherwise quiet citizens can be moved to speak up or act on their convictions—and corporations begin to feel threatened when a critical mass of perturbed consumers raises its voice. Thus a new cultural paradigm is emerging, in which the critical partnership between company and customer is managed more by the customer than by the company. JF & AS

Forkscrew Graphics, a design team committed to political and social awareness, created these "iRaq" posters, spoofing Apple's iPod campaign, to promote what the group calls a "more real" freedom.

Organizing Around Films

The Day After Tomorrow is a sci-fi action thriller in which global warming causes an extreme shift in ocean currents, generating catastrophic weather anomalies: tidal waves in New York, tornadoes in Los Angeles, and snowstorms in New Delhi. In a matter of days after the first disasters strike, the entire Northern Hemisphere is freezing over into the next ice age. The release of the film in 2004 was accompanied by a storm of its own—in the media.

Months before the film's release, a small group of shrewd online activists began writing about the movie's potential to bring the issue of global warming to the forefront of America's national agenda. Jon Stahl of ONE/Northwest (Online Networking for the Environment) suggested on his blog that climate-change activists organize around the movie because "big movie releases can be a news hook for discussion of serious issues." At first, environmental organizations hesitated to latch onto the film because they feared that the over-the top scenarios and extreme special effects would cause audiences to dismiss the reality of climate change. But the online murmurings continued (including at WorldChanging.com), and soon the mainstream press caught on. Advocacy organizations began campaigning around the movie as a wake-up call about the dangers of climate change. Before long, former vice president Al Gore was planning town hall meetings to discuss the movie and the Web site MoveOn.org was putting together thousands of volunteers to hand out leaflets to moviegoers on opening weekend. The groundswell of Netroots activity (grassroots meets the Internet) converted *The Day After Tomorrow* from trite summer movie fare into an effective tool for political activism.

Activists are increasingly using the Internet as a tool for grassroots organizing around movies. Amnesty International regularly supports films that address human rights issues through its Artists for Amnesty program. Working with filmmakers and volunteers, Amnesty International has created online resources and activism tools around many films, including *Hotel Rwanda*, *The Constant Gardener*, and *Innocent Voices*. MK

Protest Art

Feral robotic dogs roam toxic-waste sites, chasing down deadly emissions. Our sandals contain test tubes that sample the soil as we walk. Sleek, highly designed genetic mutant babies are the perfect product for the millennial lifestyle.

These are only a few of the strange realities conjured up by artists like Natalie Jeremijenko, Amy Franceschini, and Patricia Piccinini. Their work—which serves as both commentary on and protest against technology and its discontents—is part of a growing wave of art that examines the dehumanizing effects of the modern condition, using the very technologies that are subject to criticism.

"What I'm most interested in is: how do we characterize systems of which we know very little, and have very poor information?" said Natalie Jeremijenko in a 2004 Worldchanging interview. "Knowledge is very partial, very incomplete, and yet decisions are made. So, I

In her work, "F.R.U.I.T.," artist/activist Amy Franceschini raises awareness by wrapping oranges in paper printed with information about urban farming and the resources required to transport the food we consume.

specifically try to design information systems that measure urban environmental interactions."

Her "Feral Robotic Dogs" project is one such information system. The dogs—built from robotics kits and fitted with devices that measure environmental variables such as radioactivity and air quality—are let loose at Superfund waste sites and English power stations. "Because the dog's space-filling logic emulates a familiar behavior, i.e. they appear to be 'sniffing something out,'" says the project's Web site, "participants can watch and try to make sense of this data without the technical or scientific training required to be comfortable interpreting [an] EPA document on the same material."

Amy Franceschini's "Soil Sampling Shoes" and "F.R.U.I.T." also deal with environmental issues and their impact on culture. The former—a pair of sandals with an array of test tubes instead of soles—theoretically let a user surreptitiously take soil samples at Superfund sites. The latter studies how fruit gets to our local markets from farms hundreds or even thousands of miles away. Custom paper "wrappers"—with information printed on them about the resources, labor, and transit required to supply us with fruit—help to raise consumer awareness about the complex economic system involved in even the simple purchase of an orange.

Patricia Piccinini's work deals with the ramifications of genetic engineering. Her "Mutant Genome Project" conjures up the LUMP (Lifeform with Unevolved Mutant Properties), a creepy "designer baby" created solely as a lifestyle accessory. "Protein Lattice" depicts lovely nude models interacting with mice that have human ears growing on their backs (a case of art imitating a reality both fascinating and gruesome). And her recent Venice Biennale *We Are Family* installation displays a series of sculptures and digital images of human/animal hybrids, grotesque and stunning at the same time.

Like most of the best artists who deal with social themes, Piccinini is rather ambivalent about her mutants—unlike many of her contemporaries, she is less a protester than a documenter. In an introductory essay to the Biennale catalog, curator Linda Michael writes, "Though they may be in some way failed or mutant creations, her figures have a kind of innocence that makes it easy to see beauty in the grotesque. We are free to imagine new futures that are unconstrained by outworn social philosophies. Piccinini always does this in a way that makes such futures understandable in terms of what we encounter in everyday life." JE

Banksy

To call Banksy a graffiti artist does little to distinguish him from the scads of other spray-paint guerillas who illustrate the world's walls and sidewalks. Banksy has frequently been characterized as an "art terrorist," which implies a violence that is absolutely absent in Banksy's work. His one-man mission is to confront cultural failings by stealthily positioning his own art among museum collections, on walls, and anywhere in the public view.

In 2005, for instance, Banksy visited Israel's West Bank, where he spray-painted several large images on the Palestinian side of the border wall to give the appearance of a hole through

A Banksy work on a West Bank section of the security wall that separates Israel and Palestine.

which the viewer could see a tranquil beach scene full of palm trees, or a snowcapped mountain and inviting forest. The subversiveness of the "art attack" drew mixed reactions from both sides of the wall—and from around the world—but nobody would dispute the power of the act, or the convention-busting audacity of the artist.

Theatre of the Oppressed

How can an interactive theater exercise performed by slum dwellers inform government policy? Initiated by Augusto Boal in Brazil, the Theatre of the Oppressed (TO) uses diverse games and interactive theater techniques to teach oppressed citizens how to concretely transform their society. Exercises pose dilemmas and challenges to participants, tailoring content to the core social problems and power structures of particular communities and society at large. Legislative Theatre, for instance, shows how TO techniques can help transform citizen desires into laws. In a mock legislative session, participants adopt simple new laws; afterward, they put pressure on real lawmakers to approve the new laws. While Legislative Theatre has had a significant impact on the actual laws in Brazil and other countries, often participants invent laws that are already in effect. The exercise thus not only enables them to inform the law-making process, but to become better informed themselves about existing laws so that they can exercise their rights. MB

RESOURCES

Legislative Theatre: Using Performance to Make Politics by Augusto Boal (Routledge, 1998)
Founder of the international movement Theatre of the Oppressed, Boal describes in his book taking Hamlet's advice, "theater is a mirror in which may be seen the true image of nature, or reality" as a starting point for Legislative Theatre, aiming to make "theater as politics rather than merely making political theater." *Legislative Theatre* is a mandate in progress—a project still in flux and discovery—working to find new ways of involving everyone in the democratic process. This book is a unique and important guide to an alternative means of effecting social change, providing real tools for real theater that will lead to real action:

"The opening of the show for a community is an important moment, a big step. If the rehearsals are already a form of political activity in themselves (the citizens talk to one another and try to pinpoint their oppressions, to understand them by means of aesthetics), the shows are the moment of social communion, in which the other members of the community are invited to participate in the debates, still using the same theatrical language."

Smart Mobs: The Next Social Revolution by Howard Rheingold (Basic Books, 2003)
In the pivotal *Smart Mobs,* Rheingold explores the transformation of collaboration, cooperation, and community via electronic networks, mobile devices, and innovative new thinking about computer-mediated social organization. Rheingold was one of the first to comprehend and explain the new "mediasphere" we're swimming in; *Smart Mobs* should be considered required reading for worldchangers everywhere.

Rheingold writes, "The big battle coming over the future of smart mobs concerns media cartels and government agencies that are seeking to reimpose the regime of the broadcast era in which the customers of technology will be deprived of the power to create and left only with the power to consume. That power struggle is what the battles over file-sharing, copy protection, regulation of the radio spectrum are about. Are the populations of tomorrow going to be users, like the PC owners and website creators who turned technology to widespread innovation? Or will they be consumers, constrained from innovation and locked into the technology and business models of the most powerful entrenched interests?"

Direct Action

The time occasionally comes in a democracy when no amount of polite discussion will change a government or corporation's course, when rational argument and evidence make no difference, when average people are ignored in the halls of power. When that happens, it's time to act more directly, to expose the truth in ways that are impossible to ignore. Direct action is a desperate measure, but sometimes it's the only one that will jolt people awake. Sometimes, as the saying goes, when the fox is in the henhouse, the hens need to get a-flapping.

What Is Direct Action?

Direct action includes anything from arranging sit-ins (like the civil rights movement of the early 1960s) to chaining yourself to a bulldozer (like the radical Earth Liberation Front) to blocking off streets with impromptu human barriers and throwing parties where usually only cars dare to tread (like the UK-based Reclaim the Streets movement). Of course, direct action also includes firebombings, assassinations, and even harassing passers-by with low-fidelity megaphones and poorly spelled picket signs. But we're going to assume you're not evil enough to actually do any of those things.

In our media-dominated age, the most effective direct action does the following:

- attracts media attention
- amuses, engages, or otherwise positively affects passive media consumers
- avoids subjecting anyone to death, injury, or prison sentences (though a stint in the local jail is probably OK)

Attracting Media Attention

The top way to attract media attention is to not be boring. If you feel you can best engage the media with megaphones and picket signs, bake sales or drum circles, you might want to let another member of your organization handle this end of operations, or perhaps farm things out to a respectable PR firm.

Always remember: the media does not care one iota about the fine points of your ideology. The media cares about getting ratings and market share. If you can help them achieve this goal, they will give you as much airtime as you want.

The following tend to woo media attention:

- Attractive nude people running around, accompanied by provocative slogans. For example, the PETA (People for the Ethical Treatment of Animals) campaign "I'd Rather Go Naked Than Wear Fur" featured fur-eschewing supermodels who were—you guessed it—naked.

- Celebrities doing unusual things, such as actor Woody Harrelson scaling the Golden

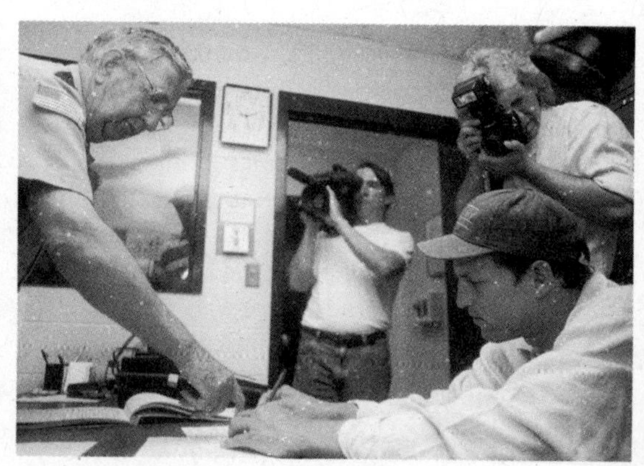

Gate Bridge in 1996 to draw attention to the plight of the Headwaters redwood forest.

- Being very, very funny. This is the strategy of the Yes Men, who pose as representatives of big corporations and government organizations and deliver hilariously absurd but horribly plausible position papers to audiences of senior trade officials and economists.

Catching the eye of the media will involve some cleverness on your part, and usually a willingness to put your most earnest tendencies on hold, at least in public. Nobody—especially not the media—likes a humorless zealot. The cult of irony rules in the Western zeitgeist right now. Chuckles, not chants, are the order of the day.

Make sure your organization's name and URL are plastered on every available surface a camera might happen to focus on—your T-shirts, your vehicles, the nude supermodel's bottom, the handcuffs you've used to lock yourself to whatever it is you're locking yourself to. This is called *brand recognition.*

Remember, as Mao Tse-tung (or maybe Andy Warhol) once pointed out: the only way to defeat the hegemony of popular media is to co-opt its tactics.

Winning Over Your Audience

Here's a little secret: pretty much every member of Western civilization believes, to some degree or another, that the Man is sticking it to them. Oddly enough, this even extends to people who—due to political or economic advantage—actually *are* the Man. We are a society of self-perceived underdogs, scrappy little guys fighting the good fight against the Big Faceless Whatever.

Another secret: most people consider activists to be ivory-tower idiot hippies who know nothing about the real world. You will have to fight this bias by being funny, clever, and most important, tuned in to the needs and desires of the common man.

For example, if you are an ecoactivist, don't demonize loggers for cutting down old-growth forests. Loggers are not evil. Loggers are people who are trying to feed their families by doing a backbreaking and dangerous job for little money, usually in areas that are economically depressed, like the rural Pacific Northwest. Instead, place yourself and the loggers on common ground, fighting the big evil lumber corporations who are both destroying Mother Earth and endangering the future livelihood of the loggers themselves by cutting down all the trees (leaving no more trees for loggers to cut down).

You are now a champion of the Average Joe, fighting the good fight against the devious machinations of the Man. Loggers will now buy you beers and clap you on the shoulder, instead of righteously kicking your idiot hippie behind.

And remember the (Michael) Moore Principle: people respond to humor, not lectures. Making the target of your opposition look like a bunch of evil morons is far more effective than simply debating them.

To paraphrase *Conan The Barbarian:* What is best in life? To crush your enemies, to drive them before you, and to hear the lamentations of their PR flacks. JE

Opposite, left: Activists in protest of Israeli policies block New York's Fifth Avenue by forming a human chain across the street, 2003.
Opposite, right: Actor Woody Harrelson signs a form acknowledging that he was read his rights following an arrest staged to promote the legalization of industrial hemp.
Left: A member of the Yes Men stands with the skeleton Gilda, "the golden skeleton in the closet," as part of a stunt meant to illustrate the risks of trusting corporate promises of socially responsible conduct.

Not Ending Up Dead or in Prison

Look, we know how it is. After a day of listening to Howard Zinn lectures on the old iPod and reading about the destruction of natural resources by greedy power mongers, there's nothing you'd rather do than head down to Starbucks with your rolling crew of concerned global citizens and get medieval on their latte-pimping asses. We understand. Really, we do.

But what will that accomplish? Starbucks' insurance will cover any damage you do. The media will portray you as a windmill-tilting hippie imbecile. You will go to prison for vandalism and destruction of private property.

Being a yardbird is much less pleasant than merely being a passive accomplice to the continuation of the capitalist power structure. Believe us. We know what we're talking about.

You will almost always find that people—and, more important, governments local and national—will place the righteousness of your cause second to their obligation to protect human life and the principle of property ownership.

Also, such activities as climbing billboards, going on hunger strikes, and standing in front of oncoming tanks do carry the risk of actually getting killed, or at least severely injured. Seriously consider the implications of such activities before partaking in them.

Luckily, most of the activities activists engage in—blocking public thoroughfares, interfering with law enforcement, and so on—will at most land you in the local pokey—and local jail, at least in the G8 countries, is usually no big deal. It's almost a rite of passage for many activists to spend a weekend in lockup. The worst thing you will probably have to deal with is your drunk cell mate.

Conclusion

As we've seen, there is a middle ground between mind-numbing picket-line boredom and full-tilt Black Bloc butt-kicking. Just keep these simple concepts in mind and you'll do fine.

And remember: if a protest happens in the forest and nobody covers it for the nightly news, well ... the forest's probably going to get clear-cut in the very near future. JE

Ruckus Society

While some present-day protesters are still content with picket signs, marches, and sit-ins, some groups go further. Take, for example, the Ruckus Society—a radical spin-off of the already radical Earth First! organization (which played a major role, alongside the Direct Action Network, in organizing the anti-WTO protests that drew massive media attention in 1999). The Ruckus Society is the strong arm of the protest movement—and whatever the issue or event, from globalization and genetically enhanced food to the 2004 Republican National Convention and the WTO, the Ruckus Society is there. Their "Action Camps" train activists in direct action methods, including the best way to chain oneself to an immovable object and to safely hang oneself from a billboard—tactics that are always sure to bring high levels of media exposure to the cause (whatever the cause may be).

What distinguishes the Ruckus Society from its predecessors is its level of practical organization and its media awareness. In a 2001 *Rolling Stone* article, writer Dan Baum noted, "When two Ruckus-trained climbers hung a

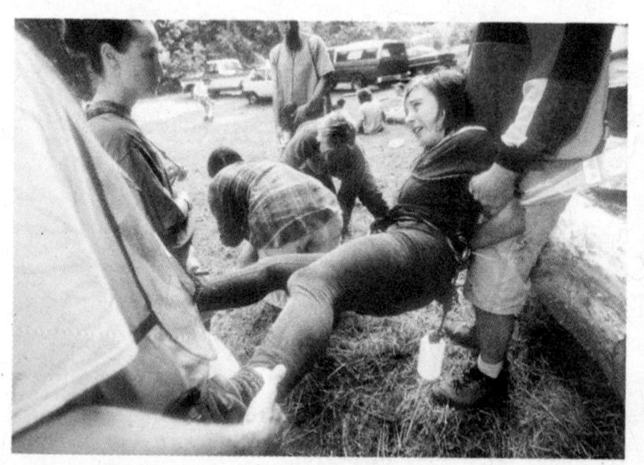

stories-high Stop Global Warming banner off the Los Angeles headquarters of the oil company ARCO a few years ago, they were only the centerpiece of a fifteen-person operation. Others directed the media to the best camera angles, delivered carefully prepared sound bites about ARCO to the cameras, monitored the police radio bands from a truck full of scanners and waited at the jail with bail money."

Ruckus Society spokespeople are always quick to point out that nonviolence is one of their core beliefs—though they seem to interpret the term as applying exclusively to acts performed on living beings. "The Boston Tea Party was property destruction," said Ruckus director John Sellers in a *San Francisco Examiner* interview (June 19, 2002). "I don't think a lot of people debate about whether it was violent or nonviolent. I think most people would say it was nonviolent." JE

Smart Mobs and Swarming Protests

Smart mob is a term coined by futurist Howard Rheingold to describe a decentralized, coordinated form of protest that is as reliant on technological tools as is the establishment it is leveled against. Smart mobs consist of individuals who may not even know one another; their protests are coordinated via the Web, the telephone, SMS messaging, and sousveillance [see Watching the Watchers, p. 448].

How do they do it? The answer is complicated, but the short version is, they *swarm.* Well trained in civil disobedience, linked by cell phones, e-mail, and online maps, operating in affinity groups and clusters, backed by ad hoc teams of independent journalists, neutral observers, and lawyers, protesters can flow across an urban landscape with a speed and coherence that leaves security forces scratching their heads. As it stands now, no city on earth can't be shut down by 10,000 serious, swarming protesters. Anywhere cell phones will work and airlines fly, activists can own the streets.

In 2003, smart mobs in San Francisco protested the American invasion of Iraq using cell phones, text messaging, and pirate radio to coordinate swarmlike movement, staying one step ahead of police. Many protesters used digital video, still cameras, and Wi-Fi connections to document the protests as they happened. Activist Lisa Rein shot video of San Francisco police as they attacked protesters; she describes the scene on her blog: "After the cops rushed the crowd, and [selected] certain individuals and [had] them put their hands behind their backs since they were going to be arrested, the crowd [began] booing and screaming. 'The whole world's watching,' it screams. ('Ha!' I thought to myself, 'I wonder if the crowd or the cops know how true that is!')"

As communications and documentation technologies become more portable and affordable, smart mobs become more and more effective. In the Philippines, smart mobs were partially responsible for the ousting of President Joseph Estrada in 2001. We have not even begun to see what other feats can be accomplished when the right people with the right technology are in the right place at the right time. JE

Opposite, left: Activists practice techniques of passive resistance at the Eco-Avenger Basic Training Camp, run by the Ruckus Society.

Opposite, right: A young labor-movement supporter joins a rally in downtown Manila, Philippines.

Nonviolent Revolution

I suppose that human beings looking at it would say that arms are the most dangerous things that a dictator, a tyrant, needs to fear. No. It is when people decide they want to be free. Once they have made up their hearts and minds to that, there's nothing that can stop them.
Archbishop Desmond Tutu

The world is lousy with dictators and repressive politicians, ruling mafias and militias. If we want to live in a better world, we have to get rid of them. Unfortunately, the hard work of actually turning them out always falls to the folks they're oppressing. Most of the time, doing away with a truly awful regime involves starting a revolution.

Revolutions sound cool to those who've never been through one. Real revolutions are terrifying and dangerous. They're actions undertaken by sensible people only when it is absolutely clear that nothing else will change a regime, when the only alternative left is to either compel change or completely overthrow the system.

Revolutions can be incredibly violent. But employing violence is often the worst way to overthrow those in power. The more despotic the government, the more violence it has at its disposal. Starting the shooting when the other side has most of the guns is bad strategy. Also, the more violent the revolution, the more likely it is that violent figures will come to power in its aftermath, just as Stalin followed the Russian Revolution, or Madame Guillotine the French. What's worse, unsuccessful violent insurrectionary groups—from the Shining Path to the IRA— often slide away from their principles and become merely well-organized thugs.

We are learning, however, how to do revolution right through the tool of nonviolent confrontation. At first glance, the idea of a "nonviolent revolution" seems counterintuitive, but nonviolent revolution works. In the last three decades, fifty out of sixty-seven successful citizen revolutions were largely or wholly nonviolent, according to the International Center on Nonviolent Conflict.

There is nothing "soft" about nonviolence. Some of the hardest-headed revolutionaries on earth have chosen nonviolence for entirely calculated, rational reasons. But one of the many advantages of nonviolent revolution is that it serves the heart as well as the head, by leaving open the possibility of postrevolutionary reconciliation and cultural renewal. Nonviolent revolution is not only more likely to succeed, but it heals instead of harms and leaves a better society in its wake. AS

Left: Archbishop Desmond Tutu, chairman of the Truth and Reconciliation Commission in South Africa, at the end of 1998.

Opposite: Mahatma Gandhi at the beginning of his fast for peace in 1948, New Delhi, India.

How Nonviolence Works

Nonviolence works by destroying the ability of those in power to use force without losing the essential support of those on whom their continued stay in power depends.

In order to undermine the power of a regime, revolutionaries need to do two things: win the public over to their cause, and convert the regime's key supporters.

To win the public's support, nonviolent revolutionaries must dramatize the issues. They must create symbolic actions—like Rosa Parks did when she refused to sit in the back of a segregated bus—that arrest the public's attention and help shatter the legitimacy of the regime in the eyes of the people. In war, the military historian Basil Liddell Hart teaches, the strategy that is best is that which allows the most flexibility in concentrating an overwhelming amount of force at a critical weak point in the opponents' forces. In nonviolence, the point is to use a small group's willingness to take a blow to concentrate the public's attention on a telling symbol of the regime's moral weaknesses.

To peel away the regime's supporters, revolutionaries have to split off elements of the ruling bloc, convincing those people who hold power, on whom the regime depends—the police, businesspeople, judges, military, and clergy—that change is both in their interests and in line with their ethics. Nonviolence is not passive: it is a form of fighting back—but fighting back, as the saying goes, with different weapons and to a different end. The end is to convert. When critical parts of the regime defect to the cause of change—when, for instance, the police refuse to break up peaceful demonstrations—the regime's days are numbered.

All of this requires intelligent hard work and extreme bravery. Nonviolence, done well, demands discipline, commitment, strategy, and leadership; it takes careful planning and training, role-playing exercises, education, and spiritual preparation. That doesn't mean that nonviolent activists need to wait for an entire movement to emerge before beginning their work. Small groups are the shock troops of nonviolence; small groups of people always take the first steps, and become the nodes around which an entire network of action coheres. But even small groups benefit from knowing how to use their tools well. AS

A Force More Powerful

First they ignore you, then they laugh at you, then they fight you, then you win.
Mahatma Gandhi

Since Mahatma Gandhi led the Indian people in rising up peacefully against the occupying English Empire, more than half a century ago, the study of nonviolence has become an art. Not only can nonviolence be taught, but the lesson plans are readily available.

One of the most impressive groups distributing the curriculum is the International Center on Nonviolent Conflict (ICNC). The ICNC does standard advocacy for nonviolence—helping the media (which tends to find gunplay a more interesting story) understand the role of nonviolence in real change; pushing for foreign policies that better support civil society groups in countries with authoritarian governments—and also provides direct access to tools for understanding and employing nonviolent confrontation directly.

To that end, the center has produced a triumvirate of great resources, collectively titled "A Force More Powerful." If you're interested in learning more about nonviolence, there is no better place to start. Of the three tools, the meatiest is the center's book *A Force More Powerful: A Century of Nonviolent Conflict,* a five-hundred-page exploration of the history and practice of nonviolence. A documentary and a video game are two additional resources that complement the book. The ICNC documentary, also called *A Force More Powerful,* covers the American civil rights movement, the Indian independence movement, the Polish Solidarity labor movement, the Chilean movement that toppled the brutal dictatorship of General Augusto Pinochet, and the South African boycotts that helped end apartheid. Perhaps most inspiringly, it also tells the story of Danish resistance to the Nazi occupation. Overwhelmed by the Nazi war machine, the Danes practiced "resistance disguised as collaboration." They refused the Nazis' demand to turn over their Jewish fellow citizens, hiding them in their homes and then smuggling them by the thousands out of Denmark to safety in fishing boats. Then the Danish resistance engaged in sabotage, strikes known as the "go home early movement" (where workers left their factories hours early each day, supposedly to work in their gardens, but in reality simply to slow down the German war effort), and constant psychological pressure on occupying troops, such as symbolic "two-minute stops" where, at a predetermined time, everyone and everything moving in the streets would suddenly freeze, stand still and silent for two minutes, and then continue. Their struggle not only undermined the Nazi war effort, but it also helped the Danes emerge from the war with their national conscience intact.

But it is *A Force More Powerful: The Game of Nonviolent Strategy* that is both the most fun, and in some ways, the most illuminat-

ing tool. Taking the role of the leader of an emerging nonviolent movement in any of ten scenarios, video game players learn to grow their movement, build alliances, and use a variety of pressure tactics to isolate and change the regime.

These "serious games" offer one of the best ways to learn. Playing with possibilities allows us to understand systems in ways intellectual study can't. The principles embodied in *A Force More Powerful* are tried and tested; in fact, a number of veteran nonviolent activists consulted on the project, and Ivan Marovic, one of the leaders of the Otpor student movement, which brought down Serbian dictator Slobodan Milosevic, played a key role in its design. AS

Ukrainian Nonviolent Protest

The Ukrainian presidential elections in November 2004 sent citizens into nonviolent protest in the street, opposing what was exposed as a fraudulent and corrupt process. The masses made a significant statement, but the cry of resistance that resounded the loudest was actually a silent and singular one—expressed through sign language via an interpreter for Ukraine's state-run television station.

On November 25, according to the *Washington Post,* as Natalia Dmytruk signed the news of the elections on TV, she took the opportunity, standing live on national television, to stop interpreting the newscaster and to tell deaf Ukrainians what she felt they needed to know: "I am addressing everybody who is deaf in the Ukraine. Our president is Viktor Yushchenko. Do not trust the results of the central election committee. They are all lies ... And I am very ashamed to translate such lies to you. Maybe you will see me again—" She then completed the newscast according to the script, unsure of what awaited her after the broadcast (April 29, 2005).

Word spread, and protesting escalated until a reelection was scheduled, and in December of that year, Viktor Yushchenko emerged victorious. Nonviolence is a strategy in which each of us can play a part. AS

Opposite, left: *A Force More Powerful: The Game of Nonviolent Strategy* is a resource that teaches players about nonviolent movement building and effecting regime change. **Opposite, right:** Revelers in Kiev, Ukraine, wave flags to commemorate the one-year anniversary of the "Orange Revolution," which brought reformer Viktor Yushchenko into power in 2004.

RESOURCES

The Art of the Impossible: Politics as Morality in Practice by Vaclav Havel (Alfred Knopf, 1997) Vaclav Havel stands as a worldchanging icon: principled yet open-minded, humble yet strong willed, subversively funny yet in deadly earnest, the playwright, human-rights dissident, and former president of the Czech Republic has been a major influence on an entire generation of Eastern European activists, artists, and writers, who've responded to his thoughtful postmodern humanism. *The Art of the Impossible* brings together many of his most powerful political essays and speeches. Together they bear witness to the fact that moral strength, intellectual courage, and political determination can change even the most hopeless situations for the better:

"I have experienced a beautiful revolt of children against the lie that their parents served, allegedly in the interest of those very children. Our antitotalitarian revolution was—at least in its beginnings—a children's revolution. It was high-school students and apprentices, adolescents, who marched in the streets. They marched when their parents were still afraid, afraid for their children and for themselves. They locked their children in at home, took them away from the cities on weekends. Then they began marching with them in the streets. First out of fear for their children, later because they become infected by their enthusiasm. The children evoked from their parents their better selves. They convinced them that they were lying and forced them to take a stand on the side of truth."

Waging Nonviolent Struggle: 20th Century Practice and 21st Century Potential by Gene Sharp (Extending Horizons Books, 2005)
Founder of the Albert Einstein Institute and the preeminent scholar of nonviolence, Gene Sharp has devoted his life to furthering the understanding and successful practice of nonviolence, and has been pivotal in some of the most important nonviolent struggles of our day. *Waging Nonviolent Struggle* is the essential compendium of his work and practice, debunking myths about nonviolence, exploring sources of power and human nature, and providing key historical examples of the overwhelming success of nonviolent techniques. We are left with no choice but to believe that "expanded applications of nonviolent struggle in the future will not only contribute to the reduction of major violence but to the expansion of democratic practices, political freedom, and social justice. The choice is ours."

Diary of a Political Idiot: Normal Life in Belgrade by Jasmina Tesanovic (Midnight Editions, 2000)
Jasmina Tesanovic writes with extraordinary moral clarity about what life feels like inside a desperate and oppressive nation. Her *Diary,* written as NATO was bombing Serbia during the Balkan conflicts, is a beautiful, vulnerable, and honest account of the mental changes an educated, liberal person undergoes as she watches her country slide over the brink and keep falling. It's also a testament to the courage it takes to turn one's back on people and institutions one can no longer support, and the emptiness that living through times that demand such courage can leave behind:

"The referendum was yesterday. One of my friends said, now they'll come and shoot us because we didn't vote. I told her she was just being paranoid. After the vote, our president gave a speech. Everything about him, his face, his voice, his words, and his emotions, was so familiar—that pathetic, patronizing tone that says he knows what's best for me, just like my father. This president, whom I know to be a corrupt liar and merciless enemy, worries for my future—that is why he is sending me to war, that is why he is fighting the rest of the world. That is why he makes me weep."

Ending Violence

Most of the worst humanitarian disasters on the planet are the result, at least in part, of war. War may be—indeed, often is—triggered or worsened by other factors: the killing fields of Rwanda were at their most terrible where hunger was most dire. But the fact remains that the immediate cause of much suffering in the world is a teenage boy with an automatic rifle.

Increasing evidence supports the notion that security and sustainability go hand in hand, that attempts to create one without the other are doomed from the start. If we are to create a more stable world, we need to do a better job of heading off conflict (or negotiating its end), of stopping genocides, and of bringing killers to justice. Curtailing the trade in small arms would be a good idea, too, in many activists' books. We also need a better understanding of the kinds of problems violence creates, in order to better relieve suffering and more quickly restore stability in war-torn areas.

The *New York Times* quoted an International Rescue Committee (IRC) report that found that only 2 percent of those dying in African war zones are violently killed. Many more die from the resulting disintegration

Opposite, left: Child soldiers of the rebel group Sudan People's Liberation Army walk away from their weapons after surrendering, Malou, Sudan, 2001.
Opposite, right: The role of the United Nations' "blue hats" is increasingly under debate, but is critical to moving toward a more peaceful future.

of the social fabric, already weak economic structures, and loss of scarce resources. Indeed, the IRC says, the primary cause of death for innocents in a time of war is their forced flight into perilous conditions. As they flee from violence, families get separated, medical supplies, food, and clean water disappear quickly, and otherwise preventable illnesses ravage whole populations.

All of this suggests that when it comes to heading off the worst effects of war, we need to adopt a multifaceted approach. Not only do we need to reinvent the way we deliver refugee aid, we need to stop the fighting in the first place, and create incentives for lasting peace. Ending the violence of war sets the stage for building bright green peace. AS

Making Better Blue Hats

To war is human. But to stop wars, prevent genocide, help spread harmony—this is the work of peacekeepers. These soldiers are willing to fight and die to keep the peace; it's a job that has only existed for a few decades.

Though they are defamed by nationalists of many stripes, derided by some globalists as being too timid ("what the UN does best is count the dead" a famous condemnation begins), United Nations peacekeepers—known as blue hats—change the world. The noblest calling a soldier can answer is to be a "soldier of peace." As the world is subject to more and more stresses and conflicts, it will have more and more need of peacekeepers of all sorts. If there is to be a bright green future, it will be built in part on the courage of peacekeeping troops.

Peacekeeping in its current form, important as it is, suffers from a huge number of problems. Romeo Dallaire is outspoken about the worst problem: simple inaction. As head of the UN peacekeeping mission that found itself powerless to check the genocide in Rwanda, Dallaire may have agonized more than any other person on earth about how the world ought to act when faced with monumental acts of evil. His conclusion? The UN must act before, not after, genocide has begun.

A growing number of experts agree with Dallaire that the international community has to learn how to stave off genocide, not just regret it. We must learn more about what kinds of conflicts lead to disastrous ruptures of the social fabric, which ones can be stopped early, and how to mitigate those that erupt full-blown. The social science of peacekeeping is a worldchanging enterprise.

But it's also clear that the UN peacekeeping forces themselves need to undergo some fundamental reforms, and that everything from their rules of engagement to their technologies and tactics must change. They must be better equipped and prepared; more attention must be focused on training units from different armies to work together; and absolute accountability must be maintained, so that soldiers don't exploit or abuse the civilian population and thus undermine local support for peacekeeping. More coordination among peacekeepers, police, and emergency workers is also needed. The list goes on, with various governmental, military, and NGO experts debating the needed reforms, and while implementation of those

reforms has been slow, the trend is clear: better peacekeeping missions are on their way. AS

Environmental Peacemaking

In a world where natural systems are strained to the limit and billions of people struggle daily to survive, what we're used to calling "natural disasters" are growing both more common and more dangerous. They are inherently more destructive than they used to be (thanks in part to climate change), but they also feed into (and their effects are made worse by) ongoing humanitarian crises and violent conflicts.

The idea that poverty, the environment, and security are linked is not a new one. Neither is the idea that disaster response needs to move beyond relief to become an opportunity to rebuild something better. Neither, for that matter, is the idea that a healthy environment can buffer the worst effects of natural disasters while providing critical ecosystem services [see Ecosystem Services, p. 486].

It all works together, or it doesn't work at all.

The good news is that these holistic approaches are already catching on, and those who respond to crises have started thinking in multiple arenas and across multiple disciplines. Winning a war doesn't always create a useful peace. Conversely, conflicting groups can move toward peaceful resolution, or at least recognize the necessity of alliances, in the face of shared environmental destruction, as occurred after the Pakistani earthquake of late 2005.

An Algorithm for Solving Civil War

Civil wars are notoriously hard to bring to a negotiated peace. In general, the sides fight until one or both are exhausted, in the process wreaking havoc on the lives of a nation's people. If only there were a remedy to help us end civil war quickly.

The Santa Fe Institute's Elizabeth Wood studies civil conflict in an effort to uncover the systems of resolution that succeed and endure. Wood's method involves using a kind of mathematics known as game theory to find the settlement most acceptable to both sides, and to prove that ending conflict more quickly will ultimately offer more gain than prolonged fighting over fragments of territory.

Her work is still theoretical, but it has serious real-world implications. And while models and figures on their own will probably never convince people to stop murdering one another, being able to concretely show warring parties why peace is in their best interest can only help the work of peace negotiators and peacekeepers.

Security in the Twenty-first Century

Like almost everything else in our lives, international security in the twenty-first century will be very different than it was in the twentieth. Warfare is evolving owing to changes in global realities.

The first big change is that state-versus-state warfare is on the way out, made obsolete by nuclear weapons and global economic interdependence. Don't be confused by the invasion of Iraq,

it was an outlier. One-on-one wars between states in the twenty-first century will be exceedingly rare.

That's good, but as interstate warfare has been declining, state-versus-nonstate warfare has surged. It started with 9/11, and it continues in Iraq, Nigeria, southern Russia, Thailand, and many other places. When we examine it in its entirety, we see that it has become a global contest (some would say an epochal war) between an alliance of states—led in some ways by the United States—and a large and amorphous community of nonstate foes. The reason for this is globalization (the same force that made state-versus-state warfare obsolete).

If we plunge below the level of the state, we see a plethora of groups, formed around the traditional moral bonds of family, clan, tribe, ethnicity, and religion (or, in the case of gangs, around manufactured loyalty). These groups have always provided a means of mutual survival. However, as technology, transportation, and trade began to connect substate groups to the global community, those groups gained skills and influence that, in combination with mindsets that are far from progressive, have made them very dangerous.

It should come as no surprise that some of these groups don't share the developed world's vision of the future. They see their religion corrupted, their economy in decline, and their environment degraded by a global system of impersonal markets, nation-states, and media. They are not unjustified in believing the goals of our world put their world at risk. Their very cultural survival is in question.

The result is war. In the past, state-on-nonstate war would have been containable to a locality (typically a backwater). However, in our new global environment, nonstate groups have gained incredible leverage. They can now wage war on states, both locally and globally, and win.

Why? These substates have developed a new method of warfare that uses nonhierarchical forms of organization very much like what we see in the open-source software community. Open source–style warfare is highly decentralized. It allows many groups, regardless of motive, to coordinate and mutually advance against common enemies. In Iraq, which is increasingly the laboratory for this century's new war, seventy-five to a hundred different insurgent groups are fighting the U.S. military, its coalition partners, and the new Iraqi government with a considerable amount of success. Not only is this type of nonorganization almost impossible to destroy (since there's no leadership cadre to decapitate), it is able to rapidly deploy innovations that meet or exceed those deployed by the U.S. military, which is finding it hard to keep up.

Another component of the new state-versus-nonstate warfare is systems disruption, strategic attacks without the use of weapons of mass destruction. This type of warfare uses our systems against us. For example, on 9/11 the transportation system was used as a weapon. In Iraq, attacks against power systems, oil pipelines, and social connections have hollowed out the newly emerging state. Not only do such disruptions render the state unable to deliver those services that are the foundation of its legitimacy, but they create an unstable environment that prompts more groups to enter the open-source war, on either side. Systems disruption is easy (we have too many networks

Opposite, left: Indians and Pakistanis work together to deploy relief tents to victims of the 2005 earthquake in Kashmir, a disputed area between the two countries.
Opposite, right: Insurgents in Fallujah, Iraq, take up position to launch mortars against U.S. troops, 2004.
Left: In what appeared to be an act of terrorism, residents of villages in Nigeria were forced to evacuate their homes when fire erupted along an oil pipeline there.

to defend), and it provides unprecedented levels of leverage. A small group that spends thousands of dollars on an attack can cause hundreds of millions of dollars in damage to the state it is attacking—as we have seen in the ongoing attacks on the Iraqi and Nigerian oil systems.

These groups self-finance their operations through transnational crime, which has limitless growth potential. Globalization has smashed the barriers between states. Transnational crime (some call it black globalization) is now a multi–trillion dollar economy and, according to Moises Naim, the editor of *Foreign Policy* magazine, it is growing seven times faster than legal trade. This shadow economy allows nonstate groups access to a global smuggling network that handles everything from arms to illegal immigrants to drugs to knockoff products to money laundering. They have quickly learned to use this network to sustain their war with the world. The ongoing attacks in Nigeria are financed through the bunkering and export of stolen oil, and the 2004 Madrid train bombings were financed through the sale of the drug ecstasy.

Over the longer term, these system-disrupting, transnational crime-fueled sons of global fragmentation will cause a radical change in how we think about informational security. As their reach spreads, they may melt the global map. Parts of states will fall under their control, and many states will hollow out. If this happens, nonstate groups will inspire more participants and see their fortunes rise as state control recedes. Unfortunately, we who live in democracies aren't prepared for this—neither our exceedingly expensive militaries, built for interstate warfare, nor our fragile economies will protect us.

What will protect us will be difficult to implement, particularly in the United States. It will require that the country change its approach to how it uses energy. Green technologies will have to become a central aspect of all of our lives. It is only through the broad adoption of green technology that we will achieve the decentralization and independence that can mitigate the shocks of global systems disruption. If al Qaeda's attack on Saudi Arabia's Abqaiq oil facility had been successful in February 2006, this would be crystal clear to all of us.

For the United States to survive the new realities of war, it must give up being a superpower. The role is too expensive to maintain, particularly considering the massive national debt. Given that we live in a world dominated by fluid markets, flirtation with bankruptcy isn't a sound long-term strategy. The United States is critical. If it can overcome these barriers (retool its defense system, abdicate as superpower, and shrink the national deficit), it could emerge intact, or even stronger than it is today, and help support the global system. If not, international violence may continue to spin out of control, putting our global system, as we understand it, at risk. For the sake of all of our futures, the world's sole remaining superpower needs to chart a new course. JR

Real National Defense

War undermines every aspect of sustainability: it destroys the environment; it upends the lives of those caught within it; it impoverishes

Rescue workers cover up casualties after a passenger train was bombed by terrorists in Madrid, Spain, 2004.

whole nations. At the same time—human nature being what it is—we all want reasonable assurances that we are safe from the aggressions of our neighbors. How do we balance these apparently contradictory needs? How do we avoid war and all the vices associated with it while keeping ourselves safe? An interesting set of answers has begun to emerge in the last couple of decades. (For an excellent overview of some of these answers, start with Amory Lovins's essay "How to Get Real Security," freely available on the Web.)

First, military personnel themselves can be one of the most effective checks to aggression. In nations where officers and troops are inculcated with democratic values, ethical practices, and cultural understanding, there is less support for foreign adventures. Those values, practices, and understandings can themselves be reinforced through cooperation with the militaries of potential rivals, via programs such as "troop swaps" (where soldiers from one nation serve in units with those from another) and joint peacekeeping missions.

Second, international treaties, like the land-mine treaty, when backed by effective monitoring and enforcement, can not only help prevent weapons of mass destruction from falling into the wrong hands, but they can cool down conflicts (and save lives) by cutting off the flow of weapons to bad actors. In addition, diplomatic efforts, when backed by strong coalitions and sanctions, can undermine the ability of aggressive nations to pose a threat, and give them real reason to play nice.

Third, nations can intentionally adopt defensive military postures. As Lovins writes, "To date, Sweden has executed the most sophisticated design of military forces for nonprovocative defense. Its coastal guns cannot be elevated to fire beyond Swedish coastal waters. It has a capable and effective air force, but with short-range aircraft that can't get very far beyond Sweden ... In every way, by technical and institutional design, they've sought to make Sweden a country you don't want to attack, but one that is clearly in a defensive posture. This approach can ultimately create a stable mutual defensive superiority—each side's defense is stronger than the other side's offense. Each has, by design, at most a limited capacity to export offense."

When such defensive nations need to project power, they can do so in coalition with other countries' militaries, under the auspices of bodies like NATO and the UN. Individually, they threaten no one; with allies, they retain the capacity to wreak havoc on rogue nations.

Finally, the ultimate way to prevent war may be what some call "leader control." Democracy itself can be a check on war: when democratic states are in good political health, they maintain checks on the military ambitions of their leaders, and hold those leaders accountable for their actions. International accountability, achieved through measures such as universal jurisdiction [see Demanding Human Rights, p. 441], provides disincentives to starting wars even for leaders in less democratic countries. Knowing a war-crimes tribunal will be their eventual reward for aggression has a marvelous tendency to make invading other nations less attractive, even to tyrants. AS

Troops attend a ceremony in Sarajevo to transfer peacekeeping duties from NATO forces to the European Union, 2004.

Truth and Reconciliation

Few countries in the world have traveled as tumultuous and inspiring a path toward racial reconciliation as South Africa has. The transition from apartheid South Africa, ostracized around the world, to Nelson Mandela's beloved Rainbow Nation is a remarkable story. Key to that story was the role of the Truth and Reconciliation Commission (TRC), chaired by Archbishop Desmond Tutu.

The TRC was a means of exacting justice without retribution. Any agent of the apartheid regime who testified fully and truthfully was granted immunity for his actions. The goal was to bring sunshine into the dark corners of South Africa's history, not wreak vengeance.

It has been argued that the greatest gift of the Commission was "a common history of the apartheid era," without which two histories would be competing for legitimacy. Regardless of how dark the past was, South African children now have a common historical view on which they can build their future. Cynthia Ngewu, a mother who lost her son during apartheid, said of the TRC process, "This thing called reconciliation ... if I am understanding it correctly ... means the perpetrator, this man who killed Christopher Piet, if it means he becomes human again, so that I, so that all of us, get our humanity back ... then I agree, then I support it all" (Rotberg and Thompson 2000.)

Values systems as diverse as Christianity and the Zulu philosophy of *ubuntu*, meaning "a person is a person through other persons," guided the commission and contributed to its success. While massive challenges remain, and the process has not been perfect, the commission surely stands as a towering testament to the capacities of the human spirit. ZH

Radio Okapi

Radio can be a force for destruction: witness the role of Radio Mille Collines in urging the Hutus of Rwanda to commit genocide. Radio Okapi—"Breath of the DRC" (the Democratic Republic of Congo)—set up by Swiss journalists as part of the reconstruction process in that country, shows the power of radio for peace.

The difficulties involved in creating an independent national radio station in a decimated country are not to be underestimated. As one report on USAid.gov puts it, "The DRC has practically no roads. It has no railway system. The river routes were closed and mail and telephone services did not work."

The phones may not be working, but it appears Radio Okapi is.

Okapi now claims the country's most listeners, and it is the most trusted and capable source of news and of educational and public-service programming. Call-in talk shows, however, are what the people tune in to hear: Okapi's talk-radio programs often include human-rights defenders and development experts, and debate is lively in all five of the languages—French, Lingala, Tshiluba, Swahili, and Kongo—in which the station broadcasts.

The plan is to make Okapi a public, listener-supported radio station after the United Nations pulls out of the DRC. Can talk radio

heal the wounds of a war that's killed 2.1 million people? Not by itself. But by providing a space where Congolese can discuss and debate, it's lending a hand in rebuilding the nation. EZ

RESOURCES

Pax Warrior
In this computer-simulation documentary about the 1994 Rwandan genocide, players take the role of a UN peacekeeping commander in Rwanda, and must make a series of increasingly difficult decisions as evidence mounts that something awful is taking place.

There isn't an obvious "right answer" in the simulation; seemingly correct choices can have unforeseen (yet utterly plausible) results: your evacuation effort may in fact put those you are trying to help at greater risk of attack from Hutu militia units. The game's goal isn't necessarily to stop the genocide (although that would obviously be an ideal outcome), but to learn more about the complexity of managing peacekeeping situations.

Blueprint for Action: A Future Worth Creating by Thomas P. M. Barnett (Putnam Adult, 2005)
Barnett's earlier book, *The Pentagon's New Map: War and Peace in the Twenty-First Century,* changed the debate about the proper role in global politics of United States military force. Here Barnett goes further and suggests that unless America uses its might to promote development, spread peace, and encourage democracy, no number of new weapons systems can keep it safe.

"Once a state exits a civil war situation, it must endure a roughly ten-year recovery period during which it builds its economy back to where it was prior to the civil war. During that ten-year period, the country has about a 4-in-10 chance of lapsing back into civil war during the first five years and a 3-in-10 chance over the second five-year period. What the data shows is that one of the biggest triggers for renewed civil war is the tendency of the surviving state to spend prodigiously on arms in the years immediately following the conflict, which in turn tends to provoke the suspicions and renewed aggression of rebel parties. What tends to dampen such spending most is the introduction of foreign military troops to keep the peace. If such troops can be offered for approximately five years on average, then the country in question typically hits a growth-recovery spurt in years 4 and 7 following the end of the conflict, and when the bulk of the country's post-conflict economic recovery is achieved, the odds of slipping back into civil war decrease dramatically."

Opposite, left. The first witness is sworn in before the Truth and Reconciliation Commission in South Africa, 1996.
Opposite, right: The head of U.N. Peacekeeping Operations, Jean-Marie Guéhenno, speaks on Radio Okapi, Kinshasa, Democratic Republic of Congo, 2006.

PLANET

In times past, people didn't travel much. Most of our ancestors were peasants, and it was extremely rare for them to venture more than a couple days' journey from their villages. Even our more footloose progenitors, like herders and sailors, by and large stuck to well-traveled paths. A rare few explorers, soldiers, merchants, and pilgrims followed their callings to far-off lands, but most folks lived and died close to home.

Today, we are a planet on the move. According to the Population Resource Center, hundreds of millions of us have left the countries where we were born to start new lives abroad. Hundreds of millions more travel long distances for recreation and business. Jet contrails crisscross the skies, signals flash through fiber-optic cables, and the planet seems to shrink every day.

With all this travel, we've lost our connection to the land around us. Few of us could match the local ecological knowledge of our most ignorant ancestors. On the other hand, we've gained a greater understanding of the wider workings of nature. We may not be able to identify the tree growing in our own backyards, but we can instantly conjure up a satellite photo of our neighborhood on our laptops. We might not be able to point south without the aid of a compass, but our cell phones can tell us our near-exact latitude and longitude. We may not be able to name the birds singing outside our windows, but we can empathize with the sorrows and joys of Antarctic penguins at our local movie theater. We are, in short, completely uninformed about the regions we call home, and yet tuned-in as never before to the workings of the planet as a whole.

What we need to do is to synthesize the two—the global and the local; the technological and the domestic. We need to use the best of the remarkable suite of environmental technologies that are emerging from labs and workshops around the world, and combine them with the kind of local ecological wisdom that comes only from a deep engagement with place. Combining the two will give us unprecedented tools for solving the planet's most dire problems.

Aside from empowering us to be locally proactive, knowing what's going on around us is simply enjoyable. We have innate connections with nature—connections that can be as easily nurtured in the heart of the city (when we know where to look) as in the middle of the woods—and acquainting ourselves with the local plants, birds, and weather helps us feel more truly at home. But learning about nature can feed our minds as well. Anyone who's ever played with an online mapping system knows that looking at the planet in new ways is endlessly entertaining.

We're not relegated to being passive observers, either. With the explosion in citizen science, we can become part of the action, adding our own observations concerning insects or whales or the climate to the larger scientific project of understanding the planet. We can work on local ecological restoration projects, applying the latest in environmental science as we cut brush and plant native trees. We can even help search for alien intelligence and hunt dangerous asteroids. Planetary knowledge can be a playground.

Planetary knowledge can also be a bummer. With our increased understanding of the little rock on which we live comes a flood of evidence that we're royally screwing things up. Ecosystems are trembling. Species that evolved over millions of years are being driven into extinction. The most troubling signs have to do with our climate. Every time we drive a car, we're participating in the largest planetary experiment ever conducted—we're changing the climate, acidifying the oceans, melting the ice caps, and generally wreaking havoc with the very systems that support life on earth.

That said, the same tools that are leading scientists to sound alarms are giving us a better understanding of how to begin tackling the problems. Not only do we know what is causing the climate crisis, but we know what to do about it. Not only do we have concrete evidence that it's real, we increasingly know what to expect locally. Not only can we stop turning our planet into a greenhouse, we can predict with increasing accuracy just what it will take to do that.

Facing the grim realities of climate change, mass extinctions, and ecosystem collapses does not make for a cheerful day. But we shouldn't lose heart.

Astronauts, seeing for the first time the tiny blue-green ball below as they travel farther and farther into the cold, empty darkness of space, report almost universally one overwhelming emotion: a profound, sudden comprehension of the beauty, fragility, and unity of the planet. Most of us will never venture into space, but all of us can learn to see the earth as they have. We can learn to see the whole planet as our home, and decide that it's about time we took care of it. AS

Knowing our planet and moving forward based on that knowledge is one of the greatest challenges we currently face, Lake Natron, Tanzania.

Placing Yourself

"Where am I?" The first question of place is universal; the need for an answer, often urgent. We each stand at the hub of a great turning wheel. How do connections radiate outward from our lives to the economy (the flows of electrons, water, materials, and signals that form the planet's industrial metabolism) and the biosphere (the flows and fluxes that power the earth as a living system)? What is the universe that starts with each of us?

Placing ourselves is place-making: an active engagement with the world that begins at our doorstep and expands outward in both space and time as we learn about and connect to our surroundings throughout our lives. Place is a mind-set that travels with us, grows with us, and helps us frame and reframe the answers to that perennial "where" question.

Since the birth of spoken language, placing has been about naming. The Kwakiutl people of Vancouver Island were not unusual in this regard. Before they came into contact with Western culture, they conferred place-names that told stories, names that could teach. "For them, a place-name would not be something that is," explains author and professor Kim Stafford, "but something that happens. They called one patch of ocean 'Where Salmon Gather.' They called one bend in the river 'Insufficient Canoe.' They called a certain meadow 'Blind Women Steaming Clover Roots Become Ducks.'" Such places are gathered into the imagination so their attributes—abundance or scarcity, hardship or surprise—can be made manifest, and human experience made easier. The Aboriginal people of Australia follow songlines across their continent, literally singing their world into existence. Neither is such intimacy with the land entirely absent in the vernacular languages and dialects of Europe or the tribal languages of Africa, Asia, and the Indian subcontinent. Names, retrieved from the past or across boundaries of culture, offer a handle on place.

Today, placing ourselves is about weaving an understanding of natural facts and artifacts into a durable fabric of identity and citizenship. Many new and powerful tools are available to help us place ourselves, from podcast birdsongs to the thematic maps streamed onto our computer screens. But even with the digital flood of data and imagery, understanding our places entails an effort to untangle our relationships with the planet. By freeing our senses to perceive the landscape around us, in city or country, we open ourselves to a wholly individual understanding of place.

Understanding place as a natural system starts with the simplest of questions: What's overhead? What's underfoot? Who's been here previously? What's that hill called? What's likely to happen here next week? Next season? The answers integrate into an intuitive first approximation of biogeography—the study of the way plant and animal communities are distributed around the earth. In discerning biogeographic patterns, we begin to perceive natural units—watersheds, biomes, ecoregions—that challenge arbitrary boundaries of political jurisdictions. We begin to detect larger patterns of migration, water systems, and seasonal change that give each place its distinctive character. We begin to anticipate the return of a certain raptor, the timing of a berry harvest, the angle of the sun on an April evening. Some of us step from this awareness directly into new allegiances, into a sense of political activism that springs from an expanded sense of community, while others just begin to feel more at home.

Another dimension of placing ourselves is having a clear-eyed grasp of the industrial metabolism in which each human life, as part of a global economy, is embedded. City dweller, farm owner, or migrant laborer, each is sustained by commercial flows of electricity, water, materials, waste, and signals so omnipresent as to be nearly invisible. To understand place, we are obligated to tease out these converging systems, to comprehend the extent of our dependencies.

In the landmark *New Yorker* article "Apartment," environmental writer Bill McKibben (*The End of Nature; Hope: Human and Wild*) describes tracing to their source the energy and water supplied to his Manhattan apartment, and visiting the terminus of the wastewater and garbage discharged from it. His journeys led him to the hydroelectric power dams of James Bay, Ontario, the uranium mine of Hack Canyon, Arizona, an offshore oil loading platform in Brazil, the freshwater reservoirs of New York's Catskill Mountains, and the garbage mountains of the vast Fresh Kills Landfill (since closed) on Staten Island in New York City. Across a hemisphere, locales so diffuse as to be largely invisible to us are a functional part of McKibben's apartment at Bleeker Street and Broadway. The "place" called New York contains even more of them. "Those wires and pipes," McKibben writes, "some of them thousands of miles long, [then] lead from my apartment to the far reaches of New York City."

Does this mean each place bleeds out into the world like the electron cloud around an atom's nucleus? Yes and no. Ecosystems are real, watersheds have physical boundaries, bioregions have a natural integrity. But connected global life is based on exchanges so numerous and furious that the "stuff" of places is in constant flux. Ecologist Raymond Dasmann distinguished "ecosystem people"—who sustain themselves and build their cultures mainly from the diversity and abundance of a local area—from "biosphere people"—whose economic relations draw materials and energy from farther afield, often from the planet as a whole. Dasmann saw this not as a dichotomy but as a cultural continuum along which people can travel.

If the latter decades of the twentieth century marked a galloping race toward a globalized economy of biosphere people, the beginning of the twenty-first century shows the glimmers of a conscious shift, by some, back toward the practices of ecosystem people. Emergent local food economies in cities around the world are signs of that shift, as are growing markets for biofuels and renewable power generated close to home. In truth, a healthy planet demands a blend of global connectivity and ecosystem attentiveness. As we enter such an era—as the industrial metabolism of society changes and natural ecoregions are reshaped by extinctions, invasive species, and unpredictable shifts triggered by climate change—we will be forced to think of "place" in new ways. In this world, place will be as fluid as the circulating atmosphere, and "living in place" will call on a capacity for innovation.

In the twenty-first century, the realities of the planet will demand that people—even people who've settled in one place for the long term—"re-place" themselves many times throughout their lives. Understanding the patterns and processes of place, even in a world homogenized by global economy and culture, will hold keys to citizenship and sufficiency. As we rename our changing places, we may discern the stories to guide the next phase of human presence on earth.

Placing yourself in the changing world is a worldchanging act. EW

From Bioregions to Bioregionalism

No sooner had biologists borrowed geographers' tools and begun to map earth's natural communities (boreal forests, temperate grasslands, tropical rain forests, and other biomes defined by their native vegetation) than some earth-centered activists began to envision new realms for civic engagement. Their "precinct" was the bioregion: a discrete area with ecological and cultural unity.

"All politics is local," said Congressman Thomas P. "Tip" O'Neill. Bioregional activists reasoned that plants, soils, and climate offered the most local politics of all, a fertile substrate in which to nurture citizenship and strengthen community. They touched a nerve in the era of the Vietnam War, offering people a new way to pledge allegiance.

San Francisco's Planet Drum Foundation, founded in 1973, set the tone for the bioregional movement. The Bay Area pioneers sought to encourage reinhabitation of earth's bioregions by taking what their Web site describes as "a grassroots approach to ecology that emphasizes sustainability, community self-determination, and regional self-reliance."

Planet Drum triggered talk of green cities, organized bioregional congresses, and inspired kindred initiatives up and down the West Coast. If such efforts never overtook the political mainstream, they at least attracted many young people into community activism and influenced a few major political figures, including Oakland mayor, former California governor, and one-time presidential candidate Jerry Brown.

The West Coast has proved to be more receptive to bioregional thinking than any other North American locale; its rugged rain-forest coast and tectonic volatility have helped shape a notion of "Cascadia"—a bioregion generally considered to be bounded by the Pacific Ocean and the Cascade Range in the Pacific Northwest. The alternative conception of place that characterizes Cascadia has found expression in visionary literature (Ernest Callenbach's *Ecotopia*), antilogging activism, and governance. The Cascadia Mayors' Council—a regional coalition composed of mayors from Oregon to British Columbia—meets intermittently to discuss opportunities to coordinate policies in transportation, economic development, and the environment. Along with this informal structure for governance, Cascadia has a Seattle-based think tank, Sightline Institute (formerly Northwest Environment Watch); an investigative newspaper, Portland-based *Cascadia Times;* and an online bioregional news portal, Tidepool.org. Even if few of the eight million residents in the mayors' corridor of concern think of themselves as "Cascadians," those predisposed to bioregional views have a rich garden of resources at hand.

Seeking to preach bioregional good news beyond the realm of the converted, Portland-based Ecotrust launched its Salmon Nation program in 2002. Through events, celebrations, education, and even product branding celebrating the iconic Pacific salmon (whose natural range encompasses the major watersheds of Cascadia), organizers promoted a new notion of "citizenship" based on an ethic of informed stewardship and a commitment to revitalizing the bioregional economy. ShoreBank Pacific, a community bank that lends to green businesses, even issues Salmon Nation credit cards.

Rooting citizenship in the natural attributes of a place continues to create avenues for engagement—too rare in modern life—with the essential and sustaining features of the planet. Pledging allegiance to place, we might at last plant politics in the realities of community and landscape that transcend partisan divides. EW

Literature of Place

How do we instill bioregional awareness in children? Why not slip the literature of place into language arts classrooms, and acquaint young readers with the best stories, poems, journal entries, essays, and songs written in and about their home region?

Minneapolis-based publisher Milkweed Editions created a colorful map delineating sixteen North American "ecoregions," and then set out to assemble an anthology for each. The result is the *Stories from Where We Live* series. Each volume in the illustrated series introduces elementary- and middle-school readers (ages nine to twelve) to writers who fold native animals, plants, cultures, weather, and landscapes into their work. Each book includes information on a region's natural history and a list of wild areas where kids and their families can encounter native species.

The first five of sixteen planned volumes include anthologies for the North Atlantic, California, and Gulf coasts, plus the Great American Prairie and the Great Lakes. The books, which feature a wide range of literature, from Willa Cather's "Country of Grass" to a Cantonese American immigrant's poetic response to Golden Gate Park, teach an attentiveness to North America's natural regions that could shape a generation's responses to place. EW

Opposite, left: Members of the U.S. Forest Service slowly make their way up Glendora Mountain Road while setting backfires along a containment line to rob a wildfire of fuel in Angeles National Forest, north of Glendora, California, 2002.
Opposite, right: Waves come ashore in Key West as Hurricane Wilma crossed the Florida Keys, headed toward the state's southwest coast, 2005.

Place Breaking

We tend to fall in love with a place through its plants, animals, and scenery. But many places also have their predictably recurrent natural disasters: wildfires, floods, hurricanes, typhoons, earthquakes, tsunamis, and volcanic eruptions. These catastrophes are every bit as native to the places we live as the birds and trees. To truly know our home places, we have to know not only how they flourish, but also how they break.

The extent to which catastrophes disrupt human lives often depends on whether the infrastructure of settlement has taken predictable catastrophes into account, or denied their existence. Unfortunately, the human propensity for denial often manifests in egregious and avoidable infrastructure failures. The insufficient engineering and reinforcement of the New Orleans levees, breached by the storm surge of Hurricane Katrina, is just one example. The dike systems in the Sacramento Delta are famously vulnerable. The homes being built in the path of wildfires at the urban-wildland interface, whether in the San Bernardino canyons of California or the lodgepole-pine forests of the northern Rockies, are practically earmarked for destruction. Such examples suggest that American-style affluence does nothing to curb folly.

Making the effort to envision—and prepare for—the plausible consequences of the unthinkable is a civic act, a necessary part of the work of placing ourselves. Some have done so brilliantly, but none in more chilling detail than the late author Marc Reisner in *A Dangerous Place*. A richly imagined "eyewitness" account of the upheaval caused by an entirely plausible (even restrained) magnitude 7.2 earthquake along the Hayward Fault, Reisner's book offers ways to think about how challenged infrastructures could fail in other places.

Literary treatments that carry important but troubling ideas into the fickle popular consciousness can also prompt—or mirror—the more rigorous scenario-building and public investment that civic authorities use to help communities withstand disaster.

A host of new uncertainties crowd the catastrophes linked closely to climate—wildfire, drought, hurricane intensity, flood frequencies. Current climate-change projections must be factored into our understanding of place, because few disaster scenarios become more benign under the erratic shifts of basic climate parameters that global warming has begun to spawn.

A sense of how our places can "break" is crucial place knowledge. Understanding the brittle systems that sustain modern lives provides no certain insurance against disaster, but the point of knowing is to survive, and to serve. A deep knowledge of place may improve our chances of surviving catastrophe; it will certainly deepen our commitment to help neighbors and community. No knowledge is more consequential. EW

RESOURCES

The Klamath Knot by David Rains Wallace (Sierra Club Press, 1983)
A wonderful work of natural history from a writer who can see not only into the natural systems that define the part of northern California in which he lives, but the epochal forces of evolution and geology that created those systems, *The Klamath Knot* is an essential guide

to understanding what placing ourselves can really mean.

The author writes, "Evolution doesn't view earth's history as a conflict between good and evil. It does essentially view it as a conflict between life and death, between increased organization and more efficient energy use on the part of life, and an opposing tendency of nonliving matter to become disorganized and lose energy—entropy. But evolution doesn't see life and death as simple adversaries: life as good and death as evil. Life cannot triumph over death in evolution. They don't fight to win. As with some of the older myths, wherein the natural dualities of light and darkness, sun and moon, male and female, performed an eternal, amoral dance of opposites, evolutionary life and death are interdependent: two halves of the world. Evolution would be impossible if organisms did not die."

Uncommon Ground: Rethinking the Human Place in Nature edited by William Cronon (W. W. Norton and Co., 1995)
William Cronon thinks that the environmentalist goal of wilderness preservation is conceptually and politically wrongheaded. The problem is that we haven't learned to live responsibly in nature—rather than trying to exclude humans, contemporary environmental activists need to help us learn to live in a sustainable relationship *with* nature: "At a time when threats to the environment have never been greater, it may be tempting to believe that people need to be mounting the barricades rather than asking abstract questions about the human place in nature. Yet without confronting such questions, it will be hard to know which barricades to mount, and harder still to persuade large numbers of people to mount them with us."

The Practice of the Wild by Gary Snyder (North Point Press, 1990)
One of the finest books on living in place ever written, poet Gary Snyder's collection of essays offers the harvest of a lifetime spent thinking about how modern people can learn to live in the kind of intimacy with the land that our native, tribal ancestors devoted their lives to learning. Snyder guides us through the back country of the history of our relationships to nature, bringing us finally to a vantage point where we can see that one of the highest aims of a prosperous, settled, and technologically advanced society must be a way of life that brings us into regular contact with our wild roots and reconnects us with the reality of the living systems on which all life depends:

"The presence of this tree signifies a rainfall and a temperature range and will indicate what your agriculture might be, how steep the pitch of your roof, what raincoats you'd need. You don't have to know such details to get by in the modern cities of Portland or Bellingham. But if you do know what is taught by plants and weather, you are in on the gossip and can truly feel more at home."

Opposite, left: Bird-watchers have collaborated with researchers, using sites such as eBird, to share sightings and field data.
Opposite, right: Amateur astronomy is another hobby that allows citizens to make a contribution to science, aiding astronomers in tracking new developments in the sky.

Citizen Science

How do we keep up the pace of technological progress in a civilization that can't train specialists fast enough? In the twentieth century, we professionalized the narrowest of fields, creating expertise within medical, engineering, and scientific disciplines—a trend that offered great benefits, but also posed serious limitations. It seems impossible for this trend to continue far into the twenty-first century—indeed, we may be bound to leap beyond professionalism. As education levels keep rising, more of us have leisure time and access to vast amounts of publicly available knowledge. The next century may bring an Age of Amateurs, when citizen exploration yields much of our greatest learning as a society.

This hearkens back to a legacy that predates big-time corporate and government-funded research, to the contributions of Benjamin Franklin, Charles Darwin, Lewis and Clark, Smithson, Audubon, Thoreau, and others. Already, amateurs, enthusiasts, and retirees contribute in many ways. Soon burgeoning technological tools will revolutionize the capabilities of citizen savants, leveling the playing field between aficionados and professional researchers, both expediting scientific advancement and democratizing countless fields of interest. Eccentrically individualist and internationally collective, citizen-driven research offers science a powerful method of escaping control by a rigid elite and remaining, instead, an adventure for us all. DB

Citizen Science and Bird-watching

During the winter holidays, while most of us are fiddling with new toys and trying on itchy sweaters, more than 50,000 people around the Western Hemisphere grab their coats and binoculars and trudge out to the woods to look for birds. They're all taking part in the Christmas Bird Count, a tradition for over a century.

Bird-watching, or birding, is one of the most popular hobbies in North America. Millions love to spend their free time hiking through mountains, woods, and swamps, ears perked and eyes peeled for any sign of a rare bird. The Christmas Bird Count allows scientists to aggregate the otherwise independent findings of all of these birders to assess migration patterns, locations, and populations. Until recently, though, gathering and processing this data was extremely daunting for researchers, whose time and funding was already stretched thin.

But a few years back, researchers, birders, and environmentalists shared an "aha!" moment—they realized they could ease the coordination of diffuse data by connecting the thousands of enthusiastic citizen birders directly with advanced technological tools.

Thus hatched eBird. Thousands of individual birders across North America have signed up, contributing their ongoing sightings and field data to a collaborative database on bird populations and behavior. In exchange, birders who use eBird (and the various other like-minded networks that are a part of this movement) get online records of the birds they've spotted, are able to access planning tools for birding trips, establish friendships with other birders, and gain a sense that they're not just watching their feathered friends, but helping to save them.

Collaborative technologies often seem abstract, and the manner in which they break huge projects down into small, even pleasurable tasks, distributed over thousands of people, can be difficult to understand. But for the thousands of folks who head outdoors on Christmas, it's a fair bet collaboration is neither abstract nor difficult to grasp. Multiply this one hobby—both fun and useful—by hundreds of others, and you start to see the potential in an Age of Amateurs. DB

Citizen Science in Action

Gardening, stargazing, bird-watching, surfing the Net: who knew our favorite hobbies could benefit science? But amateurs everywhere are indeed helping scientists, by adding their personal sightings to the mix of observational data. At the University of California, Davis, for example, home gardeners aid agricultural specialists in assessing the decline of genetic diversity in major food crops. After crash efforts got under way to collect specimens of related species from their native sources—before precious variants vanished (rare varieties of potatoes from the Andes, for instance)—researchers realized that the ambitious undertaking could be made both easier and more thorough with the assistance of people who already studied and cultivated exotic plant species as a hobby.

Stargazers, another popular breed of hobbyist, are aiding professional astronomers with the unmanageable task of tracking comets and asteroids. Amateurs of course delight in learning to operate their own sophisticated CCD scanner-telescopes, becoming faithful assistants who scan the sky patiently, in the cold, seeking undiscovered objects that they can name after themselves. What a deal!

Bird and bug lovers likewise perform priceless labor for ornithologists and entomologists. Worldwide networks of citizen collectors and observers report sightings of tagged monarch butterflies or birds in migration, making major contributions to species conservation.

Despite advances in high-tech satellites, weather agencies, too, rely more than ever on reports from widely scattered volunteer stations, both ashore and at sea. Citizen reporters help refine our models of climate change and our strategies for preserving biodiversity.

In medicine, citizen networks now offer more than just emotional support. They supply up-to-the-minute information for people who share symptoms, illnesses, or side effects. Networked patients even occasionally come up with sharp insights about their particular disease. At first daunted by this trend, medical personnel have now grown accustomed to patients who use the Internet to self-educate about their conditions, often accessing the latest journal articles, or joining worldwide support groups. Instead of making physicians' jobs

Members of the Environmental Career Organization collect data on sand foods in the Sand Dunes Recreation Area near Glamis, California.

harder, these networks tend to help by tracking outcomes, providing data, and informing patients. Informed patients are better patients. And informed citizens are making science better every day. DB

Sensors for Citizen Science

As cheap and sophisticated sensors begin to make their way from military and commercial laboratories to your local Radio Shack, neighborhoods and even whole cities will be able to detect toxins, pollutants, and various kinds of germs in their local environments. Individuals and groups of citizens will then be empowered to track water and air quality, for use in local activism and—more broadly—for adding data points to global networks. Lest we forget, many advances in robotics and open-source software are driven by loose groups of expert amateurs. Human vision and knowledge may expand even faster than it did with the invention of lenses and movable type five hundred years ago.

This trend has naturally been met with some skepticism from the scientific establishment, governments, and corporations. Not all hobbies attract sober-minded and earnest citizens. The skies that amateur astronomers gaze at nightly are also watched by thousands of fervent UFO seekers, whose florid claims nearly match the vivid intensity of their hopes that something wondrous will come down and choose them for a wild journey. New Age sects were already flourishing before the Internet arrived to fertilize the trend beyond all recognition, so we'll all have to learn to tell wheat from chaff—even if we don't have old-fashioned credentials. When Stanford geophysicist Anthony Fraser-Smith wanted to create a distributed network of sensors that might give advanced warning of earthquakes, he learned quickly that not all interested and enthusiastic individuals could be relied upon as citizen scientists. On the other hand, some volunteers impressed him immensely. In the face of deficient government funding, he found that enlisting amateurs was indispensable in achieving his research goals.

This productive and collaborative side to amateur science is only going to grow in the years to come. We can almost picture an era when even the most august experts will be constantly looking over their shoulders, because hordes of well-informed citizens will hold them accountable for every finding and report. But it will also be an era when swift and savvy citizen assistants—eager, innovative, and surprisingly well-informed—will help push the boundaries of discovery further, faster than ever before. In this environment, bureaucrats and paper-shuffling committees will lose much of their authority, because many of society's intellectual tasks will be augmented—or even taken over—by voluntary associations of passionate devotees. DB

Restoration Ecology

Some suggest that the only way to satisfy our longings for contact with nature is to head back to the woods—to go camping in the wilderness, kayaking along untamed rivers, or climbing distant peaks. We share an impractical cultural idea that the only nature worth connecting to is nature far away from humanity.

Fortunately, if we can't get out to the wild, we have other avenues for connecting to nature—and that's a good thing considering the wilderness areas we'd trample underfoot if all six billion of us decided to visit them. One such avenue is restoration ecology, the practice of bringing degraded ecosystems back to a healthier state closer to their origin. Some restoration projects are wilderness oriented—reintroducing lost species to areas where they once roamed free, such as the ongoing reintroduction of wolves in Yellowstone National Park—but many can take place in our own backyards.

Ecological restorations can range from small-scale urban park reclamations, which are happening across the United States, to huge re-creations of entire ecosystems. Whatever the scale, the goal is to restore the landscapes that once existed (like wetlands, tall grass prairies, and various river systems), or at least some of their functions. If measured by the amount of time and money Americans are willing to invest in it, restoration ecology is one of the most popular ways of being environmentally involved. For example, the cluster of projects known as Chicago Wilderness—a thirty-year attempt to return areas around Chicago to oak savanna (their pre-settlement ecosystem)—has attracted some 20,000 volunteers over its life and restored at least 17,000 acres (6,880 hectares) of a total planned 100,000 acres (40,469 hectares).

Restoration has real ecological and scientific benefits. But just as important, it gives us a chance to become reconnected with nature, to literally get our hands dirty in an active process of changing the world. Truly successful restoration projects actively encourage community participation.

The thousands of volunteers in Chicago are not only replacing exotic species with indigenous ones and bringing back native forests, but they are also becoming more actively involved in their local environment. Not all of us can do biological surveys, but most of us know how to work a shovel or pull an invasive weed.

When we volunteer on restoration projects, we form a strong attachment and commitment to the land, a relationship that can be as important as family ties and civic attachment. Sociological evidence gathered from the Chicago restorations suggests that restoration volunteers are likely to adopt a benign attitude of stewardship and responsibility toward nature as a result of such

engagement. The reasons are fairly obvious: participants learn the injurious toll human activity takes on nature by experiencing how hard it is to restore something after it has been damaged. They also gain a sense of investment in the health of local natural systems. For all these reasons, restoration can serve as a new schoolhouse for environmental responsibility, but only when we are farsighted enough to encourage voluntary public participation in these range of projects. ARL

Save a Mangrove, Save a Life: Ecological Restoration

Healthy coastal ecosystems are vital to protecting growing coastal populations from disaster. According to the UN, over half of the world's population was living within two hundred kilometers (125 miles) of a coastline in 2001, with numbers only increasing in subsequent years, and tourists compounding the problem. At the same time, climate instability is likely to render already violent cyclones, hurricanes, tsunamis, and floods even more intense and destructive. It's clearly of vital importance that we integrate ecological preservation and restoration into the process of development along the world's coasts—before the next disaster.

In the 2004 tsunami in the Indian Ocean, healthy mangroves, the leading edge of the forest, took the brunt of the great wave's energy when it hit—yet few trees were uprooted. With large areas of mangroves appearing to be relatively intact, it can be difficult to detect the extent of injury to the forest—a problem often referred to as *cryptic ecological degradation*. However, where mangroves had been weakened, even relatively slightly, by ecological harm caused by humans—like pollutants from inland development and agriculture—they were much less able to protect the inner coastline, even if they had not been thinned by strong storm winds.

Mangroves are incredibly fertile ecosystems. At the meeting point of land and sea, healthy mangroves support an amazingly biodiverse ecology. Unfortunately, development plans often involve destroying mangroves—for the sake of coastal tourism, say, or of farming shrimp for export to richer nations. But people live better with mangroves than without them, a fact that will likely become all too clear the next time a big storm hits and weakened forests have less resistance.

Mangroves have had a high rate of success in restoration projects, since they are easy to access from the coast for management and care, and generally exist in a self-sustaining environment (provided that coastal conditions don't reach extremes). Restored and well-managed mangrove forests can be extremely beneficial for local coastal populations, as they can be superior ecosystems in which to raise and harvest shellfish (a fact that gets ignored when forests instead get replaced by shrimping operations) and can supply managed timber. The work of fortifying and supporting these ecosystems promises to be one of the great preventative measures we can take to be better prepared for future marine storms. EG

As the result of a successful reintroduction program, wolves are now being spotted in Yellowstone National Park after an absence of more than seventy years. Mangroves have also thrived in reintroduction projects and are crucial to the health of coastal populations.

Restoring the Deep Past

Before the first intrepid explorers crossed over the Bering Strait from Asia, toward the end of the last ice age, North America was home to an astonishing variety of large animals: mammoths and lions, horses and camels, giant beavers, and ground sloths the size of Volkswagen Bugs. Shortly after we got here, the wildlife disappeared.

Whether or not we killed them off remains a matter of some debate (although the scales seem to be tipping against us), but the impact of this mega-extinction has become increasingly clear: the continent the Europeans saw as a savage wilderness when they arrived late in the fifteenth century was in fact something more like a garden, managed extensively by native peoples through the use of fire, and devoid of many of its largest prehistoric inhabitants.

In *Twilight of the Mammoths: Ice Age Extinctions and the Rewilding of North America*, Paul S. Martin proposes "rewilding" the continent by reintroducing, in protected locations, some of the close living cousins (particularly elephants) of the animals North America has lost. Far from being unnatural, Martin says, this would bring ecosystems—or at least experimental parts of them—closer to the healthy state they were in before we arrived.

When we think of restoring what has been lost, Martin suggests, we should aim to re-create not what was here a couple of hundred years ago, but the full suite of plants and animals that evolved here before humanity first spread out over the earth. We should practice "resurrection ecology." By bringing back mini-replicas of the world we first encountered, we'll be able to better understand what, exactly, we've made of our world—and make a better one for the future. AS

Right: Protection of the natural resources in watersheds, like this one in the Little Tennessee River near Tallassee, Tennessee, is crucial to the health of all living things. **Opposite:** As pollinators, bees are a great example of the value offered by "nature's services," which we often take for granted.

Ecosystem Services

Nature is working for you. If you don't know how, consider your faucet. You bought the plumbing, and in your monthly bill you pay for the filtration plant and pipes that deliver the water, metered, to your house. That water likely began its journey to your home scores or hundreds of miles away, percolating through a forested watershed that captured the rainfall, combed it of impurities, oxygenated it in streams, delivered it to a river, and ultimately, to a municipal drinking-water reservoir. Nature provided the water composing the ice cubes rattling in your lemonade, for free. You can't easily buy a substitute for the rainfall, the watershed, or the streamflows. It makes you wonder how much they're worth.

In similar fashion, the air we breathe, the food we eat, and the climate we depend on are all supplied in large measure by services that nature performs for free. Such "ecosystem services" underpin all life on earth, but only humans are in a position to assign a value to them. When economists first did so about a decade ago, a rough calculation showed nature's services to be worth a third more each year than all the human economic output on the planet. In the "Millennium Ecosystem Assessment

Synthesis Report," economists estimate that it would cost at least $33 trillion per year to substitute human effort for the services that nature provides for free. Of course, many of those services we simply couldn't perform at any price. That's some subsidy!

When it comes to nature, we human beings act as though we don't know the value of a buck. We don't calculate the wealth lost when our $25 trillion economy pollutes, degrades, and impairs the functions of the natural economy that sustains us.

That's slowly beginning to change. We are starting to understand that on one small planet, keeping life-support systems in good working order isn't just another expense—it's an investment in survival. EW

Pollination: The Planet's Oldest Partnership

At 100 million years, pollination may be the longest-running business partnership on the planet. Well over half of earth's quarter million flowering plants (including 80 percent of major food crops) depend on pollination by insects in order to reproduce. Insects don't have to be hired for the job. Beetles, butterflies, and honeybees are part of coevolved "pollinator complexes" that make sure each flowering species gets exactly the attention it needs. Imagine having to hire human laborers to do this work with Q-tips, in the largest agricultural jobs program ever conceived. We would fail abysmally.

"Millions of years of coevolution have finely tuned the relations between particular plants and their special pollinators," says biologist Edward O. Wilson (Buchmann and Nabhan 1997). But today, populations of wild pollinators are declining nearly everywhere on earth. Luckily, innovative efforts to restore insect habitats and cultivate pollinator gardens are under way in many countries. More are needed. EW

Millennium Ecosystem Assessment

More than 1,300 scientists from ninety-five countries collaborated on the "Millennium Ecosystem Assessment (MEA) Synthesis Report," which describes results from the first segment of a multiyear study of environmental indicators, launched by United Nations Secretary-General Kofi Annan in 2001. The report, released in 2005, evaluates the condition of earth's ecosystems, details how ecosystem changes may impact human life, and suggests tools for improving ecosystem management to alleviate poverty and enhance well-being. Overall, the story isn't good: of the twenty-four ecosystem services evaluated by the MEA, fifteen have deteriorated over the last half century. Services in decline include capture fisheries, water purification, natural control of pest species, and the capacity of mangroves, wetlands, and other natural systems to buffer against hazards like storms and tsunamis. Only four ecosystem services have improved, three of them relating to food production. Subsequent volumes in the assessment effort include a look at four scenarios for the next half century, forecasting how we might shape the future with the choices we make today. EW

Ecological Economics

Since resources are finite, we should budget their use. Since ecosystem services can be destroyed, we should create incentives to protect them. Ecological economics is about adding these insights to traditional economics, to create a broader framework that combines hard business sense with environmental and physical reality.

Why should we impose extra limits on ourselves? Because it's better for us in the long run. Just as borrowing too much on credit and getting deeper and deeper into debt hurts our financial futures, using up too much natural capital reduces our long-term wealth. The need to stay within a budget constrains every business, but it does not stop smart businesses from continually improving. Similarly, the need to stay within resource and ecoservice budgets is a limit we have to respect, but it does not stop progress in efficiency, technology, cultural endeavors, and human welfare. And we can stay within these budgets only if we properly appraise what they measure.

Let's take an example: the Canadian Boreal Initiative. Canada's vast boreal forest stretches across more than 3.1 million square miles (5 million square kilometers), covering over half of Canada's land mass. The value of this unparalleled natural resource is considered in the report "Counting Canada's Natural Capital: Assessing the Real Value of Canada's Boreal Ecosystems." First, the gross market value of natural-capital extraction—related to logging, mining, oil and gas extraction, and hydropower generation—is added up. Next, subsidies and costs of natural-capital extraction—such as tax incentives and the health costs of air pollution—are subtracted. Finally, ecosystem-service values—flood control, water filtering, biodiversity value, carbon sequestration, and nature-related recreation—are considered.

When the final value is calculated, the boreal forest is worth far more money as a living ecosystem—two and a half times more, in fact. Does this mean that the boreal forest should never be touched by commercial interests? No, but it does mean that logging, mining, and other uses should only be allowed to the extent that they do not damage the services the boreal forest gives us. Assessing the full value of natural ecosystems, and considering them as assets in our accounting, helps us develop and prosper for the long term.

In the big picture, local efficiency improvements are not enough—earth system limits as a whole have to be respected. To return to the business analogy, an efficient branch office or manufacturing plant isn't enough if a business as a whole burns through all its cash and credit. That's why the term *sustainable scale* is fundamental to ecological economics. It means ensuring that we don't overuse nonrenewable resources and get caught unprepared when they start to decline, as may happen with oil. And it means ensuring that we keep our environment healthy—setting limits on emissions of carbon and other pollutants and on the drawdown of water and soil resources.

The good news is that we've already developed many useful tools, like natural-assets accounting, efficiency and emission standards, tax shifts, and ecosystem credits. Understanding what conventional economic statistics hide or ignore can help citizens craft policies better adapted to the future, and help forward-thinking businesses gain a competitive advantage. With macrolimits and microflexibility, ecological economics can help us get our mental models in line with reality, making doing the right thing second nature. HM

Ecosystem Credits

"Ecosystem credits" are a bull market. These tradable permits are designed to protect threatened species and habitats without entirely restricting the commercial use of land. One of the elements of the Kyoto Protocol is the use of carbon "markets" as incentive for countries to reduce carbon dioxide emissions. Countries that produce CO_2 in amounts below the treaty limits, and countries not currently bound by the treaty (for now—other than the United States and Australia—mostly developing nations) can sell carbon-emission rights to countries that produce CO_2 in amounts above the treaty limits. This may just seem like a way for overproducing countries to continue spewing excess carbon dioxide, but it's actually a very good idea: few countries' emissions actually fall below the Kyoto limits, so there's considerable incentive to buy credits from nonparticipating or below-cap nations. As more countries get their emissions under control, the emissions cap will gradually be lowered, and the price of the remaining credits will go up; countries will therefore have an economic incentive to be net carbon-credit producers whose emissions fall below the cap, instead of consumers who exceed the CO_2 limits. It's an incentive for developing countries to adopt cleaner technologies sooner, so as to continue being sellers and not buyers as they grow.

The downsides to carbon trading are the complexity of the schemes, the likelihood of clever traders being able to "game" the markets, and incentive to hold out: potentially, when most other countries are operating below the cap and trying to sell remaining credits in the market, the price per unit will drop so substantially that even the most egregious emitter would be able to afford to operate unchanged.

The European Union established a system for trading carbon credits that went into effect in early 2005. The scheme involves more than 12,000 participants throughout the twenty-five European Union states. Compliance for each facility will be reported every year; therefore, noncompliance will be met not only with fines but also with disgrace. All transactions will be recorded in a common "Community Independent Transaction Log." But if you want to see who is buying and selling credits, you face a bit of a wait—detailed transaction information is being withheld for five years following initial implementation.

Emissions trading has been implemented to address various kinds of pollution, and not just in the developed world. A demonstration project in Taiwan and prototype schemes for sulfur-dioxide trading in China are promising signs for the future. For landowners in the United States (developers, utility companies, state transportation agencies, farmers, retailers, and others), ecosystem credits can help turn land-use conflicts into win-win relationships that protect both the environment and companies' bottom lines. The Endangered Species Act and the Clean Water Act both require land developers to make up for any harm they do to streams, wetlands, or habitats critical to endangered and threatened species. Simply put, a developer or other commercial occupant that compromises or destroys habitats in one location is required to protect or restore habitats of comparable scope or value somewhere else.

As with carbon credits, land developers and owners can satisfy the law through "conservation banks," which protect land containing

Vernal pools, which appear seasonally in rural areas of California, are home to rare species, yet are not currently under the protection of the government.

endangered species, and "mitigation banks," which protect wetlands and streams. In both cases, businesses putting habitats or wetlands at risk are required to purchase credits from other landowners who have legally committed to protecting their land permanently through the "banks." The "bankers" can sell these credits on the open market, sometimes at levels far exceeding traditional real-estate values. JM

Biodiversity Farming and Restoration for Profit

You've probably never heard of the California vernal-pool fairy shrimp. Don't feel bad—these tiny crustaceans live out their entire lives in little puddles and ponds that come and go, and that often dry out completely in the summer.

But therein lies the problem: these ephemeral wetlands, as they're called, have been hammered by agriculture and development to the point that fairy shrimp are now a federally protected threatened species in the United States. This creates no end of problems for people who have vernal pools on land they wish to use for farming or other business.

Enter the Dove Ridge Conservation Bank. Dove Ridge offers "mitigation banking" for fairy shrimp. Instead of trying to work around the pools on their property, developers can buy "credits" from Dove Ridge, which maintains 233 acres (94 hectares) of fairy shrimp–filled pools. By purchasing these credits, for $70,000 apiece, developers can fulfill federal requirements to mitigate the damage they'll inevitably do to the critters' habitat.

Mitigation certainly isn't a perfect fix—after all, a human-made pool full of fairy shrimp isn't anything like a wild, teeming vernal pond—but Dove Ridge may be a harbinger of better things to come. What if we reimagined wilderness as a special kind of farm, the kind of farm that grows biodiversity and other ecosystem services? In the near future, the best way to save species and protect clean air and water may be to recognize that these can be measured, given value, and taken care of.

Ecosystem services are irreplaceable. As we move into the bloom of biomimetic technology, biodiversity becomes our most important natural resource. If local people can retain the rights to the economic value of local species, they'll have powerful incentive to cultivate biodiverse farms; this will be particularly advantageous to poor people, who'll discover new ways to feed their families without destroying local ecosystems. In a similar way, if governments recognize the value of ecosystem services through tax breaks and regulations, farmers will have financial incentive to protect watercourses and restore soil. Right now it's hard to make a living in sustainable agriculture, but if there's a price tag on healthy farms, that'll quickly change. EG

RESOURCES

The Soul of Soil: A Soil-Building Guide for Master Gardeners and Farmers by Grace Gershuny and Joe Smillie (Chelsea Green Publishing, 1999)
To understand soil as the foundation on which all agricultural activities are built is the first step in "honoring our oneness with all living creatures and helping the long process of repair that a new biological era will require," argue Gershuny and Smillie. *The Soul of Soil* is indispensable to understanding soil—the texture, smell, uses, and energies inherent to it—and serves as a simple primer on the basic tenets of sustainable agriculture. In presenting the principles of ecological soil management, Gershuny and Smillie hand us a veritable tool kit of ways to care and nurture the soil under our feet.

Under Ground: How Creatures of Mud and Dirt Shape Our World by Yvonne Baskin (Island Press, 2005)
Two-thirds of the earth's biodiversity lives in its terrestrial soils and underwater sediments—*Under Ground* gives us a glimpse into this otherwise undiscovered world. More than simply weaving a tale of the fascinating and essential underground microbes and organisms busy at work making the planet habitable for humans, Baskin delivers a warning about the unseen effects of agriculture and development on the life systems under the earth's surface: "We usually take notice only when changes below ground set in motion a cascade of unwanted consequences above ground." Ultimately, Baskin acknowledges that we aren't all going to share her passions, but she nevertheless seeks to inspire: "Most of us will never

respond to microbes or nematodes with the emotional connection we muster for elephants and eagles ... [but] my hope is that we can learn to step, not only lightly, but also with wonder and awareness on the world underground."

The Future of Life by Edward O. Wilson (Alfred A Knopf, 2002)
Biologist Edward O. Wilson is one of the world's most respected scientists. He's also one of the ones ringing the loudest alarm bells. *The Future of Life,* his call to arms, is an impassioned cry for us to recognize the importance and value of the diversity of life, and to accept the position we now find ourselves suddenly thrust into: that of having to decide between preserving the amazing natural riches that nourished our success as a species and risking the loss of both nature and civilization:

"On or about October 12, 1999, the world population reached 6 billion. It has continued to climb at an annual rate of 1.4 percent, adding 200,000 people each day or the equivalent of the population of a large city each week. The rate, although beginning to slow, is still basically exponential: the more people, the faster the growth, thence still more people sooner and an even faster growth, and so on, upward toward astronomical numbers, unless this trend is reversed and growth rate is reduced to zero or less. This exponentiation means that people born in 1950 were the first to see the human population double in their lifetime, from 2.5 billion to over 6 billion. During the twentieth century more people were added to the world than in all of previous human history."

Biodiversity: How Much Nature Is Enough?

We are living, biologists like to remind us, during the Sixth Extinction: a loss of species as great as, and far more rapid than, the meteoric disaster that wiped out the dinosaurs. It is a tragic truth that we have already doomed thousands, perhaps millions, of species to extinction. Many are disappearing unbeknownst to us, before ever even being described by science. According to paleoanthropologist Richard Leakey, if we continue with business as usual, we may wipe out half of life's variety this century (Leakey and Lewin 1996).

No one knows the real implications of this. It is true that sometimes ecosystems can survive, heal, even thrive with fewer or different species than they had when we went to work on them. It is also true that sometimes killing off just one or two key species can cause a spiral of collapse, leaving a wasteland behind. Species are the rivets holding our ecosystems together, and we're knocking them out in handfuls, not knowing when the structure will collapse.

We can respond to this crisis in one of three ways. The first approach is the one every sane person believes we ought to pursue: try to save all the parts. There's no doubt that the preferred option is the preservation of what some U.S. environmentalists term the *Big Wild*—huge tracts of protected nature. But the engines driving habitat destruction and species extinction are incredibly powerful. It's probable that in even the best-case scenarios, we're going to lose a lot. Which brings us to the second response: pick and choose which parts we save, and the third: save the plans for all the parts.

The pick-and-choose approach is known as the "hotspots" strategy. The gist: identify threatened habitats with the greatest

biological diversity and use our scarce conservation funding to protect them. It's not foolish, but the problem is that scientists are discovering that ecosystems are more interconnected and complex than previously understood, and that unless hotspot reserves are really big, species there continue to go extinct. Because hotspots tend to comprise a limited area, they are also left vulnerable to changes in rainfall and temperature brought on by global warming.

The last approach involves documenting and preserving the DNA of species we know will soon be extinct. Rather than storing the parts, these frozen genetic zoos (or *zooz*, as they're increasingly termed) store the genetic instructions for the manufacture of those parts. While saving the DNA of an extinct species is certainly better than letting it slip into oblivion undocumented, huge problems remain.

Frozen zooz are not a replacement for conservation. For one thing, we're not certain we can ever bring these creatures back. As efforts to clone the extinct thylacine (or Tasmanian tiger) have shown, it's extremely difficult to bring back even an individual member of a particular species, much less a viable population. But even if you could revive a species, what you get may not be the same as what you lost. Ecosystems are by definition the interaction of species. We have no idea how to even begin reconstructing those interactions. Critters themselves may not be the same when cloned.

Preserving the world's biodiversity is not just an ethical imperative to save life for its own sake—given both our dependence on the continued health of the planet and all the lessons nature still has to teach us, preserving biodiversity may mean saving our civilization. AS

Why the Cloned Bird Doesn't Sing

Even if it becomes simple science to revive ancient animal species from genetic material, we will not have achieved true species restoration. The songbird is a poetic example. Even if we were able to clone an ancient songbird with DNA, its likeness would be betrayed by the absence of its most essential trait: song. Birds get their singing skills through a combination of instinct and learning; without the presence of feathered kin, our cloned bird's song, if it has one, will be meaningless. SR

The Culture of Extinction: Tattoos for Dodoes

Somewhere in America a group of schoolchildren is on a museum field trip, spending the day gawking and giggling at dinosaur bones. These days, dinosaurs adorn everything from coloring books to pajamas, so you would think that Americans spend a lot of time pondering extinction. But how many lesson plans ask the most important question: What does it mean to have a part in causing the extinction of a species?

The extinction crisis is one of the major problems humanity faces, yet it is hardly mentioned outside of environmental and scientific circles. The reasons for this are largely cultural. To consider

A silhouette of a village near Chernobyl—site of one of the worst nuclear disasters in history—which has become a refuge for endangered wildlife.

the extinction crisis is to visit with death, guilt, and horror. It's overwhelming and depressing.

So how do we change our cultural outlook? We need something that shifts extinction from an abstract issue to a human concern, something that moves the discussion onto the streets and into people's homes, in the same way that red and pink ribbons have helped to destigmatize HIV and breast cancer.

We might start remembering extinct species through body art. We could easily assemble and maintain a database of names, pictures, and information on species that have gone extinct or are clearly bound for extinction. People can then "adopt" dead species. The membership dues? Tattooing an image of that species (with its Linnaean name) on a visible place on your body.

By doing so, you would agree to become someone who remembers, in a very personal way, that this plant or animal no longer exists, because we killed it—someone who is willing to talk with others about it, to drag the taboo out of the closet and carry it around.

People often use tattoos to commemorate the lives of friends and loved ones. Maybe we should start using ink and skin to celebrate the species—living and dead—with whom we share the planet. AS

Involuntary Parks

The so-called demilitarized zone between North and South Korea is devoid of any human habitation or activity, and has been for about fifty years. As a result, this space—155 miles (250 kilometers) long, 2.5 miles (4 kilometers) wide—has become home to a staggering array of rare plants and animals, including the highly endangered red-crowned crane.

The DMZ is not the only government-owned wasteland of its kind. These long-forsaken, undomesticated expanses are some of the wildest remaining places on earth. They've been dubbed "involuntary parks," reflecting the accidental nature of their wildness.

Decommissioned nuclear plants seem to make good involuntary parks. Rocky Flats, a former nuclear weapons plant near Denver, Colorado, was shut down in 1989 because a slew of environmental violations—leaking storage drums, tanks, and pipelines; on-site landfills; unlined disposal trenches—had thoroughly contaminated the soil and groundwater. Today the site contains what might be the largest remaining tallgrass prairie in North America, as well as several endangered or threatened species, including peregrine falcons. The most iconic nuclear wasteland of our time, Chernobyl, is also turning into an immense wildlife preserve. Moose, wolves, and deer roam around in places where humans still can't safely walk, and more than two hundred bird species (thirty-one of them endangered) call the park home.

As pressures on more desirable lands increase, these dangerous nature sanctuaries play an ever-more-important role in preserving biodiversity. "Involuntary Parks are natural processes reasserting themselves in areas of political and technological collapse," says futurist Bruce Sterling. "An embarrassment during the 20th century, Involuntary Parks could become a somber necessity during the twenty-first." CB & JC

RESOURCE

Darwin's Ghost: The Origin of Species Updated
by Steve Jones (Random House, 2000)
Steve Jones's book is a sharp overview of current scientific thinking about evolution and biodiversity, as well as an insightful meditation on the work of Charles Darwin. "No biologist can work without the theory of evolution. Like Galileo's notion of a solar system with the sun at its center, Darwin's long argument makes sense of their subject. Ideas of origin were once, like *Moby Dick*, allegories. They helped to comprehend not the structure but the meaning of the universe. Some still hope to find symbolic significance in Darwinism. They will not: but his work turned the study of life into science rather than a collection of unrelated anecdotes."

Sustainable Forestry

Wood and paper products might seem to be an environmentalist's best friends. After all, they're made of a renewable raw material—unlike, say, plastic bags, each of which sends a dollop of petroleum on a one-way trip to the dump. But the devil is in the details. Factors such as economic desperation, lax regulation, and greed have conspired to make logging a problematic venture for both wildlife and humans.

Some environmentalists would have us believe that every logging job is as bad as a clear-cut—that parkland is the only good use of a forest. That view ignores the emergence of sustainable forestry, a full-fledged industry that aims to reap a steady harvest from the forest for generations to come, while maintaining the integrity of ecosystems and social fabrics.

When we practice sustainable forestry, we care for the forest as a whole system, made up of creatures from soil fungi to spotted owls, from lizards to loggers. In return, the forest offers a host of materials and services: not just lumber, but clean water downstream, food for animals and humans alike, recreation opportunities, and carbon storage—which means that the trees absorb CO_2 from the atmosphere, offering some potential for short-term relief from global warming.

When we adopt sustainable forestry, we commit ourselves to several principles that depart from common industrial practice: We harvest trees no faster than the forest grows back, instead of liquidating the forest and cutting trees at ever-younger ages. We log so that all the forest's native species continue to thrive, not just a few commercially significant timber varieties. We respect timber workers' labor rights, instead of busting unions whenever possible. We plow profits back into the maintenance and restoration of the forest, instead of sucking it dry and moving on to clear-cut somewhere else.

Sustainable forestry is still being defined through attentive experimentation, but it's no longer just the domain of a few concerned entrepreneurs. Loggers and environmentalists are working side by side to find responsible ways to make forests more resistant to fire; even bureaucrats in New York City have had to promote sustainable forestry to ensure the safety of their city's water supply. Whatever the motivation of the players, the outcome is the same: healthier forests equal a healthier planet. SZ

Forests: More Than Just Lumber with Leaves

In the 1990s, environmental regulators told New York City officials that they needed to begin filtering the city's municipal water supply. The water cops weren't imagining the problem: New Yorkers had already dealt with repeated orders to boil their drinking water because of microbial contamination.

Building a filtration plant to clean up the city's water would have cost $6 billion. Instead, the city decided to go directly to the source: the Catskill Mountains, a hundred miles northwest of the metropolis, where a 1,600-square-mile (4,100-square-kilometer) basin, most of it in private hands, supplies 90 percent of New York City's water. The city realized that if it could entice private landowners to take better care of the area, its water supply would be clean enough to satisfy the water cops. In 1997, New York made a pact with the towns in and around the Catskills to improve the watershed, achieving cleaner water for the city at one-fourth the cost of a new filtration plant.

What does all this have to do with sustainable forestry? Watershed-altering timber practices were part of the agreement, along with improved dairy farms and septic systems. Logging can send sediment coursing into creeks, where it fouls downstream reservoirs. The City of New York has funded more than five hundred forest-management plans covering 94,000 acres (38,000 hectares), bought and retired the development rights to forest

land, and helped loggers harvest less disruptively. It even supplies temporary bridges for creek crossings, where loggers would ordinarily muddy the stream by driving right through it. The program also backs woodworking businesses in the Catskills, on the theory that a strong forest economy will prevent the subdivision of wooded land for residential development. It all boils down to recognizing that the forest is worth vastly more than its lumber.

The Catskills isn't the only area where sustainable logging has been part of a broader effort to value the forest as a whole system. For instance, British Columbia's 2006 Great Bear Rainforest settlement put aside 5 million acres (2 million hectares) of coastal forest, while leaving 10 million acres (4 million hectares) open to selective logging that steers clear of bear dens, streams where salmon spawn, and sites sacred to native peoples. The province has also promised more than $100 million for ecofriendly development in the region. SZ

Playing with Fire

For the last hundred years, the wildfire has been the boorish cousin of the American West. No matter how hard we try to schedule family gatherings without it, eventually it crashes the party and makes a scene.

As with flesh-and-blood cousins, the wildfire problem is aggravated by our attempts to keep the troublemaker out of the picture entirely. A century of dogged efforts to extinguish all wildfires turned terrain into tinder, choked with thickets of small trees and dead wood, which explode into infernos when conditions are right.

With more people living in and around the forests of the West, those inevitable conflagrations kindled fear on all sides of the timber

Working toward sustainable forestry is essential to the health of the planet, Northern Borneo, Malaysia.

wars. Defenders of old-growth forests sat down with the loggers who'd been cutting the big trees, and all agreed they had to make the forests more fire-resistant. For a solution, they looked to the forests that greeted the first European settlers to the States: widely spaced, older trees whose thick bark protected them from scorching when the underbrush between them was cleared by frequent, low-intensity fires. They realized that fire is a natural presence in the forest, and that fires sparked by lightning and native peoples have shaped the landscape for thousands of years. But throughout the twentieth century, an aggressive forest-fire-fighting policy had allowed fuel to accumulate instead of burning off harmlessly; decades of clear-cuts, replanted with dense tree farms, added highly combustible fuel to any wildfires that escaped our attempts to contain them.

The new fire coalition did not aim to stamp out wildfire entirely. Instead, it sought to keep fires from torching people's homes. To achieve that end, coalition members decided to thin the forest, removing dead and dying trees that could carry a wildfire from the forest floor into the canopy, where it would rage out of control. Fewer, more widely spaced trees would also be less apt to set each other alight, and soil moisture would be shared among fewer trees, making them less fire-prone.

Younger forests near rural towns were the natural candidates for the first experiments. Rural workers were pleased with the promise of new jobs and raw materials, and environmentalists were relieved that the focus had shifted from protecting primeval forests exclusively to working with younger ones. These stands had already been logged at least once, and the loggers would selectively cut only the smaller trees, instead of clear-cutting it all.

In the northern California town of Hayfork, the Watershed Research and Training Center is spearheading efforts to thin the already logged national forest for small firs, pines, oaks, and other hardwoods. A local business has sprung up to turn the timber into flooring and furniture, which it sells to Whole Foods Market, among other clients. The small trees are a far cry from the behemoths that had been cut to make plywood and lumber a decade earlier, but Big Timber had been forced to close up shop in Hayfork after a fracas over threatened owls and fish limited logging in the area.

In Arizona, one of the largest projects of this kind is yearly turning the cuttings from forest thinning into 65,000 tons (59,000 metric tons) of wood pellets for home heating. This has led the project operators to an opportunity to thin another 15,000 acres (6,000 hectares) annually—unopposed by environmental groups, despite its scale. This is the kind of logging we can all agree on. SZ

Certification

Standing in a lumberyard or a home-improvement store, it's hard not to feel a little ambivalent: we need the wood stacked around us for our projects and homes, but we can't escape the realization that the planks were, until recently, trees swaying in the wind on verdant hillsides. How can we tell if the two-by-four we're about to purchase came from a muddy clear-cut or a careful, selective harvest? Did it have the tree's equivalent of a happy, cage-free, grass-fed life?

Forestry nerds spent the better part of the mid-1990s figuring out how a wood buyer could answer those questions. The result was a system of third-party certification, in which a trusted entity separate from both the buyer and the timber industry vouches for the wood. The Forest Stewardship Council (FSC)—an independent agency including environmentalists, foresters, and indigenous peoples—sets the standard. Accredited audit firms inspect forests and mills; the ones that pass may affix the FSC trademark to their wood. Globally, 168 million acres (68 million hectares) are currently certified. It may sound like a lot, but combined, that worldwide total only actually equals the size of Texas.

Originally, certification visionaries hoped the stamp of approval would result in a higher price for certified timber, to compensate forest managers for doing a more careful job. But it's turned out that most buyers are extraordinarily price conscious and are unwilling to pay more for certified lumber. Instead, sustainable suppliers gained an advantage when major retailers like Home Depot began giving preference to certified lumber. Further impetus has come from the LEED [see Big Green Buildings and Skyscrapers, p. 247] green-building program, which awards one of its coveted points for using FSC-blessed wood.

Beware of imitations: a year after FSC started up, Big Timber's main trade association, the American Forest and Paper Association (AF&PA), created its own standard, called the Sustainable Forestry Initiative (SFI). For all its slick public relations, SFI has yet to shake its reputation as an industry greenwashing group. The tip-off? Every last member of the AF&PA, from International Paper on down, has won certification, despite their widespread practices of clear-cutting and raising single-species tree farms. SZ

Investing in Sustainable Timber

Spencer Beebe and Bettina von Hagen of Ecotrust are committed to the kind of long-term ecologically and socially responsible forestry that is independently certified by the Forest Stewardship Council. With backgrounds spanning finance and conservation, Beebe and von Hagen have led the development of a new kind of forestland investment fund that derives returns not just from timber harvests, but from ecosystem services including carbon storage, non-timber forest products, and habitat provision. Unlike conventional timberland investment funds that emphasize maximum harvests followed by land disposition over a ten-year lifespan, Ecotrust Forests, LLC is managed for perpetual, multi-generation returns and long-term land ownership and management.

The Ecotrust Forests, LLC fund estimates returns competitive with conventional timberland investment funds when taking into account emerging markets in ecosystem services like those tracked by the Katoomba Group's Ecosystem Marketplace, a think tank, advocacy group, and resource portal for information about the advancement of new markets for ecosystem services. For instance, longer-term harvesting rotations (65 years or more) allow significantly more average biomass/carbon to be stored than conventional short-term industrial harvesting rotations (30 to 40 years). This increase in carbon storage can be marketed as carbon credits to entities wishing to remain Kyoto compliant. Monetizing carbon storage allows Ecotrust Forests, LLC to provide competitive returns even with an emphasis on forest restoration during the first several years after acquiring a parcel. SC

RESOURCES

More Tree Talk: The People, Politics, and Economics of Timber by Ray Raphael (Island Press, 1994)
More Tree Talk offers an excellent introduction to the human and technical landscape of forestry. In a style reminiscent of prize-winning author and radio broadcast personality Studs Terkel, Ray Raphael gives a record of his conversations with representative players in the forestry scene, from lumberjacks to forest ecologists, rangers to mill owners, in an engaging monologue format.

Towards Forest Sustainability edited by David B. Lindenmayer and Jerry F. Franklin (Island Press, 2003)
Don't be daunted: this book may be packed with information, but it's an easy read. It includes a collection of essays on the latest developments in forestry in North America, Scandinavia, and Australasia.

Ecoforestry: The Art and Science of Sustainable Forest Use edited by Alan Drengson and Duncan Taylor (New Society Publishers, 1997)
In this solid anthology, scientists and practitioners lay out a multifaceted approach to sustainable forestry.

Creating Rural Sustainability in the Global South

In much of the world today, to be poor is still to be close to the land—people still depend on small farm plots, wild game, free-flowing water, and seasonally gathered plants.

Like all of us, people in the rural Global South want to better their lives. This desire often translates into a short-term outlook: people do what they can to make money today and trust that their newfound wealth will be enough to solve the problems of tomorrow. Too often, pursuit of a better life means pushing natural systems past their capacity to heal.

Overgrazing today means less wild game tomorrow. Overfishing local waters for commercial sales means less fish and a degraded ecosystem. Cutting down trees for marketable timber means barren hillsides and polluted water. Too often, "development experts" from the outside actively encourage people in the developing world to over-exploit their resources, with promises of a better future. But by imposing market values and limiting foresight, they instead create the perfect conditions for slow destruction, not only of the environment, but also of people's ability to achieve health and prosperity in the long term.

Now, however, a new paradigm has begun to take root. It has become apparent that for those who live close to the land, the best way to prosper is to work hand in hand with nature. While the idea of working with nature is as ancient as human civilization, it's now easier to see that sustainable development done right can be truly lucrative, offering long-term mutual benefit to local populations and outside markets. Rural sustainable development is a strategy with a future.

Poverty and the Environment

Protecting the environment is one of the best ways for people to lift themselves out of poverty and protect the gains they've made. This is even more true in the Global South than it is in the Global North—as World Resources Institute president Jonathan Lash has made quite explicit: "There's never been a time when poverty has been higher on the agenda, but if we don't make the key linkages between poverty, the environment and good governance, it will be impossible to achieve the poverty target. Seventy-five percent of the world's poor are rural poor, who depend directly on natural systems for their livelihood."

Most of the poor people in the world are in direct and constant contact with the land in order to obtain food, water, medicine, and fuel. If we destroy the environment, we destroy their ability to bootstrap out of desperation, not to mention the possibility of their leapfrogging out of poverty.

Opposite, left: Rural poor, the overwhelming majority of impoverished people, rely almost exclusively on natural resources to get by, and when these resources are polluted, their survival is in jeopardy.
Opposite, right: Women in Ethiopia walk across the drought-struck land, carrying jerry cans provided by the British charity Oxfam, 2004.

Agricultural Sustainability Equals Agricultural Productivity

If we were designing a worldchanging agricultural system for the developing world, one less likely to generate the kinds of social, economic, and environmental costs we see in today's dominant system, what would we want to include? How about improved water-use efficiency; reduced pesticide use; agroforestry (both to maintain nearby forest resources and to improve carbon sequestration); conservation tillage; even aquaculture (to incorporate fish and seafood as part of a larger integrated farm system)? All wonderful ideas, but of course the reason that industrial agriculture remains dominant is that it's so much more productive, right?

Wrong.

According to a 2006 study in the February 15 edition of *Environmental Science & Technology*, a journal of the American Chemical Society, sustainable agriculture techniques like those mentioned above, when introduced to developing-world farms over the last decade, improved farm yields by an average of 79 percent over four years. And not just in a limited set of locations: the study covered 286 different projects in 57 developing countries. That's more than 12 million farms.

Of the various crops the group studied, a quarter—largely maize, potatoes, and beans—saw yields increase by 100 percent, and half saw increases of at least 20 percent. Only rice and cotton saw just minor increases or slight declines.

The research team sees three of the methods under study as particularly useful: water-efficiency practices (including collective irrigation management, rainwater harvesting to cultivate formerly degraded lands, and watershed-level conservation); soil management to improve organic-matter accumulation and carbon sequestration (including conservation tillage and no-till agriculture); and pest, weed, and disease control, emphasizing "in-field biodiversity and reduced pesticide" use.

The authors make a point of calling out the carbon-sequestration benefits of these improvements as well, pointing out that if 25 percent of developing-world farms adopted these techniques, the potential increase in carbon sequestration would amount to a hundred (plus or minus four) megatons of carbon every year. As Professor Jules Pretty of the University of Essex in England notes, "such gains in carbon may offer new opportunities to households for income generation under emerging carbon trading schemes" (Pretty et al. 2006). That is, not only does adopting sustainable-agriculture practices mean greater yields, but it could mean added income, from carbon-trading agreements. Not only are sustainable agriculture techniques better for the farmers and better for the planet, but they also lead to substantially better production—vitally important in a world where millions still go hungry. JC

Drip Irrigation

Worldwide, 70 percent of water usage goes toward agricultural irrigation, and according to the UN's World Water Assessment Program, almost 60 percent of irrigation water is wasted.

How can farmers reduce water usage? One of the best tools is drip irrigation. Running water through open ditches or canals (called surface irrigation) allows water to evaporate into

the air and soak into the ground; spraying water through the air with sprinklers is effective in humid climates, but can cause up to 45 percent evaporative loss in desert climates. Drip irrigation instead relies on pipes with tiny holes in them to allow a controlled amount of water to drip out in designated spots directly onto the ground; some systems are even buried belowground to prevent any evaporation. Drip irrigation is especially good for trees and shrubs whose soil doesn't require tilling, but it can be used anywhere. Though it can be more costly than sprinklers or surface irrigation, it can also improve uniformity of water distribution, which keeps crop quality high. Drop by drop, better irrigation systems could help solve the world's water problems.

Extending the Garden

The urban market garden has great potential for poor and hungry people in the developing world. From the adoption of mini-livestock to a reacquaintance with indigenous vegetables, the promise of gardening as a self-empowering nutritional tool has yet to be fulfilled.

Backyard animal farming, a practice that goes hand in hand with the small vegetable lot, has been overlooked as a mechanism for solving the shortfalls in protein production. The neglect of indigenous and adaptable flora and fauna has consequences that range from unsustainability to warped local economies. A rediscovery and revalidation of these methods and means will go a long way toward remedying these mistakes, filling the belly and inspiring the mind.

The Snail and the Goat

The Giant African land snail *(Achatina fulica)*—considered a delicacy in most parts of West Africa; a pest in the United States—exhibits profound livestock qualities. It eats almost any vegetable and or fruit remains, requires some calcium, and seeks out moisture and darkness. Already an export product in Ghana, the snails could conceivably turn children and adults with very limited resources into livestock breeders and traders. For those with a bit more acreage to spare, the Nigerian Dwarf goat is an excellent source of sweet milk. A star in goat-breeding circles within the United States, the miniature goat has untapped potential to provide dairy products—a rare commodity—to its countries of origin.

Urban Vegetable Platter

The success of urban gardens, such as Senegalese Doudou Diallo's triumphant small organic vegetable and fruit garden, needs to be heralded in societies "where city dwellers may spend as much as 70 percent of their income on food." Diallo's bountiful harvest is a testament to the efficacy of homegrown solutions. The reemergence of previously neglected indigenous vegetable farming continues to be another practical plank in the fight for a nutritious diet that is sustainable. "Leafy vegetables are seen as an ally in the fight against *hidden hunger,* whether they be wild or cultivated, and produced by lianas, tubers or trees. Leafy vegetables offer populations with limited access to meat or fish a rich source of protein, which is essential for pregnant women and breastfeeding mothers, as well as for young and growing children." ED

Michoacán Butterfly Reserve

Some of the most exciting models for sustainable development empower rural communities with the tools to protect their environment and fight poverty. That's the case with the Michoacán Reforestation Fund and La Cruz Habitat Protection Project, a joint effort by U.S. and Mexican conservationists who work with rural communities to improve their material and economic prospects by restoring the forests around Mexico's monarch butterfly reserves.

Every year, tens of thousands of monarchs make an amazing 3,400-odd-mile (5,472-odd-kilometer) migration from eastern Canada and the United States to the mountain-bound oyamel fir forests of the Mexican state of Michoacán, where the balance of humidity and cool temperatures is perfect for the butterflies' winter hibernation. The federal reserves, established in the mid-1980s, are only a few hours' drive from Mexico City, and attract upwards of 200,000 visitors a year to witness the clouds of butterflies.

Deforestation has devastated much of the monarchs' overwintering habitat. Poverty has driven some of the region's tree loss, as local families cut forests to clear fields for subsistence farming and grazing, gather wood for cooking fuel and construction material, and sell the logs for a meager income. Illegal logging is also a factor. There are few viable job options in the region, so trees tomorrow often seem less important than jobs today. But the spectacle of the monarch migration keeps the tourists coming back and the conservationists paying attention, which has led to some innovative ideas—and the outside funding to sustain them.

Since 1997, the Michoacán Reforestation Project has worked with Mexican conservationists and hundreds of rural residents around the reserves, supplying them with over one million pine and oyamel fir seedlings for converting subsistence agricultural fields back to forestland. Restoring forests is stemming soil depletion and erosion, protecting streams and watersheds, and attracting wildlife back to the region. Communities work together to replant the forests, and to develop and implement community forest-management plans. Trees are harvested—for firewood, construction materials, and Christmas trees—sequentially according to when they were planted, preserving habitat and ecotourism dollars, and delivering a win-win scenario for villagers and butterflies alike. EG

AIDS and the Environment

In the face of a crisis as massive and widespread as the AIDS epidemic, we sometimes fail to make the connection between catastrophic human disease and the increasing ill health of the natural world.

In rural communities throughout many African countries, one major cause of deforestation is the growing demand for coffins to bury victims of AIDS. Traditional mourning rites involve all-night vigils, which add to the problem by requiring an enormous amount of firewood for heat and light.

Aside from the trees, native medicinal plants are rapidly depleting, because of individual efforts to seek free, readily available treatments for HIV and AIDS. The threat on nontimber-forest resources compounds both the problem and the pursuit of a solution. Only the establishment of clinics, along with the availability of some conventional drugs, is offering some relief to depleted tree and plant species. AS

The Thinning of Rural Culture

The destruction of rural culture needs to be seen for what it is: a serious loss of capacity. In its current incarnation, the shift from rural to urban signifies a mislaying of traditional means of food production—meaning that more and more people are becoming increasingly dependent on industrial agriculture for sustenance.

This loss of capacity represents a steady homogenization of diverse farming cultures that evolved over the centuries in response to local conditions. Skills and knowledge built up over generations are being lost in a matter of a few years, leaving human communities—as well as the global food system—less resilient to shock and to change.

All healthy cultures, urban or rural, are thick with choices. The thicker a culture, the more resilient it is to dramatic change and catastrophe. The thinner a culture, the more brittle it is. In a thick culture, if your crop fails, you always have a handful of choices left open to you; they might be unpleasant choices, but they ensure that your family won't starve. But in declining cultures, if your crop fails, or you lose your job, there are no other choices left. The only viable option is to leave your culture—often by joining the global trend of migration from the countryside to the city—and leave behind both the personal and geographic landscapes that define you. ZH

CyberTracker

Tracking is the ancient practice—both a science and an art—of observing wildlife intimately; it requires extreme sensitivity and alertness. Louis Liebenberg, a scientist and lifelong tracker, developed the idea of the CyberTracker while hunting with indigenous Bushmen in the Kalahari. The software fuses technology and field biology, allowing users to record observations about plants or animals on a GPS-empowered handheld, and then merge those observations into a larger map representing the way habitat is changing, or animals are migrating.

"It's quite a stunning example of what fairly regular collection of data by game guards doing patrols can tell us," says Liebenberg in a 2004 *Wired* article (Lindow).

Liebenberg attests that tracking, which was the basis of the hunter-gatherer existence, represents the genesis of scientific study. For the most part, tracking is no longer necessary for survival, and only a very few elder Bushmen still possess the knowledge and experience that was once essential education for children. But Liebenberg believes the practice ought to be reestablished, this time for the purpose of conservation. "There are not enough scientists out there gathering information," Liebenberg said. "If you had trackers on the ground working with scientists, in theory you could monitor the entire global ecosystem this way."

Future of the Small Town

Just a hundred years ago, 60 percent of Americans lived in rural areas; by 1990, that number had dropped to 25 percent. In the same time period, the proportion of the American labor force employed in agriculture has dropped from nearly 50 percent to less than 1 percent, according to the National Center for Policy Analysis.

Going by these statistics, it would seem that we need to add rural life to the growing list of things that face extinction. But hidden in these numbers is a remarkable reversal—from 1990 to the end of the century, rural areas of America have grown four times faster than they did from 1980 to 1990 (Johnson and Beale 2004)—and almost all of that growth is due to an influx of transplants from urban areas.

The "rural rebound," as researchers call it, is a result of fundamental economic changes. People aren't moving to rural America to take traditional jobs like farming and mining, but to find new types of jobs made possible by improvements in infrastructure and transportation: jobs in information technology, service, manufacturing, and distribution. Retirees are leading the way for these workers, swelling the population

Right: Small-town USA is experiencing something of a reawakening as people move to rural areas to improve their quality of life and take jobs in the twenty-first-century sector. **Opposite:** The Bryant "Hay Bale" House built by the Rural Studio is the home of Shepard and Alberta Bryant and their two grandchildren. The 24-inch thick walls are stacked hay bales that were stuccoed over with concrete, providing excellent, natural, and inexpensive insulation.

of rural retirement destinations that come with beautiful scenery and low-cost living.

But there's a dark side to this migration—some of the newer jobs are in problematic industries: meat-processing plants and prisons are two growth industries in rural America. Another troubling fact of rural growth is that it is especially pronounced in counties that border metropolitan areas, as city workers trade long commutes for lower costs of living.

Many of the proposed scenarios for creating a green future focus expressly on urban America: they overlook the fact that pedestrian-friendly downtowns and public transportation aren't feasible in communities with low population densities. It will take very different scenarios to preserve small-town rural life while addressing its larger sustainability issues. EZ

Rural Studio

Rural Studio, an outreach program established by Alabama's Auburn University, was founded in 1992 by the late architect Samuel Mockbee in response to the extreme poverty of western Alabama, the former heartland of cotton farming centered around Selma. State and local authorities there had demonstrated little interest in (and possessed no resources for) addressing the community's problems resulting from endemic poverty. So Mockbee created a program that allowed students to design and build houses and community projects with their own hands. They would learn the technical aspects of architectural design, and more importantly, a respect for using raw materials in sustainable and economical ways. To date, about eighty projects have been completed by Rural Studio students, from single-family houses to community spaces, including a new fire station and town hall. Many of the structures are made from recycled and collaged materials like old tires, discarded carpet tiles from office buildings, and broken concrete donated by the highway department.

The process of working with actual clients, each presenting particular needs and unique challenges, and then building on-site within the community makes students extraordinarily sensitive to the social and ethical demands of designing for a specific context. The clients in these areas often have few resources and rely on the builder to have the insight to create practical, functional, and lasting housing.

The model of a design/build architecture program has since been replicated in many universities, but the Rural Studio—having been in the same location for fourteen years—remains unique in its grasp of place; it has become a rooted component of local networks. The students who attend Rural Studio gain not only a better understanding of their professional pursuits and the relationship of builder to client, but a true connection with a rural way of life. HL

Reaping the Wind

Farmers who are frustrated by fluctuating commodity prices are discovering a highly profitable new crop: wind [see Green Power, p. 174]. As the cost of producing electricity from wind turbines drops, developers have started to invest in wind farms, and individual farmers have begun to see the advantages of adding turbines to their farms and ranches.

Commercial wind farms require huge areas of land. Farmer and wind-energy entrepreneur Dan Juhl's Woodstock Wind Farm, established in 1993 in southwestern Minnesota, encompasses 320 acres (129 hectares) of land—it also grosses $1.6 million a year and has a twenty-five-year contract to sell power to a local utility, Xcel Energy. The success of the project convinced Juhl to start a consultancy, DanMar and Associates, to help farmers identify appropriate pieces of land, get necessary permits, and finance equipment, which can cost up to $750,000 per turbine.

While commercial wind turbines are enormous—often 200 feet (61 meters) tall and 50,000 pounds (22,680 kilograms) by weight—their "footprint" is quite modest. Farmers can plant crops almost to the base of their turbines, and ranchers report that their cattle don't seem to mind sharing a field with the towers. Plus, the most appropriate locations for wind farms, according to the U.S. National Renewable Energy Lab, are the states with large numbers of farms and ranches—the Midwest and Mountain states.

Some farmers are discovering that they don't have to invest in turbines to make money from the wind: energy companies will lease farmland in exchange for 2–4 percent of annual turbine revenue, usually two to four thousand dollars a year. This type of income can have an enormous impact on small family farms. The USDA's 2003 farm-income forecast projects that 94 percent of farm income will come from "nonfarm" activities—anything from second jobs to revenue from wind turbines. EZ

Conservation Easements

Grandma Beth died last year and left you her farm. You've got no interest in becoming a farmer, and the million dollars the local developer, Sprawlco, is offering for the land sounds mighty tempting. But you've got fond memories of sitting on the front porch, staring out over fields of corn—you want the farm to remain a farm, and to stay within the family.

It turns out there's a way to do just that, and make a few bucks in the process. Conservation easements, sometimes called conservation restrictions, are legal agreements between a property owner and a government agency or nonprofit land trust that restrict how land can be used. Your lawyer draws up a conservation restriction for Grandma's land that stipulates that the property can't be subdivided or used to build more houses, but can be used for agricultural purposes and for more agricultural buildings.

Your local assessor offers two valuations for Grandma's farm: one tells you how much you'll make if the land is used to develop condominiums, another tells you how much you'll make if the land remains farmland. Not surprisingly, the land is worth $700,000 more if you let Sprawlco take over.

But you follow your better instincts and donate the conservation easement to a local land trust. They're now responsible for ensuring the easement is enforced, and they're able to tell their supporters that they've helped conserve another forty acres of rural farmland. They write you a receipt for a charitable donation of $700,000, which goes a long way toward reducing your federal taxes for the next five years. You still own the land, and you rent it to a local farmer who continues growing corn on it, generating income.

Ultimately, your children get to visit Grandma Beth's farm and see fields of corn instead of fields of condos. And those nightmares where Grandma threatens to haunt you for eternity if you sell the family farm? They're gone for good. EZ

Whistler 2020

Every year, the town of Whistler in British Columbia draws throngs of tourists in search of outdoor recreation in a mountain resort setting. While the idea of visiting this magnificent destination is, of course, to enjoy its offerings, the line between enjoying and exploiting Whistler's natural beauty tends to blur—as it does in so many wild places that have been developed for tourism.

But we can't point the finger at tourists alone. The buildings, systems, and services that form the infrastructure of this popular resort town could easily reduce the impact of tourism—if they were set up right. Enter the Whistler 2020 plan. Recognizing the power the town held to promote and embrace sustainability, the resort owners sought to forge a plan that lays out strategies for protecting the environment, supporting the local economy, and strengthening the community.

Already one of the premier resort destinations in the world, Whistler stands to become an influential model of success in implementing sustainable principles. The planners and strategists who developed the 2020 plan also have visions for Whistler's potential by 2060: by then, if all goes well, Whistler will be not only a sustainable resort, but a fully sustainable town.

RESOURCE

The First Measured Century: An Illustrated Guide to Trends in America, 1900–2000 by Theodore Caplow, Louis Hicks, and Ben J. Wattenberg (The American Enterprise Institute Press, 2001)
An accompaniment to the PBS special of the same name, *The First Measured Century* focuses on the measurable, and myriad, aspects of human existence—from car ownership to vehicular death, religious affiliation to professional practice—and presents them in an easy-to-understand, illustrated format spanning the entire twentieth century. Fascinating and informative, this book is an invaluable resource for understanding American social trends and contemplating where we may be headed.

Local Greenhouse Forecast

In early 2006, Great Lakes ice-fishing guides took clients out in boats because the lakes remained ice free for the first winter in living memory. That same year, Johannesburg residents were thrilled by a blizzard of brown-veined white butterflies, an unprecedented hatching attributed to South Africa's exceptionally wet summer. Pacific Northwest skiers and boarders went snowless in the winter of 2005, when the jet stream took a curious shift south. Storms of unusual intensity or unexpected trajectory seem to sweep parts of the world each week, while farmers battle historic droughts in southern Africa and the southwestern United States. Many scientists say that such anomalies point to global warming.

This is how most of us will experience the climate emergency in years to come. We'll hardly notice the gradual rise in average temperatures, the slight but sustained shifts in yearly rainfall, the upward creep of sea levels. But we will notice the blizzards that arrive heralded by thunder and lightning, the Katrina-strength hurricanes, the odd sustained stretches of heavy rain or no rain at all. Consequences of a climate system energized by excess heat, such atypically heavy weather events will be a large part of the local climate forecast in the near future.

Global average temperatures have increased by 1.4 degrees Fahrenheit (0.8 degrees Celsius) during the past century, according to a recent NASA report on global warming, and the connection linking this heating trend to the release of greenhouse gases, particularly carbon dioxide from human activities that burn coal, oil, gasoline, and natural gas, is beyond dispute. *Nature* recently reported that the concentration of CO_2 is higher now than it has been at any time in the past 650,000 years—a substantial share of Homo sapiens' tenure on earth. The year 2005 was the hottest recorded since direct measurements of earth's temperature began, and there is more CO_2—and more heat—to come.

"The world we have known is history," says James Gustave Speth, former administrator of the United Nations Development Program, and dean of the Yale School of Forestry and Environmental Studies. In our new world, the local climate forecast takes on new meaning. After all, we each need to choose where to live, what kind of dwelling to live in, what kinds of local foods to eat, and what kind of recreation to engage in. We want to know what jobs may be available, and what hazards we'll have to contend with. We have a stake in whether things become wetter or drier, whether to consider mosquitoes a mere nuisance or a serious health threat. In short, we want to know what comes next.

Climate increases have accelerated the rate at which Greenland is melting.

Local forecasts are the next frontier of applied climate science. The local work starts at a global scale. Climate scientists use computer models of the global climate to test assumptions about how future increases in CO_2 and other greenhouse gases might affect the global system. (About a dozen such models constitute the "gold standard" used by the Intergovernmental Panel on Climate Change for its authoritative projections.) They create scenarios of the climate's response to different trajectories of emissions, ranging from "assume we do nothing to alter the way we use energy today" to "assume we switch to clean power everywhere as fast as technically possible." Learning from such models reveals the likely real-world effects.

Analysts working at the regional scale—focusing on, say, the U.S. Pacific Northwest—run the results from the latest global models on their own models of regional climate response, bringing things down to an intimate level of streamflows, snowpack, and water availability for forests and agricultural areas. It's a murky crystal ball, but it's the best we've got, and the regional models are becoming more reliable as their results are checked against historical records, and the calculations are refined accordingly.

In the United States, eight university-based climate centers in a consortium known as RISA (Regional Integrated Sciences and Assessments) maintain regional climate models. Although much of the work the centers do is highly technical and intended for resource managers, most of the centers have publications or Web applications intended for people without PhDs in climate science.

So, are local climate forecasts from the RISA centers coming to the Weather Channel anytime soon? Probably not. But some of the regional climate projections are robust enough to give decision makers a basis for evaluating impacts on such things as drinking water supplies, shoreline structures, hydroelectric dams, and ski resorts. You can take a look at the information these public officials are using, and plan your life accordingly. Just be sure to count on surprises. EW

Welcome to Orefornia

The Climate Impacts Group at the University of Washington, Seattle, has forecast the climate of the Pacific Northwest for the next five decades, with some confidence. In 2005, a group of economists and resource scientists used these projections to take a preliminary look at likely impacts of climate change in Oregon, with a focus on eight key sectors of the state's $128 billion economy. In a nutshell, they foresaw Oregon's climate developing an uncanny resemblance to the present-day climate of its neighbor to the south.

Climate models project a rapid rise in regional temperatures—an increase of an average of 1.9 degrees Fahrenheit (1.1 degrees Celsius) beyond late twentieth-century levels by the 2020s, and an average of nearly 3 degrees Fahrenheit (1.6 degrees Celsius) by the 2040s; in other words, more warming in the next two decades than we saw in the entire twentieth century. The mountain snowpack, the region's largest reservoir of freshwater, is expected to shrink substantially. One projection (by research at the University of Washington) shows it declining by more than half by 2040. On the coast, sea levels could rise by as much as a yard by the end of the century—an estimate that does not take the accelerated melt of Greenland and Antarctic ice sheets into account.

The economists noted some big changes: Dwindling midsummer streamflows east of the Cascades would disadvantage farmers, communities, and native fish. Ski-area operators might find themselves forced to choose between spending on snowmaking equipment and on chairlifts. Dam operators could expect to run turbines more in the winter and less in the summer, the season when regional demand to cool air and pump water would grow. Public health authorities would confront growing caseloads of West Nile virus, Lyme disease, and other diseases transmitted by insects lured by warmer temperatures.

Oregon's best economic opportunity? Public-private investment in what a U of O report refers to as a "forest of new energy technologies" for job and income growth, spawning businesses to help Oregonians adapt while generating revenues the state will need to address the impacts of climate change.

Welcome to the new state of Orefornia. Perhaps the greatest symbolic blow to the old Oregon could be the displacement of its signature wine-grape variety, as warming surpasses the variety's optimal temperature range. Goodbye pinot, hello merlot? EW

Look, It's Climate Change!

Unless you live on permafrost or in a mountain valley from which an alpine glacier is receding, it can be hard to put your finger on changes definitely linked to global warming. But such evidence is all around us, and it's growing fast.

Early in the millennium, six organizations committed to the environment got together to create a map called Global Warming: Early Warning Signs, which gathers evidence of the local consequences of global warming in every region of the world. The map, which was last updated in 2003, displays spatially referenced fingerprints—or "direct manifestations of a widespread and long-term trend toward warmer global temperatures"—and harbingers that "foreshadow the types of impacts likely to become more frequent and widespread with continued warming." Each of the featured fingerprints and harbingers is backed with a credible reference from scientific journals or news reports.

Fingerprints include heat waves, ocean warming and rising sea levels, melting glaciers, and changes in the Arctic and Antarctic. Harbingers include such phenomena as reports of early nesting (reported in twenty out of sixty-five bird species studied in England), coral-reef bleaching, and the massive drought that struck the Korean Peninsula in 2001.

In England, the evidence of global warming is being directly observed and recorded by amateur enthusiasts, thanks to one network that is gathering data by tapping the long-established tradition of local enthusiasm for natural history—and it is already telling a larger story. Many Brits are mad for phenology, which the UK Phenology Network defines as the study of periodically recurring "natural phenomena, especially in relation to climate." The network joins together some 24,000 amateur observers around the British Isles, pooling their observations online and tracking the shifting timing of natural events. The network has created Climate Change in Your Garden, an interactive component of its Web site on which a self-guided garden tour shows how the timing of common events in nature signals climate change.

Orange-tip butterflies, oak trees, great-crested newts, and hay-fever sufferers are among the advance guard of creatures impacted by climate change in Great Britain. A decade ago, Costa Rica trained cadres of "parataxonomists" to tally that country's wealth of plant and animal species. Today, we all need to take a cue from the English and become paraphenologists—marking the changing dates on nature's calendar. EW

Picturing Global Warming

In 1999, photographer Gary Braasch set out to create a photographic record of global warming. His collection of documentary images,

This photograph of beach erosion on Cape Hatteras, North Carolina, is part of photographer Gary Braasch's project *World View of Global Warming.*

World View of Global Warming (presented online, in traveling gallery exhibitions, and at universities around the world), offers a potent record of changes in sea levels, glaciers, and ice fields; of transformations in plant and animal species; and of extreme and unstable weather—and of people's efforts to adjust their lives and homes to these changes. The images will appear with commentary in the forthcoming book *Earth Under Fire: How Global Warming Is Changing the World* (University of California Press).

Similarly, *NorthSouthEastWest,* a collaborative project from ten Magnum photographers, offers "a 360-degree view of climate change" and reveals its impacts on the lives of ordinary people. Sponsored by the Climate Group in the United Kingdom, the United States, and Australia, the exhibition will travel to sixty-five nations—including Bulgaria, Italy, Russia, South Africa, and South Korea—as part of the British Council's ZeroCarbonCity campaign to raise public awareness about climate change and energy crises. EW

Taking Local Action: A Tale of Three Cities

As climate scientists' projections concerning global-warming impacts at the regional level improve, one audience is starting to tune in: local governments. The folks responsible for keeping traffic moving, maintaining bridges, dealing with floods and overflows, and clearing fallen trees, landslides, and other urban mishaps are keenly interested in likely changes in rainfall, storm intensities, and flood levels. Many of their decisions have an impact that lasts for decades, and mistakes can be costly.

Cities around the world have been quick to find ways to reduce and encourage behavior to reduce greenhouse-gas emissions. Now they recognize that whatever success they achieve in meeting or beating Kyoto Treaty targets, they will still have to adapt to a warmer, less stable climate. Few cities have yet taken a thorough look at local impacts, but early efforts in Boston, Seattle, and London point the way for other urban areas.

In 2004, three universities teamed up with local agencies to produce "Climate's Long-Term Impact on Metro Boston" (the CLIMB study). Using available climate projections, the CLIMB team assessed potential impacts on the hundred municipalities and six counties that make up the metropolitan region. They placed special emphasis on Boston's coastal zone, where 1.2 million residents in thirty-two municipalities live near present-day sea level. With a two-foot increase in sea levels and a commensurate rise in the intensity and frequency of storm surges, coastal flooding would heighten property damage, and demand for emergency services could cost $94 billion over the course of the century.

The team evaluated three adaptive strategies that Boston could use to cope with costs and impacts of this magnitude. The city could "Ride It Out" and take no adaptive steps other than rebuilding after damage; it could "Build Its Way Out" through limited preemptive measures like strengthening seawalls; or it could pursue a "Green" strategy of taking actions early to blunt the effects of rising sea levels and temperatures. Not surprisingly, early action looked likely to save the Boston area tens of billions of dollars over the century.

One city that took note was Seattle—also a port city with a shoreline vulnerable to flooding, and hilly terrain prone to landslides. There, the Office of the City Auditor (the agency responsible for gauging government effectiveness) launched a series of reports examining how climate change would affect the operations and infrastructure of various city departments. The first looked at the Department of Transportation, the bureau responsible for Seattle's roads, bridges, seawall, and urban forest.

The auditor's review made some stunning observations. A planned seawall replacement might need to be reengineered to take new estimates of sea-level rise into account. Seattle's 105 city-owned bridges, over one-third of which are already considered to be in poor condition or worse, would face additional deterioration from thermal expansion and heightened winter rainfall. Seattle's $100 million "urban forest" of trees and shrubs, maintained on city rights-of-way, already showed evidence of impact from hotter, drier summers. Even the pavement on Seattle streets, an urban asset with a replacement value of $4.1 billion, might need to be reformulated to take new temperature ranges and moisture loads into account. All told, the climate-change impacts could mean spending billions of dollars on public infrastructure.

Cities around the world are beginning to discern common vulnerabilities and to share insights on adaptive strategies. London has taken the lead with an effort called the London Climate Change Partnership, which seeks to reduce carbon dioxide emissions and to help the seven million residents of Greater London adapt to climate-change impacts. Municipalities worldwide share experience through a network known as the International Council for Local Environmental Initiatives, with a membership of 475 cities, towns, and counties representing 300 million people. Such networks may prove to be a key to exchanging lessons that will embolden local action to defeat the heat. EW

Forecasting the African Climate

Getting information about the local impact of climate change depends on having scientists dedicated to studying the issue, and having the funding to facilitate their work. This is extremely problematic in Africa, the continent likely to suffer the worst effects of global climate disruption. Europe, the United States, Brazil, China, and India all have advanced efforts under way, but Africa has few climate scientists of its own, and little money for advanced climate-science efforts.

But the Africa Network in Earth System Science, a transnational group formed in Nairobi, Kenya, in September 2005, may change that. In fact, the network could boost support for pan-African geophysical sciences in general. The creation of another academic network may seem like a fairly prosaic development, but it means that the specific interests and concerns of Africa have a better chance of receiving adequate attention, which in turn means that nonobvious effects with global implications are more likely to be discovered and studied early on. The success of the Africa Network in Earth System Science is in all of our best interests. JC

RESOURCES

An Inconvenient Truth directed by Davis Guggenheim (2006) and ***An Inconvenient Truth*** by Al Gore (Rodale Press, 2006)
If you want a crash course in what climate change is, how we know it's here, and what we can expect if we don't do something about it, there is no better resource than Al Gore's documentary and companion book *An Inconvenient Truth*. The documentary consists mostly of footage of Gore giving his now-famous lecture on how we know climate change is real, here, and serious. It's not flashy, but *An Inconvenient Truth* is an incredibly important film and the book should be mandatory reading in every household. Gore's work will change the American debate on climate change, and that will change everything.

Stop Global Warming
http://StopGlobalWarming.org
For those concerned about the issue, log on to the Stop Global Warming Virtual March founded by activist and *An Inconvenient Truth* producer Laurie David, Senator John McCain, and Robert F. Kennedy, Jr. The site includes educational information, tips on simple things that you can do in your everyday life that can make a difference, and provides the sense of a growing community of concerned individuals. Over half a million Americans have already signed on to this call to action.

Climate Foresight

How will rising temperatures and weirder weather upset our expectations of how the world works and what the future holds? No one knows precisely, but learning to nurture good answers to that question bears heavily on our ability to create effective solutions.

In business, in health, in our efforts to protect the environment, climate change is rewriting the rules. Industries whose existence our parents took as givens, like Big Oil, will be gone within our lifetimes, while new industries—perhaps Big Solar—will spring up in their places. We'll find ourselves under threat from diseases not known in our areas (like West Nile virus in the Eastern United States). Plants and animals will invade regions they never could have survived in before, as the climate in those places becomes friendlier to them. Meanwhile, our efforts to protect fragile ecosystems will have to take into account that the weather that created those places is now changing. In a warming world, the only certainty is change.

Mountains serve as metaphors in the era we're entering. Mountain ecosystems around the world tend to vary considerably from one elevation to the next: on the lowest slopes, one kind of forest may be common, with its own complement of plants and animals, while on the higher slopes (where it tends to be slightly colder) different kinds of trees may be home to a different community of critters. But as temperatures rise, the trees in the lower forests may find that their ideal growing conditions have migrated upslope, while they themselves are rooted in place. This provides a massive conservation challenge, but it also provides a pretty good metaphor for the challenges we will all face: How will the things we've taken for granted in our lives "migrate upslope"? How will our futures be different, when a shifting climate has moved the ground beneath our feet?

We're still learning the answers, but to fail to ask these questions is to leave ourselves vulnerable to some heavy weather indeed. AS

The Response So Far

Around the world, while nations and subnational jurisdictions act to combat climate change, cities are making some of the most vigorous moves to promote climate protection. The International Cities for Local Environmental Initiatives/Cities for Climate Protection Campaign supports urban areas around the world in implementing greenhouse-gas reductions, primarily in municipal operations.

In 2005, Seattle mayor Greg Nickels upped the ante by committing to meet Kyoto goals not just in the city's operations—the city made that commitment in 2000 and more than achieved it—but also in its commercial economy. Nickels

Temperatures in Point Barrow, Alaska, have risen an average of more than four degrees in the last three decades.

challenged other U.S. cities to take up the slack left by federal inaction. According to the City of Seattle's Web site (as of May 2006) 230 cities have made the Seattle commitment, and the U.S. Conference of Mayors has unanimously endorsed the concept. Some cities are closer than others; Portland, Oregon, for example, is already producing lower amounts of CO_2 than it did in 1990 (according to a report on global warming cited on SustainablePortland.org), and expects to hit its goal of reducing the 1990 figure by 5 percent soon.

Other U.S. states and regions are also taking action. In 2009, California will begin implementing the world's first limits on auto CO_2 emissions, mandating a cut in new car and light truck emissions of 30 percent by 2015, according to that state's Air Resources Board. Since numerous other states participate in the California air-pollution-control regime, the state's actions affect a large portion of the United States, including about 20 percent of the nation's automobile market. Oregon, Washington, Massachusetts, and New Hampshire have passed binding limits on power plant CO_2, while Northeast states are coming together to shape a mini-Kyoto to cap power sector emissions through the Regional Greenhouse Gas Initiative. California, Washington, and Oregon are working toward a similar framework through the West Coast Governors Climate Initiative.

Overall, about 30 percent of the American population lives in states or cities that either have adopted or are set to adopt policies in accord with the Kyoto protocols. Moreover, the Kyoto-friendly regions account for nearly 50 percent of the total U.S. gross domestic product—a total economic output greater than that of Japan, which currently has the world's second-largest economy. While the mechanisms for compliance vary considerably, environmental economists are cautiously hopeful that working at the city and state level will allow the development of flexible, locally relevant policies for fighting global warming. PM & JC

Oil companies may one day be held accountable as one of the biggest culprits in the production of greenhouse gases.

Politics of Problems

In a world still overwhelmed by conflict, terror, and economic uncertainty, it's difficult to imagine that none of these issues will be the key political issue of the first part of this century. But the earth's environment—the climate, specifically—is likely to become the greatest issue of political debate and global struggle we'll face for the next twenty years. The climate is moving to center stage in part because environmental problems can make other issues—like conflict, terror, and economic uncertainty—all the worse, and in part because the disruption of the climate caused by global warming will lead to problems on a previously unimaginable scale.

If history is a guide, one of the stickier problems of the global-warming era will be rather prosaic: assigning blame. If natural disasters with no possible human cause—such as the 2004 South Asian tsunami—can lead to accusations of negligence and corruption, any event with a much clearer human origin will become an ongoing source of debate and conflict regarding responsibility. Since the nation with arguably the greatest burden of blame is also the world's preeminent military power, these debates and conflicts are more likely to be legal and economic than martial.

We could, for example, see "carbon trials," where the leaders of the corporations deemed most responsible for spewing greenhouse gases—oil companies, automakers, and utilities with coal-fired power plants, for example—are brought before the U.S. Congress and made to face angry politicians. Imagine the cigarette hearings of the 1990s, with an even angrier citizenry.

We could see international lawsuits from communities most affected by global warming (and least to blame for it), backed by the threat of economic sanctions from bigger powers. Inuit groups have already filed suit [see Polar Regions, p. 527] against the United States for causing the melting of Arctic lands; even if that lawsuit is thrown out, it won't be the last one filed.

We could see global boycotts of products and services from companies deemed to be resisting the necessary changes to improve environmental sustainability and reduce their carbon footprints. The Internet makes global collaborative action much easier than ever before, and millions of interested citizens around the world make it harder for companies to hide environmental misbehavior. As an example, the Web site ExxonSecrets.org already tracks in enormous detail precisely how Exxon Mobil funds the small network of scientists and think tanks that continue to insist (despite overwhelming evidence and near-unanimous scientific consensus to the contrary) that climate change is just a theory; some activists have proposed keeping records on such activities as evidence of potential crimes against humanity.

None of these scenarios are certain, but all are at least somewhat likely, along with arguments and accusations that we can't yet imagine. The global-warming era is just beginning, and we all have a stake in the outcome. JC

Global Warming and Global Health

One of the ironies of the global-warming crisis is that the regions that produced the most greenhouse gases over the last century are by and large the regions that are least likely to see serious adverse results. According to research by the University of Wisconsin, Madison, and the World Health Organization (WHO), many of the poorest nations have been, and will continue to be, those hardest hit by climate disruption—particularly concerning health-related effects. These nations (with the arguable exceptions of India and China) rank at the bottom of the list when it comes to emitting greenhouse gases. The WHO now estimates that at least 150,000 deaths each year are directly attributable to climate disruption. Over the next twenty-five years, that number will rise substantially.

The causes of death fall into two categories: noninfectious health effects, caused by heat waves and crop failures, for example; and infectious diseases such as malaria, dengue fever, and even salmonella. Developing nations are hardest hit by climate-related health problems for reasons of geography and politics. As the UW/WHO study's lead author, Dr. Jonathan Patz, puts it, "Those least able to cope and least responsible for the greenhouse gases that cause global warming are most affected. Herein lies an enormous global ethical challenge."

Global development and environmental sustainability are interconnected. Efforts like the Millennium Development Goals, which aim to reduce poverty and improve health care in the world's poorest nations, are demonstrably critical tools for environmental response, as well. Similarly, the need to slow the pace of global warming and avoid its harshest results is as much an issue of humanitarian responsibility as it is one of environmental stewardship. JC

RESOURCE

Flood Maps
http://flood.firetree.net/
With many climate models suggesting that the ocean—which is already rising slowly due to global warming—could potentially get twenty feet higher by the end of the century, being able to imagine the impacts of rising seas is useful. Several Web sites out there will help you understand what sea-level rise could mean in your area; Flood.Firetree.net is our top pick.

A Personal Action Plan

I don't believe in "average people" doing anything [about the climate]. People ought to support mitigation and adaptation within their own line of work, no matter how un-average that is. I mean: if you're a butcher, baker, ballerina, banker, or a plumber, envision yourself as the post-fossil-fuel version of yourself, and get right after it.
Bruce Sterling

With problems as seemingly insurmountable as global warming, we can easily slip into a mind-set of "let the government handle it, let big corporations handle it, there's nothing I can do." But we don't have to take the easy way out. We can each make changes in our own lives that will make a difference—and little changes compound one another, so a relative few individuals taking action can have a disproportionately large effect. And there's a bonus to crafting a personal plan of action sooner rather than later: when governments and big corporations ultimately do begin to take action, we'll be ahead of the game, not taken by surprise.

So what should our personal response be? What can we do to reduce our environmental footprint? We can find ways to reduce our personal contribution to climate change, to use fewer of our planet's resources, and to consciously ward off environmental collapse. We can choose to hammer out a positive path today. In the end, learning how to live in more climate-friendly ways will benefit not only the planet, but also ourselves. AS

Making Smart Choices

Seeing the big climate picture is useful for more than national planning and response; it's also very important for making your own big life choices. Ask yourself the following questions regarding the choices you currently make. Your answers may reveal something about the kind of future you're building for yourself:

What kinds of stuff do you buy? Start thinking now about energy efficiency as fundamental to what you buy or use. Items that waste energy (appliances that draw power even when they're "off," homes that leak heat through drafty windows) will be serious liabilities in the years to come. Key idea: Efficiency is a long-term win-win strategy.

What do you do for a living? Is your job likely to see increasing or declining demand in a time of economic stress, population migration, and environmental woe? Is the company you work for apt to be seen as a cause of these problems (justifiably or not)? Key idea: Professions that help people cope with big changes are sure to see increased demand.

Where do you live? Locales that seem attractive now could become much hotter in future summers, and more prone to big storms, persistent

drought, and/or invasive plants and bugs. The key idea is, if a spot is bordering on unsustainable now, it's liable to be in trouble in a decade or two. One big hint: avoid beachfront property.

Given the fact that we live in an era of rapid technological advancement, we're not all that accustomed to thinking decades ahead. But the environment doesn't keep pace with human activity. We have a pretty good sense of how the climate is likely to change over the next decades, and even if we get smart and change our behavior on a dime, these big, slow processes will continue to apply. We are now committed to changing the climate—but by making smart choices in our own lives, we can both prosper and prevent global warming from becoming any worse than it needs to be. JC

Opposite, left: Purchasing energy-efficient compact-fluorescent lightbulbs is a small but simple way to shrink your home's footprint.
Opposite, right: A nature educator at a zoo in Philadelphia educates a class of second graders about cheetahs. Zoos act as a catalyst for connecting urban kids to their natural surroundings.
Above: Greenpeace workers in Bangkok, Thailand, hand out pamphlets to passersby, supplying them with information on solar energy and clean power.

Check Your Carbon Footprint

So you've checked your "environmental footprint" [see Questioning Consumption, p. 32] score online, and now you know your overall impact on the planet. But how do you break that number down into concrete elements? If you want to know how best to make a difference in terms of global warming, for example, it helps to know just how much of a contribution you make to that problem in the first place.

SafeClimate.net's carbon-footprint calculator asks you for information about your travel and transportation habits and your home-energy use—the two biggest sources of individual greenhouse-gas emission. Applying figures from the International Panel on Climate Change and the U.S. Department of Energy, the calculator—which can compile data for Europe; Australia and New Zealand; the United States and Canada; and Japan—works out how much carbon dioxide is typically generated by your activities, then compares that to world averages.

The Carbon Footprint Calculator (at—of all places—the BP Global Web site) takes a bit of a different approach. It asks for fewer numerical details, but focuses on questions about a variety of energy-saving appliances and practices. The calculator—which can compile data only for the United States and the United Kingdom—is a subtle but effective way of suggesting changes to behavior; you can see with a few clicks just how much of a difference it would make to put in double-glazed windows or energy-efficient lights. JC

Personal Carbon Trading

What if we could essentially get paid to reduce our contributions to global warming, in a fun and effective system that rewards our reductions? The Kyoto Treaty includes a scheme that lets companies and countries trade "carbon credits"—those members that move quickly to reduce greenhouse gases can make a bit of money off those that lag behind. But carbon-trading systems aren't just for the big players. If folks at the Tyndall Centre for Climate Change Research in the United Kingdom have their way, we all could get in on the fun. They've come up with a concept they call Domestic Tradable Quotas (DTQs)—essentially, personal carbon credits.

Under the DTQ plan, individuals would receive an annual carbon quota, one that would decline over time as broader carbon restrictions became more stringent. People who lived more efficiently would have extra credits to sell off to those who lived less cautiously, and could make a tidy sum. This would encourage everyone to be more efficient—both to save money and to get in on the game while there's still money to be made.

On the downside, the DTQ concept would definitely be complex to implement. It's one thing to put carbon taxes on energy consumption, but it's another to keep track of indirect carbon emissions—such as the greenhouse gases that come from the trucks hauling food to the market—from individual consumption. At worst, it could end up requiring a huge, and potentially very intrusive, bureaucracy.

But on the upside, the DTQ proposal does get us to consider other unconventional approaches to fighting global warming—and if that's all it ever does, it will still make a real difference. JC

Carbon Offsets

There's a limit to how much people can do to reduce the greenhouse gases they put into the air. Even the most ecoconscious people need to fly, to drive, or to do things (like buy groceries) that require *other* people to fly or drive. But that doesn't mean you simply have to accept the resulting carbon emissions without doing something to help.

"Carbon offsets" let you underwrite activities that actively reduce global CO_2 emissions, thereby balancing out your personal emissions. Some of the organizations providing carbon offsets, such as Sustainable Travel International, support renewable energy projects and distribution of high-efficiency lights and stoves in the developing world; others, like the UK-based CO_2 Balance group, focus on carbon "sequestration" efforts like planting trees. A handful, such as Drive Neutral, actually buy up carbon credits on the global market—thereby increasing the incentives for companies and countries lagging behind the Kyoto Treaty limits to get more efficient. And a few, such as Carbonfund, do all of the above.

Many of the carbon-offset groups emphasize balancing out travel-related emissions, but nearly all of them provide broad tools for figuring out your personal carbon footprint for all activities. Most people in the United States produce about seven tons of CO_2 emissions annually according to EPA stats, but your footprint could be much lower—or higher. Prices of CO_2 offset vary widely, depending in large part on the offset methods chosen by the organization; cost can be as low as six dollars or as high as twenty dollars per ton; most organizations average out at about ten dollars per ton. But whatever the cost, it's always worth investing wisely in the future of the planet. JC

RESOURCES

High Tide: The Truth about Our Climate Crisis by Mark Lynas (Picador Paperback Original, 2004)
There are plenty of books out there on the impending threat of global warming—what makes Lynas's book groundbreaking is his compelling and terrifying evidence of how global warming is affecting people's lives and our planet even now. His book is a narrative, told through the people who are living with the daily effects of global warming, but it is also a damning critique of American environmental policy under the Bush administration: "Climate change begins and ends in America. We all know that the United States is the world's largest polluter, and yet the sheer weight of responsibility that Americans now carry of the world's plight has barely touched the national psyche." Let this book be the impetus for a much-needed wake-up call.

The Discovery of Global Warming by Spencer R. Weart (Harvard University Press, 2003)
"Our response to the threat of global warming will affect our personal well-being, the evolution of human society, indeed all life on our planet," writes Weart—strong words concerning the most pressing problem the planet faces today. *The Discovery of Global Warming* tells the all-important story of how scientists confirmed and announced the very real, very human causes behind the changes affecting our planet. Engaging and balanced, Weart's narrative gives equal importance to the political and the social subtexts that contributed to the scientists' research and claims. Ultimately, Weart has given us a highly relevant work of nonfiction that is key to understanding the contemporary debate surrounding global warming.

Stormy Weather: 101 Solutions to Global Climate Change by Guy Dauncey and Patrick Mazza (New Society Publishers, 2001)
In contrast with the copious resources detailing the gloom-and-doom prospects we face as a result of global warming, Dauncey and Mazza's book provides a refreshing change of pace, offering simple and profound solutions that can be the start of serious planetary change. Organized into easy-to-read two-page spreads, the book is a wealth of information, explaining alternative fuels and energy sources as well as the major problems facing the planet today—and providing 101 real-life, actionable solutions that anyone can implement. *Stormy Weather* aims to spur action: "if there ever was a call to grow up as a species and to end our obsession with money, power and fossil fuels, this is it."

World as Lover, World as Self by Joanna Macy (Parallax Press, 1991)
There's no doubt that to remedy the environmental problems we're experiencing today will require much more than just changing economic or business practices—Joanna Macy contends that it will require a holistic approach: "We are capable of suffering with our world, and that is the true meaning of compassion." Writing with eloquence and dignity about the "green self," Macy provides a spiritual look at the interconnectedness of the planet, and provides a blueprint for how we can reverse our destructive actions. Grounded in Buddhist philosophy, Macy is a planetary advocate worth paying attention to.

Mapping

In August 1854, cholera swept through London's Soho district. The common belief held that cholera was spread through bad air—the "miasma" theory. Physician John Snow, however, had another idea: he believed that cholera was a microorganism that infected people through the water they drank. Snow suspected a single water pump to be the culprit (most of the city's victims lived within walking distance of the Broad Street pump), and by interviewing victims' families and then plotting the evidence on a map, Snow was able not only to help end the epidemic but to make his theory intelligible to the authorities—work that established Snow's place as one of the fathers of modern epidemiology and medical geography.

Maps communicate essential information in ways few other media can. Snow's use of maps and data as analytical and communication tools for science heralded a trend that has become positively epidemic 150 years later: cartography is an essential tool for researchers and advocates, policy shapers and corporations. These days, when we want to change the world, we often start by mapping it. CFD

A Revolution in Mapping

People used to make maps by hand. Now new forms of software, like Geographic Information Systems (GIS), do the job for us. As with many other democratizing technologies that emerged in the late twentieth century, faster, cheaper computing power is largely responsible for the growing use of digital mapping applications—by everyone from disaster responders and forestry-reform advocates to legal-service providers and overnight-delivery companies. Today, if we know where something is, we can easily show it on a map.

And we're learning where more and more things are. The explosion of cheap, open-source tools has been key to the spread of mapping. Data for mapping is everywhere. Data may now come from satellites that measure moisture in the leaves of trees or from radio collars around the necks of unknowing caribou as they migrate hundreds of miles; maps may draw on information relayed from temperature sensors along the ocean floor or from pollution-measuring equipment in our major cities. And the agile software that draws these maps is being developed not in the cubicles and server farms of large corporations, but among a dizzyingly productive network of open-source developers.

Like the maps of the past, the maps of the future will aid both in data's analysis and in its communication to broader audiences. Data can be gathered via remote sensors that can measure invisible parts of the electromagnetic spectrum or synthesize information over vast geographic areas—promising to reveal insights on the invisible things that are right under our noses. That is exactly what happened on the Shuttle Radar

Left: This map of Tropical Depression Alberto was created with images from Geostationary Operational Environmental Satellite (GOES-12), which is operated by the National Oceanic and Atmospheric Administration.

Opposite: The mapping of craters, such as this one on Vulcano Island in Italy, has been essential in establishing the historical trajectory of the earth.

Topographic Mission (SRTM), during which the Chicxulub Crater on the Yucatán peninsula was discovered. Scientists had long suspected that such a crater existed: there had to be some evidence, somewhere, of the asteroid impact that wiped out 70 percent of life on the planet 65 million years ago. But it wasn't until the 1990s, when the SRTM data was processed and scientists mapped the remnants of the 112-mile-wide (180-kilometer-wide) crater that they were able to confirm their theory. Maps reached deep into the past and pulled back an essential piece of our planetary history.

In the United States, we are fortunate to have widespread access to an enormous volume of data that we can use in mapping efforts. But increasingly, the growing number of satellites that produce remotely sensed data—in multiple spectra at varying resolutions, anywhere on the globe—is closing the gap between the United States and the rest of the world. Not long ago, if we wanted to see satellite-generated information on a map, it took lots of number crunching by big computers found only at places like NASA's Jet Propulsion Laboratory. Now the information is available, on your laptop, almost as events unfold on the ground.

Easy access to timely data means that useful, sophisticated applications can emerge organically as they are needed. That's what happened in 2005, when Hurricane Katrina hit. As the storm mounted and spun inexorably landward, an ad hoc team hastily built a Web site in classic smart mob [see Direct Action, p. 459] form, using flexible, open-source mapping tools (PrimaGIS and ka-Map—two of the more interesting Web-based, open-source mapping resources) to share real-time imagery and other spatial data. In a matter of days, evacuees could log on from wherever they found themselves, enter their home address, and assess their neighborhoods through high-resolution, color aerial photos. Since the spatial data was published on the site using open-source standards, more advanced users could access raw data for use on their own Web maps, combine it with other information—the locations of functioning gas stations, for instance—and customize the maps they or their friends and families needed. CFD

The James Reserve

Though it may seem counterintuitive, mapping isn't confined to showing the big picture. The revolution in the production of spatial data (information about places) extends to the micro scale—to soil patches and birds' nests, to a fox traversing a canyon at night, or to the subtle respiration of CO_2 by a patch of silver pine. These are some of the natural systems being measured and tracked at the James Reserve, a protected mountainous area in San Jacinto, California. Here scientists use an expansive network of wireless remote sensors to measure everything from barometric shifts throughout the day to the number of times a willow flycatcher returns to its nest during a given period. The scientists systematically record everything from microbes to watersheds—ten orders of magnitude in spatial and temporal scale—in an unprecedented effort to understand the workings of a single piece of land. Similar remote networks may ultimately be used to monitor whole ecosystems like forests, or rift valleys along the ocean floor. And maps are what make the data recorded by these sensors understandable.

As sensors get cheaper, we can expect to know more about more places, and this information may help us solve seemingly intractable environmental problems. What if we monitored the comings and goings of reintroduced wolf populations so that an automatic warning system could alert ranchers when livestock were in danger? Telemetry data streamed to GIS applications like those operated by the U.S. Geological Survey in Alaska may make this possible.

In Washington's Puget Sound, scientists are learning that juvenile salmon from the northern rivers may brave predators and adverse marine conditions to venture to the South Sound in preparation for their oceanic migration. Why don't the fish just swim directly out to sea? What happens to them while they dally? Mapping their behavior patterns with other information such as water temperature, coastal vegetation, and dissolved oxygen helps us understand how to help these endangered fish thrive again.

The idea that remote sensors may one day be ubiquitous opens the door to a world where maps are simply another form of media, akin to your computer screen, for delivering streaming data. NASA's Moderate Resolution Imaging Spectroradiometer (MODIS) produces satellite imagery of the entire earth roughly every two days, in thirty-six spectral bands. Using increasingly accessible tools, almost anyone can integrate the data into a personal online map. The explosion of integrating tools like Google Earth and World Wind, which let us play with powerful mapping software and various kinds of data at home, is only the earliest hint of what this could mean. Certain GIS functions based on proprietary software, currently accessible only to trained users, are being replicated in flexible, open-source code that can be configured by users with far less technical training. The ability to access and publish sophisticated mappable data puts in our hands the power to both know and show the world. We don't need to wait for another John Snow—today we can put the water pumps and cholera cases on the map ourselves. CFD

Charting the Deep Oceans

We live on an ocean planet. Most of the earth lies underwater, and has—for most of history—been a place of mystery and unknown depths. For centuries, our maps of the land got better and better, while our maps of the ocean remained replete with sea monsters and bizarre warnings. Even mariners who sailed the world had essentially no idea what went on under the sea's surface.

All of that is changing now. The oceans are the new frontier of knowledge, and we're making startling new discoveries faster than they can be written up in the scientific press. What we're finding is that the ocean is both staggeringly more interesting and gravely more troubled than we'd ever expected. It teems with undiscovered life. It is also suffering from overfishing, pollution, and global warming (which, through a complex mechanism, is making the oceans more and more acidic, threatening the health of the entire planet).

The more we learn about the oceans, the more we understand our dependence on them. Given how ignorant we remain about how nature works in the planet's marine depths, exploration of the oceans has changed from being merely a bold scientific adventure to becoming essential research for our survival.

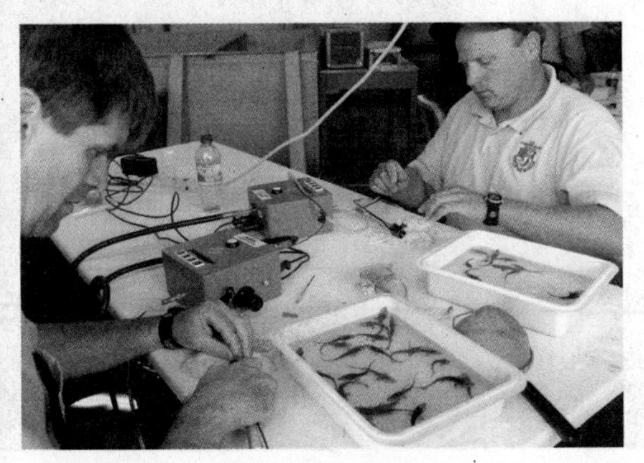

Tagging fish in preparation for release will allow scientists to track the success of reintroduction programs.

The Technology of Ocean Exploration

Robot submarines. Deep-sea sensors. Radio-tagged fish. DNA analysis. Satellite mapping. Not long ago, oceanography meant dipping a net over the side of a boat. Now, those days are long gone.

Oceanographers and ocean biologists now have technologies at their fingertips that rival those of space explorers. As oceanographer James Lindholm of the Pfleger Institute says, "For every tool we have to explore outer space—space stations, tethered missions, rovers, mapping—we have a comparable tool for ocean exploration... This suite of technologies allows us to study an environment that is equally hostile to human life" (SeaWeb 2005).

Some of what we're learning about the oceans comes as a direct result of space research—like the work of satellites that can see across the breadth of the electromagnetic spectrum, peering into the ocean's depths and identifying changes in temperature and chemistry. NASA's Aqua satellite, part of its Earth Observing System network of satellites, and the European Space Agency's Envirosat are mapping plankton levels, pollution, and ocean temperatures. Satellites, though, can only see into upper ocean levels. To plumb the deepest parts of the ocean, we need to take a dive—with sensors, submersibles, and robotic "autonomous underwater vehicles."

Underwater gliders are doing a lot of the work. These vessels use small changes in

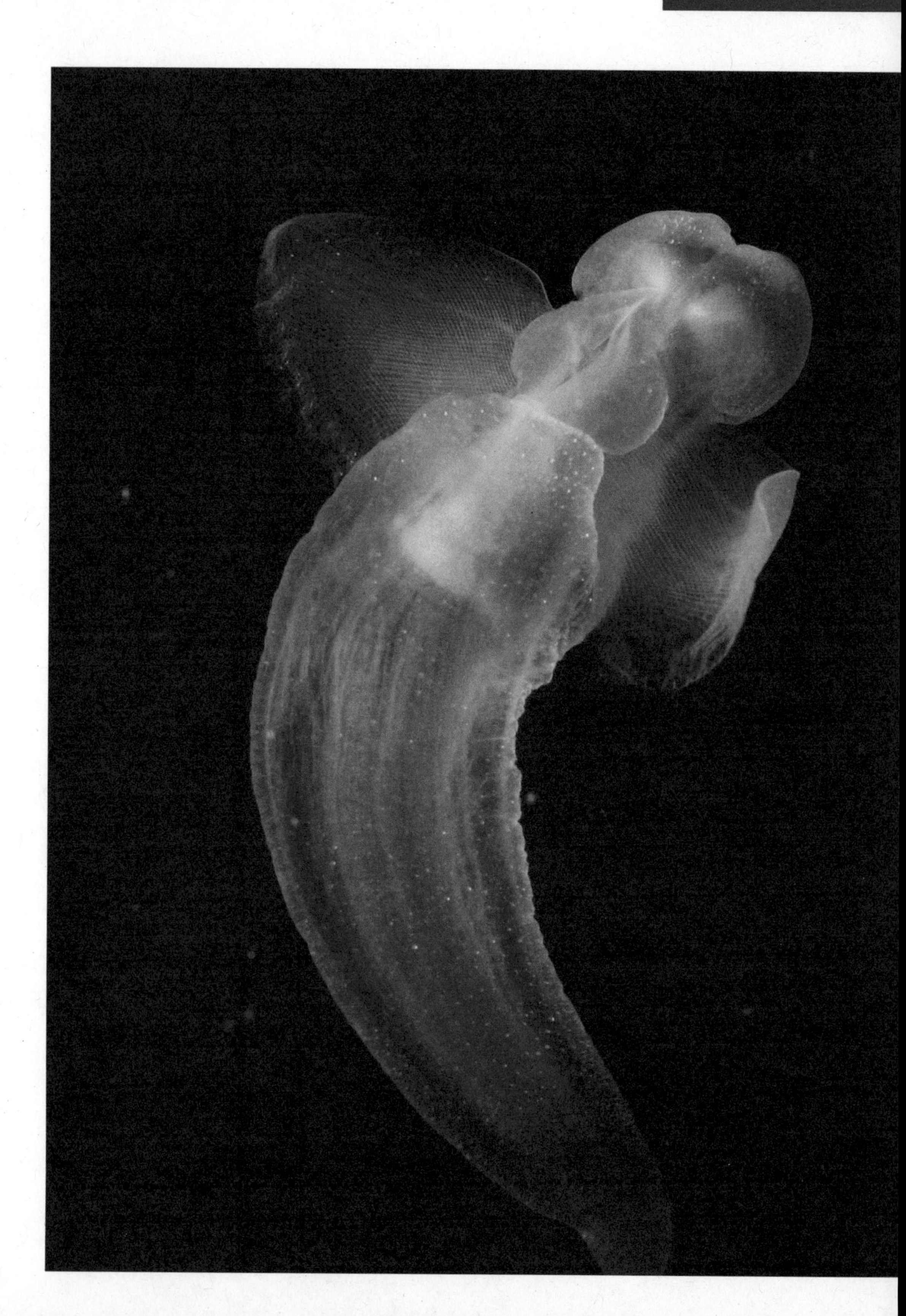

buoyancy in conjunction with their wings to propel themselves and turn vertical motion into forward movement with very little power. This lets them travel for thousands of miles and for months at a time, enabling studies of undersea regions in far greater detail than ever before possible. They're also very gentle on the environment—since they don't have external thrusters, they don't stir up sediment, and there are no exhaust fumes or dripping oils to pollute the ocean.

What we're finding, thanks to all this technology, is truly extraordinary; whole ecosystems we had no clue even existed a couple of decades ago; weird, nonsolar forests of giant worms living off the heat and sulfur around volcanic vents miles beneath the surface; and odd food chains of fish that live off falling scraps from the surface, feeding and breeding in complete darkness.

This oceanographic revolution couldn't come at a better time. Oceans are in real trouble. Pollution is causing "hypoxic" dead zones, where too little oxygen exists to support life of any kind. A quarter of all fish stocks are nearing extinction, and another half are at the limits of sustainability (Mann 2004). Coral reefs, essential to the ocean ecosystem, are dying around the world, victims of sewage, human carelessness, and rising ocean temperatures. The oceans are likely to bear the brunt of global warming, hit especially hard by a "thermal inertia" that will cause ocean temperatures to keep rising for years—even after we finally stop dumping greenhouse gases into the atmosphere. JC

Plankton, Reefs, and the Undersea "Canaries"

How do you feel about phytoplankton? You should be a big fan. These microscopic creatures swim around at the bottom of the ocean food chain, and are pretty much the closest thing on the planet to a foundation for life. They feed the ocean's animals. They also produce around half of the oxygen in our atmosphere. But phytoplankton are very sensitive to water temperature—crank up the heat, and they don't feed or breed as well. Because of this, they serve as some of the most important indicators for ocean researchers studying the effects of global warming. If the phytoplankton aren't faring well, it's a pretty good warning sign. They're the canaries in the oceanic coal mine.

The effect of global warming on coral reefs—the rain forests of the ocean—is more direct. Warming waters can cause "heat shock," bleaching reefs and quickly killing them, while increased carbon dioxide in the water makes reefs more acidic and brittle. Research suggests that by the midpoint of this century, the rate of degradation of reefs will outpace their ability to self-repair, essentially killing them off.

Fortunately, such a result isn't inevitable. Swift action to reduce greenhouse-gas emissions will help slow the acidic-reef effect, and there are several ways to encourage greater coral reef growth. The traditional way of seeding reefs is to intentionally sink a ship (or sometimes an oil-drilling platform), and wait for it to acquire mineral buildup. But clearly this is an invasive approach. More recently, the Global Coral Reef Alliance started using small electric charges to accelerate mineral accretion in a small fledgling

Opposite: This pelagic snail was collected from the deep Arctic Canada Basin. As global warming takes its toll on polar regions, the need to document these underexplored areas intensifies.

Right: Bleached coral documented by the Centre of Marine Studies in Queensland, Australia, reveals that the Great Barrier Reef has been damaged by rising water temperatures caused by global warming.

reef structure. This approach allows new reefs to form in areas too damaged to support natural reef restoration—and it's a whole lot better for the ocean than sinking oil-drilling platforms. JC

Protecting Biological Resources on the Deep Ocean Floor

The deep ocean's diversity of strange life-forms—quite different from life as we know it anywhere else on the planet—makes it of real interest to bioprospectors, people who plumb living things for materials and chemicals that may have medical or commercial viability. While commercial exploitation of deep-sea genetic resources is still a ways off, such exploitation could bring real problems, from the enclosure of genetic rights to the destruction of vulnerable deep-sea habitats.

Sam Johnston, one of the authors of a recent UN study on the issue, says that because the field is in its early stages of development, we still have time to create a legal framework for deep-sea bioprospecting: "We have a window of opportunity. The issues are much easier to deal with before commercial interests become heavily vested in the hunt for deep-sea genetic material" (Spotts, *Christian Science Monitor*, June 16, 2005).

So what can be done? The study recommends the creation of a whole new international organization (kind of like the UN of the sea) to manage the deep sea and safeguard its biological treasures. Oversight by such an institution would ensure that these resources are not appropriated for use by private companies, and are used only for peaceful purposes.

We may soon see international deep-sea eco-cops chasing bio-pirates through the hot plumes and black smokers of the ocean floor. AS

RESOURCE

The Sea Around Us by Rachel Carson (Oxford University Press, 2003)
An instant classic upon its original publication in 1951, *The Sea Around Us* is the timeless story of the oceans, told through memorable images and recounted in loving prose. However, it is also a warning that holds true today, more than half a century later, of the damage we can inflict—and have already inflicted—on the oceans. "Of all the things [Carson] labored to do, her most unintended accomplishment was to inspire us with an example of how we as individuals can strive to live." Rachel Carson inspires readers to live sustainably, consciously, and in awe of the planet on which we live.

Polar Regions

In spring 2004, Ben Saunders became the fourth person in history (and the youngest by ten years) to ski solo to the geographic North Pole. In October 2006, he sets out on the longest unsupported polar journey in history—a 1,800-mile (2,897-kilometer) journey from the coast of Antarctica to the South Pole.

Klaus Toepfer, the executive director of the United Nations Environment Program, describes the Arctic region as the "barometer of global climate change—an environmental early warning system for the world." Over the past six years, I've been lucky enough to spend several months in the high Arctic, witnessing this barometer at work firsthand.

In spring 2004, I set out to make the first-ever solo ski crossing of the Arctic Ocean, a 1,240-mile (1,996-kilometer) journey from the north coast of Siberia to Ward Hunt Island, northern Canada, via the geographic North Pole. Reinhold Messner, one of the world's most accomplished mountaineers, had attempted the same thing in the late 1990s. He was rescued after a few days on the ice and described the expedition as "ten times as dangerous as Everest." Considering that the high Arctic is home to the world's largest land-based carnivore (the polar bear) and that frostbite (which I'd contracted at fifty below during a 2001 expedition) is the least of one's worries when traveling through this region, I had no illusions about the scale of the challenge I was undertaking.

The last solo and unsupported expedition to the North Pole from Russia had been completed by Norwegian Børge Ousland in spring 1996. As I flew by helicopter to the same starting point in early March 2001, what I saw amazed me. Whereas Ousland had been able to ski from the land straight onto the frozen crust of the Arctic Ocean, I found an area of open water more than ten miles wide separating the pack ice from the northernmost tip of Siberia. I was flown to the edge of the pack and spent seventy-two days alone, battling conditions described by NASA as "the worst since records began." When you're skiing over the sea, the words *worst* and *warmest* are interchangeable; I encountered unprecedented areas of open water, and some of the highest temperatures ever recorded in the region. During past expeditions, there had been times when it felt like the Arctic was trying to kill me. This time around, it felt like it was trying to tell me something.

I had had a similarly disconcerting experience in 2005 in the Kangerlussuaq mountains of Greenland, where my teammate, Tony Haile, and I traveled to field-test equipment for our upcoming expedition. (Greenland has the largest ice cap on the Northern Hemisphere; it's the closest thing we have to the terrain of Antarctica, and therefore it was a perfect training ground.) I knew something was amiss when I caught sight of some drifting pack ice from our tiny Twin Otter ski-plane. Spotting the pack took me straight back to the Arctic and the three months I spent traveling over the shifting surface of that forgotten ocean in the spring of 2004, and I was certain I wouldn't want to pitch my tent on what I could see beneath me—it was far too fractured and weak.

The route we had chosen involved several days of ascending from near sea level to 6,500 feet (2,000 meters). I was prepared for tough climbs with the sledge forever pulling me back. I was prepared for the occasional crevasse fall. I was even bracing myself for Tony's cooking. The one thing I hadn't prepared for was to be wandering around our first campsite in the high Arctic with my shirt off. It seemed Greenland's welcome was to try and broil us alive. We spent a month trekking through this mountain range, and the temperature was never anything less than blistering. Our expedition jackets lay in our sledges, our bodies sweated gallons into our thin thermal tops—often the only clothing we wore during the day. The heat was playing hell with our rehydration calculations. Our factor-60 sun cream did not stop either of us turning a very British pink under the sun. This was Greenland; this was the Arctic, and it felt like a snow-themed Cancun.

Eventually we took to skiing during the night—the sun was still up, but we could handle

the temperature better, and we even had one or two genuinely cold nights.

Upon our return, we found that for the first time in living memory, a mountaineering expedition team even farther north than us had been able to sail right up to the coast.

The polar regions have always been in a state of climatic flux; that is one of the few certainties of earth science. In 1912, Captain Robert Falcon Scott's team unearthed fossilized ferns as they struggled back from the South Pole. In 1996, John Tarduno and his Paleomagnetic Research Group stumbled across a unique fossil find high above the Arctic Circle: the 80-to-90-million-year-old remains of fish, turtles, and champsosaurs (a semiaquatic reptile resembling the crocodile, which had never been found that far north). Tarduno's discovery implies that polar climates at that time were warm (with a mean annual temperature exceeding 57 degrees Fahrenheit [14 degrees Celsius]) rather than being below freezing. One line from the abstract of his study is particularly alarming: "Magmatism at six large igneous provinces at this time suggests that volcanic carbon dioxide emissions helped cause the global warmth."

Our civilization is mimicking those ancient volcanoes, pumping huge volumes of carbon dioxide into the atmosphere. I'm certainly not a scientist, but it is clear to me that the climates of these huge regions (Antarctica and the Arctic Ocean combine to cover an area nearly four times the size of China) are changing fast. Some of the undeniably human-made damage is less easy to spot: a fourteen-fold increase in the mercury levels of polar bears tested in Greenland over the past thirty years is one example. The poles may be incredibly remote, but they are far from untouched.

Above: Researchers say that warming temperatures in polar regions may be the cause of declining health among polar bear populations. Rising spring temperatures have led to earlier breakup of ice, leaving bears with less time to hunt for food.

Opposite, left: Ben Saunders with the flexible photovoltaic panel that produced electricity beyond "even our wildest expectations."

Opposite, right: An Eskimo whaling crew paddles through unfrozen waters, a reflection of the shifting infrastructure of circumpolar communities.

I'm not an explorer, at least not in the old-fashioned, Edwardian sense of the word. In an age of satellite, sonar, and laser, I don't exactly ski along drawing maps. For me, expeditions are a chance to explore my potential as an athlete, but I hope they're also about something bigger. At a time when taking responsibility for our impact on the planet's ecosystems is more important than ever, it strikes me that fewer and fewer young people are actually engaging with, or know anything of, the great outdoors. I certainly hope that my expeditions will fleetingly point the media's spotlight at the polar regions and the delicate balance in which the earth hangs at the start of the twenty-first century. But equally important, I hope they inspire younger people to take up adventures of their own, learning to be at home in nature, and perhaps launching themselves into the biggest adventure there is: blazing a path to a sustainable future. BS

The First Green Polar Exploration

One of my aims for the 2006 expedition is to break new ground environmentally. Every day, Antarctica receives enough solar and wind energy to fuel the entire planet's energy needs several times over, yet even contemporary expeditions have relied on fossil-fuel-burning stoves to melt snow.

I had a gnawing suspicion that there *had to* be a way to harness a tiny fraction of this raw energy to melt our modest daily requirement of drinking water and recharge camera and communications gear on an expedition. On the advice of dozens of readers on the WorldChanging Web site, my teammate, Tony Haile, and I decided to field-test flexible photovoltaic panels (for electricity production) and a number of "direct solar" systems that harness and focus the suns rays to produce heat—in our case, for melting snow.

Our experiments with solar power offered mixed results, but the flexible PV panel from Iowa Thin Film Technologies exceeded even our wildest expectations. Even in the early hours of the morning, with a low sun and overcast sky, the panel continued to charge our gear. In full sunlight, it was so powerful that we could run our Iridium sat phone (by far our most power-hungry gadget) directly from the panel, with the phone's own battery removed. BS

Circumpolar Peoples versus the Great Polar Melt

The Arctic isn't just icebergs, polar bears, and the occasional insane British explorer. People live there too, and the North Pole melt is hammering at their way of life. Things are changing so quickly that indigenous people don't even have words in their languages for some of the new things they're encountering, like new animals and insects, which are traveling farther north as forests start to grow where tundra once was.

But the long-term effects may be much worse than disorientation. Circumpolar native peoples—Samis, Inuits, Chukchis, and others—may find their communities and cultures wiped out. The infrastructure of coastal communities is already being destroyed, and recent studies have confirmed that ice-dependent animals, like polar bears, may face extinction. Wildlife research biologists from the U.S. Geological Survey and the Canadian Wildlife survey reported in 2005 that the Arctic

ice cap receded two hundred miles farther during that summer than the average recession rate two decades prior. This means that polar bears may be forced to swim distances far beyond what their energy allows. As a result, more and more polar bears face drowning from exhaustion (Iredale, *Sunday Times,* December 18, 2005).

For native peoples of the Arctic, the shock of seeing their home change so much is kindling activism. They are involving themselves in cultural survival work, political mobilization, technological and cultural innovation, even legal action.

In 2005 the Inuit Circumpolar Council (ICC)—a group representing native peoples in Canada, Alaska, Greenland, and Russia—announced that it is suing the United States to defend the Inuits' right to exist. In the petition submitted to the Inter-American Commission on Human Rights, the ICC charges that U.S. inaction on curbing greenhouse-gas emissions is a major cause of Arctic warming—which is destroying the environment that has both formed and nurtured the traditional Inuit way of life. This lawsuit has the potential to transform the politics of global warming, by shifting the issue of human rights away from uniquely local contexts and asserting that the faulty actions of one nation can be measured beyond that country's borders. EG & AS

Concordia Station

The crew at the Concordia Station—one of the most isolated permanent scientific outposts on Antarctica—is prepping for a mission to Mars.

Concordia Station is a joint project of French and Italian national research programs, with the involvement of the European Space Agency (ESA). It's situated at the top of nearly 10,500 feet (3,200 meters) of ice. The extreme temperature conditions—it can go as low as -59 degrees Fahrenheit (-51 degrees Celsius) in the summer and -120 degrees Fahrenheit (-85 degrees Celsius) in the winter—coupled with the low air pressure (only 645 hectopascals) and extreme isolation (during the winter, it is quite literally impossible to get to the station), make it an ideal setting for studying human biology under conditions similar to those astronauts would experience on a trip to Mars.

The crew will study the psychological effects of being grouped together in isolation, and continue their ongoing analysis into building self-sustaining research environments in extreme conditions. It's never too soon to start—current ESA plans call for the human exploration of Mars close to 2030.

For now, the experiments conducted at Concordia serve multiple purposes: providing information for future space exploration; building our knowledge of the only *relatively* unspoiled area left on the planet; and establishing the requirements for life in extreme environmental conditions. Climate disruption is not likely to lead to Antarctic-like conditions anywhere but Antarctica, but the lessons we learn about building materials, sustainable energy, and the dynamics of a group in isolation will be useful in other kinds of environments—and extremely valuable when survival in an extremely hostile environment (like that on Mars) is no longer a drill. JC

RESOURCES

The Two-Mile Time Machine: Ice Cores, Abrupt Climate Change, and Our Future by Richard B. Alley (Princeton University Press, 2000)
"The climate-change community is so much more confident of global warming than is the popular press," warns Greenland ice-core expert Richard Alley, one of the key scientists in the early 1990s who discovered that the last ice age ended abruptly, in a span of only three years. *The Two-Mile Time Machine* tells the story of global climate change through annual readings of the Greenland Ice Sheet, a fascinating and important story that informs the climate we live in today, and how we will proceed tomorrow.

Forty Signs of Rain and ***Fifty Degrees Below*** by Kim Stanley Robinson (Bantam, 2004)
Science fiction writer Kim Stanley Robinson's novels—parts one and two of a trilogy—tell the near-future story of a band of scientists and public servants who are battling to alert the world to the threat of sudden climate change, and to save it once the heavy weather sets in. The thinking-person's version of the movie *The Day After Tomorrow,* the books are full of interesting ideas

about science, the planet, and why it's hard for humans to act in their own long-term best interests.

Here's a taste: "Helicopters and blimps had already taken to the air in great numbers. Now all the TV channels in the world could reveal the extent of the flood from on high. Much of downtown Washington, DC, remained awash. A giant shallow lake occupied precisely the most famous and public parts of the city; it looked like someone had decided to expand the Mall's reflecting pool beyond all reason."

The Solar System: Greens in Space

Exploring space is green.

Exploring space is a crucial component of our ongoing efforts to better understand—and protect—our home planet. Some old-school environmentalists decry all the money spent on space programs; in their eyes this investment is, at best, a costly waste and, at worst, an invitation to more global irresponsibility. They envision a space program designed to allow us to just ditch our tortured planet as soon as we have space colonies ("No leaving the planet, boys," one bumper sticker reads, "until you clean up your mess."). But their fear is largely unfounded and exceptionally shortsighted. Literally and figuratively, we can't see the whole planet unless we look at it from space, and we can't really understand it unless we can see it as a whole. To really protect the planet, we sometimes have to leave it.

Over the past few decades, notions of environmental sustainability have shifted from focusing on cleaning up pollution to focusing on understanding (and, where needed, intervening in) global environmental systems. Picking up litter and reducing smog are easy concepts to understand; the dynamics between climate cycles,

These brilliant open star clusters are located about 200,000 light-years away and are roughly 65 light-years across.

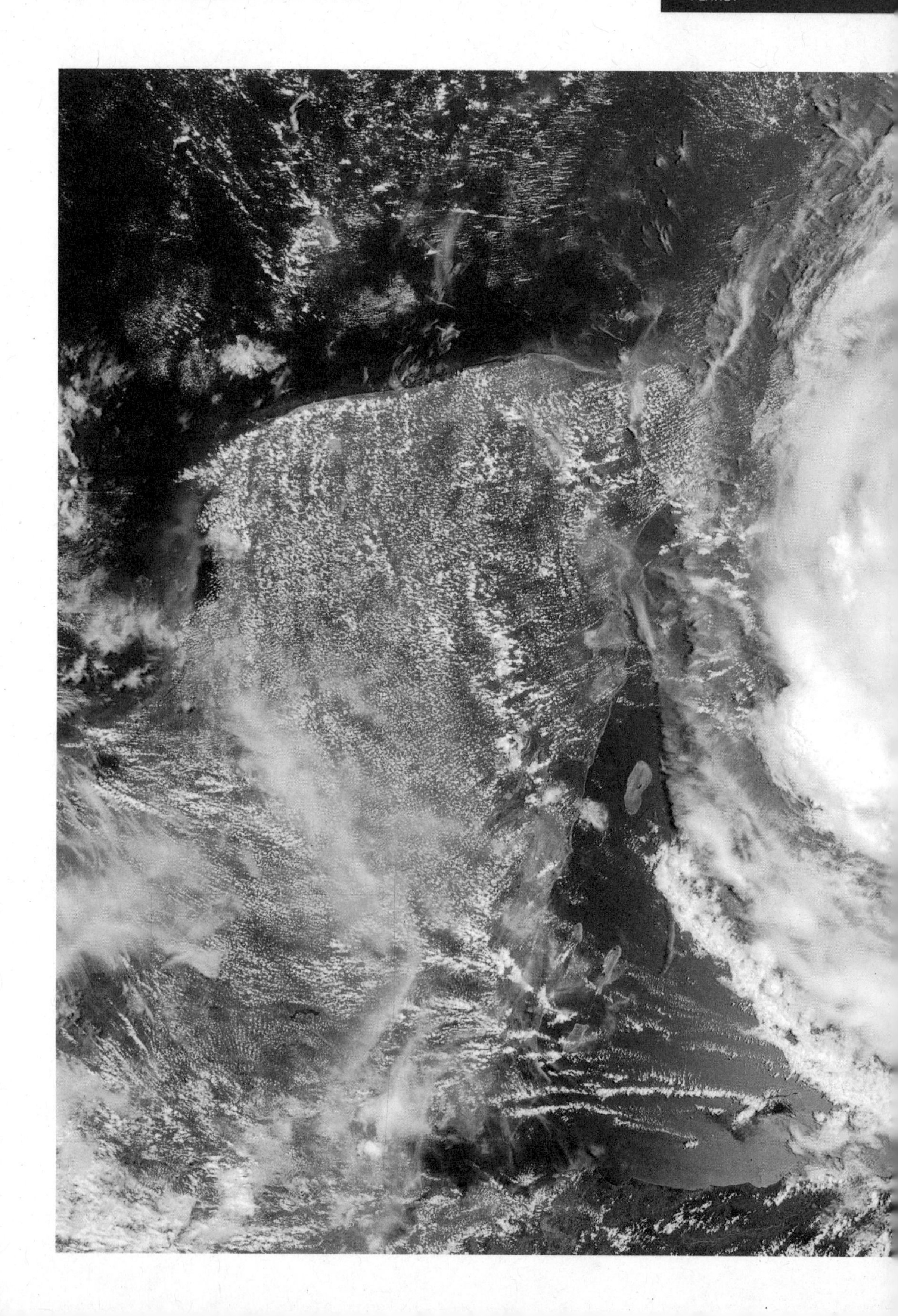

insulation, CO_2 emissions from natural and artificial sources, and solar cycles are harder to comprehend. But we'll never get a handle on how our environment truly functions without a better understanding of the larger environment in which our planet exists. It also helps to know how other planets have evolved. Turning our backs on space exploration means cutting ourselves off from a chance to really know Earth.

A space program with a planetary focus would combine current research into our planet's climate and geography (much of which can only be done from orbit) with expanded research into the workings of the rest of our solar system. Plenty of big questions about our planetary neighbors remain unanswered. Venus, Earth, and Mars all orbit within our sun's "habitable belt," and there is some preliminary research suggesting that each may have started out with similar potential for life. Why did Venus fall victim to a runaway greenhouse effect, while Mars dried up? Why did Earth alone manage to emerge from its early uninhabitable "iceball" period with the potential to support life? We can speculate, but on-site exploration will give us far better answers than will remote theorizing. Given the potential disasters associated with climate change, these are not idle questions. The better we understand how similar planets work, the better we understand our own planetology.

There are myriad connections between space research and green issues. If, for example, we were to discover life in the oceans under the icy surface of Jupiter's moon Europa, it would be our first opportunity to do truly comparative biology, which might profoundly expand our understanding of the miracle of life on planet Earth. Since all the living creatures we've ever found are related to one another, having what might be radically different life-forms to study could alert us to new understandings of life itself.

For now, and likely for the next couple of decades, a green space program would *not* mean sending people into space. Instead, it would mean mounting a much more ambitious (and well-funded) effort to send robotic explorers throughout the solar system—and beyond. Automated science missions have done remarkably well considering how little money has been made available to them. Mars Exploration Rover is the most spectacular recent example, but automated probes gathered material from a comet, monitored solar weather, dove into the crushing atmosphere of Jupiter, and found liquid water on Saturn's moon Enceladus. Robotic space exploration is relatively inexpensive—and the information we get back, its potential to help us better understand global environmental problems, is simply priceless.

Space exploration may get cheaper still. Over the next couple of decades, the field may no longer be limited to governments and big corporations. Via private space launches or via the (increasingly likely) space elevator, Earth's orbit could become as accessible as the deep ocean—not easy, not cheap, but still quite accessible to scientific research and even to the truly adventurous tourist.

Preceding pages: Both NASA and the European Space Agency (ESA) monitor climate change and natural disasters—such as this image of Hurricane Emily—with satellites. **Left:** In this photograph, American astronaut Edward H. White, a pioneer of space exploration, takes an "orbital stroll" outside of his spacecraft in zero gravity.

Author Robert Zimmerman referred to this emerging era as a "space renaissance"—a revolution in the way people on earth see and use space resources. The following are some tools that may help create that renaissance:

Microsatellites: Sending things into space will always be easier, and cheaper, than sending people into space. Small, comparatively cheap satellites with sensor kits and radios—meant to study a particular phenomenon before eventually burning up in Earth's atmosphere—could survey urban-growth patterns, monitor fisheries, look for early signs of drought or flooding, and even engage in a bit of open-source intelligence gathering.

Improved Climate Monitoring: A particularly important use of satellites—whether micro or macro—will be keeping a close watch on climate change. Both NASA and the European Space Agency (ESA) have climate-related satellite programs, and China plans to have climate satellites in orbit by 2012. But we could always use more.

Humanitarian Satellites: Satellite information aids humanitarian causes, and this use of the technology does not get the attention it deserves. Recently, satellites were invaluable in coordinating aid workers' movements in Darfur, where it can take as long as ten days to drive 75 miles (120 kilometers). The International Charter: Space and Major Disasters is a global agreement to coordinate the delivery of satellite data to rescue and relief efforts. It was most recently invoked during the 2004 Indian Ocean tsunami; satellites gave us some of the most powerful images of the tsunami's destruction, and have proved crucial in the recovery and reconstruction process. NASA has signed an agreement with the World Conservation Union (IUCN) to provide satellite data in support of a variety of conservation efforts. (The IUCN is the world's largest "environmental knowledge network," comprising members from 140 countries, 114 government agencies, and over 800 nongovernmental organizations.) These are all worthy efforts, but because there are as yet no dedicated humanitarian satellite networks, they also all require the temporary redirection of satellite resources away from their primary missions. Cheap, private launch vehicles would give humanitarian and conservation groups access to satellite technology.

But the biggest prize—and the greatest challenge—of space exploration would be to send satellites or even landers to other planets in our solar system. We may first have to build an elevator to reduce the energy costs of getting to high orbit (and, potentially, to serve as a launch "slingshot"). Even without interplanetary satellites, our understanding of how planets function may be on the verge of a revolution. Mars will undoubtedly get the most attention, given the intriguing evidence concerning life there. Already, university researchers are working on novel ideas for moving around the Red Planet, and for improving the technologies for detecting biological activity.

Of course, there are potential downsides to increased human activity in space. Within Earth's orbit, more satellites mean more chances for accidents; so-called space junk is already a concern. A proliferation of private satellite launches will only add to those headaches. Of greater long-term concern is the possibility of contaminating other planets with earthly microbes riding along on poorly handled space probes. Most earthborne bacteria would die quickly on Mars: no ozone layer means abundant ultraviolet radiation, on top of the sub-Antarctic temperatures and atmospheric density far lower than Earth's. We know all too well, however, that evolution is a hardy process. It would be appalling if our own carelessness destroyed our chances of learning whether Mars has its own native microbes because some earthly extremophile had become the martian equivalent of kudzu. JC

Space Elevator

Quite possibly the wildest idea for promoting space exploration is the Earth-to-orbit elevator.

This is one of those notions that, at first blush, sounds almost too ludicrous to be real. After all, we're accustomed to thinking of rockets as our only way into space—mixing danger and adventure. Taking an elevator into space sounds almost boring. It turns out, however, that a space elevator is not only plausible, it's potentially revolutionary.

Take a carbon nanoribbon—one a bit stronger than those we can make today, but not by much—and run it from a launch platform in the equatorial Pacific Ocean to a satellite about 62,000 miles (99,779 kilometers) up. The weight at the end keeps the ribbon taut, allowing a structure about the size of a dump truck to trundle up and down, carrying tons of satellites, construction materials, and even people. Because the launch platforms are stationed in the ocean, they can be moved around storms and remain well clear of inhabited areas.

The first elevator could cost as much as $50 billion, or as little as $10 billion, according to the LiftPort Group, one of the companies competing to build it—a lot of money, but not an outrageous amount. Many governments would be able to put that into their annual budgets without blinking. Once built, an elevator would make building *more* elevators much cheaper, since the cost of getting material into orbit would drop from the current $20,000 per pound to about $400 per pound (and likely even less, over time).

The elevator is not a new idea—and it may become reality sooner than you'd imagine. Early concepts of a "tower" reaching into orbit first popped up in the late nineteenth century, and Arthur C. Clarke imagined a more plausible version in his 1953 *Fountains of Paradise*. In the late 1990s, when NASA's Institute for Advanced Concepts first started seriously entertaining the idea, they projected that development would take until the latter part of the twenty-first century. But the LiftPort Group now says that a working version could be built by 2018. Going up? JC

Recycling Space Junk

Though we've only been exploring space since the 1960s, we've already turned the skies above us into an orbiting trash heap. More than 110,000 known, tracked chunks of litter are hurtling around the planet at more than 17,500 miles per hour (that's 28,164 kilometers per hour), according to NASA. And there may be more than a million smaller bits and pieces.

Some are shards of broken satellites (there are hundreds of whole, dead satellites zinging around up there too), some are pieces of equipment lost by astronauts, or just dumped out the window. An estimated 300,000 fragments were created by the explosion of one Pegasus rocket alone.

This stuff is a hazard. In 1983, a paint chip that was almost too small to see cracked the windshield of one of the space shuttles. Satellites are pretty routinely damaged by high-velocity grit and grime. As the BBC reminds us, "A pea-sized ball moving this fast is as dangerous as a 400-lb safe traveling at 60 mph."

Forgetting for a moment the danger posed to astronauts in the International Space Station (experts say there's a one-in-ten chance of a debris accident in the next ten years [David 1996]), these junk particles endanger the low-orbiting satellites most useful for studying Earth. A fast-moving debris ring around the planet presents some serious challenges to the kind of space science that can tackle environmental and social challenges.

But what can be done? We can't hand out trash bags to astronauts and have a litter-patrol day. Still, there are several mutually compatible options. First, we could prevent further accumulation of some space junk by using any of a variety of technologies to "deorbit" satellites at the end of their useful lives—an option that would be made more realistic by stricter laws governing the use of space. Second, we could clean up our orbit, slowly, using a laser "broom" (mounted on the Space Station, the shuttle, or special satellites) to "sweep" debris from the skies. Third—and most ambitiously—we could recycle the junk and use it as the counterweight for an orbital sling (sometimes called a skyhook): an orbiting, spinning tether, sort of a cousin to the space elevator, designed to lift and fling satellites from low orbit into higher orbit without added propellant. If none of these cleanup proposals worked, we could start exploring the *really* weird ideas. AS

Life in the Shooting Gallery

Right now, we know of 1,100 large asteroids a kilometer wide or bigger with orbits that come near Earth. Of those, scientists have studied the orbits of 700 sufficiently to determine that they will *not* pose a danger to the planet in the next century; 400 remain mysteries. But these are only the planet-killer-size rocks, the kind that wiped out the dinosaurs. If you count all the asteroids 150 meters (nearly 500 feet) or larger (which could still take out part of a continent, and perhaps trigger all sorts of climate mayhem), there are over a million nearby. In short, we live in a cosmic shooting gallery.

What can we do to lessen the odds of taking a hit? Hollywood notwithstanding, nuclear weapons don't work. Many asteroids are not very dense, and would be more likely to absorb the energy of a nuke than to be torn apart by one. Our best bet might be to *push* the asteroid. Given enough warning—about a decade—we could nudge asteroids far enough off-course to miss us. Groups like the B612 Foundation, led by Apollo astronaut Rusty Schweickart, are coming up with plans to design and test the necessary gear.

The best defense currently within our grasp is to crank up the search for asteroids that have Earth's name on them. Doing so would require adding telescopes and satellites to the effort. That takes money, but it would be cheap compared with the cost of being hit by even a minor asteroid. With funding for more observation projects, and with improved telescope technology, a smart mob of passionate amateurs could join the hunt, via distributed online efforts to snoop out likely suspects. The decline of the dinos shows that we live in a dangerous solar system, but unlike the dinosaurs, we can see trouble coming and, perhaps, avert it—if we're wise enough to look. JC

RESOURCE

Entering Space: Creating a Spacefaring Civilization by Robert Zubrin
(Jeremy P. Tarcher/Putnam, 1999)
Entering Space is engineer-visionary Robert Zubrin's manifesto for a new age of space exploration. "This is a book about creating a spacefaring civilization—the next step in the development of human society."

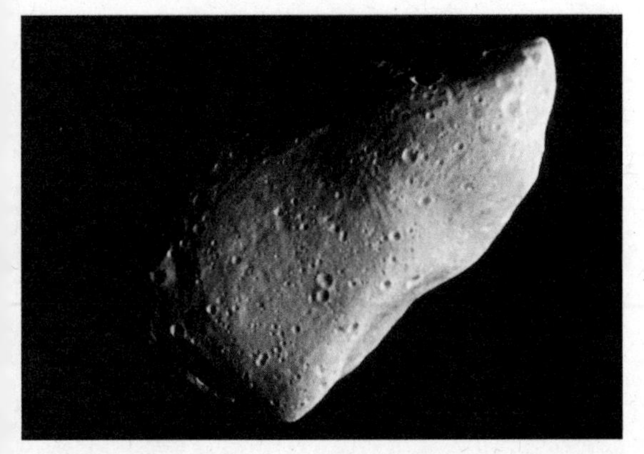

Imagining the Future

Who among us has not had at least one long, dark night where we wonder if there really is any hope for humanity at all?

To see the world we live in and not sometimes despair would mean we were less than fully human. When we hear that somewhere on the planet children are dying needlessly from hunger or disease every day, whose heart doesn't feel a pang, at least for a moment? When we read of the day's political scandals, with billions wasted on corruption and war, who doesn't sometimes wonder if it's too late to turn things around? When we learn that many wonderful things—coral reefs, polar bears—may essentially be gone by the end of the century, who doesn't wonder if things have just gone too far?

Things are bad. The problems are huge. The future, we sometimes suspect, will be unthinkable.

But despair is a trap. None of the problems we face are insurmountable. There is no reason to believe that the eight billion of us who will be here by 2050 can't live well on this planet. The technologies are possible. The design innovations are realistic. The social change is within our scope. We have the money. In fact, the biggest barrier

Left: Searching for asteroids using telescopes and satellites now will help prevent the enormous economic and human costs of being hit by an asteroid in the future.
Page 537: Nature has a way of reemerging, even in soil that has been catastrophically altered either by man or nature itself. Here a sapling grows in the ash surrounding Mount St. Helens National Volcanic Monument, Washington.

to a bright green future may be entirely in our heads—we simply can't imagine it.

On a planet with so much poverty, the worst is the poverty of imagination to which we've grown accustomed. We're used to politicians without a single new idea, journalists who dress up tired clichés and call them revolutionary, and pundits who take for granted the idea that actual change is unrealistic.

The bravest and most important thing any of us can do is to actively imagine a much better future: not to imagine it in the casual sense of daydreaming about it, but to imagine it in the way an architect imagines a house she is planning to build—to imagine it as reality, to try to see it whole, to lovingly dwell on its details, and to see ourselves walking through it one day.

We don't have to take on the task of imagining that future for the whole world—though it helps when we do. We can start with imagining our own lives, our own futures, transformed to honor our deepest beliefs about how the world ought to be. We can imagine the day when we live well with no harm at all to the planet. We can imagine the day when the work we do makes not just a profit, but a difference. We can imagine our community of friends and neighbors thriving, and the politicians we vote for being, once again, people we respect and admire. Whatever our hopes for a better world, we can imagine living lives in which those hopes are realities.

Then we can get to work building those lives. We don't need to wait for the Revolution. No charismatic leader is required. The tools for the job are either at hand, or they are things we can create together. A bright green future begins when each of us, today, decides to live as though that future were already here.

Living that way is not only the greatest adventure life has to offer, but it is also the cure for despair. "Sentiment without action," environmental author Edward Abbey warns, "is the ruin of the soul." But action inspired by deep feeling gives meaning to our lives, connects us together, and raises our spirits. It also happens to be the formula for changing the world.

If this book has a message, it is this: Imagine a better future. Find your allies. Share tools. Build it. Start now.

One final thought. It helps, when facing the future, to look from time to time toward the horizon. Too many people today assume that we are in decline, and that tomorrow is bound to be worse than today. It is a useful discipline, on occasion, to practice assuming they are all wrong, that the future will be unimaginably better, and that it will keep getting better, forever—that, as H. G. Wells puts it, "All of the past is but the beginning of a beginning; all that the human mind has accomplished is but the dream before the awakening." AS

SELECTED BIBLIOGRAPHY

Ableman, Michael, and Alice Waters. *On Good Land: The Autobiography of an Urban Farm*. San Francisco: Chronicle Books, 1998.

Ableman, Michael. *Fields of Plenty: A Farmer's Journey in Search of Real Food and the People Who Grow It*. San Francisco: Chronicle Books, 2005.

Anholt, Simon. *Brand New Justice*. 2nd ed. Oxford: Butterworth-Heinemann, 2005.

AtKisson, Alan. *Believing Cassandra: An Optimist Looks at a Pessimist's World*. White River Junction, VT: Chelsea Green Publishing, 1999.

Ausubel, Jesse. "The Environment for Future Business." *Pollution Prevention Review* 8, no. 1 (Winter 1998): 39–52.

Ayittey, George B. N. *Africa Unchained: The Blueprint for Africa's Future*. New York: Palgrave Macmillan, 2005.

Banks, Suzy, and Richard Heinichen. *Rainwater Collection for the Mechanically Challenged*. Dripping Springs, TX: Tank Town Publishing, 2004.

Barnett, Thomas P. M. *Blueprint for Action: A Future Worth Creating*. New York: G. P. Putnam's Sons, 2005.

Baskin, Yvonne. *Under Ground: How Creatures of Mud and Dirt Shape Our World*. Washington, DC: Island Press, 2005.

Baum, Dan. "Revolution 101: The Ruckus Society." *Rolling Stone*, n.d. 2001.

BBC. "Finland Tops Global School Table." BBC News, December 7, 2004. http://news.bbc.co.uk/1/hi/education/4073753.stm.

Beery, Jenny, Esther Eidinow, and Nancy Murphy. "The Mont Fleur Scenarios." *Deeper News* (Emeryville, CA: Global Business Network) 7, no. 1 (1997). http://www.arlingtoninstitute.org/future/Mont_Fleur.pdf.

Benkler, Yochai. *The Wealth of Networks*. New Haven: Yale University Press, 2006.

Benyus, Janine. *Biomimicry: Innovation Inspired by Nature*. New York: Harper Perennial, 2002.

Berger, Shoshana, and Grace Hawthorne. *Ready-Made: How to Make (Almost) Everything*. New York: Clarkson Potter, 2005.

Berkun, Scott. *The Art of Project Management*. Sebastopol, CA: O'Reilly Media, 2005.

Berley, Peter. *Fresh Food Fast: Delicious, Seasonal Vegetarian Meals in Under an Hour*. New York: Regan Books, 2006.

Berry, Wendell. *What Are People For?* New York: North Point, 1990.

Black, Richard. "Environment Key to Helping Poor." BBC News, August 31, 2005. http://news.bbc.co.uk/2/hi/science/nature/4199138.stm.

Blumenthal, Robin Goldwyn. "Good Vibes: Socially Responsible Investing Is Gaining Fans ... and Clout." *Barron's*, July 7, 2003.

Boal, Augusto. *Legislative Theatre: Using Performance to Make Politics*. New York: Routledge, 1998.

Bollier, David. *Silent Theft: The Private Plunder of Our Common Wealth*. New York: Routledge, 2002.

Bornstein, David. *How to Change the World: Social Entrepreneurs and the Power of New Ideas*. Oxford: Oxford University Press, 2004.

Bornstein, David. *The Price of a Dream: The Story of the Grameen Bank*. Oxford: Oxford University Press, 2005.

Boyd, Clark. "Estonia Embraces Web Without Wires." BBC News, May 5, 2004. http://news.bbc.co.uk/2/hi/technology/3673619.stm.

Brand, Stewart. *The Clock of the Long Now—Time and Responsibility*. New York: Basic Books, 1999.

Brower, Michael, and Warren Leon. *The Consumer's Guide to Effective Environmental Choices*. New York: Three Rivers Press, 1999.

Brown, Juanita, the World Café Community, et al. *The World Café: Shaping Our Futures through Conversations that Matter*. San Francisco: Berrett-Koehler, 2005.

Brown, Lester R. *Plan B 2.0: Rescuing a Planet Under Stress and a Civilization in Trouble*. New York: W. W. Norton, 2006.

Brownell, Blaine. *Transmaterial*. New York: Princeton Architectural Press, 2005.

Buchmann, Stephen L., and Gary Paul Nabhan. *The Forgotten Pollinators*. Washington, DC: Island Press, 1997.

Camejo, Peter. *The SRI Advantage: Why Socially Responsible Investing Has Outperformed Financially*. Gabriola, BC: New Society Publishers, 2002.

Charter, Martin, and Ursula Tischner. *Sustainable Solutions: Developing Products and Services for the Future*. Sheffield: Greenleaf Publishing, 2001.

Cooperrider, David L., Diana Whitney, and Jacqueline M. Stravos. *Appreciative Inquiry Handbook: The First in a Series of AI Workbooks for Leaders of Change*. San Francisco: Berrett-Koehler, 2004.

Crumlish, Christian. *The Power of Many: How the Living Web Is Transforming Politics, Business, and Everyday Life*. Alameda, CA: Sybex, 2004.

Datschefski, Edwin. *The Total Beauty of Sustainable Products*. Crans-Près-Céligny, Switzerland: Rotovision, 2001.

Dauncey, Guy, and Patrick Mazza. *Stormy Weather: 101 Solutions to Global Climate Change*. Gabriola, BC: New Society Publishers, 2001.

David, Leonard. "Space Debris." *New Scientist*, May 11, 1996. http://see.msfc.nasa.gov/Sparkman/Section_Docs/article_1.htm.

Demaret, Luc. "India: Hope Dawns as Women Beat Poverty." International Labor Organization Press Release, July 2004. http://www.ilo.org/public/english/dialogue/actrav/new/india04.htm.

Desai, Pooran, and Sue Riddlestone. *BioRegional Solutions for Living on One Planet*. London: Green Books, 2003.

Design Trust for Public Space. "High Performance Infrastructure Guidelines." Design Trust for Public Space, and New York City Department of Design and Construction, October 2005.

Diamond, Jared. *Guns, Germs and Steel: The Fates of Human Societies*. New York: W. W. Norton, 1997.

Drengson, Alan, and Duncan Taylor, eds. *Ecoforestry: The Art and Science of Sustainable Forest Use*. Gabriola, BC: New Society Publishers, 1997.

Drexler, Eric. *Engines of Creation: The Coming Era of Nanotechnology*. Garden City, NY: Anchor Press, 1987.

Elbaek, Uffe. *Kaospilot A–Z*. Denmark: KaosCommunication, 2003.

Elgin, Duane. *Voluntary Simplicity*. New York: Harper Paperbacks, 1998.

Estill, Lyle. *Biodiesel Power: The Passion, the People and the Politics of the Next Renewable Fuel*. Gabriola, BC: New Society Publishers, 2005.

Ferguson, Niall. *Empire: The Rise and Demise of the British Empire and the Lessons for Global Power.* New York: Basic Books, 2003.

Fraioli, James O. *Ocean Friendly Cuisine: Sustainable Seafood Recipes from the World's Finest Chefs*. Minocqua, WI: Willow Creek Press, 2005.

Frank, Thomas, and Matt Weiland, eds. *Commodify Your Dissent.* New York: W. W. Norton, 1997.

Freeman, Michael. *Space: Japanese Design Solutions for Compact Living*. New York: Universe, 2004.

Fuad-Luke, Alastair. *ecoDesign: The Sourcebook*. San Francisco: Chronicle Books, 2002.

Fukuoka, Masanobu. *The One-Straw Revolution: An Introduction to Natural Farming.* Emmaus, PA: Rodale Press, 1978.

Fuller, Donald A. *Sustainable Marketing: Managerial-Ecological Issues*. Thousand Oaks, CA: Sage, 1999.

Garreau, Joel. *Radical Evolution: The Promise and Peril of Enhancing Our Minds, Our Bodies—and What it Means to Be Human.* New York: Doubleday, 2004.

Gershuny, Grace, and Joe Smillie. *The Soul of Soil: A Soil-Building Guide for Master Gardeners and Farmers*. White River Junction, VT: Chelsea Green Publishing, 1999.

Gibson, William. *Pattern Recognition*. New York: G. P. Putnam's Sons, 2003.

Gillmor, Dan. *We the Media: Grassroots Journalism by the People, for the People*. Sebastopol, CA: O'Reilly Media, 2004.

Gipe, Paul. *Wind Power: Renewable Energy for Home, Farm, and Business*. White River Junction, VT: Chelsea Green Publishing, 2004.

Gissen, David, ed. *Big and Green: Toward Sustainable Architecture in the 21st Century.* New York: Princeton Architectural Press, 2003.

Gladwell, Malcom. *Blink: The Power of Thinking Without Thinking*. Boston: Little, Brown, 2005.

Gobster, Paul, and Bruce Hull, eds. *Restoring Nature*. Washington, DC: Island Press, 2000.

Graddol, David. "The Future of Language." *Science* 303, no. 5662 (2004): 1329–31.

Green, Judith M. *Deep Democracy.* Lanham, MD: Rowman and Littlefield, 1999.

Griffith, Nicola. *Slow River.* New York: Ballantine Books, 1996.

Groh, Trauger, and Steven McFadden. *Farms of Tomorrow Revisited: Community Supported Farms—Farm Supported Communities*. Kimberton, PA: Bio-dynamic Farming and Gardening Association, 1998.

Guenster, Nadja, Jeroen Derwall, Rob Bauer, and Kees Koedijk. "The Economic Value of Corporate Eco-Efficiency." A study by the Center for Responsible Business at the Haas School of Business, UC Berkeley, 2005.

Hargroves, Karlson, and Michael H. Smith. *The Natural Advantage of Nations: Business Opportunities, Innovation and Governance in the 21st Century.* London: Earthscan Publications, 2005.

Harrison, Owen. *Open Space Technology: A User's Guide*. San Francisco: Berrett-Koehler, 1997.

Hart, Stuart L. *Capitalism at the Crossroads: The Unlimited Business Opportunities in Solving the World's Most Difficult Problems*. Upper Saddle River, NJ: Wharton School Publishing, 2005.

Havel, Vaclav. *The Art of the Impossible: Politics as Morality in Practice*. New York: Knopf, 1997.

Hawken, Paul, Amory Lovins, and L. Hunter Lovins. *Natural Capitalism: Creating the Next Industrial Revolution*. Boston: Little, Brown, 1999.

Hawken, Paul. *The Ecology of Commerce: A Declaration of Sustainability.* New York: HarperCollins, 1993.

Higgs, Eric. *Nature by Design*. Cambridge, MA: The MIT Press, 2003.

Hochschild, Adam. *Bury the Chains: Prophets and Rebels in the Fight to Free an Empire's Slaves*. Boston: Houghton Mifflin, 2005.

Holmgren, David. *Permaculture: Principles and Pathways Beyond Sustainability.* Victoria, Australia: Holmgren Design Studios, 2002.

Homer-Dixon, Thomas. *The Ingenuity Gap: Facing the Economic, Environmental, and Other Challenges of an Increasingly Complex and Unpredictable World*. New York: Knopf, 2000.

Hunter, Linda Mason, and Mikki Halpin. *Green Clean: The Environmentally Sound Guide to Cleaning Your Home*. New York: Melcher Media, 2005.

Jackson, Wes, Wendell Berry, and Bruce Colman. *Meeting the Expectations of the Land*. New York: North Point Press, 1985.

Jacobs, Jane. *The Death and Life of Great American Cities*. New York: Vintage Books, 1961.

Jégou, François. "Making a Habit of Sustainability." Interview by Jane Szita. *Dwell* 5, no. 1 (2004): 166–68.

Johnson, Kenneth M., and Calvin L. Beale. "The Rural Rebound: Recent Nonmetropolitan Demographic Trends in the United States." *Wilson Quarterly*, Spring 1998. Revised 2004. http://www.luc.edu/depts/sociology/johnson/TheWilsonQuarterly.html.

Johnston, David, and Kim Master. *Green Remodeling: Changing the World One Room at a Time.* Gabriola, BC: New Society Publishers, 2004.

Jones, Steve. *Darwin's Ghost: The Origin of Species Updated*. New York: Random House, 2000.

Jordan, William R. III, *The Sunflower Forest*. Berkeley: University of California Press, 2003.

Kac, Eduardo. *Telepresence and Bio Art: Networking Humans, Rabbits, and Robots*. Ann Arbor: University of Michigan Press, 2005.

Kahane, Adam. *Solving Tough Problems: An Open Way of Talking, Listening, and Creating New Realities*. San Francisco: Berrett-Koehler, 2004.

Kelly, Eamonn. *Powerful Times: Rising to the Challenge of Our Uncertain World*. Upper Saddle River, NJ: Wharton School Publishing, 2006.

Kemp, William H. *Smart Power: An Urban Guide to Renewable Energy and Efficiency.* Tamworth, ON: Aztext Press, 2004.

Kemp, William H. *The Renewable Energy Handbook*. Tamworth, ON: Aztext Press, 2005.

Kent, Deidre. *Healthy Money, Healthy Planet: Developing Sustainability Through New Money Systems*. Nelson, NZ: Craig Potton Publishing, 2005.

Khan, Lloyd. *Shelter.* Bolinas, CA: Shelter Publications, 1973.

Kilarski, Barbara. *Keep Chickens! Tending Small Flocks in Cities, Suburbs, and Other Small Spaces*. North Adams, MA: Storey Publishing, 2003.

Kimbrell, Andrew. *Your Right to Know: Genetic Engineering and the Secret Changes in Your Food*. Berkeley, CA: Ten Speed Press, 2006.

Klein, Naomi. *No Logo*. New York: Picador, 2002.

Koenig, Klaus W. *The Rainwater Technology Handbook*. Dortmund, Germany: Wilo-Brain, 2001.

Kurlansky, Mark. *Cod: A Biography of the Fish That Changed the World.* New York: Penguin Books, 1998.

Kurzweil, Ray, and Terry Grossman. *Fantastic Voyage: Live Long Enough to Live Forever.* Emmaus, PA: Rodale Press, 2004.

Le Corbusier. 1923. *Towards a New Architecture.* New York: Dover Publications, 1985.

Leakey, Richard, and Roger Lewin. *The Sixth Extinction.* New York: Anchor Books, 1996.

Lewis, Helen, and John Gertsakis. *Design and Environment: A Global Guide to Designing Greener Goods.* Sheffield, UK: Greenleaf Publishing, 2001.

Lindenmayer, David B., and Jerry F. Franklin, eds. *Towards Forest Sustainability.* Washington, DC: Island Press, 2003.

Lindow, Megan. "Learning to Track Like a Bushman." *Wired News,* January 22, 2004. http://www.wired.com/news/technology/0,1282,61919,00.html.

Lipson, Elaine. *The Organic Foods Sourcebook.* New York: McGraw-Hill, 2001.

Lovins, Amory. "How to Get Real Security." *Whole Earth Review,* Fall 2002. http://www.rmi.org/images/other/Security/S02-13_HowRealSecurity.pdf.

Lynas, Mark. *High Tide: The Truth about Our Climate Crisis.* New York: Picador Paperback Original, 2004.

Macy, Joanna. *World as Lover, World as Self.* Berkeley, CA: Parallax Press, 1991.

Madison, Deborah. *Local Flavors: Cooking and Eating from America's Farmers' Markets.* New York: Broadway Books, 2002.

Makovsky, Paul. "Pedestrian Cities." *Metropolis Magazine,* August/September 2002. http://www.metropolismag.com/html/content_0802/ped/index.html.

Mann, Charles C. "The Bluewater Revolution." *Wired,* May 2004. http://www.wired.com/wired/archive/12.05/fish.

Mann, Charles C. *1491: New Revelations of the Americas Before Columbus.* New York: Knopf, 2005.

Manning, Richard. *A Good House: Building a Life on the Land.* New York: Grove Press, 1993.

Manning, Richard. *Against the Grain: How Agriculture Has Hijacked Civilization.* New York: North Point Press, 2004.

Manzini, Ezio, and François Jégou. *Sustainable Everyday: Scenarios of Everyday Life.* Milan: Edizione Ambiente, 2003.

Martin, Paul S. *Twilight of the Mammoths: Ice Age Extinctions and the Rewilding of North America.* Berkeley: University of California Press, 2005.

Maskrecki, Piotr. *The Smaller Majority.* Cambridge, MA: Belknap Press of Harvard University Press, 2005.

Mau, Bruce. *Massive Change.* London: Phaidon Press, 2004.

McDonough, William, and Michael Braungart. *Cradle to Cradle: Remaking the Way We Make Things.* New York: North Point Press, 2002.

McKibben, Bill. "Apartment." *New Yorker,* 1988.

McKibben, Bill. *The End of Nature.* New York: Random House, 1989.

McKibben, Bill. *Hope: Human and Wild.* Boston: Little, Brown, 1995.

McNeill, J. R. *Something New Under the Sun: An Environmental History of the Twentieth-Century World.* New York: W. W. Norton, 2000.

Meadows, Donella, Jorgen Radners, and Dennis Meadows. *Limits to Growth: The 30-Year Update*. White River Junction, VT: Chelsea Green Publishing, 2004.

Mega-Cities Project. "The Poverty/Environment Nexus in Mega-Cities." The Mega-Cities Project Publication, 1998. http://www.megacitiesproject.org/publications/pdf/mcp018lessons.pdf.

Menzel, Peter, and Faith D'Aluisio. *Hungry Planet: What the World Eats*. Napa, CA: Material World Books, 2005.

Mitchell, William J. *Me ++: The Cyborg Self and the Networked City*. Cambridge, MA: The MIT Press, 2003.

MoveOn.org. *MoveOn's 50 Ways to Love Your Country: How to Find Your Political Voice and Become a Catalyst for Change*. Maui, HI: Inner Ocean Publishing, 2004.

Nabhan, Gary Paul. *Enduring Seeds: Native American Agriculture and Wild Plant Conservation*. Tucson: University of Arizona Press, 2002.

National Consumer Council and Sustainable Development Commission, Sustainable Consumption Roundtable. "I Will if You Will: Towards Sustainable Consumption." http://www.sd-commission.org.uk.

Nattrass, Brian, and Mary Altomare. *Dancing with the Tiger: Learning Sustainability Step by Natural Step*. Gabriola, BC: New Society Publishers, 2002.

Newman, Robert. *The Fountain at the Center of the World*. Brooklyn, NY: Soft Skull Press, 2004.

Nworah, Uche. "Study on Nigerian Diaspora." *Global Politician*. May 4, 2005. http://globalpolitician.com/articledes.asp?ID=682&cid=8&sid=0.

Nye, Joseph S. *Soft Power: The Means to Success in World Politics*. New York: PublicAffairs, 2005.

Orr, David. *Ecological Literacy: Education and the Transition to a Postmodern World*. Albany: State University of New York Press, 1992.

Ottman, Jacquelyn A. *Green Marketing: Opportunities for Innovation in the New Marketing Age*. 2nd ed. N.p.: BookSurge Publishing, 2004.

Paine, Thomas. 1776. *Common Sense*. New York: Penguin Classics, 1982.

Paine, Thomas. 1791. *Rights of Man*. New York: Penguin Books, 1984.

Papanek, Victor. *Design for the Real World: Human Ecology and Social Change*. Chicago: Academy Chicago Publishers, 1985.

Papanek, Victor. *The Green Imperative: Natural Design for the Real World*. New York: Thames and Hudson, 1995.

Pattullo, Polly. *The Ethical Travel Guide: Your Passport to Exciting Alternative Holidays*. London: Earthscan, 2006.

Pollan, Michael. *A Place of My Own: The Education of an Amateur Builder*. New York: Random House, 1997.

Price, Dan. *Radical Simplicity: Creating an Authentic Life*. Philadelphia: Running Press, 2005.

Priest, Dana. *The Mission: Waging War and Keeping Peace with America's Military*. New York: W. W. Norton, 2004.

Propper de Callejon, Diana, Mark Donohue, and Rob Day. "Clean Technology: More Good News for the Socially and Environmentally Responsible Investor." *GreenMoney Journal* 14, no. 4 (2006). http://www.greenmoneyjournal.com/article.mpl?newsletterid=34&articleid=412.

Raphael, Ray. *More Tree Talk: The People, Politics, and Economics of Timber*. Washington, DC: Island Press, 1994.

Raymond, Eric S. *The Cathedral and the Bazaar: Musings on Linux and Open Source by an Accidental Revolutionary.* Cambridge, MA: O'Reilly Media, 2001.

Rees, Martin. *Our Final Hour—A Scientist's Warning: How Terror, Error, and Environmental Distance Threatened Humankind's Future in this Century—on Earth and Beyond.* New York: Basic Books, 2003.

Reisner, Marc. *A Dangerous Place: California's Unsettling Fate.* New York: Pantheon Books, 2003.

Rejali, Darius. "Torturing Can't Be Defended, Doesn't Even Work." *The Oregonian*, April 16, 2002. http://progressiveaustin.org/torture.htm.

Reporters Without Borders. *Handbook for Bloggers and Cyber-Dissidents* (PDF). N.p.: Reporters Without Borders, 2005. http://www.rsf.org/IMG/pdf/handbook_bloggers_cyberdissidents-GB.pdf.

Rheingold, Howard. *Smart Mobs: The Next Social Revolution.* Cambridge, MA: Basic Books, 2003.

Richardson, Julie, Terry Irwin, and Chris Sherwin. "Design and Sustainability: A Scoping Report for the Sustainable Design Forum, Design Council," 2005.

Rifkin, Jeremy. *The Hydrogen Economy.* New York: Tarcher, 2003.

Rivoli, Pietra. *The Travels of a T-Shirt in the Global Economy: An Economist Examines the Markets, Power, and Politics of World Trade.* Hoboken, NJ: John Wiley and Sons, 2005.

Robin, Vicki, and Joe Dominguez. *Your Money or Your Life.* New York: Penguin Books, 1999.

Robinson, Kim Stanley. *Forty Signs of Rain.* New York: Bantam, 2004.

Rogers, Heather. *Gone Tomorrow: The Hidden Life of Garbage.* New York: The New Press, 2005.

Rosenblatt, Gideon. "Movement as Network: Connecting People and Organizations in the Environmental Movement." January 2004. http://www.movementasnetwork.org/MovementAsNetwork-final-1.0.pdf.

Rotberg, Robert, and Dennis Thompson, eds. *Truth v. Justice: The Morality of Truth Commissions.* Princeton, NJ: Princeton University Press, 2000.

Sachs, Jeffrey. *The End of Poverty: Economic Possibilities for Our Time.* New York: Penguin Press, 2005.

Safina, Carl. *Song for the Blue Ocean: Encounters Along the World's Coasts and Beneath the Seas.* New York: Henry Holt, 1998.

Schaeffer, John. *Solar Living Sourcebook.* Gabriola, BC: New Society Publishers, 2005.

Schlosser, Eric. *Fast Food Nation.* New York: Harper Perennial, 2002.

Scott, Ridley. *Blade Runner.* Warner Brothers, 1982.

SeaWeb. "Technological Revolutions in Sensors, Robotics, and Telecommunications Allow New Views of Ocean." Physorg.com, Space and Earth Science, February 19, 2005. http://www.physorg.com/news3116.html.

Sen, Rinku. *Stir It Up: Lessons in Community Organizing and Advocacy.* San Francisco: Jossey-Bass, 2003.

Sharp, Gene. *Waging Nonviolent Struggle: 20th Century Practice and 21st Century Potential.* Boston: Extending Horizons Books, 2005.

Shawcross, William. *Deliver Us from Evil: Warlords and Peacekeepers in a World of Endless Conflict.* London: Bloomsbury, 2000.

Singh, Kanta. "Engendering Impact—Indian Scenario." Paper presented at EDIAIS Conference, Manchester, UK, November 24–25, 2003. http://www.enterprise-impact.org.uk/pdf/Singh.pdf.

Snyder, Gary. *The Practice of the Wild*. New York: North Point Press, 1990.

Spurlock, Morgan. *Super Size Me*. Goldwyn Films, 2004.

Stafford, Kim. *Having Everything Right: Essays of Place*. Seattle: Sasquatch Books, 1997.

Steffen, W. et al. *Global Change and the Earth System: A Planet Under Pressure*. Berlin: Springer, 2004.

Stephenson, Neal. *The Diamond Age*. New York: Bantam Books, 1995.

Sterling, Bruce, and Lorraine Wild. *Shaping Things*. Cambridge, MA: The MIT Press, 2005.

Sterling, Bruce. "Viridian Note: The World Is Becoming Uninsurable." N.d. Mailing List. http://www.viridiandesign.org/notes/1-25/Note%2000022.txt.

Sterling, Bruce. *Holy Fire*. New York: Bantam Books, 1996.

Sterling, Bruce. *Tomorrow Now: Envisoning the Next Fifty Years*. New York: Random House, 2002.

Stevens, William K. *Miracle Under the Oaks*. New York: Pocket Books, 1995.

Sunstein, Cass R. *Why Societies Need Dissent*. Cambridge, MA: Harvard University Press, 2003.

Suzuki, David, and Holly Dressel. *Good News for a Change: How Everyday People Are Helping the Planet*. Vancouver: Greystone Books, 2002.

Tapscott, Don, and David Ticoll. *The Naked Corporation: How the Age of Transparency Will Revolutionize Business*. New York: Free Press, 2003.

Tesanovic, Jasmina. *Diary of a Political Idiot: Normal Life in Belgrade*. San Francisco: Midnight Editions, 2000.

Thackara, John. *In the Bubble*. Cambridge, MA: The MIT Press, 2005.

Todd, Nancy Jack. *A Safe and Sustainable World: The Promise of Ecological Design*. Washington, DC: Island Press, 2005.

Trippi, Joe. *The Revolution Will Not Be Televised: Democracy, the Internet, and the Overthrow of Everything*. New York: Regan Books, 2004.

van Hinte, Ed. *Eternally Yours: Time in Design*. Rotterdam, Netherlands: 010 Publishers, 2005.

Vogel, Steven. *Cats' Paws and Catapults: Mechanical Worlds of Nature and People*. New York: W. W. Norton, 2002.

Wallace, David Rains. *The Klamath Knot*. San Francisco: Sierra Club Press, 1983.

Ward, Peter D., and Donald Brownlee. *Rare Earth: Why Complex Life Is Uncommon in the Universe*. New York: Copernicus, 2000.

Weart, Spencer R. *The Discovery of Global Warming*. Cambridge, MA: Harvard University Press, 2003.

Weber, Steven. *The Success of Open Source*. Cambridge, MA: Harvard University Press, 2004.

Wells, Spencer. *The Journey of Man: A Genetic Odyssey*. Princeton, NJ: Princeton University Press, 2002.

White, Richard. *The Organic Machine: The Remaking of the Columbia River*. New York: Hill and Wang, 1996.

Wirzba, Norman, ed. *The Essential Agrarian Reader: The Future of Culture, Community, and the Land*. Washington, DC: Shoemaker and Hoard, 2004.

Worster, Donald. *Nature's Economy: A History of Ecological Ideas*. Cambridge: Cambridge University Press, 1977.

Yeon-Ho, Oh. "OhMyNews Makes Every Citizen a Reporter." Interview by Yeon-Jung Yu. *Japan Media Review*, September 17, 2003. http://www.japanmediareview.com:80/japan/internet/1063672919_3.php.

Yue, Pan. "The Chinese Miracle Will End Soon." Interview by Andreas Lorenz. *Der Spiegel*, March 7, 2005.

N.B. Resources listed in the book are not repeated in this bibliography.

Most of the books listed in this bibliography that are currently in print are available for order at www.WorldChanging.com/book.

CONTRIBUTOR BIOGRAPHIES

Worldchanging: A User's Guide for the 21st Century was written by the contributors listed below. Initials have been placed at the end of sections to indicate individual contributions. In the cases where initials do not appear, writing and research was done by the Worldchanging team.

SUHIT ANANTULA [SA]
Suhit Anantula believes that the world is green. In past lives he has worked in financial-services outsourcing and rural development in India. Currently finishing his MBA in Australia, he is working on making organizations environmentally friendly. Suhit runs a blog at www.SuhitAnatula.com.

NICK ASTER [NA]
Nick Aster is a recent graduate of the Presidio School of Management with an MBA in sustainable management. He has a background in online media and built and helps manage www.TreeHugger.com.

ALAN ATKISSON [AA]
Alan AtKisson is president of the AtKisson Group, an international consultancy that has been advising leaders around the world in sustainable development since 1992. He also currently directs the international Earth Charter Initiative, a global effort to promote the Earth Charter, which was created by an independent high-level commission through an intensive process of international consultation, launched in the year 2000 with the help of the Dutch government, and subsequently endorsed by thousands of organizations and government agencies worldwide. See www.AtKisson.com and www.EarthCharter.org.

SHOSHANA BERGER [SB]
Shoshana Berger is the editor in chief of *ReadyMade* magazine and the coauthor of *ReadyMade: How to Make Almost Everything* (Clarkson Potter, 2005). She changed this bio five times.

SCOTT BERKUN [SEB]
A writer living in Seattle, Washington, Scott Berkun is the author of the best-selling book *The Art of Project Management* (O'Reilly Media, 2005). He writes on management at www.ScottBerkun.com.

CARISSA BLUESTONE [CB]
Worldchanging associate editor Carissa Bluestone earned her BA in journalism from New York University, after which she ended up going into publishing instead. After working for Random House, Inc., for five years, she decided to "go off the grid," trading steady paychecks and health benefits for the freelance life, and trading New York for Seattle. Her hobbies include hiking, Muay Thai kickboxing, and writing poetry. Although she's proud of her East Coast roots, she tries to hug a tree at least once a day now.

MILLE BOJER [MB]
Marianne "Mille" Bojer works in the field of design and facilitation of dialogic change processes. She is an associate of Generon Consulting (www.generonconsulting.com). She is also a cofounder and associate of Pioneers of Change (www.pioneersofchange.net), an international learning network of young people working for systemic change within their own spheres of influence. Originally from Denmark, she currently lives in South Africa.

DAVID BORNSTEIN [DNB]
David Bornstein is the author of *How to Change the World: Social Entrepreneurs and the Power of New Ideas* (Oxford Univ. Press, 2004) and *The Price of a Dream: The Story of the Grameen Bank* (Oxford Univ. Press, 2005). He lives in New York.

NICOLE-ANNE BOYER [NAB]
Nicole is a futurist and foresight specialist based in Paris. By getting people to think more wisely and creatively about the future, she believes,

we'll learn to live more sustainably in the present. Nicole is managing director of Adaptive Edge (www.adaptive-edge.com), a business-school lecturer, writer, and social entrepreneur.

DAVID BRIN [DB]

David Brin's popular science-fiction novels have been *New York Times* best sellers; have won Hugo, Nebula, and other awards; and have been translated into more than twenty languages. His 1990 ecological thriller *Earth* (Spectra) foreshadowed global warming, cyberwarfare, the World Wide Web, and Gulf Coast flooding. Other novels include *The Postman* (Spectra, 1985), on which a 1998 movie starring Kevin Costner was loosely based; *The Life Eaters*, a groundbreaking hardcover graphic novel (DC/Wildstorm, 2004); and *Kiln People* (Tor, 2003). David is also a noted scientist and speaker/consultant who makes frequent TV appearances to discuss near-future trends. He serves on advisory committees in arenas ranging from nanotechnology to national defense, and his award-winning nonfiction book *The Transparent Society: Will Technology Make Us Choose Between Freedom and Privacy?* (Perseus, 1998) deals with issues of openness, security, and liberty in the new wired age.

JAMAIS CASCIO [JC]

Jamais Cascio is a WorldChanging.com cofounder and former senior contributing editor. He specializes in the creation of plausible, compelling scenarios for what the next few decades could hold; his clients range from Hollywood producers to global nongovernmental organizations. Jamais's essays on the environment, technology, and social change regularly appear in both online and print publications, and he has spoken around the world about future possibilities.

ZOË CHAFE [ZC]

Zoë Chafe is a staff researcher at the Worldwatch Institute, where she writes for *State of the World*, *Vital Signs*, and *World Watch* magazine, and coordinates Worldwatch University, the institute's youth outreach project. She previously worked with the Center on Ecotourism and Sustainable Development, where she analyzed support for responsible tourism. Zoë has also served on the steering committee for SustainUS, the U.S. youth network for sustainable development.

DAVID CLEMMONS [DC]

David Clemmons is the editor of the *VolunTourist*, an e-newsletter for the business-tourism industry and travel trade. He specializes in developing voluntourism products and services for the meetings-and-incentive-travel and leisure-travel markets. Currently he collaborates with George Washington University's International Institute of Tourism Studies and the Educational Travel Conference to offer the annual VolunTourism Forum.

STUART COWAN [SC]

Stuart Cowan is a managing partner in Portland-based Autopoieis, LLC, which delivers design, capacity, and capital in support of a world that works for all. He is the coauthor of *Ecological Design*, a visionary overview of the integration of ecology and architecture, planning, and product design. He served as research director for Ecotrust, where he led the effort to develop a comprehensive pattern language for bioregional sustainability available at www.conservationeconomy.net. He served as transaction manager at Portland Family Funds, a sustainable investment bank, where he assisted in securing $250 million in project finance for green real estate and ecoforestry. He received his doctorate in the new science of living systems from UC Berkeley and has taught at Berkeley, Antioch University, and the Bainbridge Graduate Institute.

DAWN DANBY [DD]

Dawn Danby is a sustainable design boundary-spanner. An industrial designer by training, she collaborates with both engineers and artists, designs bridges, builds furniture, draws pictures, and works to reimagine a world for more people and less stuff. Her latest obsession is forging practical connections between the trifecta of design, business, and sustainability. She lives and creates in Toronto.

CHRIS DAVIS [CFD]

Chris Davis is cofounder and executive director of CommEn Space, the Community and Environment Spatial Analysis Center, a nonprofit organization that develops spatial applications, geographic analysis, and cartography for public-interest organizations in Montana and the U.S. Pacific Northwest. Chris is a geographer who spent six years working in the former Soviet Union before turning his attention back to conservation efforts in North America.

RÉGINE DEBATTY [RD]

Régine Debatty writes about the intersection between art and technology on www.we-make-money-not-art.com and regularly speaks at conferences and festivals about artists (mis)using technology.

PATRICK DIJUSTO [PD]

Patrick DiJusto has walked on the outer ledge of the Empire State Building, lectured in the Hayden Planetarium, and lost more than $500,000 in government currency. He is currently a contributing editor at *Wired* magazine.

CORY DOCTOROW [CD]

Cory Doctorow (craphound.com) is a science-fiction novelist, blogger, and technology activist. He is the coeditor of the Weblog BoingBoing.net; a contributor to *Wired*, *Popular Science*, *Make*, and the *New York Times*, among other publications; and the former director of European affairs for the Electronic Frontier Foundation, a nonprofit civil-liberties group that defends freedom in technology, where he advocated on copyright and related rights in venues ranging from universities to the United Nations. Cofounder of the open-source peer-to-peer software company OpenCola, he presently serves on numerous boards including Technorati's. In 2006/2007, he will serve as the Fulbright Chair at the Annenberg Center for Public Diplomacy at the University of Southern California. Cory's novels are published by Tor Books and simultaneously released on the Internet under Creative Commons licenses.

JOSHUA ELLIS [JE]

Joshua Ellis is a writer, Web designer, and musician. His writing has appeared in publications like *Mondo 2000*, *Make*, and *Wetbones*, and he has written a column for the Las Vegas *CityLife* alternative weekly newspaper for several years. He is also the creative leader and cofounder of Mperia.com, an online store for independent digital music downloads. He lives in Las Vegas and enjoys playing with synthesizers, machine guns, and espresso machines.

JEREMY FALUDI [JJF]

Jeremy Faludi (www.FaludiDesign.com) is a freelance product designer, physicist, and engineer, working to birth all these beautiful green theories into the real world. He works with Rocky Mountain Institute, among other organizations, and has spoken at conferences, schools, and businesses about biomimicry and green engineering. Before going into design, he worked in the semiconductor industry; he holds degrees from Stanford University and Reed College.

JILL FEHRENBACHER [JF]

Jill Fehrenbacher is a freelance designer and the founder of the design blog Inhabitat.com, which she created as a way to catalog her endless search for new ways to improve the world through forward-thinking, high-tech, and environmentally conscious design. Educated at Brown University, where she received a BA in art semiotics, and Central St. Martins College of Art and Design, where she received an MA in design studies, she currently resides in New York City, which so far has been good for her obsession with rooftop gardens and vegan junk-food restaurants.

PAUL FLEMING [PF]

A strategic adviser for the Seattle Public Utilities, Paul Fleming is currently pursuing an MBA at the University of Washington.

GIL FRIEND [GF]

Systems ecologist and business strategist Gil Friend is president and chief executive officer of Natural Logic, Inc., a strategy and systems development company focused on boosting companies' and communities' environmental performance; its A-list of clients has included General Mills, Hewlett Packard, and the U.S. Green Building Council. Gil holds an MS in systems ecology from Antioch University and a black belt in aikido, and is a seasoned presenter of the "Natural Step" environmental management system. He has over thirty-five years of experience in business, communications, and environmental innovation.

EMILY GERTZ [EG]

Emily Gertz has been a contributor to World-Changing since 2004. She currently lives in her native town, New York City, where she is a freelance environmental journalist.

AL GORE

Al Gore is the former vice president of the United States and chairman of Current TV. Elected to the U.S. House of Representatives in 1976 and the U.S. Senate in 1984 and 1990, Gore served as the 45th vice president for eight years. He is the author of *Earth in the Balance: Ecology and the Human Spirit* and *An Inconvenient Truth.*

VINAY GUPTA [VG]

A "biracial Hindu" living in America, Vinay Gupta's long-term goals are to continue the work of Buckminster Fuller and Mohandas Gandhi in order to create an opulent peace on Earth that can be shared by all human beings, without ecological harm. Vinay has worked on several projects at the Rocky Mountain Institute, done a lot of computer programming, and has traveled extensively throughout the United States.

ZAID HASSAN [ZH]

Zaid Hassan is a writer based in London. He works with Generon Consulting on long-term projects that bring together business, civil society, government, and communities to innovate within devolving social scenarios—incorporating action-learning into the relationship between modern cultures and cultures often labeled aboriginal or indigenous. In addition to working on projects involving child malnutrition in India and aboriginal relations in Canada, Zaid is writing a book on active responses to the destruction of cultures.

STEFAN JONES [SJ]

Stefan Jones, a resident of the Silicon Forest outside of Portland, Oregon, earns a living as a software engineer and freelance writer. He has dabbled in game design and is a curate in the Viridian Design Movement.

ROBERT S. KATZ [RK]

Robert S. Katz is a research analyst with the DC-based World Resources Institute. He explores "base of the pyramid" business approaches to poverty, and is an editor of and frequent contributor to the NextBillion.net/Development Through Enterprise Web site and blog (www.nextbillion.net). His current research involves documenting unmet human needs in low-income communities, and identifying the corresponding market power of the poor.

KEVIN KELLY [KK]

Kevin Kelly is senior maverick at *Wired* magazine, author of *Out of Control* (Perseus, 1995), publisher of the daily Cool Tools Web site, and board member of the Long Now Foundation. He can be found at www.kk.org.

MICKI KRIMMEL [MK]

Micki Krimmel is the director of Internet outreach at Participant Productions, a film company with a mission to effect social change. She is currently leading the company in building an online activist community at www.participate.net, where film lovers and activists can come together to make a difference. Micki has a long history in the film industry and the Internet and has developed a passion for exploring the areas where they intersect. She is nearing her one-year anniversary as a contributor to WorldChanging.com, where she writes about global film, new tools for production and distribution, and the democratization of the filmmaking process.

ANNA LAPPÉ [AL]

Anna Lappé is a national best-selling author, public speaker, and founding principal, with Frances Moore Lappé, of the Small Planet Institute. Coauthor of *Hope's Edge: The Next Diet for a Small Planet* (Penguin, 2002) and *Grub: Ideas for an Urban, Organic Kitchen* (Penguin, 2006), Anna is currently a Food and Society Policy Fellow in a national program supported by the W. K. Kellogg Foundation. She lives in Brooklyn, New York, where she likes to visit her local farmers' market to find good grub every week.

JON LEBKOWSKY [JL]

An author and consultant based in Austin, Texas, Jon Lebkowsky has written for BoingBoing.net, *Mondo 2000*, *Whole Earth Review*, *Whole Earth Catalog*, 21C, the *Austin Chronicle*, and *FringeWare Review*, where he was also an editor and publisher. Coeditor of the book *Extreme Democracy* (Lulu Press, 2005) and contributor to WorldChanging.com, Jon also blogs at Weblogsky.com, Polycot.com, SmartMobs.com, and Tagsonomy.com.

ANDREW LIGHT [ARL]

Andrew Light (faculty.washington.edu/alight/) is an associate professor of philosophy and public affairs at the University of Washington, Seattle. He has edited, coedited, and authored eighteen books on aesthetics, environmental ethics, and the philosophy of technology, including *The Aesthetics of Everyday Life* (Columbia Univ. Press, 2005), *Moral and Political Reasoning in Environmental Practice* (MIT Press, 2003), *Technology and the Good Life?* (Univ. of Chicago Press, 2000), *Philosophies of Place* (Rowman and Littlefield, 1999), and *Environmental Pragmatism* (Routledge, 1996). He is currently finishing a book on ethical issues in restoration ecology, tentatively titled *Restoring the Culture of Nature.*

HANA LOFTUS [HL]

Hana Loftus trained as an architect and urban designer. She is currently deputy director of General Public Agency, a regeneration consultancy in London specializing in creative master planning and regeneration strategy, social and cultural engagement, and brief setting. She spent the best year of her life in the Outreach Program at the Auburn University Rural Studio; while in Alabama, she won a state competition for her old-time fiddling.

REBECCA MACKINNON [RM]

Rebecca MacKinnon is a research fellow at Harvard Law School's Berkman Center for Internet and Society, and cofounder of Global Voices Online (www.GlobalVoicesOnline.org), an international bloggers' network. Rebecca worked for CNN in Northeast Asia for over a decade, serving as a bureau chief and a correspondent in Beijing and Tokyo.

JOEL MAKOWER [JM]

A respected strategist on sustainable business, clean technology, and the green marketplace for nearly twenty years, Joel Makower (www.makower.com) is founder of GreenBiz.com, cofounder of Clean Edge, Inc., and author of more than a dozen books. He is a Batten Fellow at the Darden School of Business at the University of Virginia and advises a variety of for-profit and nonprofit organizations. The Associated Press has called him "the guru of green business practices."

HASSAN MASUM [HM]

Hassan Masum has worked as an engineer, postdoctoral scientist, policy developer, and consultant. He is currently exploring social technologies, reputation systems, and ecological economics for their potential to make doing the right thing the natural thing. Hassan contributes to WorldChanging and other creative nonprofits as a way to do good while having fun, and to plant the seeds for the massive collaborative efforts of the future.

PATRICK MAZZA [PM]

Patrick Mazza has written on ecological sustainability issues for nearly three decades as an environmental journalist and technology policy analyst. A founding member of the Climate Solutions team, he has written a number of papers aimed at increasing public understanding of climate-change science, and accelerating clean-energy development. Patrick is coauthor with Guy Dauncey of *Stormy Weather: 101 Solutions to Global Climate Change* (New Society, 2001).

DAREK MAZZONE [DM]

Darek Mazzone is a producer, radio host, DJ, and writer focused on global music and emerging culture and technology. Since 1993, Darek has produced and hosted one of the top global-music programs in the world, *Wo-Pop*, which airs weekly on influential Seattle radio station KEXP and streams continuously on the Webby Award–winning site KEXP.org. Darek's passion for and in-depth knowledge of modern global music also inspired him to create the radio programs *Planet Beat* and *Fusion* to expose international artists to audiences in the United States. *Planet Beat* airs on National Public Radio, and *Fusion* is syndicated internationally on Voice of America and throughout the United States on various stations.

DINA MEHTA [DHM]
Dina Mehta is a qualitative researcher and ethnographer based in Mumbai, India, and founder of Explore Research & Consultancy, whose clients include MTV, Unilever, ESPN, and Pitney Bowes. In her ethnographic studies, Dina explores the impact of technology in rural markets and follows trendsetting youth in urban settings. In addition to maintaining her blog, Conversations with Dina (http://dinamehta.com/), Dina has contributed to WorldChanging, Tsunami Help, KatrinaHelp, Asia Quake Help, *Skype Journal*, and Global Voices Online. She is particularly curious about social technologies that open up new means of communication and collaboration, and her contribution to grassroots online disaster-relief efforts following the 2004 Asian tsunami have been acknowledged worldwide.

GABRIEL METCALF [GM]
Gabriel Metcalf is executive director of the San Francisco Planning and Urban Research Association, a civic-planning think tank.

MIKE MILLIKIN [MM]
Mike Millikin is the founder and editor of Green Car Congress, an online site for news, analysis, and discussion of fuels and their sources, technologies, issues, and policies for sustainable mobility. A former consultant and analyst in the IT industry, and a high-tech media executive (Interop, Red Herring), Mike believes that the most powerful catalyst we have for global change is an informed and educated citizen and consumer.

ROBERT NEUWIRTH [RN]
Robert Neuwirth is the author of *Shadow Cities: A Billion Squatters, A New Urban World* (Routledge, 2004), a work of reportage based on the two years he spent living in squatter communities around the globe. He writes regularly on cities, politics, and economic issues and is currently working on a new book on the world's informal economies. His work on squatters was supported by a grant from the John D. and Catherine T. MacArthur Foundation.

CATHERINE O'BRIEN [CO]
During the course of her doctoral research, Catherine O'Brien lived at the Barefoot College with her husband and two children. She is the author of "Barefoot College ... or Knowledge Demystified," published in UNESCO's Innovations in Education series. Her husband, Ian Murray, is the coproducer with UNESCO of the documentary *Barefoot College: Knowledge Demystified*, available through www.VideoProject.com.

EMEKA OKAFOR [EO]
Emeka Okafor is an entrepreneur and consultant, a founding partner of Caranda Ventures (parent company of Caranda Teas and Caranda Coffee), a founding member of the International Private Enterprise Group, and the founder of the publicly acclaimed blog Timbuktu Chronicles. He sits on the global advisory board for Students for the Advancement of Global Entrepreneurship, and was recently named a Sun Microsystems Participation Fellow at Pop!Tech.

ORY OKOLLOH [OO]
Kenyan Ory Okolloh recently received her JD from Harvard Law School. She is currently testing out just how many ways one can use a law degree by working in Africa on issues including corruption, technology policy, and social entrepreneurship. When she's not too busy, she tries to keep up with her blog, Kenyan Pundit, and with the outrageously large number of feeds in her RSS reader.

CHRIS PHOENIX [CP]
Cofounder and director of research at the Center for Responsible Nanotechnology, Chris Phoenix obtained his BS in symbolic systems and MS in computer science from Stanford University. From 1991 to 1997 he worked as an embedded software engineer at Electronics for Imaging; then he left the software field to concentrate on dyslexia research and correction. Since 2000 he has focused exclusively on studying and writing about molecular manufacturing. Chris is a published author in nanotechnology and nanomedical research, and maintains close contacts with leading researchers in the field.

SARAH RICH [SR]
Sarah Rich is the managing editor of World-Changing.com and the *Worldchanging* book. She has worked as a brand identity copywriter, editor,

and journalist since graduating from Stanford in cultural and social anthropology. Sarah is also the managing editor of Inhabitat.com, a Weblog covering sustainable design and green building, and has published work in *Dwell*, *I.D.*, and *ReadyMade*. She is a biodiesel driver, farmers' market devotee, and crossword addict.

JOHN ROBB [JR]

John Robb, a former counterterrorism operation planner and commander, now advises corporations on the future of terrorism, infrastructure, and markets. A graduate of Yale University and the Air Force Academy, John has been published in *Fast Company* and the *New York Times*. His book on the future of terrorism, war, and the global economy will be published by Wiley in 2007.

BEN SAUNDERS [BS]

Ben Saunders (www.bensaunders.com) is a record-breaking long-distance skier, with three North Pole expeditions under his belt. He is the youngest person ever to ski solo to the North Pole and holds the record for the longest solo Arctic journey by a Briton. From 2001 to 2004, Ben skied more than 1,250 miles in the high Arctic, and in October 2006 he sets out on the longest unsupported polar expedition in history.

CAMERON SINCLAIR [CS]

Cameron Sinclair is the cofounder and executive director of Architecture for Humanity (AFH), a nonprofit that seeks architectural solutions to humanitarian crises and brings design services to communities in need. His team has implemented programs to provide housing to returning refugees in Kosovo; mobile health clinics to combat HIV/AIDS in sub-Saharan Africa; mine clearance and playground construction in the Balkans; and disaster response to Hurricane Katrina in the Gulf States. Among other projects, AFH is currently building a sports and HIV/AIDS outreach facility for youth in Somkhele in KwaZulu-Natal, South Africa, and developing "Rethinking Tent City," a project to encourage social and economic change in long-term refugee camps and settlements through sustainable-design interventions. Cameron is coauthor, with Kate Stohr, of an upcoming book on humanitarian design, *Design Like You Give a Damn: Architectural Responses to Humanitarian Crises*.

ALEX STEFFEN [AS]

Alex Steffen cofounded and edits WorldChanging.com and is the editor of this book.

BRUCE STERLING

Bruce Sterling is a Hugo Award–winning science-fiction author and futurist. He is the author of numerous novels, short stories, as well as works of nonfiction. Three of his novels were selected as *New York Times* Notable Books of the Year, and he has been a contributing writer for *Wired* since its beginning.

PHILLIP TORRONE [PT]

Phillip Torrone is an author, artist, and engineer, and senior editor of *Make* magazine. He has authored and contributed to numerous books on programming, mobile devices, design, multimedia, and hardware hacking; he regularly writes for *Popular Science*. His projects have appeared in *Wired*, *Popular Science*, *USAToday*, the *Wall Street Journal*, the *New York Times*, and on G4TechTV, NPR, and elsewhere. Phillip also produces the *Make* audio and video content on the MakeZine.com site. Prior to *Make*, Phillip was director of product development for the creative firm Fallon Worldwide.

MIKE TREDER [MT]

Cofounder and executive director of the Center for Responsible Nanotechnology, Mike Treder is a professional writer, speaker, and policy advocate, with a background in technology and communications-company management. Mike majored in biology at the University of Washington in Seattle, after which his career in the private sector included a stint with a large telecommunications firm and stints managing radio stations in major markets. He has published numerous articles and papers, and does frequent interviews with the media. As an accomplished presenter on the societal implications of emerging technologies, he has addressed conferences and groups in North America, South America, Europe, and Great Britain.

LEIF UTNE [LU]

Minneapolis-based Leif Utne is a writer, musician, and activist. Since 1998, he has worked for the family business, *Utne* magazine, where his coverage of a wide range of topics, including globalization, democracy, film, green technology, social movements, and Latin American politics has won awards from AlterNet and Project Censored. Leif produces the UtneCast podcast and is a frequent speaker on media issues, but he secretly wants to be a rock star.

EDWARD C. WOLF [EW]
Edward C. Wolf is a writer and editor with a special interest in the natural history of global change. A former senior researcher at Worldwatch Institute and director of communications at Ecotrust, his books include *Salmon Nation* (Ecotrust, 1999) and *Klamath Heartlands* (Ecotrust, 2004). He lives with his wife, two daughters, and sundry animals in Portland, Oregon.

MOLLY WRIGHT STEENSON [MWS]
A design researcher and architectural historian, Molly Wright Steenson works within the social contexts of mobile technology and on issues of urbanism. She cut her teeth on the Web in 1994, developing more than a hundred prominent Web sites. Molly was associate professor of connected communities at the Interaction Design Institute Ivrea (now a part of the Domus Academy) in Italy and will complete her masters of environmental design (in history and theory) at the Yale School of Architecture in May 2007.

ANDREW ZOLLI [AZ]
Andrew Zolli is a foresight and global-trends consultant who analyzes critical trends at the intersection of culture, technology, and global society. His firm, Z + Partners, helps global companies and institutions see, understand, and respond to complex change. He is also the curator of the annual Pop!Tech conference (www.PopTech.com), a gathering of thought leaders who explore the social impact of technology and the shape of things to come. Andrew serves as a fellow of the National Geographic Society, and as its chief futurist.

ETHAN ZUCKERMAN [EZ]
A research fellow at the Berkman Center for Internet and Society at Harvard Law School, Ethan Zuckerman is focused on the Internet in the developing world. With Rebecca MacKinnon, he founded Global Voices, an international community of Webloggers dedicated to increasing understanding through citizens' media. He's the founding chairman of WorldChanging, cofounder of the technology NGO Geekcorps, and cofounder of the Web community Tripod. When not in countries hard to find on a map, he lives with his wife and his fluffy cat in Berkshire County, Massachusetts.

SETH ZUCKERMAN [SZ]
Seth Zuckerman writes on forests, fish, and other ties that bind human beings to the rest of the natural world. His work has appeared in numerous magazines, including *Orion*, *Sierra*, and *Whole Earth*, and in the anthology *Salmon Nation: People, Fish, and Our Common Home* (Ecotrust, 1999), which he coedited with fellow WorldChanging contributor Edward C. Wolf. He divides his time between Seattle and the Northern California coast.

ACKNOWLEDGMENTS

To the memory of my parents, George and Delores Steffen, who never stopped believing that we can make the world better for everyone.

This is a big book, full of ideas, almost none of which are purely our own. Indeed, this project has many mothers, many fathers, and a few angels.

The *Worldchanging* book simply would not exist without the astonishing hard work and sheer brilliance of my three partners on this project, Sarah Rich, Tessa Levine-Sauerhoff, and Carissa Bluestone. These three editorial ninjas cheerfully performed impossible feats again and again in the six months we had to gather, write, and edit the hundreds of pieces in this book. Throughout it all, they worked like fiends, exhibited incredible insight, and managed to make me laugh every single day (though I still do not fully understand their CuteOverload obsession). I have never had such a good time working with anyone, despite the fact that our six months on this project were incredibly hard. Together we endured long day after long day after long day; birthdays, marriages, deaths, and newborns; computer disasters, blizzards of paper and thousands of digital files, rapidly evolving demands, constant travel, inspired nights and early mornings, and the challenges of running a huge project out of my home with criminally insane deadlines. My housekeeping will remain a subject to which we won't refer. Suffice it to say that Sarah, Tessa, and Carissa's names ought to be on the cover next to mine, if not above it.

Former Worldchanging managing director Chanel Reynolds remains one of my heroes in life. Chanel can run a meeting, manage a Web site redesign, offer Buddhist insight into life, balance a bank account, oversee a team, think strategically, and raise a toddler all at the same time, while managing to keep up a steady torrent of witty and compassionate comments. Chanel put Worldchanging on its feet and kept us all human. We love Chanel.

Ed Burtynsky is another hero. His optimism about Worldchanging's possibilities is exceeded only by his good cheer and sound advice. His work as a photographer stuns me, but his qualities as a man humble me. I count among my blessings meeting him.

We also cannot thank enough our friends at TED, and Susan Dawson and Chris Anderson of the Sapling Foundation. Chris, in particular, had the willingness and generosity to see what Worldchanging is trying to achieve and to throw himself enthusiastically behind the project. He has a sharp, far-seeing mind, and both his support and his constant ability to challenge us have been crucial to our success. If every wild-eyed ambitious nongovernmental organization had an angel as astute and dedicated as Chris, we'd have these planetary problems of ours licked in time to take summer vacation.

I'm convinced there is a special place in heaven reserved for those who have served as the chairs of nonprofit boards, since they've already done their time in hell. Ethan Zuckerman is not only a great friend, but he is also the finest, smartest, most principled board chair any start-up nonprofit could hope for. If there were a Congressional Medal of Honor for NGO service, we'd be pinning it on his chest. As it is, I have to settle for saying that Ethan has left us with a moral debt we can never repay.

The whole Worldchanging crew mesmerizes me with their brilliance. Dawn Danby is one of the loveliest human beings on earth, has kept me sane, and has given me the gift of quality thinking, sound advice, and a constant invitation to play. On the whole, the Worldchanging team—Alan AtKisson, Nicole-Anne Boyer, Chris Coldeway, Régine Debatty, Jeremy Faludi, Jill Fehrenbacher, Gil Friend, Emily Gertz, Rohit Gupta, Vinay Gupta, Zaid Hassan, Micki Krimmel, Jon Lebkowsky, Joel Makower, Mike Milliken, Hassan Masum, Dina Mehta, Cameron Sinclair, and Andrew Zolli—are the finest folks on earth.

Our key supporters have our profound thanks as well. As a nonprofit, we would not exist without their generosity. Thanks to Naomi Adachi, D. Benjamin Antieau, Joshua Arnow, Charles Eric Boyd, Charles and Jane Fink, Anthony Fisk, David Foley, Jonathan Foley, Colin Glassco, Dan Goldwater, Daphna Buchsbaum, Brian Hayes, Allison Hunt, David Meyers, Caroline Rennie, Robert and Myra Rich, Alex

Shmelev, Paige West, Gregory Williams, and Morden Yolles.

A million others deserve our thanks. Inevitably, we'll miss many, but for starters: Bruce Sterling, for starting this whole ball rolling; Cory Doctorow, for encouraging this mad enterprise; Stewart Brand, for blazing the path; Bill McDonough, for being an early believer. Thanks also to Amy Novogratz, Kelly Stoetzel, Yesenia Martinez, Howard Rheingold, Bethany Yarrow, Walt Crowley, Marie McCafferty, Jay Ogilvy, Robert Smith, Robyn Kistler, Jimmy Wales, Ruby Lerner, Sam Verhovic, Joel Garreau, Joe Wetherby, Karisa Kenyon, Scott Beale, Pierre Omidyar (for his timely support), Sunny Bates (who rocks!), Jennifer McCrea (one of my oldest friends), Nic Warmernhoven, Kristen Tsiastsios, Christian French, Julie Ross, Chris Luebkeman, *Wired*'s Chris Anderson, Dave Roberts, J. C. Herz, Kelli B. Kavanaugh, Cheney Brand, Dave Bosch, Aki Namioka, Joanna Grist, Ashley Parkinson, Holly Pearson, Mark Grimes, Mark Frauenfelder, Jerry Michalski, Joi Ito, Gerritt Visser, Hugh Forrest, Tad McNulty, Ze Frank, Caterina Fake, Rachel Barenblat, Denis Hayes, Vicki Robin, Kevin Danaher, Serge de Gheldere, Kate Stohr, danah boyd, Marcus Colombano, Leslie Duss, Matt Steurwalt, Heather Moss, Jenn Byers, Wendy Sykes, Anne Galloway, Allison Gentile, Bruno Giussani, Tim and Sue Hayes-McQueen, Jose Hernando, Graham Hill, Thomas Kriese, Xeni Jardin, Julie Lasky, Pete Leyden, Margaret Lydecker, Shannon May, Cary Moon, Peter Steinbreuck, Davidya Kasperzyk, RU Sirius, Larry Brilliant, Sara Johnson, Rebecca Blood, Andrew Shapiro, Jane Byrd, Dominic Muren, Emeka Okafor, Mike Treder, Elisa Murray, Stuart Henshall, David Bornstein, Roland Piquepaille, Jeff McIntire-Strasburg, C. Sven Johnson, Kenyatta Cheese, Rob Katz, Ramez Naam, Eric de Place, Josh Dorfman, Alexandra Samuel, Lucas Gonzalez, Jon Stahl, David Hsu, Jeremy Lyon, Michael Metelits, Danny O'Brien, Meaghan O'Neill, Christopher Allen, David Weinberger, Ross Mayfield, Chris Phoenix, George Mokray, John Emerson, Dale Carrico, Thomas Barnett, Kevin Kelly, Jo Twist, David Zaks, Chad Monfreda, Zoe Chafe, Maida Barbour, Janice Cripe, Tim O'Reilly, and the whole FOO Camp crowd.

The Flick Creek group have all my love, while the Progress with Friends crew (especially Lisa Witter) have my political proxy. The Pop!Tech crowd have warped my brain for the better. Polycot has been not only our Web host, but our allies and friends.

Our editor, Deb Aaronson, has rocked beyond belief: if this book is good, you can blame her. Michael Jacobs, president and CEO of Abrams, believed in and championed this book from the beginning. The tireless work of Carrie Hornbeck, Kate Norment, Laura Tam, and Sara Runnels is here in the absence of (more) errors. Maggie Kneip, Kerry Liebling, and Melody Meyer have worked days, nights, and weekends to get the word out. And Anet Sirna-Bruder helped make the vision for this book a reality.

Stefan Sagmeister is the best kind of creative genius: an approachable, collaborative, fun guy who also happens to have a brain large enough to put significant strain on his neck. He's also smart enough to have great people working with him, including Matthias Ernstberger and Roy Rub, who both spent an awful lot of time making this book look good.

Team Sideshow, lead by Dan Tucker, and including Kate Hannibal and Kathryn Williams, has gone above and beyond, with humor, dedication, and patience, to find the images you see here in this book.

The Viridian List, in some fundamental ways, planted the seeds that grew into Worldchanging. All thanks to our colleagues there.

Finally, and most importantly, thanks to my great friend Jamais Cascio. The conversations we've had about how best to change the world have informed every page of this book and transformed my life. I have never met anyone with a more interesting mind, and I've never known anyone who feels so deeply our obligation to think clearly today so that tomorrow will be brighter. He deserves more praise than I can possibly give him.

All these people are helping to bring forward ideas that may just change the world. Whatever mistakes, errors of judgment, or total screw-ups you may find on these pages are entirely my own responsibility. I only hope I've done my colleagues as proud as they've done me. AS

PHOTOGRAPHY CREDITS

SLIPCASE

Sturnus pagodarum, Brahminy Starling. Cyril Laubscher ©Dorling Kindersley
"Blue Marble 2000": Courtesy NASA Goddard Space Flight Center Visualization Analysis Lab

FRONT MATTER

"Blue Marble 2000": Courtesy NASA Goddard Space Flight Center Visualization Analysis Lab 2

STUFF

©Thomas Hoepker/Magnum Photos 30–31
©2006 Peter Menzel/menzelphoto.com 32 (L)
©Stuart Franklin/Magnum Photos 32 (R)
AP Photo/Otago Daily Times, Stephen Jaquiery 36
Edward Burtynsky's photographs appear courtesy of Robert Koch Gallery, San Francisco; Charles Cowles Gallery, New York; Nicholas Metivier Gallery, Toronto 37
Courtesy Method Products, Inc. 38
Edward Burtynsky's photographs appear courtesy of Robert Koch Gallery, San Francisco; Charles Cowles Gallery, New York; Nicholas Metivier Gallery, Toronto 40, 42–43, 44
John Humble/Stone +/Getty Images 49
John Miller/Robert Harding World Imagery/Getty Images 52 (L)
Steve Satushek/The Image Bank/Getty Images 52 (R)
Courtesy Greenmarket 53
AP Photo/The Sun, Joe Nicholson 56
AP Photo/John Froschauer 57
Courtesy Jon Huey 58 (L,R)
©Michael Moran 59
AP Photo/Nam Y. Huh 60
AP Photo/Christina Paolucci 64 (L)
AP Photo/Victoria Arocho 64 (R)
AP Photo/Boris Heger 68 (L,R)
AP Photo/The Sun, Larry Steagal 69
Courtesy ICARDA 72 (L,R)
AP Photo/Jeff Chiu 73
AP Photo/Al Grillo 75
Courtesy [ACCELERATED|COMPOSITES] LLC., design by Steve Fambro, CAD work, modeling and rendering by Marc Walston 78 (L)
Courtesy American Honda Co., Inc. 78 (R), 80 (L,R)
Courtesy TranSglass, design by Emma Woffenden and Tord Boontje, produced by Artecnica, photo by Angela Moore 83
Courtesy Dalsouple Direct, UK 84 (L)
Courtesy Droog Design by Jurgen Bey, photo by Marsel Loermans 84 (R)
Courtesy Syndesis, Inc., USA 85
Courtesy Materia, see Material Explorer at www.materia.nl 87 (L,R), 88 (L)
Courtesy LiTraCon, copyright ©LiTraCon Bt 2001–2006, LiTraCon Bt., H-6640 Csongrád, Tanya 832, Hungary – EU, info@litracon.hu, www.litracon.hu 88 (R)
Courtesy Fa.Alulight 2004 89
Courtesy ReadyMade magazine, design by Sherif Shalaby 90 (L)
Courtesy ReadyMade magazine, design by Mark Robohm, photo by Andrew Nagata 90 (R)
Courtesy ReadyMade magazine, design by Ken Kirkpatrick, photo by Jeffery Cross 91 (L)
Courtesy gregtatedesign.com 91 (R)
Courtesy Manu Prakash, MIT's Center for Bits and Atoms, 2005 93
Courtesy Squid Labs LLC 94
Courtesy Adafruit Industries 95
Courtesy microRevolt, photo by Cat Mazza and Carrie Dashow 96
Courtesy CuteCircuit 97 (L)
Courtesy Disaffected!, an anti-advergame by Persuasive Games (www.persuasivegames.com) 97 (R)
Courtesy Future Applications Lab (Viktoria Institute) and PLAY Studio (Interactive Institute) 98
Prof. Kellar Autumn, Lewis & Clark College, Portland, Oregon 100 (L,R)
Courtesy ©W. Barthlott, University of Bonn 101 (L)
Courtesy Sto AG 101 (R)
Courtesy Alisa Andrasek/biothing 103 (L)
Courtesy Front Design 103 (R)
Courtesy Milwaukee Art Museum, photo by Timothy Hursley 104
Courtesy BIX Media facade Kunsthaus Graz, photo by Nicolas Lackner 105
©2006 Peter Menzel/menzelphoto.com 106
Courtesy Doe/NREL-Jack Dempsey 112
Courtesy Freitag 115 (L,R)
Courtesy James Patten, photo by Mariliana Arvelo 116

Edward Burtynsky's photographs appear courtesy of Robert Koch Gallery, San Francisco; Charles Cowles Gallery, New York; Nicholas Metivier Gallery, Toronto 120–121
Courtesy Sprout Design 122 (L,R)
Courtesy University of Michigan and KVA Matx 124–125
Courtesy Amy Smith, MIT's D-Lab 132, 133 (L)
Courtesy Whirlwind Wheelchair International 133 (R)
Courtesy NEC Corporation 134 (L)
Courtesy P.A. Semi, Inc. 134 (R)
©2006 Dawn Danby and Paul Waggoner, design by ©2006 Dawn Danby and Jeremy Faludi 135

SHELTER

Courtesy Integrated Architecture (www.intarch.com), photo by Laszlo Regos (www.laszlofoto.com) 140–141
Courtesy Velocipede Architects, photo by David Ericson 142 (L)
©1996 Forest Stewardship Council A.C. 142 (R)
©Michael Rosa/Dreamstime.com 143
©fernandoaguila.com, Vetrazzo, LLC 144
Courtesy Jennifer Jako/The ReBuilding Center 145 (L,R)
Courtesy Zoka Zola, Architecture + Urban Design (www.zokazola.com) 148
Courtesy Duke University, Pratt School of Engineering 149
Courtesy Michelle Kaufmann Designs, photo by MKD 150 (L)
Courtesy Andrew Maynard Architects 150 (R), 151
Courtesy Knight Frank and Abito 152
©Andrea Zittel, image courtesy Andrea Rosen Gallery, NY 153 (L)
Courtesy Brave Space Design 153 (R)
©Raf Makda/VIEW & Bill Dunster Architects 156
AP Photo/IKEA/Helmut Stettin 157
Courtesy Scrapile 158 (L)
Photo courtesy Kirei USA, design by AAA Design, Mike Slattery, Chicago 158 (M)
Courtesy Plywood Office 158 (R)
Courtesy Philips Lighting Company 161 (L)
Courtesy Parans Daylight AB 161 (R)
AP Photo/John Gaps III 163
©iStockphoto.com/Photo-Dave 165 (L)
Reza Estakhrian/Taxi/Getty Images 165 (R)
Courtesy Light Up The World Foundation 167 (L)
©Rolex Awards/Xavier Lecoultre 167 (R)
Fred Hoogervorst/Panos Pictures 169 (L)
©Rolex Awards/Tomas Bertelsen 169 (R)
AP Photo/Mike Derer 171
DOE/NREL-General Services Administration (GSA) 172
DOE/NREL-Bill Timmerman 173
REUTERS/Andrew Wong 174
DOE/NREL-Warren Gretz 175
Courtesy Elsam Kraft A/S 176 (L)
Illustration by Chris Radisch, courtesy Magenn Power, Inc. 176 (R)
©Dreamstime.com 177
©Ian Murray 178
AP Photo/Robert F. Bukaty 180 (L)
Courtesy Renewable Devices Swift Turbines, Ltd. 180 (R)
Courtesy Wim Jonker Klunne (www.microhydropower.net) 182
AP Photo/Greg Baker 184 (L)
David McNew/Getty Images 184 (R)
Jenny Matthews/Panos Pictures 186
Courtesy reHOUSE/BATH, design by Fulguro and Thomas Jomini architecture workshop, ©Geoffrey Cottenceau 188
©Paul Bardagjy 191
Courtesy Q Drum, Ltd., photo by Pieter Hendrikse 192 (L)
Courtesy KickStart 192 (R)
Courtesy FogQuest, photo by Virginia Carter 193
Courtesy Stephan Augustin, Watercone 194 (L)
Photo by Rebecca Drayse, courtesy TreePeople 195
Courtesy City of Vancouver 196 (L)
©Living Designs Group, LLC 196 (R)
Noel Harding, "Elevated Wetlands," 1997–98, Taylor Creek Park, Toronto, Ontario, Canada; photo by Michel Boucher 197
©Fritz Haeg 198 (L,R)
Courtesy Alexandra Lichtenberg (www.ecohouse.com.br) 201, 202 (L,R)
Mikkel Ostergaard/Panos Pictures 204–205
©Stuart Franklin/Magnum Photos 206
AP Photo/Heiko Junge 207
Courtesy Nutriset SAS 208 (L)
Courtesy World Food Programme (2005) 208 (R)
AP Photo/Andrea Booher 211 (L)
REUTERS/Zohra Bensemra 211 (R)
Photo by Dave Warner, MD, PhD 212
Image with permission of SkyBuilt Power ©2005 213
AP Photo/Tomas Munita 218 (L)
Courtesy MineWolf Systems 218 (R)
AP Photo/Amr Nabil 220
©Steve McCurry/Magnum Photos 221
AP Photo/Eugene Hoshiko 222

CITIES

David Hanson/Stone/Getty Images 227
©iStockphoto.com/Nicholas Belton 228
AP Photo/Chuck Stoody 232
Courtesy DA/MCM+LMN Architects 233
©iStockphoto.com/sarent 235
Jim Wark/Lonely Planet Images/Getty Images 236 (L)
Courtesy Electronic Arts™ 236 (R)
Baldomero Fernandez/Stone/Getty Images 238
GOH Chai Hin/AFP/Getty Images 239
Courtesy City of Mountain View, CA 241
Courtesy Western Recon Aerial Photography 242 (L,R)
Courtesy Wal-Mart Stores, Inc. 243
AP Photo/Women's Wear Daily 245 (L)
Courtesy TR Hamzah and Yeang 245 (M,R)
Bilderberg/Photonica/Getty Images 246
AP Photo/MoMa/Foster and Partners 247
©dbox for Cook + Fox Architects LLP 248 (L, ML, MR)
Courtesy GBD Architects, Inc. 248 (R)
Photo by Max Donoso Saint for Mathews Nielsen Landscape Architects, PC 250
Paul S. Howell/Liaison/Getty Images 251
©2001 Mark D. Phillips (www.markdphillips.com) 252 (L,R)
©Matthew Plexman Photography 255
©Roofscapes, Inc. Used by permission. 257 (L)
©iStockphoto.com/marcin szmyd 257 (R)
©Colleen Coombe/Dreamstime.com 258
Photo by Signe Nielsen for Mathews Nielsen Landscape Architects, PC 259 (L)
Courtesy Ellen Bruss Design 259 (R)
REUTERS/Shannon Stapleton 260
AP Photo/Alastair Grant 263 (L)
Courtesy Breakthrough Technologies Institute 263 (R)
Courtesy Hamilton-Bailie, illustrated by Dixon Jones Architects 265
Courtesy Atlantic Station 266 (L)
AP Photo/MTI, Peter Kollanyi 266 (R)
Courtesy Atlanta Community ToolBank 268
Courtesy Christian De Sousa 269
Edward Burtynsky's photographs appear courtesy of Robert Koch Gallery, San Francisco; Charles Cowles Gallery, New York; Nicholas Metivier Gallery, Toronto 271, 272–273
AP Photo/Stringer 274 (L)
Courtesy ©William McDonough + Partners 274 (R)
Edward Burtynsky's photographs appear courtesy of Robert Koch Gallery, San Francisco; Charles Cowles Gallery, New York; Nicholas Metivier Gallery, Toronto 277
©Stuart Franklin/Magnum Photos 279 (L,R)
©Stuart Franklin/Magnum Photos 280
Sven Torfinn/Panos 281
Courtesy ©Arup 282
Chris de Bode/Panos 283 (L)
AP Photo/Javier Galeano 283 (R)
Courtesy Goa 2100 team 285 (L)
AP Photo/Eduardo Verdugo 285 (R)
Crispin Hughes/Panos 286
©Stuart Franklin/Magnum Photos 287 (L)
Andy Johnstone/Panos 287 (R)
©Raghu Rai/Magnum Photos 288–289
Sven Torfinn/Panos 292 (L)
Courtesy Grameen Foundation, photo by Tamara Plush, Smudge Productions, LLC 292 (R)
AP Photo/Khalil Senosi 294 (L)
AP Photo/Khalil Senosi 294 (R)
ISSOUF SANOGO/AFP/Getty Images 296 (L)
Courtesy Geekcorps, photo by Matt Kuiken for IESC Geekcorps 296 (R)
Courtesy Geekcorps, photo by Ian Howard for IESC Geekcorps 297
Courtesy One Laptop per Child, MIT Media Lab 299 (L)
Courtesy Free Geek 299 (R)
AP Photo/Alexander Meneghini 300
iStockphoto/Christoff B. 302
Courtesy StarSight 304

COMMUNITY

©Keren Su/Corbis 308–309
Photo by Pat Lanza 311
Photo by Richard Koman 314
©Abbas/Magnum Photos 317
©Stuart Franklin/Magnum Photos 318–319
AP Photo/Pankaj Nangia 321
©Caroline Penn/CORBIS 322
AP Photo/Greg Baker 325 (L)
©Karen Kasmauski/CORBIS 325 (R)
©Gideon Mendel/CORBIS 326
©Chris Steele-Perkins/Magnum Photos 328
©Mujahid Safodien/Reuters/Corbis 330
©Stuart Franklin/Magnum Photos 332
Courtesy GeoEye and www.globalsecurity.org 335
©Michael Brennan/CORBIS 339
©James Chase 340
Photo by Nic Paget-Clark (www.inmotionmagazine.com) 342
©David Alan Harvey/Magnum Photos 343
©Peter Turnley/CORBIS 347

©Keith Dannemiller/CORBIS 349
AP Photo/Pavel Rahman 353
AP Photo/Indianapolis Star, Matt Kryger 355
©Heifer International 356 (L,R)
AP Photo/Rusty Kennedy 357
©Ian Murray 359, 361
©Ian Berry/Magnum Photos 363
©David Samuel Robbins/CORBIS 365
©Martin Parr/Magnum Photos 366 (L)
AP Photo/Alex Brandon 366 (R)
©Natalie Behring/Panos 369
Yann Layma/ImageBank/Getty Images 370–371
©Igloolik Isuma Productions, photo by Norman Cohn 372
AP Photo/Reed Hoffmann 373
©Aubrey Wade/Panos 374
AP Photo/Ng Han Guan 375 (L)
AP Photo/Chitose Suzuki 375 (R)

BUSINESS

Courtesy Aveda Corporation 380–381
AP Photo/Pat Sullivan 383
AP Photo/TVA, Ron Schmitt 386
AP Photo/Andy Klevorn 387
Courtesy Philips Lighting Company 390 (L,R)
AP Photo/Damian Dovarganes 391
Courtesy www.adbusters.org 394
AP Photo/Ben Margot 398 (L)
AP Photo/Bullit Marquez 398 (R)
Scott Olson/Getty Images 404
Courtesy Amazon.com, Inc. or its affiliates. All rights reserved. 405

POLITICS

©Stuart Franklin/Magnum Photos 410–411
AP Photo/J. Pat Carter 412
AP Photo 413
AP Photo/Tony Harris 416 (L)
AP Photo/Elise Amendola 416 (R)
AP Photo/Charles Krupa 420
Qilai Shen/Panos 421
AP Photo/Lauren Burke 423
Sven Torfinn/Panos 427 (L)
Marie Dorigny/Editing/Panos 427 (R)
Sean Sutton/MAG/Panos 428
Courtesy America Speaks 432 (L)
AP Photo/Lefteris Pitarakis 432 (R)
AP Photo/Brennan Linsley 433
AP Photo/NIPA 434
Corruption Perception Index, 2005, Transparency International (www.transparency.org) 435
AP Photo/Khalil Senos 437
AP Photo/Karel Prinsloo 438
AP Photo/Gregory Bull 441 (L)
AP Photo/UNMIL, Mathew Elavanalthoduka, HO 441 (R)
AP Photo/James Nachtwey/VII 442
AP Photo/Amel Emric 444 (L)
Ryan Kautz/WITNESS 444 (R)
Chris Hondros/Getty Images 448
Montgomery County Sheriff's Office 451
Courtesy Forkscrew Graphics 452 (L,R)
"F.R.U.I.T." ©2005 Amy Franceschini 453
Courtesy Banksy 454
AP Photo/Gregory Bul 456 (L)
AP Photo/Lexington Herald-Leader, Michelle Patterson 456 (R)
Courtesy The Yes Men (www.theyesmen.org) 457
Gilles Mingasson/Getty Images 458 (L)
David Greedy/Getty Images 458 (R)
AP Photo/Adil Bradlow 460
AP Photo 461
Courtesy York Zimmerman, Inc. 462 (L)
AP Photo/Sergei Chuzavkov 462 (R)
AP Photo/Sayyid Azim 465 (L)
UN Photo/Sophia Paris 465 (R)
AP Photo/Achmad Ibrahim 466 (L)
AP Photo/Bilal Hussein 466 (R)
AP Photo/George Osodi 467
AP Photo/Paul White, Files 468
AP Photo/Hidajet Delic 469
AP Photo/Mike Hutchings, Pool 470 (L)
LIONEL HEALING/AFP/Getty Images 470 (R)

PLANET

©Stuart Franklin/Magnum Photos 474–475
AP Photo/Ric Francis 479 (L)
REUTERS/Carlos Barria 479 (R)
©Sandro Vannini/CORBIS 481 (L)
©Gabe Palmer/CORBIS 481 (R)
AP Photo/Imperial Valley Press, Kevin Marty 482
AP Photo/Yellowstone National Park 484 (L)
©Daniel76/Dreamstime.com 484 (R)
AP Photo/The Daily Times, Wes Hope 486
AP Photo/American Press, Shawn Martin 487
AP Photo/U.S. Fish and Wildlife Service 489
©Gueorgui Pinkhassov/Magnum Photos 492
©Peter Marlow/Magnum Photos 495
AP Photo/Suzanne Plunkett 499 (L)

AP Photo/Liz Lucas, Oxfam 499 (R)
©Craig Aurness/CORBIS 502
©Timothy Hursley, Rural Studio 503
AP Photo/John McConnico 506
©Gary Braasch 508
AP Photo/Laura Rauch 511
Edward Burtynsky's photographs appear courtesy of Robert Koch Gallery, San Francisco; Charles Cowles Gallery, New York; Nicholas Metivier Gallery, Toronto 512
AP Photo/ Ed Betz 514 (L)
AP Photo/Jacqueline Larma 514 (R)
AP Photo/Greenpeace, Vinai Dithajohn, HO 515
©NASA 518
©Jeremy Horner/CORBIS 519
AP Photo/Will Kincaid 521
AP Photo/National Oceanic and Atmospheric Administration 522
AP/Ove Hoegh-Guldberg, Centre for Marine Studies, The University of Queensland, HO 523
©W. Perry Conway/CORBIS 526
Courtesy Ben Saunders 527 (L)
David A. Harvey/National Geographic/Getty Images 527 (R)
©NASA 529, 530–531, 532
©CORBIS 535
©Gary Braasch/CORBIS 537

INDEX

Page numbers in *italics* refer to illustrations.

A

Abito, *152*, 153
Ableman, Michael, 56, 61–62
Accelerated Composites, *78*, 78–79
Access to Knowledge ("A2K"), 338
acid rain, 274, 473
"Acoustic Survival Kit," 98
ActionAid, 299
activism, 450–59, *451–54*, *456–58*, *460–62*, 460–64
 movement building, *412*, 412–17, *413*, *416*
 through citizen science, 481–83, *482*, 513
 See also boycotts; *specific causes*
Acumen Fund, 329
Addison Center (Texas), 240
advocacy, 441, 453, 462
 and blogs, 419, 420, 421–22
 grassroots, 260, 453
 Netroots, 453
 and networking, 420, 426–30, *427*, *428*, 453
 See also specific causes
Africa, 20, 192, 333, 350, 502, 510
 and agriculture, 46, *68*, 68–69, 192, 501
 and AIDS, 325–26, 330, 354, 501
 debt relief advocacy, 420
 and government corruption, 436–39, 440
 and leapfrogging technology, 292
 Nollywood (film industry), 368, 372
 war in, and sustainability, 464–65
 and water supply, 186, 192, *192*
 and women's rights, 429–30
 See also specific countries
Africa Unchained (Ayittey), 440
Against the Grain (Manning), 51
Age of Amateurs, 481, 482, 508, 535
 See also citizen science
agriculture
 industrial, versus sustainable, 52, 66, 67, 499, 501
 See also farming
AIDS, 19, 253, 307, 313, 323, 325–29, *326*, *328*, 354, *427*, 501
air conditioning. *See* cooling technology
airplane emissions, 364
air pollution, 256, 311
 airplane-related, *364*
 automobile-related, 74–75, 147, 232, 262–63, 274, *277*, 277–78, 473, 512, 515
 cooking-related, 168, 223
 in ecological economics, 488
 See also greenhouse gases
air purification, and living walls, *255*, 256
Airtecture structures, 89
Alabama, 503–4
Alaska, *75*, 76, *163*, *511*, 528
AlbanianLiterature.com, 374
Alexander, Christopher, 261
algae, hydrogen-producing, 111–12, *112*
Alias (software developer), 123–24
Alley, Richard B., 528
Alternative Urban Futures (Pinderhughes), 230
Altomare, Mary, 388–89
Alulight (aluminum), 89, *89*
aluminum, recycling, 48
Amandla Waste Creations (AWC), 343, 344
Amazon.com, 403, *405*
AMD (microchip company), 298
American Association for the Advancement of Science, 444
American Forest and Paper Association (AF&PA), 497
American Society of Landscape Architects, 200
American Water Works Association, 187
American Wind Energy Association, 175
AmericaSpeaks, 432, *432*
AmmanNet (online radio), 424
Amnesty International, 429, 446, 453
Anholt, Simon, 395
animal farming. *See* farming, livestock
Ansip, Andrus, *434*
Antarctica, 527, 528
antennae design, and biomimicry, 100–101
antibranding, 393, *394*, 394–95
anticipation (business strategy), 397–98
anticircumvention laws, copyright, 337
antibiotics, and aquaculture, 55–56
apartheid, 415–16, 417
"Apartment" (McKibben), 477
apartments, compact, *152*, 153
apparel. *See* clothing
Apple computer, 338, 353, 374, 400, 406
appliances, 389–90
 and energy use, 48–50, 165, 166–67, 167, 514
 for Global South, 168–69, *169*
 and smart grid technology, 184–85

Appreciative Inquiry Handbook (Cooperrider/Whitney/Stravos), 433
Aprovecho Research Center, 168
Aptera (hybrid car), *78*, 78–79
aquaculture. *See* fishing industry
architecture, 249, 503–4
  biomorphic design, 105–6
  See also home design; skyscraper design
Architecture for Humanity, 327
architecture of participation (Internet), 419
Arctic Circle, 527–28
  See also polar regions
Aresa Biodetection, 219
Argentina, 444–45
Arizona, 496
Ars Electronica, 99
art, and protest, *452*, 453–55
art, and technology, *83*, *96*, 96–99, *97*, *98*, 343
  and biomorphic design, 103–4, *105*, 105–6
  and bio-utilization, 110
  "Elevated Wetlands," 196, *197*
  and living walls, *255*, 256
Art of the Impossible, The (Havel), 463
Art of the Start, The (Kawasaki), 406
Arup (consultancy firm), 131, 276
Asia, 20
  and bioengineered crops, 68–69, 70, 72
  and international trade, 41, 45, 119, *120–21*
  See also specific countries
asphalt, *238*, 240, 256
  alternatives to, 258
Asset-Based Community Development Institute, 344
asteroid monitoring, 534–35, *535*
asthma, and air pollution, 311
astronomy, and citizen science, *481*, 482, 483
Athens (Greece), *227*
Atlanta, 266, *266*, 269, *269*
"A-Z Living Units," 153, *153*
A to Z Textiles, 329
Auburn University, 503
audio books, online global, 374
Aurelle "candles" (Philips), *161*
Australia, 107, 191, 476
  coral reefs, and global warming, *523*
Austria, *105*, 105–6
Ausubel, Jesse, 397–98
automakers, and accountability, 118–19, 513
automobiles, 74–81, *89*, 244
  and air pollution, 74–75, 147, 232, 262–63, 274, 277, 277–78, 452, 473, 512, 515
  biofuels and synthetic fuels, for, 75, 76–77, 78, 80–81, 112
  car-share programs, 155, 237, 268, 270
  and congestion, 262–63, *263*, 274, 279
  Dymaxion (Fuller), 104
  electric-powered, 75, 78–79, 278, 391
  End-of-Life Vehicle Directive, 119
  flexible-fuel, 76
  fuel-cell, 75, *78*, 79–80, *80*, 277
  fuel-conservation tips, 76, 264–65
  hybrids, 75, 76, 78–79, 81, 270, 277–78, *391*, 391–92, 398
  hydrogen-powered, 75, 78, 79–80, *80*, 81, 108, 111, *112*, 277
  technology advances, in, 75, 76, 78–79, 108
  in urban design, 228, 230, 231–32, 241, *265*, 265–66
  Wuling SunShine, 277–78
Aveda (company), *380–81*, 383
Ayittey, George B. N., 440

B

backyard animal farming, 500
backyard biodiversity, 199–200
Bahrain, 422
bamboo (wood), 37, 143, *143*, 158, *158*, 159
Banco des Palmas (Brazil), 345
Bangladesh, 293, 347, 348, 352–53, 354, 356
Banker to the Poor (Yunus), 354
Bank of America Tower (New York), 248, *248*
Banks, Suzy, 196–97
Banksy (artist), *454*, 454–55
Barefoot College, *359*, 359–62, *361*
Barnett, Thomas P. M., 471
Barreto, Alexandre, 290–91
Baskin, Yvonne, 490–91
bathroom cleaners, 47–48
Battelle Pacific Northwest National Laboratory, 175
batteries, and biotechnology, 113–14
B100 biodiesel fuel, 77, 81
Beatley, Timothy, 229–30
beauty salons, and outreach programs, 321–22, 327
BedZED (zero energy development), 155–56, *156*
Beijing, 269, 270, 274, 277, 278, *410–11*, 412
Beijing Consensus, The (Ramo), 278
Belmar Urban Center (Denver), 242, *242*
Benetech, 445–46
Benfield, F. Kaid, 243
Benyus, Janine, 99, 100, 102, 110
Berelowitz, Lance, 234

Berger, Shoshana, 92
Berji Kristin (Tekin), 290
Berley, Peter, 52–53
Berlin
Reichstag, *246*, 246–47
Berry, Wendell, 70
BetterHumans.com, 108
Bey, Jurgen
designed garden chairs, *84*
Bezos, Jeff, 403
Bicitekas, 267
bicycles, *95*, 263, 266–67, 268, 278, *283*, 340
big-box stores, 32, 54, *238*, 239, 240, 242–43, *243*
Big & Green (Gissen), 249
Bill, MV, 290–91
bio-art, 99
bio-assistance, 110, 113
biodegradable
plastics, 110, 112, 113
products as, 38, 48
biodiesel fuel, 76–78, 80–81
Biodiesel Power (Estill), 80
biodiversity
backyard, 199–200
in ecological economics, 488, 490–91
and genetic science, 71–74, 482, *482*
preservation methods, 71–74, 491–93, *492*
bioengineered crops. *See* genetic science
biofilters, living walls as, *255*, 256
biofuels, 168, 223, 247, 333, 334, 364
automobile, 75, 76–77, 78, 80–81, 112
biogeography, 476
biological engineering. *See* genetic science; synthetic biology
biological nutrients, and recycling, 110–11
biomass
as energy source, 275, 360
and sustainable forestry, 497
biomechanics, 102
biomimicry, 99–102, *100*, *101*, 110, 144, 196, 282
in battery-making, 113
and collaborative design, 124–25
Biomimicry (Benyus), 99, 100, 102
biomorphism, 102–6, *103*, *105*, 110
in skyscraper design, 247, *247*
Biopaver (pavement), 194–95
Bio-PDO, 112–13
biopiracy, 334–35
bioplastics, 112–13
bioprospecting, deep-sea, 524
BioRegional Development Group, 156–57
BioRegional Solutions for Living on One Planet (Desai/Riddlestone), 156–57
bioregions, 156–57
defined, 477–78
bioremediation, 196, *197*, *250*, 250–51, *251*, 253, 257
biotechnology, 110, 335
industrial, 109–14
See also neobiological industry
Biothing (R&D lab), 103, *103*
biotoxins, 450
bio-utilization, 110
bird-counting (citizen science), *481*, 481–82
birds
and biodiversity preservation, 492
black globalization. *See* transnational crime
Blackspot (footwear brand), *394*, 394–95
Blair Watch Project, 449
Blender 3D, 123
blogs, Internet
as advocacy tool, 419, 420, 421–22
disaster-related, 214–15
political-related, 418–21, 421–23, 422, 423, *423*, 424–25, 425–26, 459
and software, 423
Blue Angel (eco-label), 166
blue hats (UN), *465*, 465–66, 469
Blue Hill Restaurant (New York), 59, *59*
Blueprint for Action (Barnett), 471
Boal, Augusto, 455
Bogotá (Colombia), *283*, 283–84
Bollywood, 375
BookCrossing.com, *373*, 373–74
books, online. *See* literature, online
Boontje, Tord
designed TranSglass, *83*
boreal ecosystem, 477, 488
See also forestry, sustainable; logging industry
Bornstein, David, 352–53
Bosnia, 20
Boston (CLIMB study), 509
Botswana, 19, 325
Bowling Alone (Putnam), 270
boycotts, 393, 394, *398*, 398–99, 414–15, 452, 462, 513
BP Global, 516
Braasch, Gary, *508*, 508–9
Brand, Stewart, 110
branding, product, 393–97, *394*, 405
manipulative, 396
Brand New Justice (Anholt), 395

brand recognition (activism), 457, 478
Braungart, Michael, 100, 101, 115, 117–18, 160, 195
Brazil, 286, 335, 337
bus rapid transit (Curitiba), *263*, 263–64
community development, 282, 345, 435, 455
Rio de Janeiro, 290–91
São Paulo, as megacity, 280, *280*
and South-South collaboration, 334, 335, *335*
telecentros (São Paulo), *300*, 300–301
Brazilian Microbe Bank, 335
bribery, and corruption, 434, 436, *437*
"bridge blogging," 422
British Columbia, 231–32, 233, 495, 505
Vancouver, urban design in, 194, 228, 231–34, *232*, 233, 263
The Brookings Institution, 239
Brower, Michael, 39
Brownell, Blaine, 88–89
"brownfields," restoring, 155, *250*, 250–53, *251*, *252*
Bryant, Shepard and Alberta, *503*
B612 Foundation, 535
Buchanan, Peter, 249
building materials. *See* materials, building
Burke Brise Soleil (roof), *104*, 105
Bury the Chains (Hochschild), 414–15
business, socially-responsible, 118–22, 379, *380–81*
and brand identity, 393–97, *394*
growth strategies for, 397–401
indicator measurements for, 382, 402–3
insurers, advocating, 384
and investments, 382–85, 385, 387, 400, 497 (*see also* microfinance)
marketing strategies for, 389–92, *390*, *391*, 400–401
and profitability, 379, 382, 383–84, *386*, 398, 400
and "shareholder activism," 383, 384
start-up tips for, 403–6
sustainability benefits, from, 382, *386*, 386–89, *387*, 391
sustainability steps, for, 387–88
"systems models" for, 402–3
and transparency, 383–84, 393, 396, 397, *398*, 398–400, 406, 425, 448
bus systems, 241, 242, 252, 268, 283
bus rapid transit, 242, *263*, 263–64, 283

C

Cabeça de Porco (Pig Head) (Barreto), 291
Café Slavia (Prague), 431
Calcutta, 280
California, 65, *73*, *184*, 269, 341, 348, *479*
CityPlace (Long Beach), 241
"conservation banks," *489*, 489–90
The Crossings, 240–41, *241*
emissions standards, 512
environmental monitoring, 479, 519–20
James Reserve, 520
Los Angeles, 194, 256, 459
and "place"-making, 476–80
solar farm, 174
sustainable forestry, in, 496
and urban design, 241, 341
See also San Francisco; University of California
California Open Source Textbook Project, 315
Callenbach, Ernest, 478
Cambodia, 299, 330
Camejo, Peter, 385
Canada, 166, 193, *255*, 505
automobile programs, 263, 270
"Elevated Wetlands" (Toronto), 196, *197*
Inuit people, 126, 528
LEED Gold standard, 232, 233, 248
net metering (surplus energy), 180
and sustainable forestry, 488, 495
Toronto, 196, *197*, *255*, 270
Whistler 2020 plan, 505
See also British Columbia
cancer, 47, 51, 110, 115, 158, 253
Capitalism at the Crossroads (Hart), 389
Caplow, Theodore, 505
carbon credits, 172, 364, 489, 497, 499, 516
carbon dioxide emissions, 41, 111–12, 512, 515–16, 532
and biotechnology, 80, 105, 113, 247
and "carbon offsets," 516
and global warming, 506–7
and Kyoto Treaty, 489, 497, 509, 511, 512, 516
and marine ecosystems, 523
monitoring technology, 98, 450
and polar regions, 526
and sustainable forestry, 494, 497
See also zero-energy
carbon footprints, 513, 514–16
"carbon offsets," 516
carbon sequestration, 499, 516
carbon trading, 489, 516
See also carbon credits
Caribbean, 214–15, 286, 290, 350
carpeting, *87*, *88*, 144
Carson, Rachel, 524

Cathedral and the Bazaar, The (Raymond), 130
Cats' Paws and Catapults (Vogel), 102
celebrities, as protesters, *456*, 456–57
cell phone technology, 84–85, 119
as activist tool, 443–44, 447, 448, 449, 450, 459
in leapfrogging process, *292*, 292–94, *294*, 296, 297, 302, 304, 334
and phone-based currency, 295
See also text messaging
Center for Democracy and Technology, 436
Center for Effective Philanthropy, 359
Center for Maximum Potential Building Systems, *191*, *194*
Center for Resource Solutions (San Francisco), 172
Center for Responsible Nanotechnology, 108
Centre for Economic Policy Research (London), 292
Centre for Indian Knowledge Systems (CIKS), 72
Centre on Housing Rights and Evictions, 290
certification programs, 55, 115
Cradle to Cradle (C2C) protocol, 115–16, 160
Energy Star, 48–50, 160, 166, 228
Euroflower eco-label, 166
Fair Trade in Tourism, 365
fish farming, 65
Green-e Renewable Energy, 172
hardwood, *142*, 143, 496–97
See also LEED certification
chairs
garden, *84*
office, 159–60
Chamoiseau, Patrick, 290
charities. *See* philanthropy
chemicals, and toxins, 158, 195
and biofilters, 250, 257
and cancer, 51, 110, 115
in carpeting, 144
in household cleaners, 47–50
in industrial farming, 51
volatile organic compounds (VOCs), 159, 160
Chen, Donald D. T., 243
Chernobyl, *492*, 493
Chery (automaker), 278
Chevrolet (automaker), 452
Chicago, *60*, *257*, 267
Zero Energy House, 147–48, *148*
Chicago Wildnerness project, 484–85
Chicxulub Crater (Yucatán Peninsula), 519
Child Help International, 354
Childline (India), 353–54
children, 307
AIDS orphans, 313, 325–26, *326*, 354
community-support programs, for, 310–13, *361*, 362
and educating girls, 316–21, *317*, *318–19*, 362
and education, in refugee camps, *221*, 221–22
and maternal health programs, 326–27, 357
nutrition education, 57–58, *58*, 60
Children's Parliament (Barefoot College), *361*, 362
Chile, 192, *193*, 462
chilled beams (radiant cooling), 248
China, 119, *271*, *274*, 315, *370–71*, 375, 403, *421*
automobile industry, 274, 277, 277–78
Beijing, 269, 270, 274, 277, 278, *410–11*, 413
and climate monitoring, 489, 510, 513, 533
and currency valuation, 41, 44
ecological footprint, 32, 33
and green building, 174, *271*, 271–78, *272–73*, *274*, *275–76*, *277*
gross domestic product (GDP), 274, 275, 278
and human rights, 29, *42–43*, 44, *410–11*
and international trade, *37*, 41–44, 45, 46, 174
public health programs, *325*
Shanghai, *272–73*, 276, 277, *277*
and South-South collaboration, 333, 334
and sprawl, 238, *239*
Chinadaily.com, 274
China-U.S. Center for Sustainable Development, 275–76
Chinese Housing Industry Association, 275–76
Chipotle Mexican Grill, 59–60
Chiras, Dan, 243–44
"choice fatigue," 33
cholera, 518
Christmas Bird Count, 481–82
circuit board production, 95, 136
CiteULike, 129
cities. *See* megacities; urban design
citizen journalism, 423
citizen media, 422–23, 424–25
citizen science, 473, 481–83, *482*, 508, 535
City Comforts (Sucher), 260–61
CityPlace (Long Beach), 241
civil rights movement, *413*, *451*, 456, 461, 462
civil wars, ending, 466, 471
Clarke, Arthur C., 534
cleaning products, 38, 47–48
clean tech investment, 384
Clean Water Act, 489
climate change, 17, *506*, 506–10, *508*
and citizen science, 482
computer models, for, 203, 479, 507

effect of, on planet, 203, 473, 513, 525–29, *526*, *527*
from global warming (*see* global warming)
government initiatives, to address, *511*, 511–13, *512*, 517
personal initiatives, to address, *514*, 514–17, *515*
satellite monitoring of, 533
and space exploration, 529–35, *530–31*, *532*, *535*
ClimatePrediction.net, 129
CLIMB study (Boston), 509
cloning, and biodiversity, 492
"closed-loop" design, and sustainability, 111, 117, 119, 245, 275–76
clothing industry, 36–38, *37*, 44, 46, 98, 112, 117
labor practices, 36, 41, 44–45, *96*, 97
coal (fossil fuel), *132*, *133*, 168, 172, 244
and pollution, 233, 274, 506
(*see also* greenhouse gases)
Coalition for a Livable Future (Portland), 235–36
Coalition for the Homeless, 353
coastal populations, and climate change, 485, 488, 509, 514–15, 527
coconut, 159, 329–30
Cod (Kurlansky), 66
coffee table (Syndecrete), *85*
collaboration, distributed, 128–29, 400
collaboration, South-South, *332*, 332–36, *335*
collaborative design, humanitarian, 216–17
collaborative product design, 123–26, 399–400, 406
for Global South, 131–34
open-source, *95*, 127–30, 133, 399–400
colleges, organic foods at, 58, *58*
Colman, Bruce, 70
Colombia, *283*, 283–84, 286, 445
Bogotá, *283*, 283–84
Colorado, 493
Denver, 241–42, *242*
Columbia University, 103
Commodify Your Dissent (Frank/Weiland), 396
communication
disaster-relief, 210–13, *212*, 214–15, 217, 222, *222*
See also telephones; Internet
community, 268, 342–46, 477–78
and capital, types of, 343, *343*
and "enabling philanthropy," 355–57
holistic approach to, 310–13, 339–41
"Imagine" movement, 433
increasing global awareness, of, 368–76, *369–75*
and microfinance, 346–51, *347*
"place"-making, in, *259*, 259–61, 265
product-service systems (PSS), *268*, 268–71, *269*
social entrepreneurship, in, 352–54, *353*, 403–6, *404*, *405*, 415
"Travelers' Philanthropy" movement, *363*, 363–67, *365*, *366*
"community asset map," 344
community capital (concept), *342*, 342–46, 351, 419
community gardens, 60, 61
community programs
for education, 310–13, 313–16, *314*, 316–22, *317*, *318–19*, 320, *321*, *322*, *359*, 359–62, *361*
for public health, 321–22, *322*, 323–31, *325*, *326*, *328*, *330*
community-supported agriculture (CSA), 54–55, 57
compact fluorescent lightbulbs (CFL), 23, 164, *390*, 390–91, *514*
compact spaces, living in, 152–54
composting, 50
in organic farming, 52, 195
and product design, *84*, 110–11, 136, 191–95
tent city, 209
toilets, 312–13
computers, 29, *134*, 136, 104, 302
and green design, 134–37
ICT4D movement, *296*, 296–300, *297*, *299*, *300*
laptops, 29, 298, *299*, 301, 302, 314–15
Linux (operating system), 127–28, 130, 298, *300*, 314, 399
and literacy programs, 298, *299*, 301, 314–15, 435
recycling, 136–37, 298, 299, 301
See also software
computing centers. *See* cybercafes; telecentros
Concordia Station (Antarctica), 528
concrete, *88*, 110
fly-ash, 233
light-colored, 258
Concrete Canvas tent, 209
congestion charges, automobile, 262–63, *263*
Congress for the New Urbanism (Web site), 237–38
"conservation banks," 489–90
"conservation bomb," and energy, 165, 183
conservation easements, 504–5
consumerism, 33, 81–83, 185, 398
and boycotts, 393, 394, *398*, 398–99, 414, 452, 513
and company branding, 393–97, *394*, 405
and company "greenwashing," 38–39, 389, 497
and company marketing, 389–92, *390*, *391*, 400–401
and product ecodesign, 83–86
for reducing climate change, 514, *514*
Consumer Project on Technology, 339

Consumer's Guide to Effective Environmental Choices, The (Brower/Leon), 39
consumption. *See* energy consumption; product consumption
container shipping, *40*, *44*, 45
content-management software, 420, 423
Conversation Cafés, 431–32
Cook + Fox Architects, 248
cooking technology, in Global South, 132, 168–69, 168–70, *169*, 222–23
cooling technology
air conditioning, 147, 166, 202, 234, 247, 256, 258
biomimetic methods, 282
chilled beams, 248
cool roofs, 258, *258*, 340
green facades, 148, 257, *257*
green roofs, 148, 190, 202, 231, 233, *233*, 256–57, 257, 270, 277, 340
light-colored concrete, 258
and solar panels, 282
ventilation, 202, 234, 245
cool roofs, 258, *258*, 340
Cooperrider, David L., 433
copyleft licensing, 338
copyright law, 336–39
coral reefs, and global warming, *523*, 523–24
The Core (Eden Project), 105
corn-based plastics, 112, 136
Cornell University, 113
Corpo Nove (Italian company), 88
Corporate Fallout Detector (CFD), 116, *116*
corporations. *See* business, socially-responsible
corruption, government, 20–21, 278, 374, *383*, 409, 434, 436, 463
Corruption Fighters' Toolkit (Transparency International), 439
Corruption Perception Index, *435*, 437
Costa Rica, 348, 350, 508
"cost of living"
in China, 44
and product consumption (U.S.), 35
Cotton Board, 36
cotton industry, 35–37, 46, 396, 503
Council on Foundations, 358–59
countertops, recycled-glass, 144, *144*
cows, and sustainable farming, 63, 64, 297
Crabgrass Frontier (Jackson), 243
Cradle to Cradle (McDonough/Braungart), 29, 100, 101, 115, 117–18, 160, 195, 275
"craft-it-yourself." *See* DIY products
Cranfield University (UK), 219
craters, mapping, 519, *519*
Creative Capital Foundation, 99
Creative Commons licensing, 217, 315, 337–38
credit loans (microcredit), 281, 293, 345, 352
credits, environmental. *See* carbon credits; ecosystem credits
crime, 441, 442, 468
See also bribery; corruption
Critical Mass (CM), *266*, 266–67
Cronon, William, 480
The Crossings (Calif.), 240–41, *241*
Crumlish, Christian, 430
cryptic ecological degradation, 485
Cuba, 41, 61, 403
culture jamming (protest), 393, 452
currencies, in economic development, 39–41, 44–45, 345, 385
Cute Circuits, 97, *97*
cybercafes, *296*, 297, 298, 299
See also telecentros
CyberTracker, 501–2
Czech Republic, 431, 435, 463

D

DakNet, 298
Daley-Harris, Sam, 351
Dallas, suburban-retrofitting in, 240
Dalsouple flooring tiles, *84*
D'Aluisio, Faith, 70
Dancing with the Tiger (Nattrass/Altomare), 388–89
Dangerous Place, A (Reisner), 479
Darfur (Sudan), 223–24, 533
Darwin's Ghost (Jones), 493
databases, 335, 492–93
Corporate Fallout Detector (CFD), 116, *116*
disaster-relief, 212–13, 214–15, 217
product material, 87–88
Datschefski, Edwin, 86
Dauncey, Guy, 517
Davis, Mike, 291
Dean, Howard, 419–21, *420*, 430
Death and Life of Great American Cities, The (Jacobs), 238
debt relief advocacy, 420
Declaração de Guerra (Bill), 290–91
deforestation, 132, *132*, *133*, 142, 158, 168, 500–501
See also logging industry

Delhi, 280
democracies, networked, 418–21
democracy, 20, 21, 409, 436, 446, 469
  corporate, 383–84
  and human rights, *410–11*, 415–16
Democratic National Convention (2004), *423*
Democratic Republic of the Congo, *204–5*, *470*, 470–71
Denmark, 175, 462
  Kalundborg industrial park (Denmark), 111
  KaosPilots (business school), 407
Denver, 493
  greyfield redevelopments, in, 241–42, *242*
Desai, Pooran, 156–57
Design and Environment (Lewis, Gertsakis, et al.), 89–90
Design for the Real World (Papanek), 126
Design Trust for Public Space (New York), 258
Detroit
  EarthWorks Urban Farm, 60–61
developers, land, 229, 232, 241, 252, 495
  and conservation easements, 504–5
  and ecosystem credits, 489–90
"developing-nations license," 217
Development as Freedom (Sen), 345
developments, housing. *See* suburbs, retrofitting; urban design
Dhaka, 280
diabetes, 71
Diary of a Political Idiot (Tesanovic), 464
DiCaprio, Leonardo, *391*
diesel-electric hybrid, 78–79
diesel fuel, 75, 77, 78–79
  and biodiesel, 76, 77, 80–81
Digital Bookmobile, 314, *314*
Digital Divide network, 215
digital product design, 123–24
Dimagi (company), 331
Direct Action Network, 458
direct action (protest), *456*, 456–59, *457*, *458*
disabilities
  and technology, *133*, 133–34, 148–49, *149*
Disaffected! (video game), *97*, 97–98
Disarmco, 219
disassembly, product design for, 119, 122, *122*, 136–37
disaster reconstruction, 210, 216–17, 220–23, 466, 479, 509
disaster relief
  communications tools, 210–13, *212*, 214–15, 217, 222, *222*, 533
  refugee camps, *207*, 207–10, 216–17, 220–23, 465
disasters, natural
  and coastal population, 485, 488, 509, 514–15
  See also specific type
Discovery of Global Warming, The (Weart), 517
diseases, 19–20, 51, 208, 325, 332, 482, 518
  avian flu, 324, 331
  cancer, 47, 51, 110, 115, 158, 253
  malaria, 279, 313, 324, 329–30, 332, 513
  and pandemics, 323–25, 518
  public health initiatives, 208, 311, 323–25, *325*, 327
  in refugee camps, 207–8
  related to climate change, 274, 511, 513, *526*
  SARS, 324, *325*, 363
  smallpox, 323, 324
Dish Stirling solar-power system, *172*, 174
distributed collaboration, 128–29, 400
DIY ("do-it-yourself") products, *90*, 90–92, *91*, *94*, *95*, 95–96, 271
  See also "engineer-it-yourself" products
D-Lab, 132
DNA, 100, 103, 445
  and biodiversity preservation, 492
  and biological engineering, 113
  and nanotechnology, 107
DNA computers, 114
Doctors Without Borders, 193, 207
Domestic Tradable Quotas (DTQs), 516
Dominguez, Joe, 34–35
"doomsday" vault (Norway), 72–73
DoubleSpace Kitchenette, 153–54
DoubleVision (camera), 448
Dove Ridge Conservation Bank, 490
downcycling, 118, 137
Dragon (mine-clearing machine), 219
Dream City (Berelowitz), 234
Drengson, Alan, 497
Drexler, Eric, 106
drip irrigation, 360, 499–500
Droog Design, *84*
drought, 20, 72, 186, *499*, 508, 515
  and organic farming, 552–53
Drupal (open-source system), 420
dry cleaning products, 50
dryers, clothes, 48–50
Duany, Andres, 244
Duchamp, Marcel, 92
Duisburg-Nord (Germany)
  brownfield restoration in, 251–52
Duke University, 149, *149*
DuPont, 112–13, 400

Durapalm (wood), 159
Dwell (magazine), 151–52
Dymaxion car (Fuller), 104

E

Earth First!, 458
"Earth Phone," 449–50
Earth Policy Institute, 313
earthquakes, 479, 483
  and blogs, 214–15
  Kashmir (2005), 211, *211*, 215, *427*, 466
  and prefab homes, 149, 216
  See also disaster reconstruction; disaster relief
EarthTrends Environmental Information Portal (Web site), 403
Earth Under Fire (Braasch), 509
EarthWorks Urban Farm (Detroit), 60–61
Eastgate development (Zimbabwe), 282, *282*
East Timor, 446
eBird, *481*, 481–82
ecoactivism, 457, 478, *484*, 484–85
  See also activism
Eco-Avenger Basic Training Camp, 458
ECO CIRCLE (recycling), 38
"eco-cities," described, 276
ecodesign, product, 81–86, *83*, *84*, *85*, 122–24
ecoDesign (Fuad-Luke), 86
Ecoforestry (Drengson/Taylor), 497
EcoHouse (Brazil), *201*, 201–2, *202*
eco-labeling systems, 166
  See also Energy Star label
ecological degradation, cryptic, 485
ecological economics, 343–44, 486–87, 488, 490, 495
"ecological footprint," 18, *32*, 33, 45, *514*
  calculating, 32–33, 515
  defined, 15–16
ecology
  and citizen science, 481–83, *482*
  ecosystem services, value of, 486–91, *487*, 497
  and forestry, 494–97, *495* (*see also* logging industry)
  grassroots approach to, 477–78
  industrial ("closed-loop" system), 111
  "place"-making, in nature, 476–80, *479*
  preserving biodiversity, in, 491–93
  restoration, *484*, 484–85, 494
economic capital (concept), 343–44, 351
economy
  effect of global warming, on (Oregon model), 507–8
  effect of international trade, on, 41, 363
  effect of malnutrition, on, 19
  and Kuznets curve, 19
  and leapfrogging technology, 294, *294*
  See also poverty
"ecoregions" (North America), 476, 477, 478
ecosystem credits, 364, 489–90, 497
Ecosystem Marketplace, 497
"ecosystem people," 477
ecosystems, MEA report on, 386, 486–88
ecosystem services, value of, 486–91, *487*
  and ecological economics, 486–87, 488, 490, 495, 497
  and ecosystem credits, 364, 489–90, 497
  and pollination, 487, *487*
Ecotopia (Callenbach), 478
ecotourism, *365*, 365–66, 501
Ecotrust, 146, 478
Ecotrust Forests, LLC, 497
Ecuador, 327
Eden Project (England), 105
Edible Estates, *198*, 198–99
Editt Tower (Singapore), *245*, 245–46
education, 310, 354, 478
  of AIDS orphans, 313, 325–26, *326*
  and Barefoot College, *359*, 359–62, *361*
  on climate change, 515
  community programs for, 310–13, 313–16, *314*, 316–22, *317*, *318–19*, *321*, *322*
  computers provided for, 301, 314–15, 435
  and effect on economy, 20
  Finnish school system, 311–12
  for girls, 316–21, *317*, *318–19*, 362
  and learning journeys (tourism), 365–66
  literacy programs, 297, 298, 301, 313–16
  night schools, 362
  online, 130, 313–16, *314*
  in refugee camps, *221*, 221–22
  and school feeding programs, 320
E85 fuel, 76–77
e-Government Handbook, 436
Egypt, 314, 315
Elbaek, Uffe, 406
elderly, technology for, 269, *269*
election campaigns. *See* political elections
electricity, 304–5
  access to, in Global South, 19, *167*, 167–68, 303
  and blackouts, *171*, 183

cost comparisons, 171, 172
energy sources, for (*see specific type*)
home energy measurements of, 166–67
smart grid technology for, 183–85, *184*, 210
Electrolux, 389–90
Electronic Frontier Foundation (EFF), 419
electronics, and vampire power, 165, 167
"Elevated Wetlands" (Harding), 196, *197*
Elgin, Duane, 34
Ella Baker Center (Oakland), 341
eMachineShop, 93
"embodied energy" (building construction), 145
Emerging Green Builders program, 150
Emerging Markets Handset program, 293
emissions. *See* greenhouse gases
emissions trading. *See* ecosystem credits
Empire (Ferguson), 412
Empire State Building (New York), *171*
employment, community programs for, 300, 310–11, 354
"empowered place" (community), 268
Endangered Species Act, 489
end-of-life options (products), 38
End of Life Vehicle Directive, 119
End of Poverty, The (Sachs), 345–46
End of Suburbia, The (Greene), 244
End of Nature, The (McKibben), 477
Enduring Seeds (Nabhan), 74
energy, nonrenewable. *See* fossil fuels
energy, renewable, 17, 64, 170–72, 268
carbon credits, to support, 364
and clean tech investment, 384
cost of, versus nonrenewable, 170–71, 172, 175, 244
in disaster-relief plans, 213–14
and environmental impact, 23, 171–72
in Global South strategy, 210
home systems for, *178*, 179–83, *180*, *182*
"intermittency" issue, 172, 177
main sources of, 179
and nanotechnology, 108
off-grid options, 172, *178*, 179–83, *180*, *182*
and smart grids, 183–85, *184*, 210
suppliers of (databases), 171, 183
See also solar energy; tidal power; wind power
energy audit, home, 165–66, 170
energy conservation, tips for, 164–67, *165*, *167*
energy consumption, 162, 165, 166–67
by appliances, 48–40, 165, 166–67, 184–85, 228, 514
by buildings, 147
in "closed-loop" system, 111
and environmental costs, 23, 162, *163*
and smart grids, 183–85, *184*, 210
and smart home technology, 148–49, *149*
"Energy Curtain" (window shade), 161–62
energy efficiency. *See* energy conservation, tips for
energy rebates, 187
energy-saving measures, and home value, 181
Energy Star label, 48–50, 160, 166, 228
"engineer-it-yourself" products, *93*, 93–96
See also DIY products
Engineers Without Borders, 358
Engines of Creation (Drexler), 106
English, as global language, *375*, 375–76
Enron scandal, *383*
Entering Space (Zubrin), 535
entrepreneurship, social, 352–54, 353, 403–6, *404*, *405*, 415
Environmental Career Organization, *482*
environmental monitoring. *See* monitoring, environmental
environmental refugees, 203, 206, *211*
"The Environment for Future Business" (Ausubel), 397–98
EQUITAS (forensic group), 445
Essential Agrarian Reader, The (Wirzba), 69–70
Estill, Lyle, 80
Estonia, *434*, 435–36
Estufa Justa stove, 168
Eternally Yours Foundation, 86
Eternally Yours (van Hinte), 86
ethanol fuel, 76–77, 364
Ethiopia, 445, *499*
Euroflower eco-label, 166
Europe
product recycling, 136–37
as urban design source, 229–30, 235, 236
See also European Union; *specific countries*
European Space Agency (ESA), 521, 528, *530–31*, 533
European Union, 320
carbon credit trading system, 489
economic size of, 41
End-of-Life Vehicle Directive, 119
and Euroflower eco-label, 166
toxic chemical initiatives, 115
See also Europe; *specific countries*
Evergreen State College (Wash.), *58*
evolution, species
and biomimetic design, 99, 100–101
exports, and Global South economy, 46
extinction, species, 71, 485–86, 489

and biodiversity, 491–93
and "conservation banks," *489*, 489–90
from global warming, 526, 527
public awareness of, 482, 492–93
Sixth Extinction, 16, 491
Exxon Mobil, *163*, 513

F

"fabbing" (quick fabrication), 123
"Fab Lab" (fabrication laboratory), 93, *93*
fabrics, 35–38, *36*, 111
facades, green, 257, *257*
Fahamu (nonprofit), 429–30
Fair Trade, 46, 55, 395, 414
farmers' markets, 53, *53*, 54, 55, 56, 57, 58, 60
farming, 51–52, 219, 488
dairy, 64, *64*, 500
and fair-trade movement, 55
industrial, 51, 52, 66, 68
livestock, 51, 63–64, 65–66, 456, 500
monocultural, 51, 52, 66, 68
prairielike methods, 67, 68
sustainable methods for, 52, 66–70, 261, *498*
urban, *60*, 60–62, 69, 199, *274*, *340*, 500
wind power systems, for, 175
farms, rooftop, *274*, *340*
Farms of Tomorrow Revisited (Groh/McFadden), 57
fast food, and nutrition, 51–52, 57–58, 59–60, 62
Fast Food Nation (Schlosser), 62
Fast Runner, The (film), *372*, 372–73
Favela Rising (Web site), 290
FCX hydrogen fuel-cell car, 75, *78*, 79–80
"Feral Robotic Dogs" (activist art), 454
Ferguson, Niall, 412
fertilizers, 17, 55–56, 66
organic, 52
fiber-optic cables, lighting with, 161
fibers, renewable, 36–37
Fibonacci sequence, and biomorphic architecture, 105
Fields of Plenty (Ableman), 56
Fifty Degrees Below (Robinson), 528–29
film industries, global, 368–73, *369–72*, 375
films, organizing activism around, 453
filters. *See* biofilters
Filtrón (water purifier), 193
Find Solar (Web site), 182–83
Finland, 97, 311–12
First Measured Century, The (Caplow/Hicks/Wattenberg), 505
First Mile Solutions, 298
fishing industry, *308–9*
fish farms, 63, *64*, 64–65, 69, *69*, 488
reintroduction programs, 251, *521*
and salmon, 69, 186, 478, 495, 520
and shellfish, 65, 69, 485, *489*, 490
fistula, obstetric, 357
Flex Your Power (Web site), 169–70
Flickr (Web site), 215
Flood.Firetree.net, 513
flooding, 217
and climate change, 203, 485, 488, 506, 507, 509, 513
and organic farming, 52–53
and permeable pavement, 190, 194–95, *196*
flood maps, 513
flooring, *84*, 142–44
floor wax, ecofriendly, 47
Florida, *479*
flour making, design for, 133
fly-ash concrete (building material), 233
Flying Electric Generator (windmill), 176
fog-catching (water collection), 192–93, *193*
FogQuest, 193
food, tips for purchasing, 53–56
See also fast food; organic foods
food banks, *58*
foodborne illness, 51
food co-ops, 54
Food Force (video game), *208*, 208–9
food production. *See* farming
Force More Powerful, A (ICNC), *462*, 462–63
Ford Motors, 77
The Foreign Policy Centre (UK), 278
forensics, and human rights, 444–45
forestry, sustainable, 494–97, 495
See also logging industry
Forest Stewardship Council (FSC), *142*, 143, 496–97
Fortune at the Bottom of the Pyramid, The (Prahalad), 299–300
Forty Signs of Rain (Robinson), 528–29
fossil fuels, 17, 19, 109
and cars, 74–75
"clean three" strategy, 164–65
conservation strategies, 164–66
cost of, versus renewable power, 171, 172
environmental effects from, 162–64, 506

and farming, 51, 52
and global warming, 506
(*see also* greenhouse gases)
"sustainable scale," for, 488
See also specific type
Foster, Norman, *245*, 246–47
Fountains of Paradise (Clarke), 534
Fraioli, James O., 66
France, 208, 528
Franceschini, Amy, 453, 454
Frank, Thomas, 396
Franklin, Jerry F., 497
Freecycle.org, 33
Free Geek (Portland), 299, *299*
Freeman, Michael, 154
Freeplay Lifeline Radio, 222
Freitag (bag company), 114, *115*
"The French Democracy" (film), 424
Fresh Food Fast (Berley), 52–53
Fritz Institute, 211
Front (design firm), *103*, 103–4
"F.R.U.I.T." (activist art), *453*, 454
F-T (Fischer-Tropsch) synthetic fuels, 77
Fuad-Luke, Alastair, 86
fuel, airplane, 364
fuel, automobile
biodiesel, 76, 77, 80–81
conservation tips for, 76, 264–65
diesel, 75, 76, 77, 78–79
electric, 75, 78–79, 278, 391
ethanol (E85), 76–77
fossil fuels, 74–75
F-T synthetic fuels, 77
gasoline, 75, 76
hydrogen, 75, 78, 79–80, *80*, 81, 108, 111, *112*, 277
solar, 76
SVO (straight vegetable oil), 77
Web site for, 81
fuel, cooking, 132, *132*, *133*
Fukuoka, Masanobu, 70
Fuller, Buckminster, 99, 102, 104, 247
Fuller, Donald A., 392
fungus, used for mycoremediation, 250–51, 253
FUPROVI (Foundation for Housing Promotion) (Costa Rica), 348, 350
furniture, ecofriendly, 84, *85*, *157*, 157–60, *158*, 496
convertible, 153–54
furniture polish, ecofriendly, 47
Future of Life, The (Wilson), 491
Future of Ideas, The (Lessig), 338

G

Gaiam (company), 182
games, environmental-related, 455
computer simulation, 209, *236*, 237
video games, *97*, 97–98, *208*, 208–9, *462*, 462–63
Gandhi, Mahatma, 344, *461*, 462
Gap (clothing manufacturer), 44, 46
Gapminder (Web site), 403
garbage disposals, and environment, 50
garden chairs, *84*
gardens, 482, 490, 500
adding compost to, 50, 209, 312–13
biodiversity, in, 60, 199–200
community, 60, 61
Edible Estates, *198*, 198–99
green roofs, as, 257
land mine-detection, 219
rain, 257
teaching children about, 57–58, *58*, 60, 199
urban, 270, 500
water-conserving, 189
garment trade (clothing). *See* clothing industry
gas, natural, 77, 80, 506
See also greenhouse gases
gasoline, 51, 75, 113, 506
with ethanol, 76
See also greenhouse gases
Gates Foundation, 326
gay rights, *416*
gecko tape, *100*, 101, 102
Geekcorps, *296*, *297*, 298
Geely Motors (automaker), 278
genetic algorithms, and biotechnology, 100, 103, *103*
genetic science, 55–56, 67–69, 73, 112, 113, 492, 524
and biodiversity, 71–74, 482, *482*, 492
genetic zoos, 492
gentrification. *See* urban revival
Genware (software), 103
Georgia, 266, *266*, 269, *269*
Geostationary Operational Environmental Satellite (GOES-12), *518*
geothermal heat pump, 149
Germany, 41, 166, 180
brownfield restoration, 251–52
End-of-Life Vehicle Directive, 118–19
rainwater harvesting, 191–92
Reichstag (Berlin), *246*, 246–47
Gershuny, Grace, 490
Gertsakis, John, 89–90

G8 (Group of Eight), *357*, 450, 458
Ghana, *93*, *132*, *133*
Giant African land snail, 500
Gibson, William, 295, 301, 396–97, 398
Gillmor, Dan, 423, 424–25
Gipe, Paul, 177
girls, educating, 316–21, *317*, *318–19*, 362
GIS (geographic information system) mapping, 211, 285, 518, 519, 520
Gissen, David, 249
giving, charitable. *See* philanthropy
glass, recycled, 48, *83*, 144, *144*
glass cleaners, 47
Global Call to Action Against Poverty (GCAP), 430
Global Coral Reef Alliance, 523–24
Global Crop Diversity Trust (GCDT), 72, 73
Global Development Research Center, 321
Global Exchange, 46
Global Footprint Network, 33
Global Fund for Women, 322
Global MapAid (NGO), 211
Global North
 "choice fatigue," in, 33
 manufacture outsourcing, 46
 and megacities, 280
 and product design, 131
 reliance on fossil fuels, 19, 162
 and trade, 41
Global Public Health Information Network (GPHIN), 324
Global Report on Human Settlements (UN), 291
Global South, 17–18, 19, 210, 217
 and corruption, 20–21, 434
 and disease (*see* diseases; public health)
 illiteracy rates, 20, 297
 and leapfrogging (*see* leapfrogging, in Global South)
 life expectancy rates, 41, 326–27
 megacities, in, 225, 280
 rural environments, and sustainability, *498*, 498–502, *499*
 and trade, 41, 44–45, 46, 54–55
Global System for Mobile (GSM), 214
Global Voices Online, 422
global warming
 and activism, 453
 and climate models, 507–8, 509–10
 corporate accountability, for, 396, 512–13
 effect of carbon dioxide, on, 494, 526
 and government initiatives, *511*, 511–13, *512*, 517
 health problems, from, 513, *526*
 linked to greenhouse gases, 506–7
 and marine ecosystems, 521, *522*, 523, *523*
 photographic record of, *508*, 508–9
 in polar regions, *506*, 525–29, *526*, 527
 See also climate change
"Global Warming: Early Warning Signs" map, 508
Globish (language), 375–76
Goa (India), 284–85, *285*
Goldman Sachs, 384
Gone Tomorrow: The Hidden Life of Garbage (Rogers), 35
Good House, A (Manning), 151
Google (search engine), 123, 265, 316, 373, 425, 520
Gore, Al, 440, 453, 510
government
 and climate-change strategies, 401, 509–10, 511–13, *512*, 517
 and corruption, 20–21, 278, 409, 434, 436, 463
 transparency, in (*see* transparency, in government)
 in urban planning, 234–35, 236–37
 See also political elections
Gowanus Canal (Brooklyn, New York), *252*, 252–53
GPS (Global Positioning System) technology, 264–65, 285, 303, 501
Grameen Bank, 347, 348, 351, 352–53, 354, 356
Grameen Phone, *292*, 293
grants, for microenterprise, 356–57, 358–59
grapefruit-seed extract, 48
grassroots advocacy, 453
 methods for, 260, 477–78
G/Rated Home Remodeling Guide (G/Rated), 146
"gray water," recycling, 186, 188, 201, 233, 248, 276
grazing, animal, 16, 63, 498
Great Bear Rainforest (British Columbia), 495
Great Britain, 168
 abolition movement, in (1800s), 412, 414–15
 car-share programs, 264–65, 270
 climate-change studies in, 508
 environmental monitoring, in, 454, 508
 London, 155–56, *156*, 247, *247*, 263, *263*, *269*, 269–70, 518
 See also United Kingdom
Great Plains Restoration Council, 64
Greece, *227*
Green Car Congress, 81
Green Clean (Hunter/Halpin), 50
Greene, Gregory, 244

Green-e Renewable Energy Certification Program, 172
green facades, 257, *257*
greenfield development, 243
GreenFuel (company), 112
Green Home Remodel (guide), 146
greenhouse gases, 147, 274, 400–401, 515, 523
  from farming, 51, 63
  and global warming, 506–7, 513, 528
  initiatives to reduce, 155, 166, 509, 511–12
  and oil company accountability, *512*, 513
  See also carbon dioxide emissions
Green Imperative, The (Papanek), 126
Greenland, 126, *506*, 525, 526, 528
"green" manure (fertilizer), 52
Green Marketing (Ottman), 392
Greenpeace, *515*
green products
  and design (*See* product design)
  identifying, 114–18
  and producer responsibility, 118–22
Green Remodeling (Johnston/Master), 145–46
green roofs, 148, 190, 231, 233, *233*, 256–57, 270, 277, 340
green tags. *See* renewable-energy certificates (RECs)
Green Urbanism (Beatley), 229–30
"greenwashing," 38–39, 389, 497
  and organic certification, 55
greyfield redevelopment, 240–43, *241*, *242*, *243*
GridPoint (company), 180
grids, energy
  off-grid options, 172, *178*, 179–83, *180*, *182*, 213, *213*, 248–49
  photovoltaic (PV) systems, 172–74, *173*
  and smart grids, 183–85, *184*, 210, 303
GridWise project, 185
Griffin, Peter, 215
Griffith, Nicola, 196
grocery stores, community, 54
Groh, Trauger, 57
Groningen (Dutch city), 230
Groove (computer network), 212
gross domestic product (GDP), and open-source concept, 128
Guantang Chuangye Sustainable Development project (China), 276
Guayaquil (Ecuador), 327
Guéhenno, Jean-Marie, *470*

H

Halpin, Mikki, 50
Handbook for Bloggers and Cyber-Dissidents, 425
Hargroves, Karlson, 401
Harlem Children's Zone, 310–11, *311*
Harrelson, Woody, *456*, 456–57
Harrison, Owen, 433–34
Hart, Stuart L., 389
Hasbrouck, Edward, 367
Havel, Vaclav, 463
Hawaii, *143*, 212–13
Hawthorne, Grace, 92
"Hay Bale" House, *503*
Hayes, Brian, 305
HBLEDs (high-brightness light-emitting diodes), 125
health, public. *See* public health
HealthStore (Kenya), 330–31
Healthy Money, Healthy Planet (Kent), 385
heating, water, 80, *80*, 150, 181, 202, 248
heating technology, 258
  forced-air solar, 180–81
  and green roofs, 233
  radiant, 147, 160, 172, *172*, *174*, 258
  solar-thermal, 172, *172*, *174*
  through passive solar design, 202
heat pump, geothermal, 149
Heifer International, 356, *356*
heirloom plants, 71, 74
hemp (fabric), 37–38
Henya stove, 168–69
heritage breeds, preserving, *73*, 73–74
Heritage Foods USA, 73
Herman Miller (furniture maker), 158, 159–60
Hertz, David
  designed coffee table, *85*
Hewlett-Packard (HP), 122, 136, 137
Hicks, Louis, 505
High Tide (Lynas), 517
Hinglish (language), 375–76
Hinte, Ed van, 86, 269
Hippo Water Roller, 192
hitchhiking (transportation), 264, 268, 269, 270
HIV/AIDS. *See* AIDS
Hochschild, Adam, 414–15
Holl House, 150, *151*
home design, *140–41*, 143, 147–52, 166, 276, *503*
  for compact spaces, 152–54
  and "embodied energy," 145
  and growth flexibility, 150, *151*

insulation, *165*, 166, 233, *503*
and landscaping, 147, *198*, 198–200
lighting, 160–62, *161*
and McMansions, 239
modular, 150, *151*, *152*, 152–54
open-source humanitarian, 216–17
passive solar, 147–48, 155, 201–2, 209
for "place"-making, 261
and prefab homes, 149–50, *150*, 152, 216
Rural Studio (design/build program), *503*, 503–4
and smart technology, 148–49, *149*
water fixtures, 187–88
windows, 155, 161–62
for zero energy consumption, 147–48, *148*, 155–56, *156*
See also remodeling; roofs; shelter design
home energy audit, 165–66, 170
Home Energy Station (HES), 80, *80*
home-energy systems, *178*, 179–83, *180*, *182*, 515
and smart grids, 183–85, *184*, 210
Homer-Dixon, Thomas, 336
home value, and energy-saving measures, 181
Hong Kong, 333, 425
Hope (McKibben), 477
hormones, in meat production, 55–56
Hospital in a Box, 214
hospitals, organic foods in, 58
Hospitals for a Healthy Environment, 58–59
Hotchkiss, Ralf, 133
hot water systems, 80, *80*, 181, 202, 248
H2Ouse (Web site), 189
Housing by People (Turner), 291
housing developments, green, 155–56, *156*
See also urban design
Houston, *211*, *251*
HOV (high-occupancy vehicle) lanes, 263, 264
How to Change the World (Bornstein), 352
Hoyt Street Properties, 252
Huangbaiyu (China), 275–76
Hub (London), *269*, 269–70
Hug Shirt, 97, *97*
Huichol people (Mexico), 125–26
human capital (concept), *343*, 343–44, 404
humanitarian design, open-source, 216–17, 336–39
See also disaster reconstruction; disaster relief
Humanitarian Logistics Software, 211
humanitarian satellites, disaster coordination by, 533
human rights, 316, 414–16, *416*, 435, 439, 441–47, *444*
and civil rights movement, *413*, *451*, 456, 461, 462
and forensic anthropology, *444*, 444–45
and Inuit lawsuit, 513, 528
and Liberia, 441
and torture, *442*, 442–43, 444
and tourism industry, 365
universal jurisdiction, and, 441–42, 469
and war crimes, 441–44, 469
watchdogs, and technology, 441, 445–46, 447–50, *448*
See also labor rights
human rights advocacy, 41, 425, *441*, 441–47, *442*, *444*, 453
Human Rights Data Analysis Group (HRDAG), 445–46
Human Rights Watch (HRW), 428, *441*, 446
Hungry Planet (Menzel/D'Aluisio), 70
Hunter, Linda Mason, 50
hurricanes, 211
Emily (2005), *530–31*
Katrina (2005), 203, 206, 210, *211*, 214, 215, 479, 506, 519
Wilma (2005), 211, *479*
See also disaster reconstruction; disaster relief
hybrid vehicles. *See* automobiles, hybrid
Hyderabad (India), *318–19*
hydrocarbon gas, fuels from, 77
hydrogen fuel, 75, 78, 79–80, *80*, 81, 108, 111, *112*, 278
hydrogen-producing algae, 111–12, *112*, 195
Hydrogen Economy, The (Rifkin), 80
hydrokinetic-power systems, 177, *177*, 179, 181–82, *182*, 244

I

IBM, 185
ice caps, and global warming, *506*, 513, 525, 527–28
ICT4D (development technologies), *296*, 296–300, *297*, *299*, *300*
Ideal Bite (Web site), 50
IKEA, *157*, 158
Illinois, *60*, *257*, 267
Chicago Wildnerness project, 484–85
Zero Energy House, 147–48, *148*
"Imagine" movement, 433
Immaculate Computers, 137
imports, and Global South economy, 41, 44–45, 54–55
Inconvenient Truth, An (Gore), 510
India, *167*, 168, *179*, 275, 375–76, 462, 476
children's helpline, 353–54
community development, 287–90, *288–89*, 291, *359*, 359–62, *361*, 434

DakNet, 298
education programs, 314, 315, *318–19*, 320, *359*, 359–62, 361
and international trade, 37, 46, 333
and microfinance, 347–48
Mumbai, 279, 280, 287–90, *288–89*, 291, 353–54
and public health programs, 324, 331, 513
seed-bank programs, 72
and South-South collaboration, 333, 334
traditional knowledge, preserving, 72, *308–9*
Udaipur as a Learning City (ULC), 344
urban planning (Goa), 284–85, *285*
womens' self-help groups, in, 321, *321*
indicators, business, 382, 402–3
Indonesia, 211, 280, 286, *317*, 333
"industrial ecology" (Kalundborg industrial park), 111
infill housing, in urban design, 229, 232, 241
infrastructure, urban
and disaster reconstruction, 210, 216–17, 466, 479, 509
effect of sprawl, on, 229
effect of war, on, 464–65
government role, in (U. S.), 236
greening methods for, 254–58, *255*, *257*, *258*, *259*, 340
and inner-ring suburbs, 238–39, 240, 339–41, 503
squatter communities, *286*, 286–91, *287*, *288–89*, 339
technologies for, in leapfrogging nations, *302*, 302–5, *304*, 333
Infrastructure (Hayes), 305
Ingenuity Gap, The (Homer-Dixon), 336
Inhabitat (blog), 160
inner-ring suburbs, 238–39, 240, 339–41, 503
Institute for OneWorld Health, 331
Instructables.com, 93, 95, *95*
insulation, home, *165*, 166, 233, *503*
Integrated Architecture, *140–41*, 143
intellectual-property protection, 217, 315, 339, 400
and open-source design, 217, 400
Interactive Institute (Sweden), 161
Inter-American Commission on Human Rights, 528
Interbrand, 395
Interface, Inc., 144, 388
Intergovernmental Panel on Climate Change, 507
International Campaign to Ban Landmines (ICBL), 218, 428–29
International Center for Agricultural Research in the Dry Areas (ICARDA), 72, *72*
International Center on Nonviolent Conflict (ICNC), 460, 462–63
International Cities for Local Environmental Initiatives/Cities for Climate Protection Campaign, 511
International Council for Local Environmental Initiatives, 510
International Ecotourism Society, 365
International Futures, 402
International Panel on Climate Change, 515
International Rescue Committee (IRC), 464–65
International Sustainable Urban Systems Design competition (Tokyo, 2003), 284
International System for Total Early Disease Detection (INSTEDD), 325
Internet, 482
advocacy-group networking, through, 409, 420, 421–22, 426–30, 452
in disaster relief, 212–14, 214–15, 217
forum facilitators, 418–19, 432
forum filters ("bozo filters"), 418
in Global South, 214, *296*, 296–300, *297*, *299*, *300*, 300–301, 304
ICT4D (development technologies), *296*, 296–300, *297*, *299*, *300*
networking politics, 418–21, *421*, 421–25, *423*, 431–34, 459
NGO networks, 426–28
and online education, 301, 313–16, *314*, 435
pandemic-tracking, through, 324–25
and smart mobs, 459
and tags (categorizations), 420
See also blogs
Internet Archive (library), 314
In the Bubble (Thackara), 83, 85–86, 268, 269
Inuit Circumpolar Council (ICC), 528
Inuit people, 126, *372*, 372–73, 513, 527–28
Inveneo (communications technology), 214, 299
investments
and socially-responsible business, 382–85, 387, 400, 497
"involuntary parks," *492*, 493
iPod, *90*, 91, *452*
Iran, 334
Iraq, 423, 442, *442*, *452*, 459, *466*, 467
"iRaq" posters (protest art), *452*
irradiation, food, 55–56
irrigation, 245
drip method, 360, 499–500
fog collection, for, 193
and permeable pavement, 194–95
rainwater-harvesting, for, 194–95, 201, 233, *233*, 245, 499
Super-MoneyMaker water pump, 192, *192*

Israel, *454*, *456*
Israeli-Palestinian School for Peace, *432*, 432–33, *433*
Istanbul (Turkey), 290, 291
Italy, 263, 269, 270, *519*
Ivory Coast, 46, *296*

J

Jackson, Kenneth T., 243
Jackson, Wes, 67, 70
Jacobs, Jane, 238
Jakarta (Indonesia), 280, *317*
Jamaica, 350
James Reserve (California), 520
Japan, 41, 70, 79–80, 180, 270, 333
 ECO CIRCLE (clothing reprocessing), 38
 Muji ("unbrand"), 394
 Tokyo, 225, 279, 280, 284
jeans, and Global South labor, 44, 46
Jégou, François, 268–69, 270–71
Jiko stove, 168, 169
Jo'burg Memo, 19, 274
Johnston, David, 145–46
Jones, Steve, 493
Jones, Van, 341
Jordan (Amman), 423–24

K

Kac, Eduardo, 99
Kafunda Learning Village (Zimbabwe), 312–13
Kahane, Adam, 417
Kalundborg industrial park (Denmark), 111
Kansas, *175*, 199
KaosPilot A–Z (Elbaek), 406
Karachi (Pakistan), 280
Kashmir earthquake (2005), 211, *211*, *427*, *466*
Kawasaki, Guy, 406
Keep Chickens! (Kilarski), 65–66
Kemp, William H., 182
Kent, Deidre, 385
Kentish Flats wind farm (UK), 175, *176*
Kenya, 286, *286*, *287*, 291, 372, 375, 510
 cooking technology, 168–69
 and corruption, *437*, 437–39
 leapfrogging technology, 292, *292*, 293–95, *294*
 and microfinance, 350
 Nairobi, *281*, 286, *286*, *287*, 291, 292, *328*, 427, *437*, 510
 public health programs, *328*, 330–31, *427*
 squatter communities, 286, *286*, *287*
Khmer Software Initiative, 299
KickStart (nonprofit organization), 192
Kilarski, Barbara, 65–66
Kill A Watt energy-measurement system, 166
King, Martin Luther, Jr., 341, *413*
Kinko's, and *Disaffected!* game, 97
Kinnisand (company), 86, *88*
Kirei board, *158*, 159
Kitchen, Angelica, 52
kitchen cleaners, 47
kitchens, modular, *153*, 153–54
Kiva (microfinance group), 348
Klamath Knot, The (Wallace), 479–80
knitPro (Web application), *96*, 97
Kodama, Sachiko, 89
Kolam Partnership (UK), 304
Kolkata (Calcutta), 280
Kosovo, 446
Kunsthaus Graz (Austria), *105*, 105–6
Kurlansky, Mark, 66
Kuznets curve, 18
Kyoto Treaty, 509
 and ecosystem credits, 489, 497, 516
 U.S. city initiatives, 511, 512

L

labor rights, 36, *42–43*, 44, 55, 114, 393, 394–95, *398*, *459*
Lagos (Nigeria), 225–26, *279*, 279–81, *287*
Lake Chad (Africa), 186
Lama Concept (company), *87*
land development
 governmental authority over, 234–35, 236–37
 See also developers, land
landfills, 48, 77
 and building material waste, 145, 150, 252, 269
 and product life cycle, 28–29
 See also recycling
Land Institute, 67, 70
land mines, *218*, 218–20, *428*, 428–29, 469
land pollution, 66–67
 and brownfield remediation, 155, *250*, 250–53, *251*, *252*
 See also farming, industrial
landscaping

biodiversity, in, 199–200
energy savings, from, 147–48, 256
and irrigation, 233, 245
and lawn design, *198*, 198–200
See also gardens
language trends, global, *375*, 375–76
Las Vegas, *236*
Latin America, 68–69, 193, 329–30
See also specific countries
laundry detergents, 47
lawn design, *198*, 198–200
leapfrogging, in Global South, 174, 277, 292–93, 498
cell phone technology, *292*, 292–94, *294*, 296, 297, 302, 304
and disaster relief plans, 209–10
and infrastructure technologies, *302*, 302–5, *304*, 333
and Internet availability, *296*, 296–300, *297*, *299*, *300*, 300–301, 302, 331
and South-South collaborations, *332*, 332–36, *335*
learning journeys (tourism), 365–66
LED (light-emitting diode), 125, 160–61, *161*, 162, *167*, 168, 210, 232, 303, 304, *304*
LEED (Leadership in Energy and Environmental Design) certification, 150, 231, 232, 233, *248*, 248–49, 252, 275, 496
Legislative Theatre (Boal), 455
Leon, Warren, 39
Lesonsky, Rieva, 406
Lesotho, 44–45, 46
Lessig, Lawrence, 338
Lewis, Helen, 89–90
Lewis and Clark College, 101
Liberia, 20, 441, 442
libraries, 232, 269
online, 128, 316, 335, 373
Traditional Knowledge, 335
LibraryThing.com, 373
LibriVox.org, 374
Lichtenberg, Alexandra, 201, *201*
LifeStraw, 208
LiftPort Group, 534
lighting systems, 160–62
access to, in Global South, *167*, 167–68
and albedo value, 258
energy-saving tips, for, 164, 165–66
Portable Light, *124*, *125*, 125–26
StarSight (streetlight system), 304, *304*
sunlight-transport systems, 89, *161*, 161–62, 247, 276
Light Up the World Foundation, *167*, 167–68
Limits To Growth (Meadows), 17
Lindenmayer, David B., 497
linoleum (flooring), 143–44
Linux (operating system), 127–28, 130, 298, *300*, 314, 399
Lipson, Elaine, 53
literacy programs, 297, 316
community programs for, 313–16, *314*
and Internet availabilility, 297, 298
One Laptop per Child project, 298, *299*, 314–15
See also education
literature, online global, *373*, 373–74, 404, *405*
Litracon transmaterial, *88*
liveCD for Disaster IT, 214
Live/Work (carsharing service), 270
Living Machines (water purification), 195–96, *196*
living walls (air purifiers), *255*, 256
loans, community business. *See* microfinance
LOCA: Location Oriented Critical Arts (monitoring network), 97
Local Flavors (Madison), 53
logging industry, 142, 158, 494
and ecological economics, 488
in Global South, 498, 500–501
and protests, 457
and sustainable forestry, 494–97, *495*
logo-free brands, *394*, 394–95
London, 510, 518
BedZED, 155–56, *156*
and congestion charges, 263, *263*
The Hub, *269*, 269–70
Swiss Re building, 247, *247*
London Array (wind power project), 175
long-term planning (business strategy), 397, 400
Los Angeles, 194, *195*, 256, 459
Lotusan (paint), 101, *101*, 102
Lovins, Amory B., 304, 469
low-flow water fixtures, 187–88
Lynas, Mark, 517
Lyocell (fabric), 37

M

Machinima movies, 424
Macy, Joanna, 517
Madison, Deborah, 53
Madrid train bombings (2004), 468, *468*
Magenn (windmill company), 176, *176*
MakePovertyHistory.ca, 430
makezine.com, 95

malaria, 279, 313, 329–30, 332, 513
Malaysia, 282, 333–34, *495*
Mali, 46, *297*
malnutrition, 19
  and Plumpy'nut, 208, *208*
Mama Mike's (remittance method), 350
Mandela, Nelson, 415, *416*, 417, 470
mangroves, in restoration ecology, *484*, 485, 488
manipulative branding, 396
Manji, Firoze, 429, 430
Manning, Richard, 51, 151
manure, as power source, 64, 293
manure, "green," 52
Manzini, Ezio, 268, 270–71
Mao Xianghui, *421*
MapAid, 211
mapping, 237, 303, 521
  environmental changes, 449, 501, 508, *518*, 518–20, *519*
mapping tools
  GIS, 211, 285, 518, 519, 520
  open-source, 237, 518, 519, 520
  satellites, *518*, 519–20, 521
marine ecosystems, *521*, 521–24, *522*, *523*
  and bioprospectors, 524
The Marine Stewardship Council (MSC), 65, 66
marketing, green, 389–92, *390*, *391*, 400–401
Marmoleum Click (flooring), 143
Marotte (company), *87*
Martin, Paul S., 486
Martus software, 445
Massachusetts, 197, *416*, 512
  Boston, 459, 509
Massachusetts Institute of Technology (MIT), 93, 95, 113, 124–25, 130, 131, 298, 314–15
  OpenCourseWare, 315
mass transit. *See* transportation
Master, Kim, 145–46
Material ConneXion (online resource), 88
Material Explorer (online resource), *87*, 88, *88*
materials, building, 37, *88*, 89, 145, *145*, 233, 269, 271
  See also remodeling
materials, product
  databases for, 87–88
  and sustainability, 87–90
Maynard, Andrew, 150
Mazza, Patrick, 517
MBDC (design firm), 111
Mbonise Cultural Concepts, *366*
McConnell, Brian, 180–81
McDonough, William, 110–11, 123, 275–76
  Cradle to Cradle, 29, 100, 101, 102, 115–16, 117–18, 160, 195, 275
McFadden, Steven, 57
McKibben, Bill, 477
McLachlan, Sarah, *357*, 357–58
Meadows, Dana, 17
meat, 63
  heritage breeds, preserving, 73–74
  organic, 55–56, 60
  See also farming, livestock
media, and activism, 415, 462
  and direct action (protest), *456*, 456–59, *457*, *458*
  and NGOs, 428
  and political censorship, 421, *421*, 423–24, 425
  and smart mobs, 459
  through Internet, 418, 423, 439
medicine
  and citizen science, 482
  See also public health
Meeting the Expectations of the Land (Jackson/Berry/Colman), 70
meetup.com, 419
megacities, 280
  growth in number of, 203, 225, 280, *280*
  innovations for, *282*, 282–86, *283*, *285*
  and slums, *286*, 286–91, *287*, *339*
  See also specific cities
Mega-Cities Project, 307
Menzel, Peter, 70
merino wool (fabric), *36*, 37
Metalworks, Inc., *387*
Metcalf Federal Building (Chicago), *173*
Method Home (cleaning products), 38
Metropolitics: The New Suburban Reality (Orfield), 244
Mexico
  Huichol people, 125–26
  Mexico City, 267, 280, *285*, 285–86
  Michoacán Butterfly Reserve, 500–501
  microfinance, in, *349*, 351
Michigan, 126, *387*
  Detroit, 60–61
  energy-efficient housing, *140–41*, 143
Michoacán Reforestation Project, 500–501
microfinance, 346–51, *347*
  and grants, 356–57, 358–59
  and microloans, 281, 293, 345, 346–48, 352–53, 354
  and remittances, *349*, 350–51

micro-hydro systems, 181–82, *182*
micromanufacturing. *See* nanotechnology
microsatellites, for environmental monitoring, 533
microturbines, 179–80, 213, *213*, 304–5
migration, and climate change, 511, 527–28
migration, human, 261, *271*, 350, 473, 476
and "rural rebound" (U.S.), 501, *502*, 502–3
See also refugees
Milan, 263, 269, 270
Milicevic, Myriel, 97
Milkweed Editions, 478
Millennium Development Goals (UN), 307, 316, 317, 351, 513
"Millennium Ecosystem Assessment (MEA) Synthesis Report," 386, 486–88
Milwaukee Art Museum (MAM), *104*, 105
Minewolf, *218*, 218–19
Minnesota, 243, 504
Mirra chair, 159–60
"mitigation banks," 490
mobile phones. *See* cell phone technology
Mobile Power Station (MPS), *213*, 213–14
Modal (fabric), 37
Moderate Resolution Imaging Spectroradiometer (MODIS), 520
modular design
for flooring, 144
for home, 149–50, *151*, *152*, 152–54
molecular manufacturing, 108–9
monetary system, and currencies, 39–41, 44–45, 295, 345
monitoring, environmental, 97
by citizen scientists, 450, 483
and mapping, 520
of oceans, 521–23
and space exploration, 532–33
through activist art, 454
through CyberTracker, 501–2
through sensors, 448, 450, 483, 519–20
monocultural farming, 51, 52, 66, 68
Mont Fleur Scenarios (South Africa), 415–16
Moodle (software), 315–16
Moore's Law, 168
More Tree Talk (Raphael), 497
"Movement as Network" (Rosenblatt), 420
movement building, *412*, 412–17, *413*, *416*
MoveOn.org, 440, 453
Muji (company), 394, 395
Multi-Fiber Arrangement (MFA), 44
Mumbai
Childline program, 353–54
as megacity, 279, 280
squatter communities, 287–90, *288–89*, 291
mushrooms, used for mycoremediation, 250–51, 253
music, global, *369*, *374*, 374–75
music creation system, 98, *98*
Myanmar, 158, *369*, 445
Mycelium Running (Stamets), 253
mycoremediation, 250–51, 253
Mystery of Capital, The (De Soto), 291

N

Nabhan, Gary, 74
Nairobi (Kenya), *281*, *292*, *328*, *427*, *437*, 510
squatter communities, 286, *286*, *287*, 291
Naked Corporation, The (Tapscott/Ticoll), 399
nanotechnology, *106*, 106–9, 114
and nanotubes, 107
NASA, 112, 506, 520, *530–31*, 533
spaceship antenna design, 100–101
Water Recovery System, 214
National Center for Family Literacy, 316
National Center for Policy Analysis, 502
National Oceanic and Atmospheric Administration, *518*
National Recycling Coalition, 48
National Renewable Energy Laboratory (NREL), 111, *112*, 177, 504
native (indigenous) plants
and biodiversity, 71–74
and biopiracy, 334–35
in medicine, 501
and urban gardening, 500
native peoples
and Great Bear Rainforest settlement, 495
Native American, 71, 74
polar-region, *527*, 527–28
Native Seeds/SEARCH (NS/S), 71
NATO, 469
Nattrass, Brian, 388–89
natural-assets accounting, 488
natural gas. *See* gas, natural
Natural Step (business framework), 388
Natural Advantage of Nations, The (Hargroves/Smith), 401
nature, human "place," in, 15–16, 476–80, *479*, *484*, 484–85, 491, 498
NatureWorks PLA, 112, 113
Navdanya (seed bank organization), 72

NEC (company), 136
Neighbourhood Satellites (Milicevic), 97
neobiological industry, 109–14, 335
  and hydrogen-producing algae, 111–12, *112*, 195
  RUrbanism (Goa), 284–85
Nepal, *167*, 168, 181–82, 192, 298
net metering (surplus energy), 180–81
"net primary productivity" (consumption), 16
NetRelief Kit, 214
Netroots advocacy, 453
network advocacy, 426–30, *427*, 428
networking politics, *423*, 459
  "long tail" effect, 427
  networked democracies, 418, 419, 420–21
  Reed's law, for, 419
  and social networks, 419
  tools for, *421*, 421–25, *423*, 431–34
  (*see also specific tools*)
Netzpolitik.org, 425
Neuwirth, Robert, 291
Nevada, *236*
New Alchemy Institute (Cape Cod), 197
New Hampshire, 512
New Rice for Africa (NERICA), *68*, 68–69
New World Order, 20
New York City, 260, 280, *448*, 448–49, *456*
  Bank of America Tower, 248, *248*
  energy consumption, 228–29
  environmental strategies, 258, 263, 494–95
  Gowanus Canal (Brooklyn), *252*, 252–53
  Harlem Children's Zone (HCZ), 310–11, *311*
  Peripheral City art series, 253
  Sustainable South Bronx (SSB), *340*, 340–41
  water supply, 477, 494–95
Nex-G (company), 304
The Next American City (Web site), 229
NGO-in-a-Box program, 427–28
NGOs (nongovernmental organizations), 168, 348, *398*, 399, 426–29, *427*, 465
  and boycotting, 414
  and disaster relief, 211, 212, 217, *427*
  emergence of, 414
  and intellectual property rights, 337, 339
  and networking, 426–29, 431
Nigeria
  and cooking technology, 169, *169*, 334
  Lagos, 225–26, *279*, 279–81, *287*
  Nollywood (film industry), 368, 372
  and war, 467, *467*, 468
Nigerian Dwarf goat (milk supplier), 500
Nike, 388, 393, 394–95
Nollywood (film industry), 368, 372
nonrenewable energy. *See* fossil fuels
nonviolent revolution, *460*, 460–64, *461*, *462*
North America
  consumption mentality, in, 34–35
  and product recycling, 119, 122
  species extinction, in, 485–86
  and toxic substance restrictions, 115
  See also Canada; United States
North Dakota, 175, 364
NorthSouthEastWest (photography project), 509
Northwest Environment Watch. *See* Sightline Institute
Norway, 72–73, *207*
Norwegian Refugee Council, *207*
nuclear plants, as "involuntary parks," *492*, 493
Nutriset SAS (company), 208
nutrition
  and malnutrition, 19, 208, *208*
  teaching children about, 57–58, *58*, 60–61

O

Oakland (Calif.), 269, 341
Ocean Friendly Cuisine (Fraioli), 66
oceans. *See* marine ecosystems
off-gassing, standards for, 115
off-grid operations, energy, 172, *178*, 179–83, *180*, *182*, 213, *213*
office chairs, 159–60
offices, shared, *269*, 269–70
off-site home fabrication, 150, 152
OhMyNews.com, 422
oil, 17, 19, 75, 488
  oil spills, 45, *163*, 251
  production peak, 74, 79
  See also greenhouse gases
Olympics
  China (2008), 276–77, 278
  London (2012), 156
  Vancouver (2010), 231, 233, *233*, 234
Olyset (mosquito net), 329
O'Naturals (restaurant), 60
Once There Were Greenfields (Benfield/Raimi/Chen), 243
One by One (giving network), *357*
100-Mile Diet, 54
One Laptop per Child (OLPC) project, 298, *299*, 314–15
ONE/Northwest (Online Networking for the

Environment), 420, 453
One Small Project (Web site), 290
One Straw Revolution, The (Fukuoka), 70
OneWorld International, 294
On Good Land (Ableman/Waters), 61–62
OpenCourseWare (MIT), 130, 315
open-source content-management systems, 420
open-source databases, 331, 332, 445–46
open-source education, 130, 313, 315–16
open-source humanitarian design, 216–17, 336–39
open-source mapping tools, 237, 518, 519, 520
open-source political campaign, 421
open-source software, 127–30, 133, 137, 399, 400
access to, in Global South, 299, *300*, 300–301, 302, 332, 337, 338
and citizen scientists, 483
and collaborative design, 127–30, 133, 399, 400
for mapping, 237, 518, 519, 520
NGO-in-a-Box, 428
in *telecentros*, 300, *300*
Translate.org, 299
and urban planning, 237
open-source style warfare, 467–68
Open Space Technology (Harrison), 433–34
opinion-sharing systems, 129
Opportunity International, 348, 356
"Orange County, China," 238, *239*
orange-rind plastic, 113
Orcelle ("green flagship"), 45
Oregon, 184, 512
climate change projections, 507
Portland (*see* Portland)
Oregon Health and Science University (OHSU), *248*, 248–49
Orfield, Myron, 235, 244
organic foods, 55, 72, 404, *404*
certification of, 55–56
and farming techniques, 52–53, 53–56, *64*, 66–70, 199, 261
fruits and vegetables, *52*, 52–53, 341
"greenwashing," 55–56
in institutions, 57–58, *58*
in restaurants, 59–60
teaching children about, 57, 57–58, *58*, 60–61
Organic Foods Sourcebook (Lipson), 53
Ottman, Jacquelyn A., 392
outreach programs. *See specific type*
ovens, solar, 169, *169*
Oxfam (charity), 499

P

P. A. Semi, 136
Pacific Northwest, climate projections for, 507–8
packaging, product, 48, 113
Paine, Thomas, 446–47
paint
Lotusan, 101, *101*
VOC-free, 160
Pakistan, 46, 280
ecological footprint, 18, 33
See also Kashmir earthquake (2005)
Palestine, 423, *454*
Israeli-Palestinian School for Peace, *432*, 432–33, *433*
palm oil industry, 333–34
Panamax ships, 45
pandemics, 323–25, 331, 518
Papanek, Victor, 124–25, 126, 269
paper mills, by-products from, 48
paper packaging, toxins in, 48
paper products, and environment, 48
Parans lighting system, 161, *161*
parks, landscape, *250*
in brownfield restoration, 251–52
farms, as, 261
"involuntary," *492*, 493
in urban design, 231, 252, 276
Parks, Rosa, *451*, 461
"Participatory Panopticon," 447
particleboard, toxins in, 158
passive energy consumption (vampire power), 165, 167, 514
passive solar home designs, 147–48, 155, 201–2
Patagonia (company), 38
patent law
and biopiracy, 334–35
and Global South development, 337
and "prior art," 335
PATH (public-health organization), 326–27
Pathways Out of Poverty (Daley-Harris), 351
Pattern Language, A (Alexander et al.), 261
Pattern Recognition (Gibson), 396–97
pavement, asphalt, *238*, 240, 256, 258
See also greyfield redevelopment
pavement, permeable, 190, 194–95, *196*
Pax Warrior (computer-simulation documentary), 471
pay-as-you-go plans, automobile, 262
PDO (polymer ingredient), 112–13
Peabody Trust, 156

peacekeepers, military, *465*, 465–66, 469, *470*, 471
"peak oil" projections, 74, 79
Pearl District Development (Portland), 252
pedestrians, and urban design, 241, 242, 252, 263, *265*
Pennsylvania, 267
Pentagon's New Map, The (Barnett), 471
People's Grocery (Oakland), 341
Peripheral City art series (New York), 253
permeable pavement, 190, 194–95, *196*
Persuasive Games, 97, 97–98
pesticides, 17
    and cotton, 36, 396
    and Global South development, 61, 499
    and industrial farming, 51, 66, 67
    and organic foods, 55–56, 61, 72
    synthetic, 17, 55–56
PETA (People for the Ethical Treatment of Animals), 456
petroleum, 400–401
    and derivatives, 47, 48
    and industrial farming, 51
    and plastics, 112
    substitutes for, 75
    waste generation from, 400–401
Pew Center on Global Climate Change, 274
pharmaceutical companies
    and copyright, 337
    and HealthStore (Kenya), 330–31
    Institute for OneWorld Health, 331
Phase-change Incubator, 132
Phenology Network (UK), 508
philanthropy, *355*, 355–59, *356*, *357*, 362
    and advocacy networks, 427
    giving circles, in, 357
    through carbon credits, 364
    in travel and tourism, *363*, 363–67, *365*, *366*
    and volunTourism, 366, *366*
Philips, and cell phone manufacture, 293
Philips, and light-bulb manufacture, 161, *390*, 390–91
photovoltaics, 105, 108, 172–74, *173*, 179, 181, 183, 527, *527*
    grid-connected, 172–74, *173*
    off-grid, 172
    Portable Light project, 125, 126
phytoplankton, and global warming, 523–24
Pinderhughes, Raquel, 230
Pittsburgh, 267
Placebrands, 395
"place"-making (concept), 259–61, 265, *265*, 476–80, *479*
Place of My Own, A (Pollan), 151
planet, 473–74
    "ecological footprint," on, 15–16, 18, 32–33
    environmental projections for, 17
    global approach, to sustainability, 254
    See also climate change
Planet Drum Foundation, 477–78
Planet of Slums (Davis), 291
plant oils, 47, 110
plants
    in green facades, 257, *257*
    in green roofs, 148, 190, 231, 233, *233*, 256–57, 270, 277, 340
    and living walls, *255*, 256
    migration of, and climate change, 511
    rain gardens, 257
    seed preservation, 71, 74
    smart breeding, 67–68
    used for bioremediation (*see* bioremediation)
    and water filtration, 257
    See also farming; native (indigenous) plants
plastics, 17
    biodegradable, 110, 112–13, 136
    and bioplastics, 112–13
    and recycling, 48, *49*
Plater-Zyberk, Elizabeth, 244
Pledge 25 Club (Botswana), 325
plumbing
    and Global South development, 303
Plumpy'nut, 208, *208*
Plyboo (plywood alternative), *158*, 159
Plywood Office (company), *158*, 159
polar bears, and global warming, 527–28
polar regions, 525–29, *526*, *527*
    and native peoples, 527–28
polio, 323–24, 325
political elections, 20, 293–94, 434, *434*, 436, 438, 449, 463
political movements, building, *412*, 412–17, *413*, *416*
political networking
    and advocacy networks, 426–30, *427*, *428*
    "encounter programs," *432*, 432–33
    "long tail" effect (finding allies), 427
    through film, 424
    through Internet, 418–27, *423*, 431–32, 459
    through radio, 423–24
politics, and activism, 409, *410–11*, 412–17, *421*, 421–25, *423*, *448*, 448–49, *460*, 460–64, *461*, *462*
politics, and human rights, *441*, 441–47, *442*, *444*, 447–50, *448*, 459
politics, transparency in, 21, 246–47, 294, 421, 430, 434, *434*, 434–40, 435, *435*, 436–39, *437*, *438*, 440

Pollan, Michael, 151
pollination (ecosystem service), 487, *487*
pollution. *See specific type*
population, world, 109
  and family-planning availability, 281
  growth in, 16, 109, 491
  and megacities, 225
  and migration, 350, 473
  and suburban sprawl, 238
Portable Light, *124*, *125*, 125–26
Portland, 142, *145*, 146, 234–38, *235*, 252, 299, *299*, 478, 512
Pot-in-Pot refrigerator, 169, *169*, 334
Potters for Peace, 193
poultry, and heritage breeds, *73*, 73–74
poverty, 307, 488, *499*
  community outreach programs, 222, 360, 503–4
  and corruption, 436
  and debt relief advocacy, 430
  and education rates, 20
  and microfinance, 347–51, 352–54, *353*
  and Millenium Development goals, 351
  and philanthropy, 300, 355–59, *356*
  related to hunger, 19
  and sustainability, 72, 498, 500–501
  world rates of, 18, 41
Power Integrations, 167
Power of Many, The (Crumlish), 430
power plants, and greenhouse-gas emissions, 175, 183, 274, 512, 513
Power Scorecard (Web site), 171–72
Practical Nomad, The (Hasbrouck), 367
Practice of the Wild, The (Snyder), 480
Prague, 431
Prahalad, C. K., 299–300
prairielike farming methods, 67, 68
Precautionary Principle, and nanotechnology, 107–8
PRé (Dutch company), 88
prefab homes, 149–50, *150*, 152, 216
Price, Dan, 95–96
Price of a Dream, The (Bornstein), 352–53
primary metals, waste generation from, 400–401
Prius (hybrid car), 79, 391–92, 398
Procter & Gamble, 38
produce, organic. *See* organic foods
produce imports, and Fair Trade, 55
producer responsibility (manufacturing), 118–22
product certification, 115–16
  See also certification programs
product consumption, *30–31*, *32*, 32–35, 35–39
  and ecological footprint, 15–16, 32–33
product design, 83–86
  and biomimicry, 99–102
  and biomorphism, 102–6
  and biotechnology, 109–14
  "closed-loop" system of, 111
  collaborative, 123–26, 127–30, 131–34, 399–400
  and consumerism, 81–83
  for development, 131–34
  for disassembly, 119, 122, *122*, 136–37
  and "do-it-yourself" (DIY) movement, *90*, 90–92, *91*, 95–96
  and "engineer-it-yourself" movement, 93–96, *94*, *95*
  for Global South, 131–34
  and "greenwashing," 38–39
  identifying green, 114–18
  and materials, 87–90
  and nanotechnology, 107–9
  and Precautionary Principle, 107–8
  and producer responsibility, 118–22
  for repair, 122
  software for, 88
  and sustainability, *83*, 83–86, *84*, *85*, 122–24
  3-D tools for, 123–24
  and transparency, 398
product life cycle
  "closed-loop," 111, 117, 119, 137
  end-of-life return policies, 38
product packaging, and sustainability, 48, 113
products, household
  toxins in, 47–48
product-service systems (PSS), *268*, 268–71, *269*
product sustainability software, 88
profitability, and sustainability, 379, 383–84, *386*, 398, 400
Project for Public Spaces, 252
Project Gutenberg, 129, 337
protest, 67–68, 450–55, *451*, *452*, *453*, *454*
  and antibranding, 393
  boycotts, 393, 394, *398*, 398–99, 414–15, 452, 462, 513
  culture jamming, 393, 452
  direct action, *456*, 456–59, *457*, *458*
  nonviolent, 456, 459, *460*, 460–64, *461*, *462*
  Ruckus Society, 458–59
  and smart mobs, 455, 459
protest art, *452*, 453–54
Protrude, Flow sculptures (Kodama/Takeno), 89
public health, 32, 58–59, 323–31, 340, 482
  community programs, 310, 323–31, *325*, *326*, *328*, *330*, 341, 361–62

HIV/AIDS prevention programs, 222, 313, 327–29, *328*
initiatives, 208, 311, 323–25, 325, 327
and toxic household products, 47–48
See also diseases
Public Library of Science, 128
pulp and paper industry, waste generation from, 400–401
Putnam, Robert D., 270
PWRficient processor, *134*

Q

Q-Drum (water container), 192, *192*
QinetiQ (company), 193

R

radiant heat/cooling systems, 160, 248, 258
Radical Simplicity (Price), 95–96
radio
in activism, 423–24, 459
in Global South development, 222, 296, 298
Radio Okapi, *470*, 470–71
Raimi, Matthew D., 243
rain forests, deforestation of, 142, 158
rain gardens, 257
Rainwater Collection for the Mechanically Challenged (Banks/Heinichen), 196–97
rainwater harvesting, 186, 190–92, *191*, *194*, *195*, 196–97, 201, 248, 499
at Barefoot College (India), 360, 361
for irrigation, 190, 499–500
and permeable pavement, 190, 194–95, *196*
through green roofs, 233, *233*, 340
Ramo, Joshua Cooper, 278
ranching. *See* farming, livestock
Raphael, Ray, 497
raw materials, and production, 46, 48, 110
and bio-utilization, 110
and nanotechnology, 107
Raymond, Eric, 130
ReadyMade (Berger/Hawthorne), 92
readymade (Duchamp), 92
Real Goods (store), 182
ReBuilding Center (Portland), 145, *145*
reciprocity, generalized (concept), 270
reclaimed ecosystems. *See* restoration ecology
reclaimed wood (flooring), 142–43
Reclaim the Streets movement (UK), 456
recycling, 35, 48, *84*, *158*, 158–59
building materials, 142, 145, *145*, 252
and closed loop system, 119, 137
clothing, 38
computers, 136–37, 298, 299, 301
ECO CIRCLE (technique), 38
End-of-Life Vehicle Directive, 119
energy used in, 48
and Freecycle.org, 33
glass, 48, *83*, 144, *144*
paper, 48, 158
plastic, 48, *49*
printers, 119, 122
product content certification, 115
sewage, 201, 245
of space junk, 534
and technical nutrients, 110–11
using embedded ID tags, for, 117
See also water recycling
Red Cross, 203, 211, *211*
Red Dive (artists' group), 99, 253
Reed's law (networking), 419
reforestation, 156, 500–501
See also forestry, sustainable
RE:FORM Studio (Sweden), 98, *98*, 161–62
refrigerators, *93*
and energy consumption, 50
Pot-in-Pot, 169, *169*, 334
Refugee Convention (1951), 206
refugees, 203, *206*
economic, 206, *206*
environmental, 203, 206
and refugee camps, *207*, 207–10, 212, 216, *220*, 220–23, *221*, *222*
See also disaster relief
reHOUSE/BATH, *188*, 188–89
Reichstag (Berlin), *246*, 246–47
reinsurance companies, 247, *247*, 384
Reisner, Marc, 479
ReliefWeb (United Nations), 218
religion, and civil war, 20
remittances, and microfinance, *349*, 350–51
remodeling, *142*, 142–46, 150, 248
EcoHouse (Brazil), *201*, 201–2, *202*
Renewable Devices (wind turbine maker), 179, *180*
renewable energy. *See* energy, renewable
renewable-energy certificates (RECs), 171
Renewable Energy Handbook, The (Kemp), 182
repair, designing for, 122

Reporters Without Borders, 425
Republican National Convention (2004), *448*, 448–49, 458
reputation systems, 129
ResearchChannel, 130
respiratory disease, and pollution, 144, 311
restaurants, *59*, 59–60
  and fast food, 51, 59–60
restoration ecology, *484*, 484–85, 494
"resurrection ecology," 486
Reticulars (curtain product), 103, *103*
reuse centers, building material, 144–45, *145*
revolution, nonviolent, *460*, 460–64, *461*, *462*
Revolution Will Not Be Televised, The (Trippi), 420–21
RFID (radio frequency identification), 117
Rheingold, Howard, 455, 459
rice, 70
  and biodiversity preservation, 72, 74
  genetic modification of, *68*, 68–69
Riddlestone, Sue, 156–57
ride-matching (carpooling), 264–65
Rifkin, Jeremy, 80
rights. *See* copyright law; intellectual-property protection; patent law
Rights of Man (Paine), 446–47
Rio de Janeiro, 290–91
Rio Grande river, 186
RISA (Regional Integrated Sciences and Assessments), 507
Rivoli, Pietra, 45–46
Robin, Vicki, 34–35, 431
Robinson, Kim Stanley, 528–29
robotics, 99, 454
  carbon-dioxide detection, 98
  and citizen scientists, 483
  for ocean monitoring, 521–22
  for space exploration, 532
Rocket Stove, 168, 169
Rocky Flats (Colorado nuclear plant), 493
Rocky Mountain Institute, 136, 209, 247, 304
Rodale Institute Farming Systems Trial, 52
Rogers, Heather, 35
roof design (Burke Brise Soleil), *104*, 105
roofs, cool, 258, *258*, 340
roofs, white, 258, *258*
roof shingles, 147, 179, 258, *258*
rooftop farms, 274, *340*
Rosenblatt, Gideon, 420
Roundabout Play Pump (water pump), 192
Royal College of Art (London), 188, 209
RSS technology, and networking, 420, 423, 430
Ruckus Society (protest group), *458*, 458–59
rural environments, and sustainability, *498*, 498–502, *499*, *502*, 502–5, *503*
  in Global South, *498*, 498–502, *499*
  in U.S., *502*, 502–5, *503*
"rural rebound" (U.S.), *502*, 502–5, *503*
Rural Studio (design/build program), *503*, 503–4
RUrbanism (Goa), 284–85, *285*
Russia, 41, 403, 467, 528
  See also Soviet Union
Rwanda, genocide in, 20, *204–5*, 446, 465, 470, 471

S

Sachs, Jeffrey D., 345–46
Sacramento delta, 479
Safaricom (electronic bank), 295
Safe and Sustainable World, A (Todd), 197
SafeClimate.net, 515–16
Safina, Carl, 66
salmon industry, 69, 186, 478, 495, 520
Salmon Nation program, 478
Salon.com, 265
Sambaza (phone-based currency), 295
Sandia National Laboratories, 100
San Francisco
  community outreach programs, 322
  Planet Drum Foundation, 477–78
  and protests, 456–57, 459
sangams (farming groups), 72
São Paulo, 280, *280*
  telecentros, in, *300*, 300–301
Sarajevo, 20
satellites, 97, 521, 534
  and asteroid monitoring, 534–35, *535*
  and climate monitoring, 533
  and Global South, 303, 334, 533
  and mapping, 303, *518*, 519–20
  used in disaster relief, 214, 533
Saudi Arabia, 468
Saunders, Ben, 525–27, *527*
Schlosser, Eric, 62
schools
  nutrition in, 57, 57–58, *58*
  See also education
school-stipend program (World Bank), 320
SciDevNet.com, 174
The Science and Development Network, 335–36

Scrapile (furniture line), *158*, 158–59
Sea Around Us, The (Carson), 524
seafood. *See* fishing industry
Seafood Watch, 65
sea levels
    and climate change projections, 506, 507, 509, 513
Seattle, 146, 267, 478, 507
    climate change strategy, 509, 511–12
seed banks, 71–73, 74
Seed Savers Exchange, 74
Sen, Amartya, 313, 345
Sen, Rinku, 417
Senegal, 500
9/11 terrorist attack, 467
Serbia, 463, 464
sewage management, 19, 50
    Living Machines, 195–96, *196*
    and recycling, 201, 245
Shadow Cities (Neuwirth), 291
Shanghai (China), *272–73*, 276, 277, *277*
Shaping Things (Sterling), 117
"shareholder activism," 383, 384
sharing, product. *See* product-service systems (PSS)
Sharp, Gene, 464
Sharp (roof tile manufacturer), 179
shellfish, 485, *489*, 490
    and seafood farming, 65, 69
    See also fishing industry
Shell oil company, 277
shelter design, 95–96, 104
    compostable tent city, 209
    and disaster-reconstruction, 216–17, 220–23
    field hospitals, 209, 214
    See also home design; remodeling
shelving, modular, 154
shingles, roof, 147, 179, 258, *258*
ship-breaking (scrapping ships), 45
shipping, container, *40*, *44*, 45
ships, Panamax, 45
short-message service (SMS). *See* text messaging
shower, water-saving, 188
Sierra Club, 228
Sierra Leone, *374*, 442
Sightline Institute, 231, 478
silver pot (water purifier), 193
SimaPro software, 88
SimCity (computer game), 209, *236*, 237
Simputer Trust, 298
Singapore, 304, 333
    Editt Tower, *245*, 245–46
Sixth Extinction, 16, 491
Siyathemba (AIDS outreach program), 327–28
SketchUp (3-D software), 123
SkyBuilt (MPS), 213, *213*
Skype technology (voice-over Internet system), 215
skyscraper design, *245–48*, 245–49
    LEED standards for, 247–48
Sky WindPower, 176
Slashdot.org, 418
Slow Food Movement, 51–52, 73
Slow River (Griffith), 196
slugging (carpooling), 264
slums, 286–91, *287*, *339*, 455
Slum/Shack Dwellers International, 290
Small is Profitable (Lovins et al.), 304
smallpox, 324
smart breeding, of plants, 67–68
smart grids, and energy, 183–85, *184*, 210, 303
smart home technology, 148–49, *149*
SmartHouse project (Duke Univ.), 149, *149*
smart mobs, 519, 535
    and protest, 455, 459
Smart Mobs (Rheingold), 455
Smart Power (Kemp), 182
Smillie, Joe, 490
Smith, Amy, 131–33
Smith, Michael H., 401
SMS messaging. *See* text messaging
snails, and backyard animal farming, 500
Snyder, Gary, 71, 480
social capital (concept), 343–44
social entrepreneurship, 352–54, *353*, 403–6, *404*, *405*, 415
Socially Responsible Investment (SRI) movement, 382–83, 385, 387, 400
social networking, 418, 419, 427, 449
social programs. *See* community programs
social software movement, 419, 427, 430, 431
"social tagging" (Internet), 420
soft path method (water resource management), 190–92, 194–95, *196*
software
    content-management, 420, 423
    copyleft licensing, of, 338
    and copyright, 337
    open-source (*see* open-source software)
    3-D, 123–24
    See also specific applications
soil erosion, 67, 203
soil management, for sustainability, 52–53, 195, 490, 499

soil pollution, 250
and industrial farming, 66–67
See also bioremediation
solar cells, 108, 174
solar energy
and albedo value, 258
in bioremediation, 196
as renewable resource, 17, 172–74, 527
solar energy systems, 172
"active," 147
for airplanes, 364
and Barefoot College (India), 360–61, 362
and China economy, 174
environmental impact of, 171
global markets for, 174
for home use, 147, 155, 179–81, *180*, 182–83, 202, *202*, 282
for institutional use, 248–49
and "intermittency" issue, 172
in leapfrogging infrastructure, 303
and LEDs (light-emitting diodes), 125, 168
and "net metering," 180–81
and Parans lighting system, 161, *161*
"passive," 147–48, 155, 201–2
and photovoltaics, 172–74, *173*, 179
solar-thermal, 172, *172*, 174
StarSight (streetlight system), 304, *304*
types of, 172
used in disaster-relief plans, 210, *213*, 213–14
water purification, 303
See also photovoltaics
Solar Energy Systems Infrastructure, 213
solar heaters, forced-air, 180–81
Solar Living Source Book (Schaeffer), 182
solar ovens, 169, *169*
solar-powered appliances, 180–81, 193, 222, 390
solar system. *See* space exploration
Solving Tough Problems (Kahane), 417
Song for the Blue Ocean (Safina), 66
"Sonic City" (music system), 98, *98*
Soto, Hernando de, 291
Soul of Soil, The (Gershuny/Smillie), 490
sousveillance, and watchdogs, *448*, 448–49, 459
South Africa, 44, 192, 299, 335, 354, 434
and AIDS, *326*, 327, 329, 330, 337
and apartheid, 415–16, 417, 444, 462, 470
South-South collaboration, 334
tourism, 365, 366
Truth and Reconciliation Commission (TRC), 444, 460, 470, *470*
South America, 333
See also specific countries
South-East Asia Earthquake and Tsunami Blog (SEA-EAT blog), 214–15
Southern California Edison (utility), 174
South Korea, 333, *398*, 422, 436, 493
South-South collaboration (Global South), *332*, 332–36, *335*
Soviet Union, 41, 61, 73
See also Russia
Soweto Mountain of Hope (SoMoHo) (South Africa), *342*, 342–43
soybeans
biofuels, from, 76
organic-farming techniques, 52
space elevator, Earth-to-orbit, 533–34
space exploration, 358, *529*, 529–35, *530–31*, *532*, *535*
and Earth monitoring, 521, 532–33
Space (Freeman), 154
space junk, recycling, 534
"space renaissance," 533
spaceship antennae, and biomimicry, 100–101
Spaceship Earth (Fuller), 104
SpaceshipRadio.com, 374
Spain, 468, *468*
Speck, Jeff, 244
SpokePOV (electronics kit), *95*
sprawl, 228–30, 231, *239*, 243, 252
alternatives to, 244
government control over (Portland), 234
and government subsidies, 244
and sustainability, 250
and transportation, 239, 266, *266*
and urban design, 261
Sprout Design (ecodesign studio), 122
Spurlock, Morgan, 62
squatter communities, *286*, 286–91, *287*, *288–89*, 339
Squid Labs, 94
SRI Advantage, The (Camejo), 385
Sri Lanka, *167*, 168, 214–15, 335
Stamets, Paul, 250–51, 253
Starbucks, 388, 393
StarSight (streetlight system), 304, *304*
start-ups, business, 403–6, *404*, *405*
Start Your Own Business (Lesonsky), 406
state-on-nonstate warfare, *466*, 466–68
steel, 100, 107
recycled, 48
Sterling, Bruce, 20, 84, 117, 262, 493, 514
Stir It Up (Sen), 417

Stirling solar-power system, *172*, 174
Stop Global Warming (Web site), 510
Stories from Where We Live (Milkweed Ed.), 478
Stormy Weather (Dauncey/Mazza), 517
stoves
  and air pollution, 223
  See also cooking technology, in Global South
Straphangers Campaign, 260, *260*
Stravos, Jacqueline M., 433
Streetcar (carsharing service), 270
street design, 241, 242
  for "place"-making, *265*, 265–66
streetlight system (StarSight), 304, *304*
Strong Angel (hands-on lab), *212*, 212–13
subsidies, government, 232
  for cotton, 46
  for energy, 172
  for sprawl, 244
Suburban Nation (Duany/Plater-Zyberk/Speck), 244
suburbs, 228, 229, 261
  inner-ring, 238–39, 240, 339–41, 503
  newer, versus inner-ring, 238–39
  See also sprawl
suburbs, retrofitting, 238–44, 248
  and big-box store reuse, 242–43, *243*
  greyfield redevelopment, 240–43, *241*, *242*, *243*
  inner-ring, 238–39, 240, 339–41
  and transportation issues, 239, 241, 242, 243, 244
subway systems, 260, *260*, *285*, 285–86
Success of Open Source, The (Weber), 130
Sucher, David, 260–61
Sudan, 20, *465*
sugarcane, cooking fuel from, 132
"Sunlight Table," 162
sunlight-transport systems (lighting), 89, *161*, 161–62, 247, 276
Sun Microsystems, 136, 137
SunPower (roof tile manufacturer), 179
Sunset Breezehouse (prefab home), 151
Sunstein, Cass R., 446
Superbia! (Chiras/Wann), 243–44
Superfund waste sites, 454
Super-MoneyMaker water pump, 192, *192*
SuperPlus (supermarket chain), 350
Super Size Me (Spurlock), 62
superstores. *See* big-box stores
sustainability, 307
  and company profitability, 379, 382, 383–84, *386*, 398, 400
  cost benefits of, 340
  and country stability, 464–66, 468–69
  shifts in focus, of, 529
  time-money transition costs, of, 285
  and wealth (Kuznets curve), 19, 22
  See also specific topics
Sustainable Everyday (Manzini/Jégou), 270–71
Sustainable Forestry Initiative (SFI), 497
Sustainable Marketing (Fuller), 392
SustainablePortland.org, 512
"sustainable scale," in ecological economics, 488
Sustainable South Bronx (SSB), 340, *340*
Sustainable Tourism Stewardship Council, 365
Sustainable Travel International, 516
SVO (straight vegetable oil) fuel, 77
sweatshops, and human rights, 36, 41, 44–45, *96*, 97
Sweden, 98, *98*, 161, 329, 388, 390, 469
  Front (design firm), *103*, 103–4
  Gapminder (Web site), 403
  RE:FORM (research studio), 98, *98*, 161–62
Swift windmill, 179
Swiss Re (reinsurer), 247, *247*, 384
Switzerland, 180
  chocolate industry, 46
  Freitag (bag company), 114, *115*
Syndecrete (composite), *85*
synthetic biology, 113
synthetic fuels, automobile, 75, 76–77, 78, 80–81, 112
synthetic pesticides, 17, 55–56
systems disruption, and war, 467–68
"systems models," for business, 402–3

T

Tactical Technology Collective, 428
tagging systems (reputation systems), 129
tags, categorization (Internet), 420
tags, green. *See* renewable-energy certificates (RECs)
Taiwan, 333, 489
take-back laws, product, 122, 136–37
Takeno, Minako, 89
Tangye New Town (China), 275–76
"Tank Man" (China), *410–11*
Tanzania, 329, 375, 474–75
Tapscott, Don, 399
Tate & Lyle, 112
Tateni Home Care Services, 354
tax, parking, 263
tax credits, 320, 348, 490
  energy, 170, 172

Taylor, Charles, 441, *441*, 442
Taylor, Duncan, 497
teak wood, and deforestation, 158
technical nutrients, and recycling, 110–11
Technorati.com (Web site), 373, 420, 425
Teijin (Japanese company), 38
Tekin, Latife, 290
telecentros, 222, *300*, 300–301
telephone-based currency, 295
telephones, 527
  "Earth Phone," 449–50
  Grameen Phone, *292*, 293
  VoIP (voice over Internet protocol), 215, 299
  See also cell phone technology; text messaging
Telepresence and Bio Art (Kac), 99
television, 447
  and Global South, 334, *370–71*, 373
  and power consumption, 165, 167
Tennessee, *386*, *486*
Ten Shades of Green (Buchanan), 249
tent city, compostable, 209
terrorist attacks, 443, 467, 468
  See also state-on-nonstate warfare
Tesanovic, Jasmina, 464
Tetris Shelves, 154
Texaco (Chamoiseau), 290
Texas, 74, 107, *191*, 197, *211*, 240, *251*
Texas Instruments, 136
Texxi taxis (England), 264–65
textbooks, open-source, 315
textiles. *See* fabrics
text messaging, 215, 264, 269, 294, 325
  and smart mobs, 459
  used for advocacy, 429–30, 459
Thackara, John, 83, 85–86, 268, 269
Thailand, 214–15, 315, 333, 467, *515*
Theatre of the Oppressed (Brazil), 455
thermal mass (passive solar technique), 147
"thermosiphon" water system, 202
ThinkCycle, 124–25
"Third World" cities
  in Global North, 307
  See also Global South
3-D software, and product design, 123, 128, 424
3-D tools, for green design, 123–124
Ticoll, David, 399
tidal power (renewable energy source), 177, *177*, 179, 244
Tidepool.org, 478
tiles, recycled-glass, 144
Tilonia, 362
time factor, and sustainability, 285
Todd, Nancy Jack, 197
Toginon (Japanese company), 122, *122*
Tohono O'odham Nation, 71
toilet bowl cleaners, 47–48
toilets
  closed-loop system, 245
  composting, 233, 312–13
  latrine design (VIP), 131
  water-saving, 187–88
Tokyo, 225, 279, 280, 284
Tomorrow Now (Sterling), 20
tool-sharing programs, 269, *269*
Toronto, 196, *197*, *255*, 270
  bioremediation, in, 196, *197*
torture, and human rights, *442*, 442–43, 444
Total Beauty of Sustainable Products, The (Datschefski), 86
tourism. *See* travel and tourism
Towards Forest Sustainability (Lindenmayer/Franklin), 497
toxic chemicals, 400–401
  and biofilters, 250, *255*, 256, 257
  bioremediation, for, 196, *197*, 250, *250*, *251*
  "body burden," from, 47, 51
  and certification standards, 115
  in household products, 47–48
  mycoremediation, for, 250–51, 253
Toyota, 277
  Prius (hybrid car), 79, 391–92, 398
trade, global
  container shipping, 45
  copyright law, 337
  and currency valuation, 41, 44
  and ecosystem credits, 489
  establishment of, post-World War II, 39–41, *40*, *44*
  Fair Trade, 46, 55, 395, 414
  garment trade, 41, 44–45, 46, *96*, 97
  transnational crime, 468
  travel and tourism industry, 363
  See also World Trade Organization (WTO)
trade, local
  alternative currencies for, 345
  radio-for-guns (Nigeria), 222
traditional knowledge, preserving, 72, *308–9*, 335, 360, 501
"traffic calming," 265–66
Traficando Informação (Bill), 290–91
Trans-Alaska Oil Pipeline, 75, 76

Trans-Fair USA (certification), 55
transgenic genetically modified (GM) crops. *See* genetic science
TranSglass, *83*
"transitional" organic produce, 56
transit-oriented development, 236, 239, 241, 242, 252, 262–67, *263*, *265*, 266, *266*, 276, 282, 283, 340
Translate.org, 299
Transmaterial (Brownell), 88–89
transnational crime, 468
transparency, 185, 445
in business, 115, 166, 383–84, 393, 396, 397, *398*, 398–400, 406, 425, 448
in government, 246–47, 294, 409, *434*, 434–40, *435*, *437*, *438*
and "greenwashing," 38–39, 389, 497
in microfinance, 348
in politics, 21, 246–47, 294, 421, 430, 434, 435, 436–39, 440
Transparency International (TI), 21, *435*, 437, 439
transportation, 231–32
automobiles (*See* automobiles)
bicycles, 266–67, 268, 278, 283, 340
and brownfield development, 252
bus rapid transit, 242, *263*, 263–64, 283
commuter rail, 241, 242, 252, 263
pedestrian-oriented, 241, 242, 252, *265*
and product-service systems (PSS), 268
subways, 260, *260*, *285*, 285–86
and urban design (*see* urban design, transit-oriented)
Transport for London (Web site), 263
trash
production approach to, versus consumption, 35
See also landfills; recycling
travel and tourism, *363*, 363–67, *365*, *366*, 485
and "carbon offsets," 516
ecotourism, *365*, 365–66, 501
learning journeys, 365–66
volunTourism, 366, *366*
Whistler 2020 plan (for managing tourism), 505
Travels of a T-Shirt in the Global Economy, The (Rivoli), 45–46
TreePeople (organization), 194
trees, in passive solar design, 194, 256
T.R.E.E.S. (water collection project), 194
Trees for a Green LA program, 256
Trickle Up (grants program), 356–57
Trippi, Joe, 420–21
Tropical Depression Alberto, *518*
tropical hardwoods, and deforestation, 142, 158
Trust Banks, 356
Truth and Reconciliation Commission (South Africa), 444, 460, 470, *470*
tsunami, Indian Ocean (2004), 211, 212–13, 214–15, 222, 485, 512, 533
Tunisia, 314
Turkey, 286, 290, 291
girls' education programs, 320–21
turkeys, and heritage breeds, *73*, 73–74
Turner, John F. C., 291
Tutu, Desmond, 460, *460*, 470
Twilight of the Mammoths (Martin), 486
Two-Mile Time Machine, The (Alley), 528

U

ubuntu (Zulu philosophy), 128, 470
Udaipur as a Learning City (ULC), 344
Uganda, 132, *186*, 350, 372, 375
Digital Bookmobile, 314, *314*
and Grameen Phone, *292*, 293
Inveneo system, 299
solar ovens, 169, *169*
Ukraine
Chernobyl, *492*, 493
Orange Revolution, *462*, 463
Uncommon Ground (Cronon), 480
Under Ground (Baskin), 490–91
unemployment, 307
See also employment, community programs for
UNESCO, 124, 317
UNICEF, 320
Unilever, 38
United Kingdom, 193, 200, 219, 304, 516
and carbon "sequestration," 516
and China trade, 41
compact-apartment design, *152*, 153
Domestic Tradable Quotas, 516
and political activism, 449
Reclaim the Streets movement, 456
wind energy systems, 175, *176*, 179
See also Great Britain
United Nations, 64–65, 363, 499, 525
Campaign to End Fistula, 357
Food Force (video game), *208*, 208–9
Global Report on Human Settlements, 291
Human Settlements Programme (UN-HABITAT), 279
on land mines, 218, 219–20

Millennium Development Goals, 307, 316, 317, 351, 513
"Millennium Ecosystem Assessment (MEA) Synthesis Report," 386, 486–88
peacekeepers, *465*, 465–66, 469, *470*, 471
War Crimes Tribunals, 442, 444
WIPO, 315, 419, 501
United Nations University, 203
United States, 478
ecological footprint, 18, 33
and international trade, 41, 44–45, 46
and terrorist attacks, 467, 468
top waste-producing industries, 400–401
and waste exports, 45
See also specific states
universal jurisdiction, and human rights, 441–42, 469
University of British Columbia, 233
University of California (Berkeley), 58, 101, 111, 168, 383, 430
University of California (Davis), 482
University of California (Santa Barbara), 324–25
University of Essex (England), 499
University of Guelph-Humber (Toronto), *255*
University of Michigan, 126
University of North Dakota, 364
University of Syiah Kuala (Indonesia), 211
University of Texas (Dallas), 107
University of Washington (Seattle), 507
University of Wisconsin (Madison), 513
urban design, 225–26, 252
and brownfield development, *250*, 250–53, *251*, *252*
and China, *271*, 271–78, *272–73*, *274*, *277*
and compact homes, *152*, 152–54, *153*
density of, and efficiency, 228, *228*, *236*, 254
ecofriendly requirements for, 261, 276
and ecovillages, 105, 275–76
and gentrification, 240
governmental authority over, 232, 234–35, 236–37
and green initiatives, 232
and housing developments, ecofriendly, 105, 155–56, *156*
infill housing, 229, 232, 241
LEED standards for, 248
"place"-making, in, *259*, 259–61, 265
and planning, 232, *236*, 236–37
Portland, 234–38, *235*, 252, 299
and skyscrapers, *245–48*, 245–49
for temperature control, 256–57, *257*, 258, 277
transit-oriented, 236, 239, 241, 242, 252, 262–67, *263*, *265*, *266*, 276, 282, 283, 340
(*see also* street design)
Vancouver, 231–34, *232*, *233*
See also infrastructure, urban
urban farming, *60*, 60–62, 69, 199, *274*, *340*, 500
urban-heat-island effect, 256, 258, 277
urban living, advantages of, 226, 281, *281*
urban revival, 240, 261, 501
urban sounds, music from ("Sonic City"), 98, *98*
"urban village," creating, 260–61
U.S. Centers for Disease Control and Prevention (CDC), 51, 331
U.S. Congress, 234–35
U.S. Department of Agriculture (USDA), 55–56, 504
U.S. Department of Defense, 77
U.S. Department of Energy, 515
U.S. Environmental Protection Agency (EPA), 50, 80, 166, 171, 516
U.S. Forest Service, *479*
U.S. Green Building Council (USGBC), 150, 228, 231, 234, 247–48
U.S. Supreme Court, 443
U.S. Agency for International Development (USAID), 46, 470
utilities. *See* power plants; *specific utility type*
Uzbekistan, 46

V

vampire power (energy consumption), 165, 167, 514
Vancouver, 54, 194, 228, 231–34, *232*, *233*, 263
Vancouver Island, 476
Vavilov, Nikolai Ivanovich, 73
vegetable oil, as biofuel, 77, 247
vegetables, organic. *See* organic foods
vegetarianism, 52–53, 63
Velcro, 99, 100
Velocipede Architects, *142*
Venice Biennale, 454
Ventilated Improved Pit (VIP) latrine, 131
ventilation systems, 202, 234, 245, 247, 256
vernal-pool fairy shrimp (California), 490
Vestal Design, 153
Vetrazzo (ceramic aggregate material), 144, *144*
VIA (company), 136
video, music ("World on Fire"), *357*, 357–58
video games
Disaffected!, *97*, 97–98
Food Force, *208*, 208–9
A Force More Powerful (ICNC), *462*, 462–63
The Movies, 424

video technology, 222
as watchdog tools, 428, 447, 448–49, *449*, 450, 459
Video Vote Vigil, 449
Viktoria Institute, 98, *98*
violence, country. *See* war
"Viridian Design Manifesto" (Sterling), 84
virtuous circles (philanthropy), 355–57
viruses, making batteries from, 113–14
vision (business strategy), 397, 400–401
Vodafone (mobile phone company), 292
Vogel, Steven, 102
VoIP (voice over Internet protocol), 215, 299
volatile organic compounds (VOC), 159, 160
Volkswagen (automaker), 277, *277*
Voluntary Simplicity (Elgin), 34
volunteers, and activism. *See* citizen science
volunTourism, 366, *366*
voting. *See* political elections

W

Wackernagel, Mathis, 15–16, 32
Waging Nonviolent Struggle (Sharp), 464
Wallace, David Rains, 479–80
Wallenius Wilhelmsen (maritime corporation), 45
wall panel (Woven Wood), *87*
wallpaper design, and biomorphism, 104
walls, living (air purifiers), *255*, 256
Wal-Mart, 54
and big-box store reuse, 242–43, *243*
and imports, from China, 41, 44
and labor practices, 393
Wann, Dave, 243–44
war, 423
civil wars, ending, 466, 471
effect on sustainability, 464–66, 468–69
global warming effect, on, 512
nonprovocative defense strategies, 468–69
open source-style, 467–68
state-on-nonstate, *466*, 466–68
and systems disruption, 467, 468
using game theory, to end, 466, 471
war crimes, 441–44, 469
and forensic anthropology, 444–45
washing machines, energy-efficient, 48–50
Washington (state), *69*, *142*, 512, 520, 535, *537*
Seattle, 146, 191, 267, 507, 509, 511–12
waste, art from, 343
waste management. *See* sewage management
waste-producing industries, top (U.S.), 400–401
wastewater recycling. *See* "gray water"
watchdog advocacy
and technology, 441, 443–44, 445–46, 447–50, *448*, 459
water, as energy resource
hydrokinetic-power systems, 177, *177*, 179, 181–82, *182*, 244
micro-hydro systems, 181–82, *182*
water-collection methods, 186, 189–93, *191*, *192*, *193*, 194
fog catching, 192–93, *193*
permeable pavement, 190, 194–95, *196*
Q-Drum (water container), 192, *192*
rainwater harvesting (*see* rainwater harvesting)
T.R.E.E.S. project, 194
Watercone (water purifier), 193, *194*
water conservation, 186, 187–89, 283, *387*, 499–500
water desalinization, 193
water filtration, 250
for gray water, 201
and LifeStraw, 208
and New York City, 494–95
using plants, for, 257 (*see also* bioremediation)
water fixtures, low-flow, 187–88
water heaters, 184
Home Energy Station (HES), 80
solar, 181, 202
tankless, 150
water pollution, 186
and bioremediation, 196, *197*, 250, 257
in China, 274, *274*
from shipping, 45
water pumps, 202
cholera, spread from, 518
Super-MoneyMaker, 192, *192*
water purification, 193, *194*, 201, 245, 488
and desalinization, 193
in disaster-relief plans, 193, 213, *213*
LifeStraw, 208
Living Machines, 195–96, *196*
and rain gardens, 257
reverse-osmosis, 213, *213*
solar-powered, 303
water quality, 19, *186*, *287*
testing, 132, 362
water recycling, *188*, 188–89, 201, 233, 248, 275, 276
"gray" water, 186, 188, 201, 233, 248, 276
water resource management, 52–53, 186, *186*, 476, 508
Clean Water Act, 489
hard-path versus soft-path, 189–90, 190–92, 194–95, *196*

for leapfrogging nations, *287*, 303
and local control, benefits of, 361
See also irrigation
Waters, Alice, 58, 61–62
watersheds, 229, 476, 494, 499–500
value of, to ecosystem, 486, *486*
Water Use It Wisely (Web site), 189
waterwheels, for micro-hydro systems, 181–82
Wattenberg, Ben J., 505
Wattson energy-measurement system, 166–67
wax, furniture/floor, 47
wealth, and happiness (concept), 32
We Are Family (activist art), 454
Weart, Spencer R., 517
weather. *See* climate change
Weber, Steven, 130
Weblogs. *See* blogs
Web 2.0 (participation concept), 419
Weiland, Matt, 396
West Africa, 333–34, 500
We the Media (Gillmor), 424–25
wetlands, 196, *197*
and Clean Water Act, 489
and "mitigation banks," 490
"Wexelblat disaster," 211
Wharton School of Business, 384
wheelchair design, *133*, 133–34
Where's George (Web site), 325
Whirlpool, 185
Whirlwind Wheelchair International, *133*, 133–34
Whistler 2020 plan (British Columbia), 505
White, Edward H., *532*
white-box systems (computers), 137
"white goods," 389–90
See also appliances
white roofs, 258, *258*
Whitney, Diana, 433
Whole Earth Catalog (Brand), 110
"whole earth" perspective, and Internet, 419
Whole Foods Market, 404, *404*
Why Societies Need Dissent (Sunstein), 446
Wi-Fi. *See* wireless networks (Wi-Fi)
Wikibooks, 315
Wikipedia, 129, 130, 215, 315, 373
wildfire, and sustainable forestry, *479*, 495–96
wildlife-gardening.org.uk, 200
Wilson, Edward O., 487, 491
windmills, flying, 176, *176*
windows, 166
and temperature control, 155, 162, 234, 248
window shade ("Energy Curtain"), 161–62
wind power, as energy resource
environmental impact of, 23, 171
and polar region resources, 527
wind power energy systems, 174–76, *175*, *176*, 177
cost of, versus fossil-fuel, 172, 175
for home use, 179–80, *180*, 182
and "intermittency" issue, 172
in leapfrogging infrastructure, 303
wind farms, 23, 170, 172, 174–75, *175*, 176, *176*, 504
wind turbines, 168, 174, 175, 176, *176*, 179, 304–5
Wind Power (Gipe), 177
Windsave (wind turbine maker), 179
WIPO (World Intellectual Property Organization), 315, 419, 501
WirelessMoment.com, 449
wireless networks (Wi-Fi), 214, *297*, 297–98, 299, 304, *304*, 435, 459
Wirzba, Norman, 69–70
Witness (human rights organization), 443–44, *444*, 445
wolves, tracking, 484, *484*, 520
women, and empowerment, 307, 317, 320
AIDS orphans program (Zambia), 326
domestic-violence prevention, 316, 321–22
education programs, 20, 312–13, 360, 361
Grameen Phone, *292*, 293
and human rights, 316, 429, 441, *441*, 442
maternal health programs, 326–27, 357
and microfinance, 347, 350, 352, 356
Millennium Development Goals (UN), 316
Night for Women (Bogotá), 283, *283*
self-help groups, 321, *321*
and urban living, benefits from, 44, 281, *281*
See also girls, educating
wood
certification programs, *142*, 143, 496–97
flooring, 142–44
furniture, 158
reclaimed, 142–43
renewable, 158, *158*, 159 (*see also* bamboo)
See also deforestation
Woofenden, Emma
designed TranSglass, *83*
wool, merino (fabric), *36*, 37
woonerf ("street for living"), 265
World as Lover, World as Self (Macy), 517
World Bank, 167, 181, 271, 320, 350
World Cafés, 432
WorldCat (online catalog), 316
WorldChanging.com, 24, 25–26, 527

World Computer Exchange, 298
World Conference on Education for All (UNESCO), 317
World Conservation Union (IUCN), 533
World Economic Forum, 311, 314–15
World Food Program, 320
"World Game" (Fuller), 104
World Health Organization (WHO), 274, 324, 513
World Intellectual Property Organization (WIPO), 315, 419, 501
"World on Fire" (music video), *357*, 357–58
World Refugee Survey, 203
World Resources Institute, 403, 498
World Shipping Council, 45
World Summit on Sustainable Development (2002), 274
World Tourism Organization, 363
World Trade Organization (WTO), 450, 458
World View of Global Warming (Braasch), *508*, 508–9
World War II, 246
  post-war economic programs, 39–41
  post-war refugee rights, 206
  and post-war suburban development, 238–39
Worldwatch Institute, 139, 174, 175
World Watch (magazine), 20
World Water Assessment Program (UN), 499
World Wind, 520
Wuling SunShine (car), 277–78
Wyoming, 242–43

X

X Prize Foundation, 358

Y

Yahoo.com, 373
Yes Men (activist group), 457, *457*
Your Money or Your Life (Robin/Dominguez), 34–35
"youth bulge" (population growth), 281
Yunus, Muhammad, 351, 352, *353*, 354

Z

Zambia, 326, 327
ZeroCarbonCity campaign, 509
"zero-carbon" design (housing development), 155–56, *156*
zero-energy homes, 147–48, *148*, 155–56, *156*
Zero Energy House (Chicago), 147–48, *148*
Zimbabwe, 19, *182*, 282
  Eastgate development, 282, *282*
Zittel, Andrea, 153
Zope (content-management system), 420
Zubrin, Robert, 535

EDITOR: Deborah Aaronson
PHOTO RESEARCH: Sideshow Media
PRODUCTION MANAGER: Anet Sirna-Bruder

FOR WORLDCHANGING:
EDITOR: Alex Steffen
MANAGING EDITOR: Sarah Rich
ASSOCIATE EDITOR: Carissa Bluestone
ASSISTANT EDITOR: Tessa Levine-Sauerhoff

Library of Congress Cataloging-in-Publication Data

Worldchanging : a user's guide for the 21st century / edited by Alex Steffen.
p. cm.
Includes bibliographical references and index.
ISBN 978-0-8109-3095-7 (hardcover)
ISBN 978-0-8109-7085-4 (paperback)
1. Environmentalism. 2. Sustainable development. 3. Green movement. I. Steffen, Alex.

HC79.E5W676 2006
333.7—dc22

2006021673

This book was originally published by Abrams in hardcover in 2006.

Printed and bound in the United States
10 9 8 7 6 5 4 3 2 1

harry n. abrams, inc.
a subsidiary of La Martinière Groupe
115 West 18th Street
New York, NY 10011
www.hnabooks.com